DYNAMICS OF FLEXIBLE MULTIBODY SYSTEMS: THEORY AND EXPERIMENT

NGINEERS

ISION, ASME

MECHANICAL ENGINEERS
ring Center ■ New York, N.Y. 10017

Statement from By-Laws: The Society shall not be responsible for statements or opinions advanced in papers . . . or printed in its publications (7.1.3)

ISBN No. 0-7918-1076-3

Library of Congress
Catalog Number 92-56539

FOREWORD

The study of dynamics and control of flexible multibody systems has become increasingly important in recent years due to the problems encountered in the design of proposed space stations, large spacecrafts, lightweight robots performing rapid maneuvers and machine components subjected to high speed operating conditions. In order to provide a forum for exchange of recent developments on the subject, the Applied Mechanics Division and the Dynamic Systems and Control Division of ASME jointly sponsored a symposium on "Dynamics of Flexible Multibody Systems: Theory and Experiment" at the Winter Annual Meeting held in Anaheim, California, November 8–13, 1992.

This book of proceedings contains all the twenty-seven papers presented in six sessions on November 12 and 13, 1992 at the symposium and covers various issues related to the problem of flexible multibody dynamics and control. The papers appear in the same order as they were presented.

A successful organization of a symposium is a considerable task, particularly for someone like myself who can hardly claim any expertise on this topic. First, I would like to thank all the authors for promptly submitting their valuable contributions and meeting the deadlines, some of which were not very convenient due to some reason or the other. I would also like to thank Professor Wayne Book, Professor Ahmed Shabana, and Dr. Ramen Singh for organizing various sessions for the symposium and Dr. Henry Waites for sharing the burden. Last, but not the least, my sincere thanks are due to Ms. Marilyn Swaim of the Mechanical Engineering Department, Auburn University for providing secretarial assistance and Ms. Barbara Signorelli at the ASME Headquarters for making sure of its publication on time.

S. C. Sinha
Auburn University

CONTENTS

AMD-Vol. 141/DSC-Vol. 37, Dynamics of Flexible Multibody Systems:
Theory and Experiment
ASME 1992

COOPERATIVE MANIPULATION OF FLEXIBLE OBJECTS: INITIAL EXPERIMENTS

David W. Meer and Stephen M. Rock
Aerospace Robotics Laboratory
Stanford University
Stanford, California

Abstract

The vast majority of the work done on multiple manipulator systems has focused on manipulating rigid objects, both in free-space motion and contact tasks. Not all objects encountered in potential robotic applications are rigid, however. Spring-loaded parts, lightweight space structure members, and heavy cabling provide just a few examples of flexible objects that robots may need to manipulate in the future.

This paper presents a testbed for the study of cooperative manipulation of flexible objects. It discusses some of the important characteristics required for this study. Using this testbed, the limitations of two control strategies used for manipulation of rigid objects are demonstrated when applied to flexible objects. The results justify the validity of the testbed as well as providing motivation for further study.

Introduction

The advantages of using multiple manipulators include increased payload capability, improved dexterity with larger objects, and expanded functionality. Most previous research, however, focused on developing control strategies for multiple robotic arms manipulating a single, rigid body. What happens when the manipulated object is flexible? Various potential robotic applications, from the assembly of spring loaded parts in a manufacturing environment to the servicing of satellite solar arrays in orbit, will involve the manipulation of flexible objects by multiple manipulators.

One of the most promising and general approaches to cooperative manipulation is object-level control. This technique allows the operator to issue task level commands, such as "capture this object" or "insert this connector into that fixture". The controller takes care of the details of the operation, drawing upon a library of task primitives, freeing the user to perform other tasks. This capability has been developed and demonstrated successfully on a wide variety of experimental platforms. [1] [2] [3]

The goal of this research is to extend object-level control to flexible objects. This paper presents some preliminary findings. First, an experimental testbed is described. It consists of a pair of arms and a flexible object. Next, two attempts to apply previously developed control strategies to a flexible object using this experimental testbed are discussed. The first, Object Impedance Control (OIC), developed for cooperative manipulation of rigid objects, performed poorly in attempts to regulate the free space motion of the object. In fact, this controller was unstable for higher object stiffnesses. The second control strategy, a coordinated PD control, was stable and could perform free space manipulations without undue excitation of the object's flexibility. The coordinated PD controller, however, proved insufficient for tasks involving deformation of the object. These results show that current controllers, designed for manipulation of rigid objects, perform poorly when applied to a flexible object and that the experimental testbed embodies the problem of interest.

Related Work

Some work has been done on the control of flexible objects with robotic manipulators. This body of work addresses various aspects of the problem, including trajectories and task formulation. It does not, however, focus on the interaction between the flexibility and the controller.

Zheng and Luh [4] used a flexible object to eliminate kinematic redundancy problems in their early work on coordinated control of multiple manipulators. These results seem to indicate that, in some cases, flexibility in the object may make the task of controlling such systems easier.

Recently, Dauchez, et al, presented experimental results for a pair of 6 dof arms deforming a spring and transporting the spring in the deformed state [5]. They used symmetric hybrid position/force control. The principal contribution of the work was the method they used to describe the task with "virtual sticks".The algorithm used, however, was so computationally complex that the controller ran at 20 Hz. Also, the hybrid control approach requires task dependent control mode switching. This can be a disadvantage when performing complex tasks.

Zheng, Pei, and Chen [6] have also done work on assembly of deformable objects. The assemblies involved sliding a long, flexible beam into a hole with a single manipulator. The principal contribution of this work was determining the proper trajectory for the arm to follow based upon the beam properties and the tightness of fit.

The goal of this research is to explore the interaction between object flexibility and the system controller.

Design Objectives

Several criteria helped shape the design of the experimental apparatus used to study cooperative manipulation of flexible objects. First, the arms should be as "ideal" as possible. An "ideal" arm would produce specified forces and accelerations at the endpoint exactly. This allows the experiments to focus on the problems introduced by flexibility in the object rather than those caused by friction at the robot joints, flexibility in the drive train, etc. The goals in designing the flexible object included (1) introducing the flexibility at a frequency of interest within the bandwidth of the control system, (2) providing the capability to change the natural frequency and stiffness of the flexible element in order to study the effect of varying these parameters, and (3) creating an object that was deformable with the available actuators in order to study assembly operations requiring deformation of the object. Finally, the testbed should not be so geometrically complex that the computational speed of the control computers severely limits the algorithms that can be applied to the system.

Experimental Apparatus

Addressing the last of the design criteria, the experimental testbed is limited to 2 dimensions, simplifying the computational complexity significantly. The flexible object has 4 degrees of freedom (DOF) and each manipulator has 2 DOF. The flexible object floats on air bearings over a granite surface plate, eliminating the effects of gravity on the system and simulating the drag-free environment of space. These simplifications enable the research to focus on the problem of interest: how flexibility affects the control of an object grasped by multiple manipulators.

Cooperating Manipulators

Figure 1 depicts one of the pair of experimental arms. Each manipulator is a direct-drive, SCARA two-link arm, with revolute "shoulder" and "elbow" joints. At the tip of each arm is a two-dimensional force sensing pneumatic gripper. These grippers fit into ports on the manipulation objects. The connection is mounted on a bearing pin joint, so the manipulator cannot apply torque at the connection. The system thus provides frictionless two-dimensional motion.

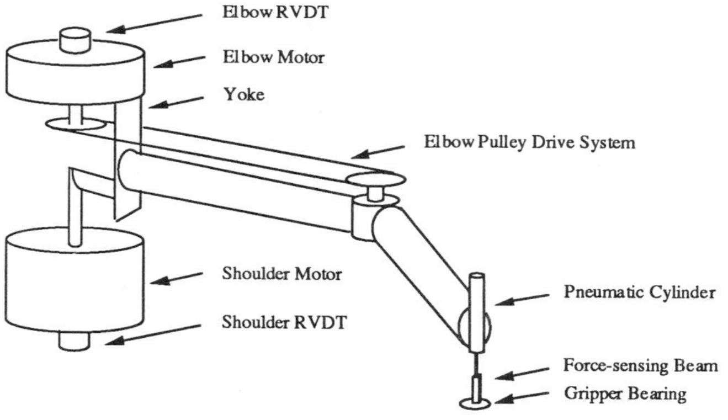

Figure 1: **Arm Schematic**

A schematic view of one of the two link, SCARA, arms used in the experimental testbed.

The manipulators have a reach of 0.65 meters and are separated by 0.60 meters at the shoulder hub. The motors on the manipulator are DC limited-angle torquers. A motor located at the shoulder transmits torque to the elbow joint through a steel cable. Joint angles are measured by a rotary variable differential transformer (RVDT) mounted on each motor shaft. Each arm also has a vision target located over the gripper for use with an overhead vision system. See [7] for a detailed description of the manipulators.

Many factors, including the direct drive nature of the arms, the vision and force sensors at the endpoint, the two dimensional nature of the experimental system that eliminates the need for gravity compensation, and accurate calibration of the motors bring these manipulators close to the "ideal".

Flexible Object

The flexible object consists of two pads that float on an air cushion over the granite surface plate. These pads are joined by a six bar linkage. The linkage is designed to add a single flexible degree of freedom to the object. Figure 2 shows the object in both the nominal configuration (solid lines) and the deformed configuration (dashed lines). The circles represent pin joints in the mechanism while the thicker lines show the two sections of steel wire that give the object its flexibility. These segments can easily be switched out to change the stiffness of the flexibility in the object. Each pad also has two gripper ports and a target for tracking by the overhead vision system (not shown in the drawing).

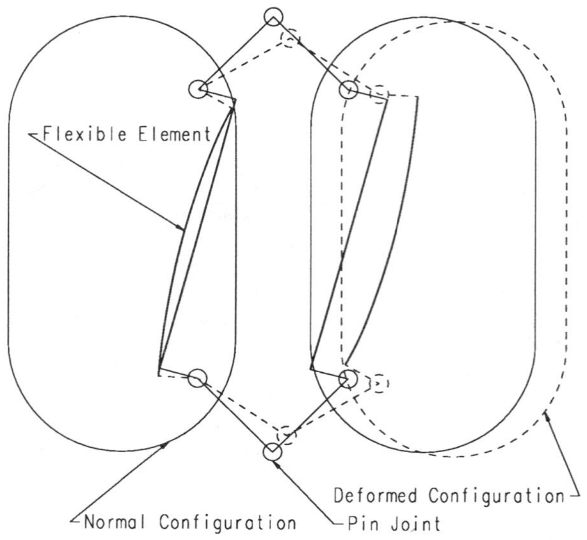

Figure 2: **Flexible Object**

This flexible object, which floats on a granite surface plate, uses a six bar mechanism with 2 flexible elements. The object thus has three rigid degrees of freedom and one flexible degree of freedom.

As Figure 3 shows, the free vibration of the flexible object, with the particular stiffness used in these initial experiments, has a natural frequency of 3.06 Hertz. This plot also shows the very lightly damped characteristic of the linkage. The linkage gives the object a range for the distance between the pads of between .025 meters and .09 meters, with a nominal separation of .064 meters.

Current Control Strategies

This section briefly outlines the two control strategies applied to the experimental system. The motivation for these experiments was to demonstrate that deficiencies exist with current control strategies when applied to flexible objects and to test the validity of the experimental testbed.

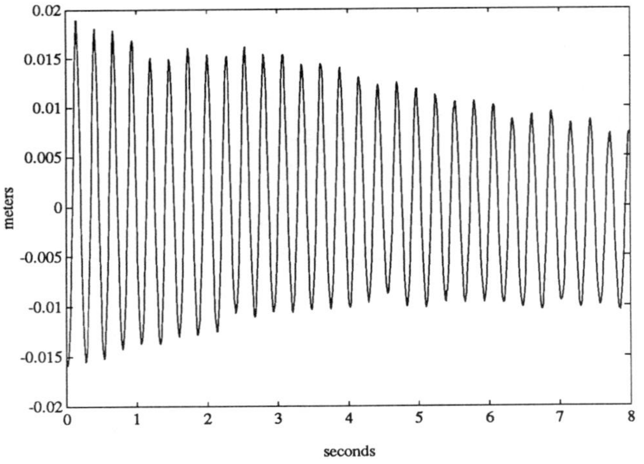

Figure 3: **Free Vibration of Flexible Object**

This plot shows the deflection from the nominal separation of the two pads of the flexible object for a nonzero initial condition. Note the lightly damped response.

Object Impedance Control

The Object Impedance Control (OIC) strategy enforces a controlled impedance of the manipulated object [8]. Equation 1 contains the particular impedance relationship chosen for this controller.

$$m_d(\ddot{x} - \ddot{x}_{des}) + k_v(\dot{x} - \dot{x}_{des}) + k_p(x - x_{des}) = f_{ext} \tag{1}$$

Here, x is the coordinate of any one DOF of an arbitrary frame fixed relative to the object's frame. The constants m_d, k_p, and k_v can be specified independently for each degree of freedom. x_{des} represents the desired position of the chosen frame and $\ddot{x}_{des}$ is acceleration feedforward. The derivation of the control equations based on the desired object behavior specified in Equation 1 is fully contained in Schneider [8]. Basically, the controller attempts to cancel the actual object dynamics and make the object behave according to Equation 1. This produces a desired acceleration, $\ddot{x}_{des}$, and force, f_{des}, at each arm endpoint. Then, if M and J represent the mass matrix and Jacobian for a given arm and q is the vector of joint angles, the arm kinematics yield:

$$\ddot{q}_{des} = J^{-1}(\ddot{x}_{des} - \dot{J}\dot{q}) \tag{2}$$

Combining this with the arm equations of motion, where C contains the nonlinear coriolis and centrifugal terms,

$$\tau = M\ddot{q}_{des} + C(q, \dot{q}) + J^T f_{des} \tag{3}$$

produces the desired torques for each arm, τ. These equations are for the simplified planar case.

This technique requires an accurate model of the dynamic behavior of the object. It also uses the location of the object in the feedback loop. Endpoint feedback techniques are generally not very robust to the introduction of unmodelled modes of vibration. Consequently, the unmodified Object Impedance Controller applied to a flexible object was expected to perform poorly.

Object PD Control

The second control strategy tested was a very simple coordinated PD control. Coordinated control refers to an approach that uses the desired motion of the center of mass of the object to calculate the desired motion of the the grip points. This control makes no attempt to compensate for dynamic forces, relying on the strictly kinematic relationship between the grip points and the object center of mass. This approach treats the arms as a simple force source, calculating the force that each arm should apply using a PD control law on the gripper port. This yields

$$f_{arm} = k_p(p_{des} - p) + k_v(\dot{p}_{des} - \dot{p}) \tag{4}$$

where f_{arm} is the desired arm endpoint force, k_p and k_v are specifiable position and velocity gains, p represents the endpoint position of the arm, and p_{des} is the desired arm endpoint location. The desired arm endpoint location, p_{des}, comes from the kinematic relationship between the desired object center of mass and the grip point on the object. The controller simply runs the desired endpoint force, f_{arm}, through the Jacobian, J, to produce the torques at the joints, τ.

$$\tau = J^T f_{arm} \tag{5}$$

This control should be stable regardless of the object's dynamics, since it essentially treats the motion of the object, including the flexibility, as a disturbance to the arm endpoint.

Experimental Results

Object Impedance Control

Experimentally, the Object Impedance Control strategy proved unstable for motions that provided sufficient excitation to the flexible object. Figure 4 shows a time history of the spring mechanism compression for a slew of 0.15 meters in 1.0 seconds in the upper plot. The lower plot shows the desired and actual center of mass X position. The slew begins at about 1.5 seconds. Clearly, the interaction between the object's flexibility and the controller is leading to instability. Also note that, despite the excitation of the flexibility, the object's X position does not deviate significantly from the desired.

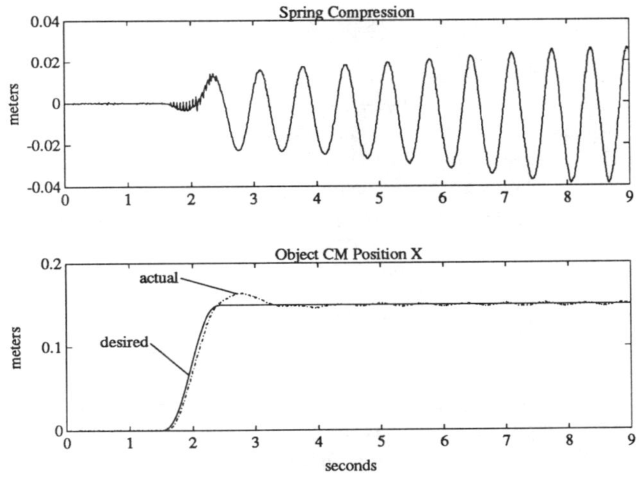

Figure 4: **Object Impedance Slew**

As this plot of the spring compression during a slew shows, the Object Impedance Controller can excite the flexible mode in the object.

Object PD Control

As Figure 5 shows, the gripper point PD controller was stable for the same slew that caused the object impedance controller to go unstable.

So, the simple PD controller works well for free space motions. While the flexibility was excited somewhat, the PD controller quickly damped it out once it began regulation.

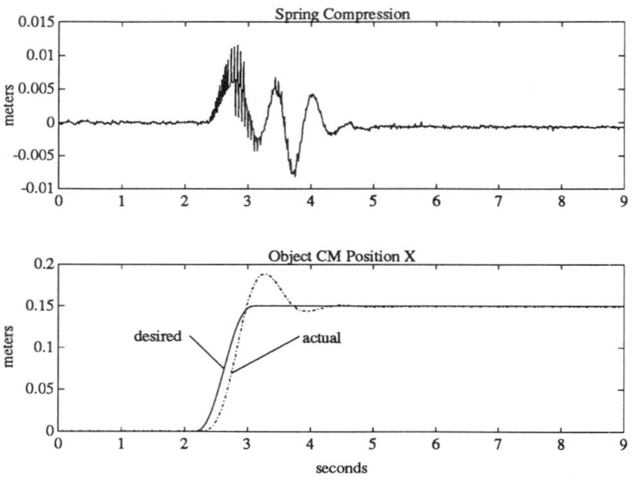

Figure 5: **PD Slew**

While the PD controller applied to the flexible object exhibits significant overshoot, it does damp out any excitation of the flexibility.

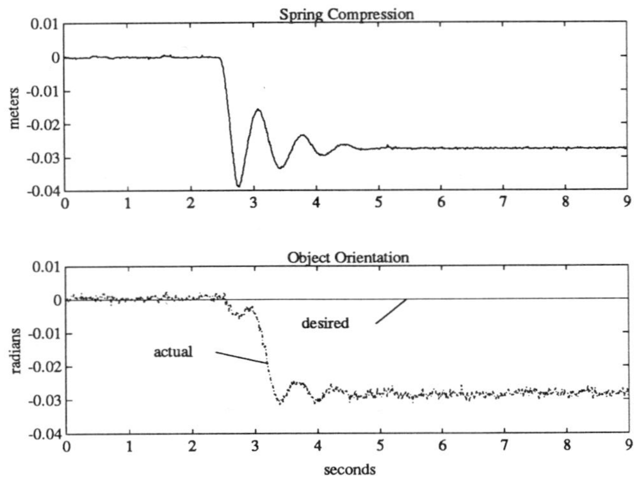

Figure 6: **PD Spring Compression**

When commanded to deform the flexible object, the PD controller rotated the object as well as compressing it.

The next test involved directly manipulating the flexibility in the object. Figure 6 shows the distance between the ports for a desired compression of -0.12 meters beginning at about 2.5 seconds. The actual compression achieved, -0.0275 meters in this case, depends upon the proportional gains used in the controller. Also note that the object rotates as it compresses. So, while this simple coordinated PD controller can damp out the object vibrations when the spring is unstretched, it does not control the object adequately for manipulations involving deformation.

Conclusions

This paper describes a hardware testbed developed to study the problem of manipulation of flexible objects. It discusses the criteria used to design a testbed to study the problem of interaction between the system controller and the flexibility in the object. A flexible object composed of two rigid bodies coupled by a mechanism designed to be flexible in one dimension is the principal difference between this system and systems studied in previous work. To validate the testbed and demonstrate that a problem does exist, two controllers were applied to the experimental system. For the first strategy, Object Impedance Control, the results demonstrated a sensitivity to modelling error. The controller performed marginally, at best, and sometimes proved unstable, depending upon the stiffness of the object. The second approach, a coordinated PD controller was stable and did exhibit reasonable performance in free motion. However, it did a poor job of controlling the object for tasks that involved deformation.

References

[1] S. Schneider and R. H. Cannon. Experiments in cooperative manipulation: A system perspective. In *Proceedings of the NASA Conference on Space Telerobotics*, Pasadena, CA, February 1989. NASA.

[2] M. A. Ullman. *Experiments in Autonomous Navigation and Control of Multi-Manipulator Free-Flying Space Robots.* PhD thesis, Stanford University, Stanford, CA 94305, (February) 1992. To be published.

[3] Lawrence Pfeffer. *Experiments in Control of Cooperating, Flexible Robotic Manipulators.* PhD thesis, Stanford University, Stanford, CA 94305, (March) 1992. To Be Published.

[4] Y. F. Zheng, J. Y. S. Luh, and P. F. Jia. A real-time distributed computer system for coordinated motion control of two industrial robots. In *Proceedings of the International Conference on Robotics and Automation*, pages 1236–1241, Raleigh, NC, April 1987. IEEE.

[5] Pierre Dauchez, Xavier Delebaree, Yann Bouffard, and Eric Degoulange. Task description for two cooperative manipulators. In *Proceedings of the American Control Conference*, pages 2503–2508, Boston, MA, June 1991. IEEE.

[6] Yuan F. Zheng, Run Pei, and Chichyang Chen. Strategies for automatic assembly of deformable objects. In *Proceedings of the International Conference on Robotics and Automation*, pages 2598–2603, Sacramento, CA, April 1991. IEEE.

[7] S. Schneider. *Experiments in the Dynamic and Strategic Control of Cooperating Manipulators.* PhD thesis, Stanford University, Stanford, CA 94305, September 1989. Also published as SUDAAR 586.

[8] S. Schneider and R. H. Cannon. Object impedance control for cooperative manipulation: Theory and experimental results. *IEEE Journal of Robotics and Automation*, 8(3), June 1992. Paper number B90145.

AMD-Vol. 141/DSC-Vol. 37, Dynamics of Flexible Multibody Systems:
Theory and Experiment
ASME 1992

NUMERICAL ANALYSIS OF NONMINIMUM PHASE ZERO FOR NONUNIFORM LINK DESIGN

Douglas L. Girvin and Wayne J. Book
George W. Woodruff School of Mechanical Engineering
Georgia Institute of Technology
Atlanta, Georgia

ABSTRACT

As the demand for light-weight robots that can operate in a large workspace increases, the structural flexibility of the links becomes more of an issue in control . When the objective is to accurately position the tip while the robot is actuated at the base, the system is nonminimum phase. One important characteristic of nonminimum phase systems is system zeros in the right half of the Laplace plane. The ability to pick the location of these nonminimum phase zeros would give the designer a new freedom similar to pole placement.

This research targets a single-link manipulator operating in the horizontal plane and modeled as a Euler-Bernoulli beam with pinned-free end conditions. Using transfer matrix theory, one can consider link designs that have variable cross-sections along the length of the beam. A FORTRAN program was developed to determine the location of poles and zeros given the system model. The program was used to confirm previous research on nonminimum phase systems, and develop a relationship for designing linearly tapered links. The method allows the designer to choose the location of the first pole and zero and then defines the appropriate taper to match the desired locations. With the pole and zero location fixed, the designer can independently change the link's moment of inertia about its axis of rotation by adjusting the height of the beam. These results can be applied to inverse dynamic algorithms currently under development at Georgia Tech and elsewhere.

INTRODUCTION

Controller design for collocated systems has been heavily researched and is well understood compared to controller design for noncollocated systems. In noncollocated systems, uncertainties from model inaccuracies and modal truncation present fundamental problems with system performance and stability [18]. The fundamental difference between collocated and noncollocated systems is the presence of these RHP zeros. To advance controller design for noncollocated systems, research needs to be conducted into the factors that affect the location of these RHP zeros. This research targets the relationship between RHP zeros and structural design.

Although research on RHP zeros is limited, there has been some notable research done in the past. In 1988, Nebot and Brubaker [13] experimented with a single-link flexible manipulator. In 1989, Spector and Flashner [19] investigated the sensitivity effects of structural models for noncollocated control systems. In 1990, Spector and Flashner [18] again studied modeling and design implications pertinent to noncollocated control. Also in 1990, Park and Asada [15],[14] investigated a minimum phase flexible arm with a torque actuation mechanism. In 1991, Park, Asada, and Rai [1] expanded their previous work on a minimum phase flexible arm with a torque transmission device.

The underlying issue in noncollocated control is how to deal with the RHP zeros in the control algorithm. A major step in solving the problem is understanding what design parameters can be used to change the location of these RHP zeros. This research targets the relationship between RHP zero location and structural design. Specifically, how do changes in the shape of the structure (link) affect the location of these zeros?

Traditionally links are designed with uniform properties along the length because analytic solutions to this problem exist. A link with variable cross-section cannot be solved analytically, but with aid of a computer a numerical approximation can be found. The key to an accurate numerical solution is a good model of the system.

The research presented in this paper models a single-link flexible rotary manipulator as a pinned-free beam. Transfer matrix theory was used to generate a beam with variable cross-section. FORTRAN code was written to generate the model and evaluate the system for the location of RHP zeros. The program was used to examine the relationship between link shape and RHP zero location. This relationship can be directly applied to controller design using the inverse dynamics approach researched at Georgia Tech and elsewhere.

TRANSFER MATRIX THEORY

Transfer matrices describe the interaction between two serially connected elements. These elements can be beams, springs, rotary joints, or many others. In 1979 Book, Majette, and Ma [6] and Book [4] (1974) used transfer matrices to develop an analysis package for flexible manipulators. They used transfer matrices to serially connect different types of elements to model the desired manipulator. Of interest in this paper is how to connect similar types of transfer matrices (beam elements) to model a beam with different cross-sectional area. Pestel and Leckie [16] provide an in depth discussion of transfer matrix derivations and applications.

Transfer matrices can be mathematically expressed by Equation 3.1. The state vector u_i is given by the state vector u_{i-1} multiplied by the transfer matrix B.

$$u_i = [B_i]u_{i-1} \tag{3.1}$$

When elements are connected serially, the states at the interface of two elements must be equal. By ordered multiplication of the transfer matrices, intermediate states can be eliminated to determine the transfer matrix for the overall system.

The concept of state vector in transfer matrix theory is not to be confused with the state space form of modern control theory. The state equation in modern control theory relates the states of the system as a function of time. In transfer matrix theory the state equation relates the states at various points along the serial chain of elements. The independent variable in a transfer matrix is the Laplace or Fourier variable with units of frequency, not time. The elements of the matrix B depend on the frequency variable and therefore the states will change as the system frequency changes. The transfer matrix B essentially contains the (Laplace or Fourier) transformed dynamic equations of motion that govern the element in analytic form. Therefore, analytical solution of the transfer matrix alone does not involve numerical approximations to the partial differential equation modelling the beam. This is desirable since numerical approximations introduce error into the solution.

A single-link manipulator as pictured in Figure 3.1 can be thought of as a beam with torque applied at one end and free at the other end. There are several steps to determine the RHP zeros and imaginary poles of this system. First, develop a model for the beam. Second, determine the appropriate boundary conditions. Third, determine the system input and output. Fourth, solve for the system zeros. The following sections will discuss each of these steps in more detail.

A link with nonuniform cross-sections can be modeled as a series of discrete elements. While the shape of these elements is similar, the size can vary to allow for changes in cross-section. The appropriate element to model a flexible link is an Euler-Bernoulli beam element. The Euler-Bernoulli model neglects the effects of rotary inertia and shear deformation in the element. [11]. This assumption is generally valid for modeling beams whose length is roughly ten times the thickness. Flexible manipulators have long, slender links which are appropriately modeled under the Euler-Bernoulli assumption.

Transfer matrices are derived from the equation of motion for a given element. For a uniform Euler-Bernoulli beam element, the equation of motion transformed to the frequency domain has the form:

$$\frac{d^4 w(x,\omega)}{dx^4} = \frac{\mu\omega^2}{EI}\, w(x,\omega)$$

where,

μ = mass density per unit length

ω = frequency in radians/second

E = Young's modulus

I = Cross sectional area moment of inertia

Notice the equation is fourth order thus requiring four states to describe the solution in transfer matrix form. The state vector for the Euler-Bernoulli element is:

$$u = \begin{bmatrix} -w \\ \psi \\ M \\ V \end{bmatrix} = \begin{bmatrix} \textit{displacement} \\ \textit{slope} \\ \textit{moment} \\ \textit{shear force} \end{bmatrix} \tag{3.3}$$

The first two elements of the state vector are displacements (w and ψ) while the last two elements are forces (V and M). This arrangement of states is characteristic of transfer matrix theory.

An analytic solution to Equation 3.2 can be found when the element has uniform properties (ie. constant cross-section, mass density, and stiffness). Equation 3.4 gives the transfer matrix for a uniform Euler-Bernoulli element. Each element of Equation 3.4 is a function of frequency and must be reevaluated as the frequency of interest changes.

$$TM = \begin{bmatrix} C_0 & lC_1 & aC_2 & alC_3 \\ \frac{\beta^4 C_3}{l} & C_0 & \frac{aC_1}{l} & aC_2 \\ \frac{\beta^4 C_2}{a} & \frac{\beta^4 l C_3}{a} & C_0 & lC_1 \\ \frac{\beta^4 C_1}{al} & \frac{\beta^4 C_2}{a} & \frac{\beta^4 C_3}{l} & C_0 \end{bmatrix} \tag{3.4}$$

where,

$$C_0 = \frac{1}{2}(\cosh\beta + \cos\beta) \quad (3.5)$$

$$C_1 = \frac{1}{2\beta}(\sinh\beta + \sin\beta) \quad (3.6)$$

$$C_2 = \frac{1}{2\beta^2}(\cosh\beta - \cos\beta) \quad (3.7)$$

$$C_3 = \frac{1}{2\beta^3}(\sinh\beta - \sin\beta) \quad (3.8)$$

and

$$\beta^4 = \frac{\omega^2 l^4 \mu}{EI} \quad (3.9) \qquad a = \frac{l^2}{EI} \quad (3.10)$$

With the transfer matrix for the fundamental beam elements, one can combine these elements serially to generate a model for the link. Figure 3.3 illustrates how a simple model can be constructed for a tapered beam. Although only two elements are considered here, more elements can be added to better approximate the shape of the link. Since the states at interface u_1 are the same for both elements, u_1 can be eliminated to obtain an overall transfer matrix for the beam:

$$u_2 = [B_2][B_1]u_0 \quad (3.13)$$

Eliminating one state simply illustrates the point that this multiplication can be carried out to eliminate all intermediate states in a model with more elements.

As previously mentioned, transfer matrices themselves are not numerical approximations. The transfer matrix for a Euler-Bernoulli beam contains the analytic solution for a uniform beam element. It is not an assumed modes solution. The approximation made in using transfer matrix theory involves the modeling of the beam and solution of the equations. To generate the model of a link with variable cross-section, the size of the elements must vary. The interface of two different size elements will be discontinuous. In Figure 3.3, interface 1 is discontinuous between elements A and B. These discontinuities are the major approximation when using transfer matrices to model a beam. This approximation can be minimized by using more elements to model a nonuniform beam. As more elements are added to the model, the discontinuities between elements will decrease thus reducing the effects of this approximation on the results.

Transfer matrix theory as used to represent a variable cross section is similar to Finite Element Analysis (FEA). In FEA, first the system must be discretized. Then an appropriate interpolation function must be selected to describe each element (ie. element stiffness). Next the system matrices must be assembled to produce a set of linear algebraic equations. Finally the linear equations are solved to get an approximate solution to the system under consideration. These boundary conditions are applied to the overall transfer matrix for the system and the appropriate state variables are set to zero.

$$\begin{bmatrix} -w \\ \psi \\ 0 \\ 0 \end{bmatrix}_{x=L} = \begin{bmatrix} B_{11} & \cdots & B_{14} \\ \vdots & \ddots & \vdots \\ B_{41} & \cdots & B_{44} \end{bmatrix} \begin{bmatrix} 0 \\ \psi \\ 0 \\ V \end{bmatrix}_{x=0} \quad (3.14)$$

Since this research targets the location of RHP zeros the system output is tip position, and the system input is joint torque. Considering the system input and output, the overall system transfer matrix will have the form:

$$\begin{bmatrix} -w \\ \psi \\ 0 \\ 0 \end{bmatrix}_{x=L} = \begin{bmatrix} B_{11} & \cdots & B_{14} \\ \vdots & \ddots & \vdots \\ B_{41} & \cdots & B_{44} \end{bmatrix} \begin{bmatrix} 0 \\ \psi \\ \tau \\ V \end{bmatrix}_{x=0} \quad (3.15)$$

In the above equation, w_L is the system output which corresponds to tip position, and τ is the system input corresponding to joint torque at the base of the manipulator.

With the system input and output chosen, Equation 3.15 can be simplified to relate system input to system output:

$$N = B_{12}B_{44}B_{33} - B_{12}B_{34}B_{43} + B_{13}B_{34}B_{42} - B_{13}B_{44}B_{32} + B_{14}B_{43}B_{32} - B_{14}B_{33}B_{42}$$

$$w_L = -\frac{N}{B_{34}B_{42} - B_{44}B_{32}} \quad (3.16)$$

Where B_{ij} are elements of the overall transfer matrix in Equation 3.15. When the frequency is found which renders the function inside the brackets zero the output at that frequency will always be zero regardless of the input; therefore, the zeros of the bracketed term are the system zeros.

To search for RHP zeros, one must consider what type of frequency to input into Equation (3.16). Using the relationship which defines the Laplace variable, s

$$s = j\omega \quad (3.17)$$

one can easily determine ω should have the form:

$$\omega = 0 - jb \qquad where\ \ 0 \le b \le \infty \quad (3.18)$$

That is, imaginary negative values of ω will result in purely real positive values of s. Thus searching Equation 3.16 with frequencies of the form of Equation 3.17 one can find the location of the RHP zeros on the real axis.

Although the location of RHP zeros is of primary concern in this research, knowledge of pole location will help in analysis of the results. Since the system damping is ignored, the poles will lie on the imaginary axis of the s-plane in complex conjugate pairs. The location of these poles can be determined by simply searching the positive imaginary axis of the s-plane. Considering the applied boundary conditions, one can extract two homogeneous equations from Equation 3.14 to get the homogeneous system:

$$\begin{Bmatrix} 0 \\ 0 \end{Bmatrix} = \begin{bmatrix} B_{32} & B_{34} \\ B_{42} & B_{44} \end{bmatrix} \begin{Bmatrix} \psi \\ V \end{Bmatrix} \qquad (3.19)$$

The poles (eigenvalues) of the system are those values of ω which make the determinant of the sub-transfer matrix in Equation 3.19 equal to zero (see reference [6] for a detailed explanation). For a two by two matrix this determinant is simply:

$$g(\omega) = B_{32}B_{44} - B_{34}B_{42} \qquad (3.20)$$

Referring to Equation 3.17, one finds that Equation 3.20 is the denominator of the input/output transfer function which is to be expected. To find the values of the purely complex poles, one must search Equation 3.20 for its roots. According to the definition of s, ω must have the form:

$$\omega = b + j0 \qquad (3.21)$$

Searching over a range of values for b will give the poles in that range. With the zero and natural frequency functions determined, the problem remains to implement a computer solution to find the RHP zeros and imaginary poles.

RESULTS

Unless otherwise specified, several dimensions remain the same from one study to the next (referred to as nominal dimensions). The overall length of the beams is 40 inches, and the height (which remains constant over length) is 1 inch. The material properties are selected to be those of aluminum: modulus of elasticity, E, is 10E6 psi, and the density is 9.55E-2 lbm/in^3.

Although the model was limited to uniform elements, there were any number of combinations one can find to represent the system. This study examined two different methods for modeling a linearly tapered beam. As shown in Figure 4.1 the link was tapered along the length in the width dimension while the height was held constant. The taper was described by two dimensions: the width at the base, A, and the width at the tip, B. The degree of taper, R=A/B, was used to compare different designs.

Using Method 1 to model the tapered link, the beam was divided into elements of equal length. For a three element model with length L, each element will have length L/3. The height of each element was the same, while the width of each element changed linearly as a function of x. Figure 4.2 presents modeling Method 1.

Using Method 2 to model the tapered link, the beam was divided into elements so the first and last element have length one-half of the intermediate elements. For a three element model with length L, the first and last elements will have length L/4 and the middle element will have length L/2. Again the height of each element was the same, while the width of each element changed linearly as a function of x. Figure 4.3 presents modeling Method 2.

Figures 4.2 and 4.3 illustrate the main difference between the two modeling methods. Method 2 compensated the elements at each end for meeting the specified end widths A and B. In both methods the width of intermediate elements was determined by the width of the tapered beam at the midpoint of each element. Since the end elements meet the specified A and B, the tapered link will not pass through the midpoint of these two elements. Method 2 compensates for this exception by making the end element lengths one half the length of the other elements.

To compare these two different modeling methods for a linearly tapered beam, a beam with nominal dimensions and A=0.75 inches and B=0.25 inches was studied. This corresponds to R=3. The number of elements was increased with each method until the zeros and poles converged. Table 4.3 presents the results from Method 1 where all elements were of equal length, and Table 4.4 presents the results from Method 2 where the end elements were half the length of all other elements. Although only two methods are considered in this research, there are many different ways to discretize a nonuniform link.

The two methods were evaluated based on an error function. When the tapered beam was modeled with 80 elements, both methods converged to nearly identical values for the poles and zeros. These values, when NE=80, were taken to be the "correct' values and other cases were compared to this case. The error, ϵ, was defined for the zeros as:

$$\epsilon = \left| \frac{z_{80,i} - z_{NE,i}}{z_{80,i}} \right| \qquad (4.2)$$

where i refers to the ith zero

A similar definition was used for the poles. The value of ϵ at the top of each column represents the maximum of all individual errors in each column. As the tables show, Method 2 provided better results for the same number of elements. In each table, one column was shaded to distinguish it as the number of elements needed to get the error under 1%. For Method 2, this column corresponded to NE=10 as opposed to NE=20 for Method 1. Thus, compensating the end elements did provide a better model of a linearly tapered beam, and this method was used in the following studies unless specified otherwise.

When comparing different link designs to evaluate pole/zero location as a function of link shape, it was necessary to keep some parameter constant to aid in the evaluation. For a single-link manipulator rotating in the horizontal plane, the link's mass moment of inertia about its axis of rotation, I_y, was of importance. This parameter directly affected the dynamic equations of motion and was an important design parameter in terms of motor selection. In the following studies, several link designs were evaluated for a given value of I_y. A tapered link's moment of inertia about its axis of rotation in terms of the links parameters: L, A, B, H, and ρ is found to be:

$$I_y = \frac{\rho H}{48}(A^3 + A^2B + AB^2 + B^3 + 4AL^2 + 12BL^2) \tag{4.3}$$

For a given tapered link design, one can use Equation 4.3 to determine I_y. Knowing I_y, one can change the value of A and solve Equation 4.3 for B. Since the equation was cubic in B, the commercial package *Mathematica* was used to solve for B. Following this method, a group of tapered link designs were generated all with the same I_y.

The first study investigated several tapered link designs with nominal dimensions and all designs having I_y=764.05 in-lb-sec^2. Table 4.5 presents the raw data for each of these designs. Even with I_y held constant, it was still difficult to interpret the data. To aid in developing a relationship between zero location and link shape, the zeros were normalized with respect to the first pole for each design. The first pole is an important parameter in control system design, and normalizing the zeros with respect to the first pole aided in the interpretation of the results. Table 4.6 presents the normalized data for those designs with I_y=764.05 in-lb-sec^2. The second study presents data for several link designs with nominal dimensions and I_y=1528.1 in-lb-sec^2. Table 4.7 shows the raw data for these link designs and Table 4.8 shows the normalized data for these designs. Figures 4.4 and 4.5 show pole/zero maps for selected values of R for I_y=764.05 and I_y=1528.1 respectively.

Several patterns were evident by examining the raw data. First as a general rule, both the poles and zeros increased (moved away from the origin) as the taper on the beam increased. Increasing the taper effectively moved more of the link mass closer to the base. Increasing the value of the poles is often desirable to push them out of the system bandwidth and increase system response speed. The ordering of poles and zeros was the second pattern recognized. In a collocated system, the poles and zeros will both lie on the imaginary axis in complex conjugate pairs and in an alternating order. This means, along the imaginary axis, the poles and zero are found in the order p_1, z_1, p_2,z_2, etc. or vice versa. Previous research [18] has found this alternating order of poles and zeros does not hold for nonminimum phase systems. Referring to Table 4.5, notice the order of the magnitude of poles and zeros was: $z_1,p_1,p_2,z_2,p_3,z_3,p_4,p_5,z_4....$ p_2 jumped in front of z_2, and the same occurred for p_5. This reordering of poles and zeros can be critical as accurate knowledge of the pole/zero order is important for control system design.

Important information was learned from examining the relationship between the taper ratio, R, and the values of the normalized zeros. Figure 4.8 better illustrates this point showing both polynomial fits on the same graph. Even though the coefficients were different for each polynomial fit, the curves were nearly identical.

This illustrates an important relationship in the design of tapered links. For a given ratio R, the normalized zero will always remain the same. The designer can choose the location of the first pole and zero, determine the normalized zero, and then using Figure 4.8 find the appropriate taper ratio R. Of course there are constraints on this process. A ratio less than one corresponds to a taper with B greater than A, which is usually undesirable. At the other end, R is limited by the value of H. If A is larger than the value of H, the link will be wider at the base than it is tall, and the assumption that the link is stiff in the vertical plane will no longer be valid. Although the designer can choose the pole/zero relationship, the values of normalized zeros are limited to approximately 0.72-0.82 (according to Figure 4.8).

A simple verification of the above relationship is the uniform beam which has no taper. According to the stated relationship, the normalized first zero should be the same for all uniform beams. Table 4.9 presents the results for several uniform beam designs. All cases had nominal dimensions. The normalized zero in all cases was 0.726 which confirmed the normalized zero will not change as long as R is constant.

Previous studies demonstrated how the designer can choose the pole/zero relationship and then determine the appropriate taper design from the ZERO results. This study presents the designer with another freedom. Once the taper is chosen, the designer can change the link to independently adjust the value of I_y. Table 4.10 presents the results of a study performed on designs with L=40 inches, and all designs have the same taper. The height of the link was changed to adjust the value of I_y.

One should notice that the pole and zero locations of all designs in Table 4.10 were the same, yet the value of I_y changed with adjustments in link height. Since the adjustment of H is out of the plane of motion, it had no effect on the location of poles and zeros. Combining this with the results from the previous study, the designer can effectively choose the location of poles and zeros and independently adjust the links moment of inertia about its axis of rotation to meet the needs of the particular system.

CONCLUSIONS

Program ZERO was developed as a tool to locate the poles and zeros of a single-link manipulator modeled as a pinned-free Euler-Bernoulli beam. The program used transfer matrix theory to allow for variable cross-sections granting the designer new freedom in analysis of nonuniform link designs. The results were shown to be very accurate when system pole location was compared to analytic solutions for uniform beams. Several results from previous studies were confirmed with this research.

First, the reordering of poles and zeros was confirmed for nonminimum phase systems. Accurate knowledge of pole/zero order is critical for proper control system design. In conjunction with this, Tables 4.3 and 4.4 show that even for very few elements in the model, the program still predicts the proper order of poles and zeros.

Second, the studies presented suggested the nonminimum phase characteristics could not be eliminated by changing the structural design of the link. The system will be nonminimum phase above a finite frequency dictated by the location of the first nonminimum phase zero. It may be possible that this frequency is out of the operating range and not of concern to the designer.

The major contributions of this research are the development of the ZERO program to determine zero and pole location for a single-link nonuniform flexible manipulator, and formulation of a design procedure to place the first pole and zero and independently change the value of the link's moment of inertia about its axis of rotation to meet the needs of the system.

Program ZERO was set up specifically for pinned-free boundary conditions of the model and determines pole and zero location based on a frequency range entered by the user. Linearly tapered beams were studied in this research, but any type of nonuniform beam can be analyzed by program ZERO. Slight modifications would also allow for different boundary conditions.

The design procedure for tapered beams allows the designer to choose the first pole and zero subject to certain physical constraints. These physical constraints only allow for approximately 25% variation in R according to Table 4.6. This zero to pole ratio defines a particular taper ratio according to the collected data. Keeping the ratio the same, the size of the taper can be changed to get the proper magnitude of the pole and zero. With the pole and zero placed, the height of the beam can be changed to adjust the link's moment of inertia about its axis of rotation. This procedure can be used to design tapered links to meet the particular requirements of the system.

Program ZERO was designed to model a single-link manipulator modeled with pinned-free boundary conditions. This is a simplified model, but it was necessary to show transfer matrices yield good results for this case before progressing to more complicated problems. Now that transfer matrices have proven useful to solve for zero location, future work exists to extend the results of this research.

First, the program could be modified so the user could input the desired boundary conditions which best represent the system. This could include hub inertia or end-point mass. Second, the program could be extended to multi-link designs to predict pole and zero location for different configurations. Transfer matrices have been derived for rotary joints and many other elements. The DSAP package developed by Book, et. al. [6] handles multi-link models and would be a good reference. Finally, the results for tapered link designs could be applied to the inverse dynamic algorithm developed by Kwon and Book [9]. This method requires mode shapes for the assumed modes and uses pinned-pinned boundary conditions, which can also be found using transfer matrix techniques as shown in Book, et al.[6].

ACKNOWLEDGEMENTS

This work was performed with partial support from grant NAG 1-623 from the National Aeronautics and Space Administration. NASA bears no responsibility for its content.

BIBLIOGRAPHY

[1] Asada, H., Park, J.-H., and Rai, S., "A Control-Configured Flexible Arm: Integrated Structure/Control Design," *Proceedings of the 1991 IEEE International Conference on Robotics and Automation*, Sacramento, California, April, 1991, pp. 2356-2362.

[2] Bayo, E., "A Finite Element Approach to Control the End-Point Motion of a Single-Link Flexible Robot," *Journal of Robotic Systems*, Vol. 4, No. 1, 1987, pp.63-75.

[3] Beer, Ferdinand, and Johnson, Russell, Jr., *Vector Mechanics for Engineers, Statics and Dynamics*, Third Edition, McGraw-Hill, New York, 1977.

[4] Book, W. J., *Design and Control of Flexible Manipulator Arms*, Ph.D. Thesis, Massachusetts Institute of Technology, April, 1974.

[5] Book, W. J., and Kwon, D.-S., "Contact Control for Advanced Applications of Light Weight Arms," *Symposium on Control of Robots and Manufacturing*, Arlington, Texas, 1990.

[6] Book, W. J., Majette, M., and Ma, K., *The Distributed Systems Analysis Package (DSAP) and Its Application to Modeling Flexible Manipulators*, NASA Contract NAS 9-13809, Subcontract No. 551, School of Mechanical Engineering, Georgia Institute of Technology, 1979.

[7] Churchill, R. V., and Brown, J. W., *Complex Variables and Applications*, Fifth Edition, McGraw-Hill Publishing Company, New York, 1990.

[8] Kwon, D.-S., *An Inverse Dynamic Tracking Control for Bracing A Flexible Manipulator*, Ph.D. Dissertation, Georgia Tech, Woodruff School of Mechanical Engineering, June, 1991.

[9] Kwon, D.-S., and Book, W. J., "An Inverse Dynamics Method Yielding Flexible Manipulator State Trajectories," *Proceedings of the American Control Conference*, June, 1990, pp. 186-193.

[10] Majette, M., *Modal State Variable Control of a Linear Distributed Mechanical System Modeled with the Transfer Matrix Method*, Master's Thesis, Georgia Tech, Woodruff School of Mechanical Engineering, June, 1985.

[11] Meirovitch, L., *Elements of Vibrational Analysis*, McGraw-Hill, New York, 1986.

[12] Misra, Pradee, "On The Control of Non-Minimum Phase Systems," *Proceedings of the 1989 American Control Conference*, 1989, pp. 1295-1296.

[13] Nebot, E. M., Lee, G. K. F., and Brubaker, T. A., "Experiments on a Single Link Flexible Manipulator,", *Proceedings from the USA-Japan Symposium on Flexible Automation Crossing Bridges: Advances in Flexible Automation and Robotics*, 1988, pp. 391-398.

[14] Park, J.-H., and Asada, H., "Design and Analysis of Flexible Arms for Minimum-Phase Endpoint Control," *Proceedings of the American Control Conference*, 1990, pp. 1220-1225.

[15] Park, J.-H., and Asada, H., "Design and Control of Minimum-Phase Flexible Arms with Torque Transmission Mechanisms," *Proceedings of the 1990 IEEE International Conference on Robotics and Automation*, 1990, pp. 1790-1795.

[16] Pestel and Leckie, *Matrix Methods in Elastomechanics*, McGraw-Hill, New York, 1963.

[17] Rao, Singiresu S., *Mechanical Vibrations*, Addison-Wesley Publishing Company, Reading, Massachusetts, 1986.

[18] Spector, V. A., and Flashner, H., "Modeling and Design Implications of Noncollocated Control in Flexible Systems," *Journal of Dynamic Systems, Measurement, and Control*, Vol. 112, June, 1990, pp. 186-193.

[19] Spector, V. A., and Flashner, H., "Sensitivity of Structural Models for Noncollocated Control Systems," *Journal of Dynamic Systems, Measurement, and Control*, Vol. 111, December, 1989, pp. 646-655.

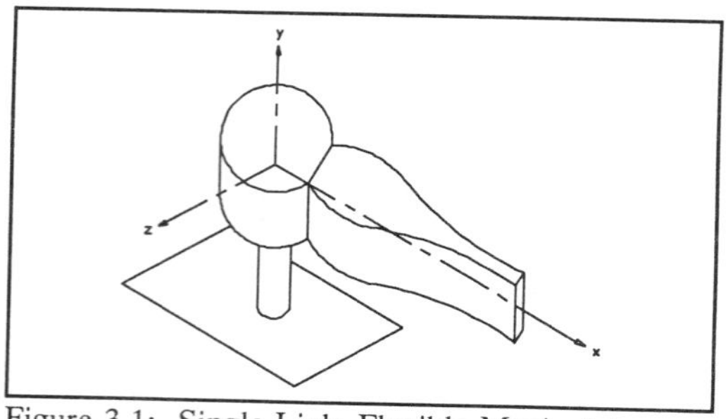

Figure 3.1: Single-Link, Flexible Manipulator

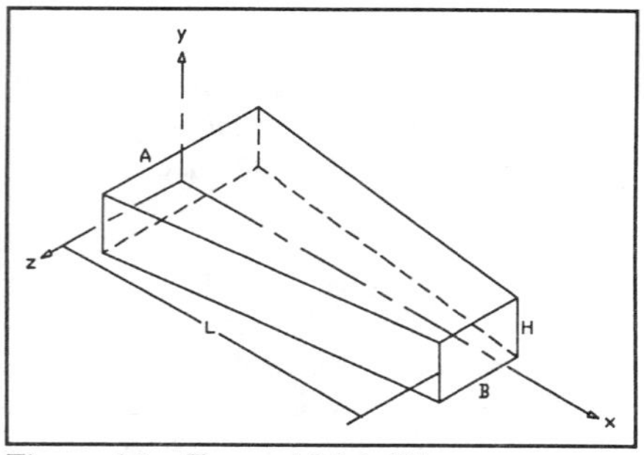

Figure 4.1: Tapered Link Diagram

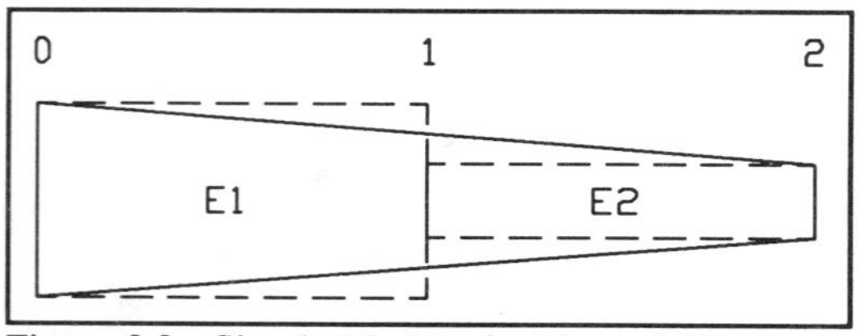

Figure 3.3: Simple Model of a Tapered Beam

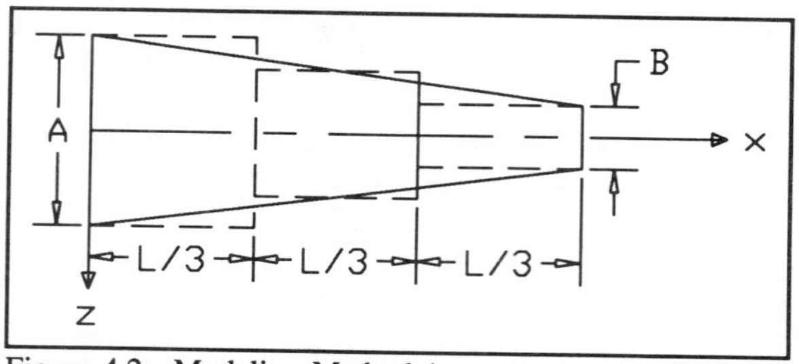

Figure 4.2: Modeling Method 1

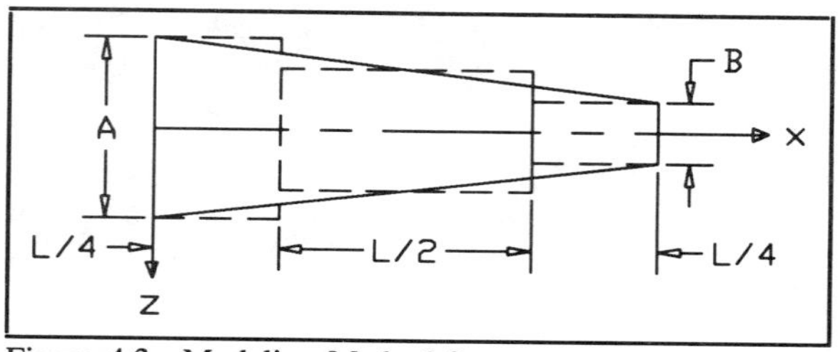

Figure 4.3: Modeling Method 2

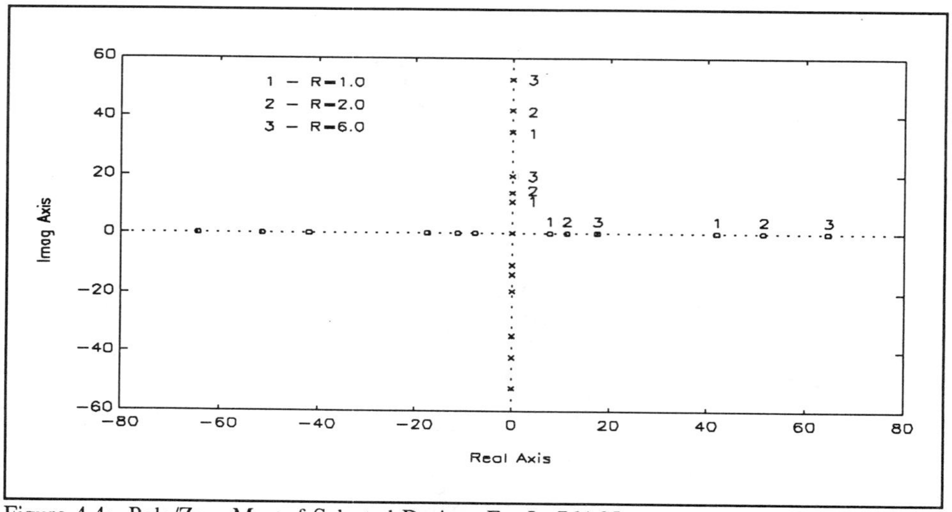

Figure 4.4: Pole/Zero Map of Selected Designs For I_y=764.05

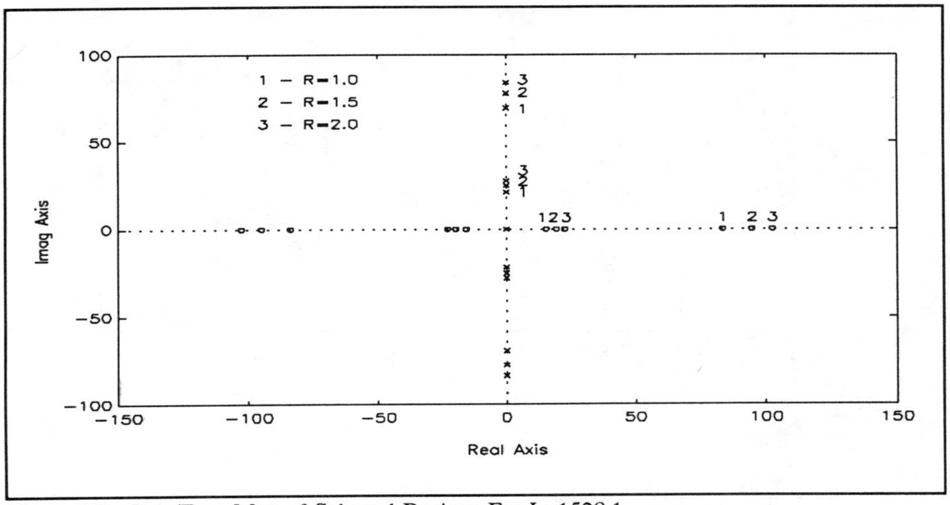

Figure 4.5: Pole/Zero Map of Selected Designs For I_y=1528.1

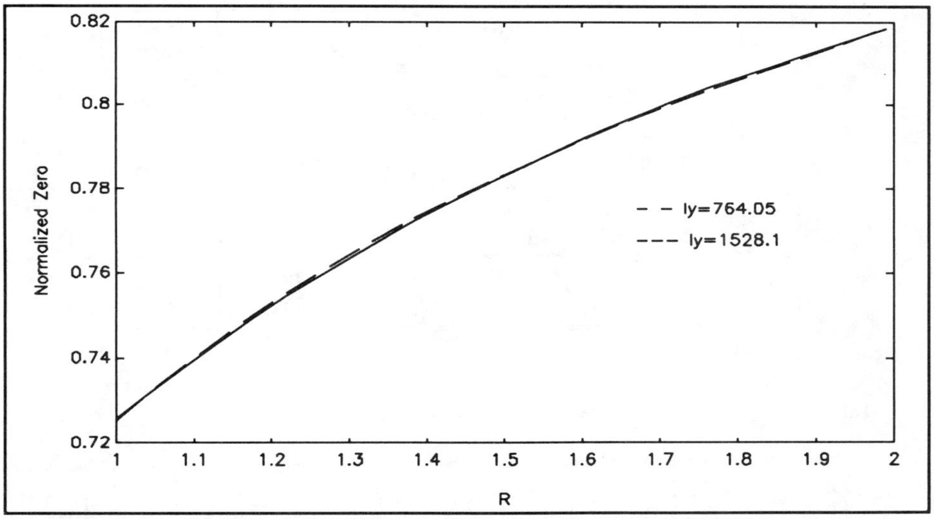

Figure 4.8: Comparison of Polynomial Curve Fits

Table 4.3: Results From Method 1

Zero Pole	NE=3 (ε≤20%)	NE=5 (ε≤8.6%)	NE=10 (ε≤1.9%)	NE=20 (ε≤0.3%)	NE=40 (ε≤0.1%)	NE=80
1	13.91 13.64	13.99 15.95	13.73 16.05	13.69 15.96	13.68 15.92	13.68 15.91
2	45.57 38.08	55.99 43.24	57.28 46.19	56.91 46.21	56.84 46.14	56.83 46.11
3	121.2 88.16	122.0 85.31	134.2 92.52	133.8 93.20	133.6 93.13	133.5 93.09
4	210.8 137.5	223.2 147.5	242.9 154.9	244.7 157.0	244.2 157.0	244.1 157.0
5	357.8 219.5	383.8 234.7	382.1 233.3	389.4 237.7	388.9 237.9	388.7 237.8

Table 4.4: Results From Method 2

Zero Pole	NE=3 (ε≤16%)	NE=5 (ε≤4.0%)	NE=10 (ε≤0.4%)	NE=20 (ε≤0.1%)	NE=40 (ε≤0.0%)	NE=80
1	13.09 15.57	13.49 15.82	13.64 15.89	13.67 15.90	13.68 15.91	13.68 15.91
2	53.77 38.66	56.12 45.82	56.62 46.03	56.78 46.09	56.82 46.10	56.83 46.10
3	120.4 85.88	135.1 93.17	133.0 92.90	133.4 93.03	133.5 93.06	133.5 93.06
4	233.6 154.4	234.7 148.3	243.4 156.6	243.9 156.9	244.1 156.9	244.1 156.9
5	360.6 220.3	384.6 231.1	388.2 237.4	388.3 237.7	388.6 237.7	388.6 237.8

Table 4.5: Tapered Beams With I_y=764.05

Zero Pole	A=.375 B=.375	A=.4 B=.367	A=.5 B=.333	A=.6 B=.3	A=.7 B=.267	A=.8 B=.233	A=.9 B=.2	A=1 B=.167
1	7.745 10.68	8.153 11.04	9.762 12.46	11.34 13.84	12.90 15.21	14.44 16.60	15.98 18.03	17.50 19.52
2	41.85 34.59	43.15 35.48	47.38 38.80	51.37 41.87	55.05 44.73	58.45 47.41	61.60 49.94	64.51 52.36
3	103.4 72.18	105.9 73.88	115.0 80.17	123.1 85.75	130.2 90.75	136.4 95.19	141.7 99.14	146.2 102.6
4	192.2 123.4	196.6 126.2	212.7 136.5	226.6 145.5	238.6 153.4	248.7 160.1	257.1 165.9	263.6 170.6
5	308.4 188.3	315.3 192.6	340.5 208.0	362.0 221.2	380.3 232.6	395.5 242.3	407.8 250.3	416.9 256.5

Table 4.6: Normalized Data For I_y=764.05

Zero	R=1.00	R=1.09	R=1.50	R=2.00	R=2.62	R=3.43	R=4.50	R=5.99
1	0.7252	0.7385	0.7835	0.8194	0.8481	0.8699	0.8863	0.8965
2	3.919	3.909	3.803	3.712	3.619	3.521	3.417	3.305
3	9.682	9.592	9.230	8.895	8.560	8.217	7.859	7.490
4	18.00	17.81	17.07	16.37	15.69	14.98	14.26	13.50
5	28.88	28.56	27.33	26.16	25.00	23.83	22.62	21.36

Table 4.7: Tapered Beams With I_y=1528.1

Zero Pole	A=.75 B=.75	A=.8 B=.733	A=.9 B=.7	A=1.0 B=.667	A=1.1 B=.633	A=1.2 B=.600
1	15.49 21.35	16.31 22.08	17.92 23.51	19.52 24.92	21.11 26.30	22.68 27.68
2	83.71 69.16	86.03 70.95	90.50 74.35	94.76 77.60	98.83 80.73	102.7 83.74
3	206.7 144.4	211.7 147.7	221.2 154.2	230.1 160.3	238.4 166.1	246.2 171.5
4	384.4 246.8	393.2 252.5	409.9 263.1	425.4 273.1	439.9 282.4	453.2 291.0
5	616.7 376.7	630.6 385.1	656.8 401.1	681.0 415.9	703.3 429.6	724.0 442.4

Table 4.9: ZERO Results For Uniform Beam Designs

Zero Pole	W=0.25"	W=0.5"	W=0.75"
1	5.163 7.116	10.33 14.23	15.49 21.35
NZ	0.726	0.726	0.726
2	27.90 23.06	55.80 46.12	83.71 69.19
3	68.90 48.12	137.8 96.23	206.7 144.3
4	128.1 82.28	256.2 164.6	384.4 246.8
5	205.6 125.6	411.1 251.1	616.7 376.7

Table 4.10: Variable Height Designs

Zero Pole	H=1.0"	H=1.5"	H=2.0"
1	11.34 13.84	11.34 13.84	11.34 13.84
2	51.37 41.87	51.37 41.87	51.37 41.87
3	123.1 85.75	123.1 85.75	123.1 85.75
4	226.6 145.5	226.6 145.5	226.6 145.5
5	362.0 221.2	362.0 221.2	362.0 221.2
I_y	764.05	1146.1	1528.1

AMD-Vol. 141/DSC-Vol. 37, Dynamics of Flexible Multibody Systems:
Theory and Experiment
ASME 1992

END-POINT CONTROL OF A TWO-LINK FLEXIBLE ROBOTIC MANIPULATOR WITH A MINI-MANIPULATOR: COUPLING ISSUES

W. L. Ballhaus and Stephen M. Rock
Aerospace Robotics Laboratory
Stanford University
Stanford, California

Abstract

Previous experiments have established the benefits and feasibility of an end-point based control scheme using a mini-manipulator at the tip of a two-link flexible manipulator [1]. The result was quick and accurate position control within a localized workspace. The control approach incorporated into those experiments involved a partitioned structure. A main arm controller operated to carry the mini-manipulator within range of the desired end-point position. A low-gain mini-manipulator controller then provided precise end-point control. The mini-manipulator end-point position controller was able to maintain small end-point errors long before the main arm controller brought the main arm to rest.

This research investigates the dynamic coupling that exists between the mini-manipulator and the main arm. This two-way coupling ultimately limits the performance achievable with a partitioned control approach. Specifically, instability can occur as the gains of the mini-manipulator end-point controller are increased. This paper demonstrates experimentally a coordinated slew of the mini-manipulator/main arm system, and the instability which results from increasing the end-point controller gains.

1 Introduction

In a number of robotic applications, it is necessary to achieve high-bandwidth position control within a localized workspace at the manipulator end-point. In the case of large, lightweight manipulators, which are common in space robotic applications, inherent structural flexibility and actuators far removed from the manipulator end-point limit the speed and accuracy with which tasks in a localized workspace can be performed.

Recent research has demonstrated that the use of a mini-manipulator and end-point sensing eliminates the accuracy problem which results from having actuators far removed from the robot end-point. Further, it significantly reduces the bandwidth limitations imposed by inherent structural flexibility for tasks within the localized workspace [1]. These benefits of the mini-manipulator come at the expense of possible complexities, however. Adding the mini-manipulator to the end of a flexible manipulator results in a two-way dynamically coupled system. When the gains of the mini-manipulator end-point controller are low, the coupling effects are insignificant, and the mini-manipulator controller is able to maintain small errors even in the presence of relatively large main arm motion. Increasing the end-point controller gains yields significant coupling between the mini-manipulator and the main arm. In addition, the effects of coupling are compounded by the presence of delay in the vision sensor. Because the mini-manipulator end-point controller utilizes measurements provided by the vision sensor, the presence of vision sensor delay reduces the overall system performance. In fact, the vision sensor delay and the two-way dynamic coupling can lead to instability as the end-point controller gains are increased.

The goal of this research is to demonstrate experimentally the instability of the mini-manipulator/main arm system which results from increasing the mini-manipulator end-point controller gains.

2 Experimental Apparatus

An experimental apparatus has been developed to allow for investigation of the dynamic coupling issues associated with a mini-manipulator mounted at the tip of an extremely flexible multi-link manipulator. In particular, the hardware has been designed to enhance the issues and problems associated with inherent structural flexibility. A schematic of the experimental apparatus is shown in Figure 1. The apparatus consists of two major subsystems: the flexible main arm and the mini-manipulator.

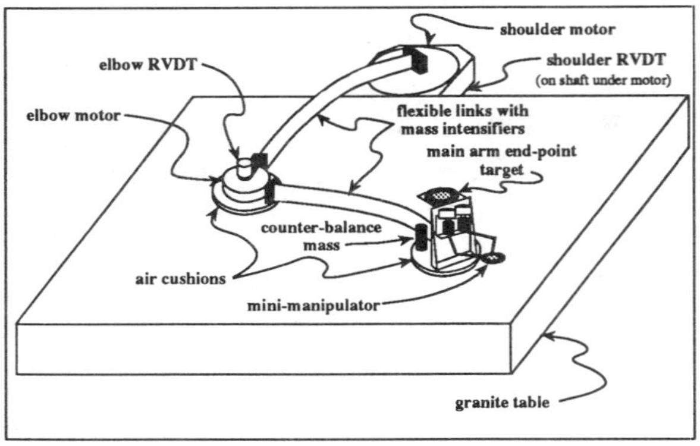

Figure 1: **Experimental Hardware Schematic**

Shown in this figure is the Stanford Multi-Link Flexible Manipulator with a mini-manipulator mounted at the tip.

2.1 Main Arm

The Stanford Multi-Link Flexible Manipulator consists of a two-link flexible manipulator which operates on air cushions in the horizontal plane of a 1.2 m by 2.4 m granite table. Each of the flexible links is 0.52 m in length, and consists of an aluminum beam (cross section 1 mm by 38.1 mm) with discrete masses evenly spaced along its length. The discrete masses, termed "Mass Intensifiers", increase the overall beam mass without changing its flexural rigidity. These Mass Intensifiers also lower the natural frequencies of vibration. The flexible links exhibit significant bending in the horizontal plane due to their narrow cross section and orientation. Air cushion supports at the elbow and end-point provide torsional stiffness. In addition to exaggerating the link flexibility, the structure was designed to be lightly damped. As a result, damping must be provided to the system through active control. The shoulder motor, mounted on the side of the granite table, can provide a peak torque of 5.43 N-m. The elbow motor, mounted on the elbow air-cushion pad, can provide a peak torque of 1.06 N-m. Both actuators are direct-drive, DC limited-angle torquers. Rotary variable differential transformers (RVDT's) are located at each of the motor shafts and provide joint angle measurements. A vision sensor, fully described in [2], is available for end-point measurements. It consists of a CCD television camera that tracks a special variable reflectivity target located at the manipulator end-point. The vision system has the capability to track multiple targets at a sample rate of 60 Hz with a resolution of approximately 1 mm over the roughly 1.5 m^2 workspace [3].

2.2 Mini-Manipulator

A two degree of freedom mini-manipulator is mounted on the tip of the two-link flexible main arm. The mini-manipulator, shown in Figure 2, consists of a five-link, closed-kinematic chain which operates in the horizontal plane. The base link, inner links, and outer links of the mini-manipulator are approximately 5.1 cm, 7.6 cm, and 10.2 cm in length respectively. To allow for quick, precise motion of the tip, the four moving links are made of hollow tubes, and are actuated by two small DC electric motors located at the base of the mini-manipulator. The motors have a peak torque of 0.35 N-m. Rotary encoders mounted on top of the motors provide angular position and velocity information. The vision system tracks a reflectivity target to provide end-point position information of the mini-manipulator.

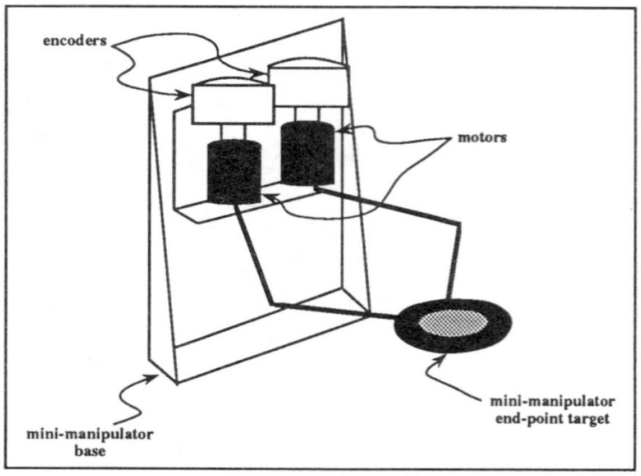

Figure 2: **Mini-Manipulator Schematic**

This figure illustrates the mini-manipulator and its various components.

3 Control Strategy

A block diagram of the system control strategy is shown in Figure 3. The system controller consists of two components: the main arm controller and the mini-manipulator controller. The mini-manipulator controller is designed to control the mini-manipulator end-point position in Cartesian space. The goal of the main arm controller is to position the base of the mini-manipulator at a location from which the target is attainable. In this initial research, the two controllers are designed independently.

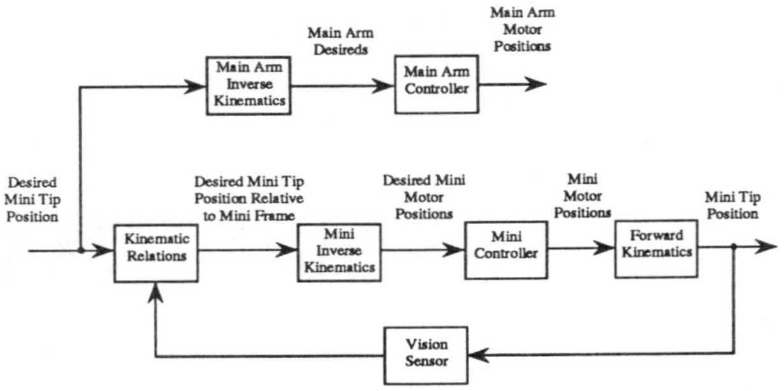

Figure 3: **System Control Strategy**

Shown in this figure is a block diagram of the combined mini-manipulator and main arm system control strategy.

3.1 Main Arm Controller

Desired joint positions and velocities are the inputs to the main arm controller. Desired final shoulder and elbow motor angles are generated from the reference mini-manipulator tip goal position using inverse kinematics (assuming rigid links). A fifth order trajectory from the initial motor positions to the final motor positions generates the desired main arm motor positions and velocities for a given slew time.

Although a high performance control strategy has previously been implemented on the main arm [4], it did not account for the additional dynamic effects of the mini-manipulator. Consequently, a PD controller is used in this initial research. The PD controller runs at 400 Hz, and the gains were chosen to maximize the closed-loop bandwidth of the main arm while maintaining sufficient damping.

3.2 Mini-Manipulator Controller

The mini-manipulator controller is an end-point based controller. The vision sensor provides end-point position measurements which are utilized to express the desired tip position in the mini-manipulator reference frame. If this desired relative tip position is within the mini-manipulator's workspace, the corresponding desired motor positions are determined from the mini-manipulator inverse kinematic relations given in [1]. If the desired relative tip position is not within the workspace, the desired motor positions are set to a nominal value. The desired motor positions serve as a reference to the mini-manipulator PD controller. The PD controller feedback gains were experimentally determined to yield a high-bandwidth, well damped response.

Typical closed loop motor responses of the mini-manipulator left motor are shown in Figures 4 and 5. Figures 4 and 5 correspond to lateral and longitudinal motion of the mini-manipulator end-point relative to its base. With an ideal vision sensor (without delay) and no dynamic coupling interaction between the mini-manipulator and the main arm, these figures would represent a mini-manipulator end-point controller with closed loop bandwidths of approximately 9.2 Hz and 7.2 Hz in the lateral and longitudinal directions respectively.

However, the actual vision system contains a delay which degrades the performance of the mini-manipulator end-point controller. This delay is evident in Figure 6. Figure 6 displays the transfer function from the actual rotation angle of the mini-manipulator frame to the rotation angle measurement provided by the vision sensor. The magnitude response is essentially a unity gain. Because of the vision sensor delay, the phase plot shows a significant decrease in phase with frequency. To obtain this data, the mini-manipulator was mounted at the end of a single link rigid manipulator. The actual rotation angle was measured by the RVDT located at the hub of the manipulator.

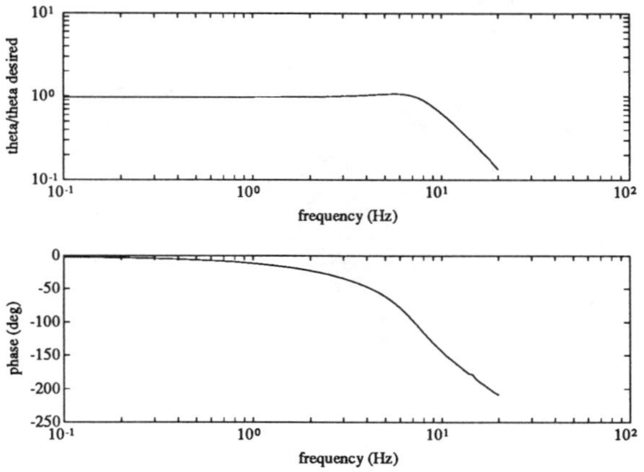

Figure 4: **Closed Loop Motor Response**

This figure shows an experimental mini-manipulator closed loop motor response for the left motor. This response corresponds to lateral motion of the mini-manipulator end-point relative to its base.

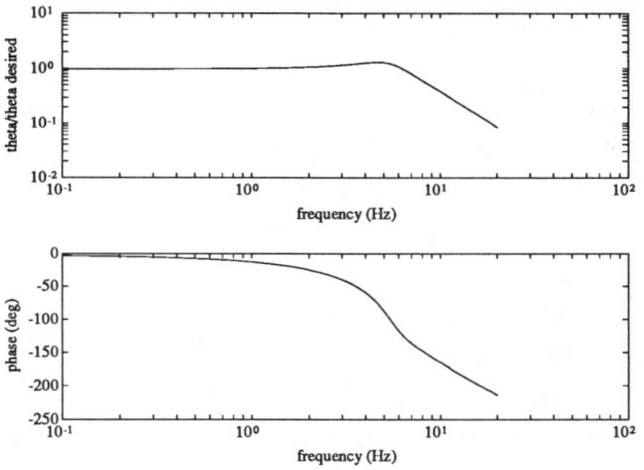

Figure 5: **Closed Loop Motor Response**

This figure shows an experimental mini-manipulator closed loop motor response for the left motor. This response corresponds to longitudinal motion of the mini-manipulator end-point relative to its base.

4 Experimental Results

4.1 Coordinated Slew

Figures 7 and 8 show the position responses of both the mini-manipulator tip and the end-point of the main arm during a coordinated slew of the mini-manipulator/main arm system. These figures illustrate the advantages of speed and precision offered by the mini-manipulator. Once the main arm brings the mini-manipulator within range of the target (desired mini-manipulator end-point position), the mini-manipulator end-point controller is able to maintain small end-point errors long before the main arm PD controller brings the main arm to rest.

Without end-point sensing, the mini-manipulator would maintain its nominal configuration, neglecting to compensate for the motion of its base. As a result, the displacement of the mini-manipulator tip would equal that of the main arm end-point.

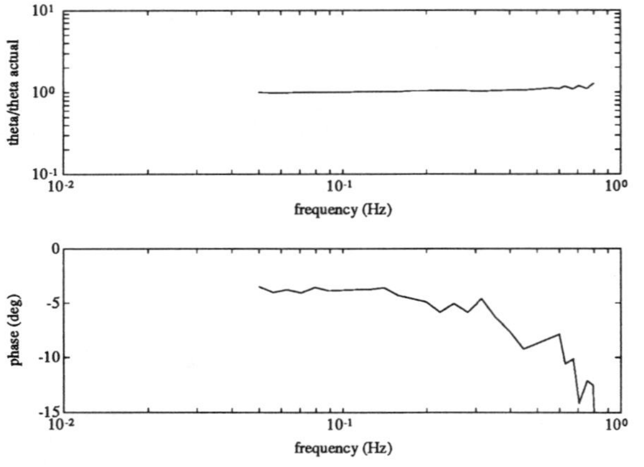

Figure 6: **Vision Sensor Transfer Function**

This figure shows magnitude and phase frequency responses for the transfer function from the actual rotation angle of the mini-manipulator frame to the rotation angle measurement provided by the vision sensor.

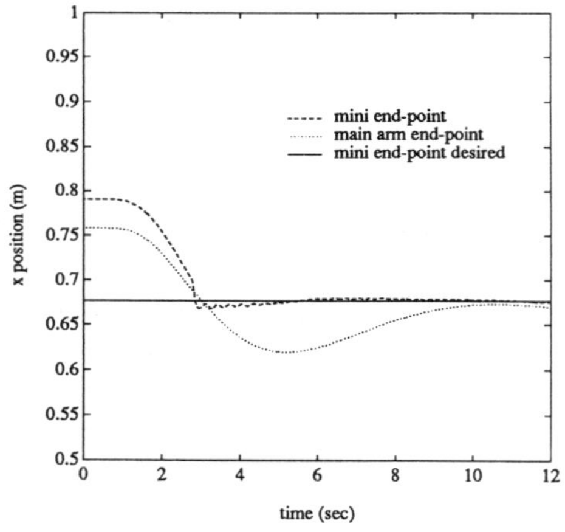

Figure 7: **Coordinated Slew**

End-point position responses for a coordinated slew of the mini-manipulator/main arm system are displayed in this figure.

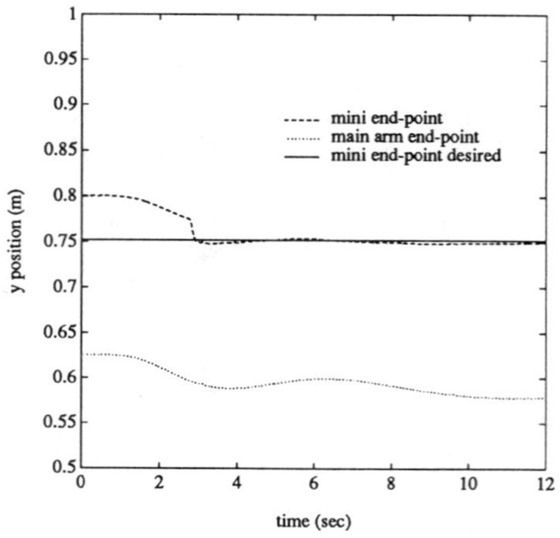

Figure 8: **Coordinated Slew**

End-point position responses for a coordinated slew of the mini-manipulator/main arm system are displayed in this figure.

4.2 Dynamic Coupling

Dynamic coupling between the mini-manipulator and the main arm is evident even for the low gain end-point controller responses shown in Figures 7 and 8. The coupling effects are most noticeable in the plot of the rotation angle of the main arm end-point shown in Figure 9. At approximately five seconds, the main arm has positioned the mini-manipulator such that the desired end-point position is within reach. As the end-point controller attempts to zero the end-point error, reaction forces are imparted by the mini-manipulator on the main arm. These reactions induce a 3 Hz oscillation in the main arm, and result

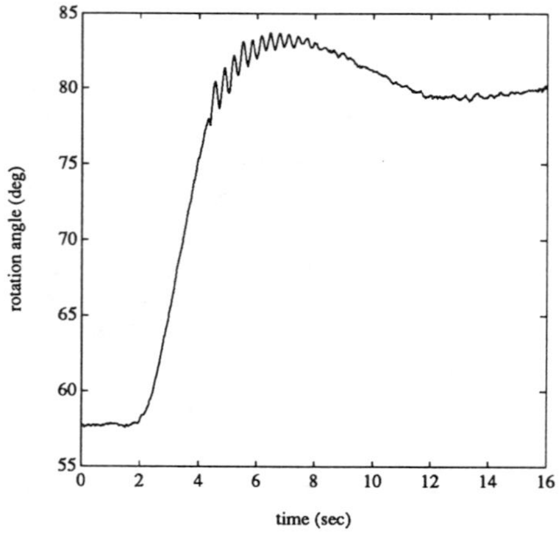

Figure 9: **Tip Rotation of the Main Arm**

The tip rotation angle of the main arm is displayed in this figure. Dynamic coupling between the mini-manipulator and the main arm is evident even for a low gain end-point controller.

in a slight rotation of the main arm end-point.

For increased end-point controller gains, the coupling becomes more significant as illustrated in Figure 10. Increasing the gains further can result in instability. As the stabilizing effects of the main arm PD controller become negligible compared to the coupling interaction caused by the mini-manipulator end-point controller, the system becomes unstable. Figure 11 demonstrates this instability for the case where the PD controller gains are zero.

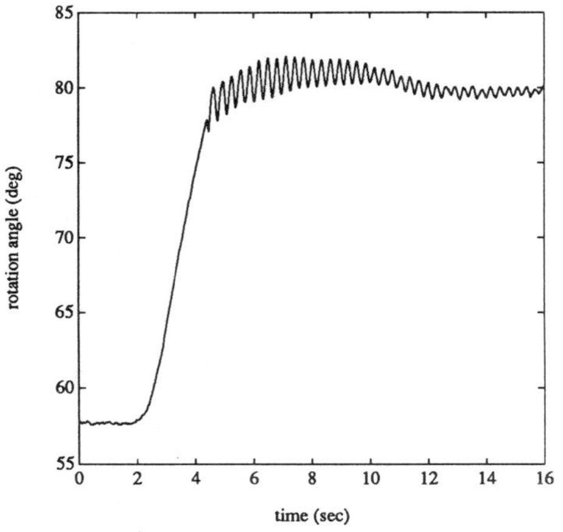

Figure 10: **Tip Rotation of the Main Arm**

This figure illustrates that for increased end-point controller gains, the dynamic coupling becomes more significant

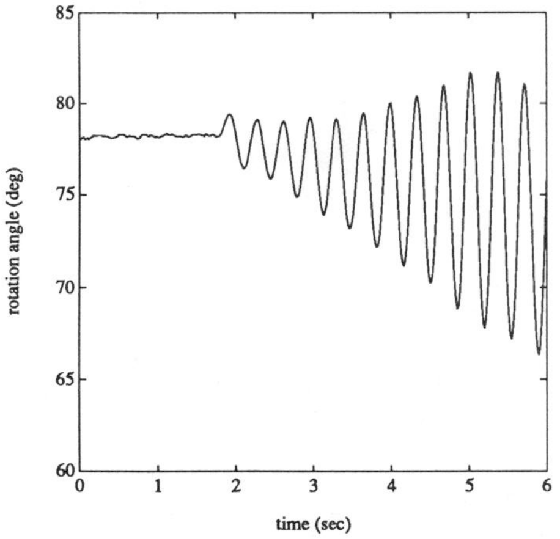

Figure 11: **Tip Rotation of the Main Arm**

Unmodelled coupling effects can lead to instability. This figure demonstrates this instability for the case where the PD controller gains are zero.

Figure 12 shows the collocated transfer functions of the main arm with the mini-manipulator mounted on its tip. The first system mode, shown in the elbow transfer function plot, occurs at approximately 3 Hz. Because of the relatively large tip and hub masses, the second link closely resembles a beam with pinned-pinned boundary conditions. As a result, the tip motion corresponding to the first system mode is primarily a rotation of the tip of the second link. The response shown in Figure 10 indicates that the unmodelled coupling effects destabilize the first system mode.

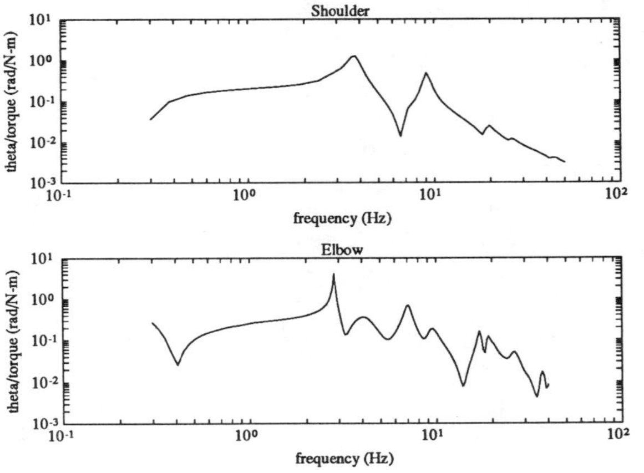

Figure 12: **Experimental Frequency Response**

This figure shows the experimental collocated transfer functions for the main arm with the mini-manipulator mounted at its tip.

5 Conclusions

This research has identified many of the issues associated with the two-way coupled mini-manipulator/main arm system. Experiments have shown that the two-way coupling limits the performance achievable with a partitioned control approach. Because of vision sensor delay and dynamic coupling, increasing the gains of the mini-manipulator end-point controller decreases the overall system performance. Specifically, experiments demonstrated a destabilization of the first system mode due to these unmodelled coupling effects.

Acknowledgements

The research reported here has been funded by the Air Force Office of Scientific Research (AFOSR). The authors would like to thank Professor Robert H. Cannon Jr. and all of the students and technical staff at the Stanford Aerospace Robotics Laboratory for their support, assistance, and suggestions.

References

[1] W. L. Ballhaus and S. M. Rock. End-point control of a two-link flexible robotic manipulator with a mini-manipulator: Initial experiments. In *Proceedings of the American Control Conference*, Chicago, IL, June 1992.

[2] S. Schneider. *Experiments in the Dynamic and Strategic Control of Cooperating Manipulators.* PhD thesis, Stanford University, Stanford, CA 94305, September 1989. Also published as SUDAAR 586.

[3] Celia M. Oakley. *Experiments in Modelling and End-Point Control of Two-Link Flexible Manipulators.* PhD thesis, Stanford University, Department of Mechanical Engineering, Stanford, CA 94305, April 1991.

[4] W. L. Ballhaus, S. M. Rock, and A. E. Bryson, Jr. Optimal control of a two-link flexible manipulator using time-varying controller gains: Initial experiments. In *Proceedings of the American Astronomical Society Guidance and Control Conference*, Keystone, CO, February 1992.

AMD-Vol. 141/DSC-Vol. 37, Dynamics of Flexible Multibody Systems:
Theory and Experiment
ASME 1992

DYNAMICS AND CONTROL OF FLEXIBLE MULTI-BODY SPACE SYSTEMS

Y. P. Kakad
College of Engineering
University of North Carolina, Charlotte
Charlotte, North Carolina

ABSTRACT

In this paper, the dynamics and control of a large space structure consisting of a chain of articulated flexible substructures are developed. The problem of maneuver control of base platform together with vibration control of flexible payloads and the individual payload pointing requirements are addressed. Although the analytics are developed for the dynamics and control of a configuration consisting of a rigid platform with articulated flexible appendages, the method can be extended to more general multi-body space systems.

INTRODUCTION

Many advanced spacecraft applications in areas such as communications, space research, astrophysics, the earth related research etc. would utilize significantly large, flexible and very complicated structures in outer space. Some of these complicated structures would consist of a number of different payloads to accomplish different tasks. Thus, a generic spacecraft would consist of a chain of articulated flexible substructures with stringent pointing accuracy due to pointing requirements of various sensing and measurement systems. Also, in order to ensure vibration free observations and measurements it would be necessary to impose strict vibration suppression requirements on the spacecraft.

The dynamical equations of a generic articulated flexible multi-body space system are derived by using an inertial space coordinate system and individual body-fixed frames for the subsystems thereby the control forces and torques could be expressed in these axes. The control strategy is proposed to be of hierarchical type whereby the attitude control of the base platform, the pointing and vibration suppression of flexible subsystems are developed as separate subsystems. The dynamical model for developing control laws is further recast in terms of subsystems with interconnecting variables at one level and as a two-point boundary value problem for the entire multi-body space system at another level. An optimal controller is obtained for the total system by solving this multilevel system as a trajectory optimization problem.

The method is applied to NASA Langley Research Center's Spacecraft COntrol Laboratory Experiment (SCOLE) [28] test article to perform arbitrary rotational slew maneuvers with simultaneous vibration suppression of the flexible appendage. The application with the details will be discussed at the end.

LIST OF SYMBOLS

C_0	Inertial frame to platform body-fixed frame transformation
C_i	Inertial frame to *ith* appendage body-fixed frame transformation
$\underline{F}_o(t)$	Force applied at the platform mass center
$\underline{G}_o(t)$	Moment applied about the platform mass center
J_0	Mass moment of inertia matrix of the platform
J_i	Mass moment of inertia matrix of the *ith* appendage
K	Stiffness matrix
L_i	The Length of the *ith* appendage
M_0	Angular velocity vector transformation for the platform
M_i	Angular velocity vector transformation for the *ith* appendage
m_0	Mass of the platform
m_i	Mass of the *ith* appendage
N	The total number of subsystems
n	The maximum number of modes considered
$\underline{Q}$	The generalized force vector
q_i	Generalized coordinates for *ith* appendage
$\underline{R}_0$	Position vector of the mass center of the platform in the inertial frame
$\underline{R}_i$	Position vector of the point of attachment of *ith* appendage in the inertial frame
$\dot{\underline{R}}_0$	Velocity vector of the mass center of the platform in the inertial frame
$\dot{\underline{R}}_i$	Velocity vector of the point of attachment of *ith* appendage in the inertial frame
$\underline{r}_i$	Position vector from the platform mass center to the

This work was supported by NASA Grant NAG-1-1414.

	point of attachment of *ith* appendage
T_0	Kinetic energy of the platform
T_i	Kinetic Energy of the *ith* appendage
T	Total Kinetic Energy
U	Total Potential Energy
U_i	Total Potential Energy of the ith appendage
$\underline{u}_i(p_{i,t})$	The *ith* appendage 3-dimensional deflection referred to its body-fixed frame
u_i	Control vector of ith system
$\underline{V}_0$	Velocity vector of the mass center of the platform in the body-fixed frame
$\underline{V}_i$	Velocity vector of the point of attachment of *ith* appendage in the body-fixed frame
$\underline{x}_i$	State vector of ith system
$\underline{z}_i$	Vector of interconnecting variables
ρ_i	Mass per unit length of the flexible appendage
$\underline{\theta}_0$	The attitude of the platform in the inertial frame
$\underline{\theta}_i$	The attitude of the *ith* appendage mass center in the inertial frame
$\dot{\underline{\theta}}_0$	The angular velocity of the platform in the inertial frame
$\dot{\underline{\theta}}_i$	The angular velocity of the ith appendage in the inertial frame
ξ	Slew Angle
$\underline{\omega}_0$	The angular velocity of the platform in its body-fixed frame
$\underline{\omega}_i$	The angular velocity of *ith* appendage in its body-fixed frame
$\Phi(\underline{\lambda})$	Dual functional for two-point boundary value problem
$\underline{\lambda}$	Vector of Lagrange multipliers

ANALYTICS

The specific multi-body space system considered for this paper comprises of a rigid-platform from which a number of flexible payloads are deployed through gimbaled joints as shown in figure 1. The nonlinear dynamics of the flexible appendages together with the rigid platform are developed and these nonlinear dynamics are expressed in terms nonlinear ordinary differential equations with coupling between rigid-body modes of the entire system and the elastic modes of the appendages or substructures. The flexible appendages are considered to be beams for this analysis, but this method can be applied for other types of flexible appendage geometry.

In developing the dynamical equations for the arbitrary maneuver of the rigidized body with articulated appendages, three different sets of coordinate systems are used. A fixed inertial coordinate system is utilized together with a body-fixed frame at the mass center of the rigid platform and additionally separate body-fixed frames are used for the appendages at their points of attachment to the rigid platform. The elastic deformations of the appendages are expressed in terms of these individual coordinate frames. The general direction cosine matrix to relate the position vector from inertial frame to body-fixed frame is represented by matrix C with appropriate substrict to refer to a specific body-fixed frame. Similarly, the transformation matrix M is used to relate angular velocity vector from inertial to body-fixed axes.

Thus the corresponding transformations for the rigid-platform fixed coordinate system is,

$$\underline{\omega}_0 = M_0^T \dot{\underline{\theta}}_0 \qquad (1)$$

and

$$\underline{V}_0(t) = C\dot{\underline{R}}_0 \quad . \qquad (2)$$

The velocity of the point of attachment for *ith* appendage in the body-fixed frame of the platform can be given as

$$\underline{V}_{0i} = \underline{V}_0 + \underline{\omega}_0 \times \underline{r}_i \quad . \qquad (3)$$

The position vector of point p_i in the *ith* appendage is given as

$$\underline{R}_{p_i} = \underline{R}_i + \underline{a}_i + \underline{u}_i(p_i,t) \quad . \qquad (4)$$

Here, $\underline{a}_i$ is the position vector from the point of attachment to p_i. The velocity vector of p_i is

$$\dot{\underline{R}}_{p_i} = \dot{\underline{R}}_i + \dot{\underline{a}}_i + \dot{\underline{u}}_i \quad . \qquad (5)$$

This can be simplified as

$$\dot{\underline{R}}_{p_i} = \underline{V}_i + \underline{\omega}_i \times [\underline{a}_i + \underline{u}_i] + \dot{\underline{u}}_i \quad .$$

Here, $\underline{V}_i = C\dot{R}_i$ and

$$\underline{u}_i = \begin{bmatrix} u_x(p_i,t) \\ u_y(p_i,t) \\ u_z(p_i,t) \end{bmatrix}$$

Also, $\underline{u}_i = \Phi_i(p_i)\, q_i$. The matrix Φ_i is a matrix of admissible functions (shape functions).

Kinetic Energy

The kinetic energy in the *ith* flexible appendage is given as

$$T_i = (1/2)\, m_i \underline{V}_{oi}^T \underline{V}_{oi} + (1/2)\, \underline{\omega}_i^T [J] \underline{\omega}_i - m_i \underline{V}oi^T [\underline{\tilde{c}}] \underline{\omega}_i + (1/2) \int \dot{\underline{u}}_i^T \dot{\underline{u}}_i dm_i$$

$$+ \underline{V}_{oi}^T \int \dot{\underline{u}}_i dm_i + \underline{\omega}_i^T \int \tilde{\underline{a}}_i \dot{\underline{u}}_i dm_i \qquad (6)$$

where the vector $\underline{c}_i$ is from the point of attachment to the mass center of the flexible appendage and using the skew symmetric form for the vector cross product for any two vectors $\underline{c}$ and $\underline{\omega}$ (in the same reference frame) as

$$\underline{c} \times \underline{\omega} = [\,\underline{\tilde{c}}\,]\, \underline{\omega}$$

$$\underline{\tilde{c}}_i = \begin{bmatrix} 0 & -c_{zi} & c_{yi} \\ c_{zi} & 0 & -c_x i \\ -c_{yi} & c_x i & 0 \end{bmatrix} \quad . \qquad (7)$$

This expression for kinetic energy can be simplified as

$$T_i = (1/2) m_i \underline{V}_{0i}^T \underline{V}_{0i} + (1/2) \underline{\omega}_i^T J_i\, \underline{\omega}_i - m_i \underline{V}_{0i}^T \underline{\tilde{c}_i}\, \underline{\omega}_i$$

$$+ (1/2) m_i \sum_{k=1}^{n} \dot{q}_{ik}^2 + \underline{V}_{0i}^T \underline{\alpha}_i + \underline{\omega}_i^T \underline{\beta}_i \quad . \qquad (8)$$

The vectors $\dot{\alpha}_i$ and $\dot{\beta}_i$ depend upon the vector $\dot{q}_i$ and are also functionally related to matrix Φ_i. The kinetic energy function can be further simplified as

$$T_i = (1/2)m_i(\underline{V}_0 + \underline{\omega}_0\times \underline{r}_i)^T (\underline{V}_0 + \underline{\omega}_0\times \underline{r}_i)+(1/2)\underline{\omega}_i^T J_i \underline{\omega}_i - m_i(\underline{V}_0 + \underline{\omega}_0\times\underline{r}_i)^T \underline{\tilde{c}}_i\underline{\omega}_i + (1/2)\, m_i \sum_{k=1}^{n} \dot{\underline{q}}_{ik}^2 + (\underline{V}_0 + \underline{\omega}_0\times\underline{r}_i)^T\dot{\underline{\alpha}}_i + \underline{\omega}_i^T \dot{\underline{\beta}}_i \quad . \tag{9}$$

The kinetic energy of the rigid platform, T_0 , is given as

$$T_0 = (1/2)\, m_0\underline{V}_0^T\underline{V}_0 + (1/2)\, \underline{\omega}_0^T[J_0]\underline{\omega}_0 \quad . \tag{10}$$

The total kinetic energy of the entire system is

$$T = T_0 + \sum_{i=1}^{N-1} T_i \tag{11}$$

including all N-1 appendages.

For the subsequent derivation, only one flexible appendage is used without loss of generality as additional similar terms would be added for each additional appendage. Rewriting equation (11) with one flexible payload,

$$T = T_0 + T_1$$

$$T = (1/2)m\underline{V}_0^T\underline{V}_0 + (1/2)\, \underline{\omega}_0^T[\, J]\underline{\omega}_0 + \underline{V}_0^T[\, A_1]\, \underline{\omega}_0 + (1/2)\, \underline{\omega}_1^T[\, J_1]\underline{\omega}_1 + \underline{V}_0^T[\, A_2]\, \underline{\omega}_1 + \underline{\omega}_0^T[\, A_3]\, \underline{\omega}_1 + \dot{\underline{q}}_1^T[\, A_4]\dot{\underline{q}}_1 + \underline{V}_0^T[\, A_5]\dot{\underline{q}}_1 + \underline{\omega}_0^T[\, A_6]\dot{\underline{q}}_1 + \underline{\omega}_1^T[\, A_7]\dot{\underline{q}}_1 \tag{12}$$

where,

$$m = m_0 + m_1 \ ,$$
$$J = J_0 - m_1[\tilde{\underline{r}}_1]^2 \ ,$$
$$A_1 = -m_1[\tilde{\underline{r}}_1] \ ,$$
$$A_2 = -m_1[\tilde{\underline{c}}_1] \ ,$$
$$A_3 = m_1\, [\tilde{\underline{r}}_1]\, [\tilde{\underline{c}}_1] \ ,$$
$$A_4 = m_1\, [I] \ ,$$
$$\underline{V}_0^T \dot{\underline{\alpha}}_1 = \underline{V}_0^T\, [A_5]\dot{\underline{q}}_1 \ ,$$
$$\underline{\omega}_0^T = [\tilde{\underline{r}}_1]\dot{\underline{\alpha}}_1 = \underline{\omega}_0[A_6]\dot{\underline{q}}_1$$

and

$$\underline{\omega}_1^T\dot{\underline{\beta}}_1 = \underline{\omega}_1^T\, [A_7]\dot{\underline{q}}_1 \ .$$

The potential energy in each flexible appendage or substructure is given as

$$U_i = (1/2)\, \underline{q}_i^T\, K_i\, \underline{q}_i \quad . \tag{13}$$

The stiffness matrix K_i depends on the substructure geometrical and material properties. The stiffness matrix for our case here is K_1 and the corresponding expression for potential energy is

$$U_1 = (1/2)\, \underline{q}_1^T\, K_1\, \underline{q}_1$$

Equations of motion

Lagrange's equations of motion for the case of independent generalized co- ordinates q_k are

$$\frac{d}{dt}\frac{\partial T}{\partial \dot{q}_k} - \frac{\partial T}{\partial q_k} = Q_k - \frac{\partial U}{\partial q_k} \qquad (k=1,2,......,n) \tag{14}$$

where, $T = T(q,\dot{q})$ is the kinetic energy
$U = U(q)$ is the potential energy, and
Q_k are the generalized forces arising from nonconservative sources.

The generalized co-ordinates are:
R_{0X},R_{0Y},R_{0Z} – position of platform mass center relative to inertial frame origin.
$\theta_{01},\theta_{02},\theta_{03}$ – roll, pitch and yaw angles of platform.
R_{1X},R_{1Y},R_{1Z} – position of point of attachment of appendage number 1.
$\theta_{11},\theta_{12},\theta_{13}$ – roll, pitch and yaw angles of the appendage number 1.
q_1 – modal deformation co-ordinates for the appendage number 1.

The previous kinetic energy expression developed in equation (12) is given in terms of nonholonomic velocities $\underline{V}_0$, $\underline{\omega}_0$, and $\underline{\omega}_1$, and generalized velocities $\dot{q}_1$. Using the notation $\bar{T}(\underline{V}_0,\underline{\omega}_0,\underline{\omega}_1,\dot{q}_1)$ for this kinetic energy expression and T for kinetic energy expression in terms of generalized velocities, the equations of motion are developed.

(a) Translational Equations of the Entire Space System

From the chain rule applied to equation (12) using equation (2), one gets

$$\begin{bmatrix} \dfrac{\partial T}{\partial \dot{R}_{0X}} \\ \dfrac{\partial T}{\partial \dot{R}_{0Y}} \\ \dfrac{\partial T}{\partial \dot{R}_{0Z}} \end{bmatrix} = C_0^T \begin{bmatrix} \dfrac{\partial \bar{T}}{\partial \underline{V}_{01}} \\ \dfrac{\partial \bar{T}}{\partial \underline{V}_{02}} \\ \dfrac{\partial \bar{T}}{\partial V_{03}} \end{bmatrix} \tag{15A}$$

Also, the generalized forces are $C_0\underline{F}_0(t)$ where $\underline{F}_o(t)$ represents the force applied at the orbiter mass center. From Lagrange's equations

$$\frac{d}{dt}\left[\frac{\partial \bar{T}}{\partial \underline{V}_0}\right] + C_0\dot{C}_0^T \left[\frac{\partial \bar{T}}{\partial \underline{V}_o}\right] = \underline{F}(t) \tag{15}$$

and from equation (12)

$$\left[\frac{\partial \bar{T}}{\partial \underline{V}_0}\right] = m_o\underline{V}_0 + A_1\underline{\omega}_0 + A_2\underline{\omega}_1 + A_5\dot{\underline{q}}_1 \quad . \tag{16}$$

Substituting equation (16) in (15),

$$m_o\dot{\underline{V}}_0 + A_1\dot{\underline{\omega}}_0 + A_2\dot{\underline{\omega}}_1 + A_5\ddot{\underline{q}}_1 = -C_0\dot{C}_0^T(m_o\underline{V}_0 + A_1\underline{\omega}_0 + A_2\underline{\omega}_1 + A_5\dot{\underline{q}}_1) + \underline{F}_0(t) \quad . \tag{17}$$

This can be rewritten as

$$m_o\dot{\underline{V}}_0 + A_1\dot{\underline{\omega}}_0 + A_2\underline{\omega}_1 + A_5\ddot{\underline{q}}_1 = \underline{N}_0 + \underline{F}_0(t) \tag{18}$$

where the nonlinear term $\underline{N}_0$ is given as

$$\begin{aligned}\underline{N}_0 &= -C_0\dot{C}_0^T(m_o\underline{V}_0 + A_1\underline{\omega}_0 + A_2\omega_1 + A_5\dot{\underline{q}}_1) \\ &= -\tilde{\underline{\omega}}_0(m_o\underline{V}_0 + A_1\underline{\omega}_0 + A_2\underline{\omega}_1 + A_5\dot{\underline{q}}_1)\end{aligned} \tag{19}$$

Here, $\tilde{\underline{\omega}}_0 = C_0\dot{C}_0^T$.

(b) Rotational Equations of the Entire Space System:

From equation (1), and again using the chain rule

$$\left[\frac{\partial T}{\partial \underline{\theta}_0}\right] = M_0\left[\frac{\partial \bar{T}}{\partial \underline{\omega}_0}\right] \quad . \tag{20}$$

Also

$$\begin{bmatrix}\dfrac{\partial T}{\partial\theta_{01}}\\ \dfrac{\partial T}{\partial\theta_{02}}\\ \dfrac{\partial T}{\partial\theta_{03}}\end{bmatrix} = \begin{bmatrix}\dfrac{\partial \underline{V}_0^T}{\partial\theta_{01}}\\ \dfrac{\partial \underline{V}_0^T}{\partial\theta_{02}}\\ \dfrac{\partial \underline{V}_0^T}{\partial\theta_{03}}\end{bmatrix}\left[\frac{\partial \bar{T}}{\partial \underline{V}_0}\right] + \begin{bmatrix}\dfrac{\partial \underline{\omega}_0^T}{\partial\theta_{01}}\\ \dfrac{\partial \underline{\omega}_0^T}{\partial\theta_{02}}\\ \dfrac{\partial \underline{\omega}_0^T}{\partial\theta_{03}}\end{bmatrix}\left[\frac{\partial \bar{T}}{\partial \underline{\omega}_0}\right] \quad . \tag{21}$$

It can be shown that

$$\left[\frac{\partial T}{\partial \underline{\theta}_0}\right] = \begin{bmatrix}\underline{V}_0^T C_0\dfrac{\partial C_0^T}{\partial\theta_{01}}\\ \underline{V}_0^T C_0\dfrac{\partial C_0^T}{\partial\theta_{02}}\\ \underline{V}_0^T C_0\dfrac{\partial C_0^T}{\partial\theta_{03}}\end{bmatrix}\left[\frac{\partial \bar{T}}{\partial \underline{V}_0}\right] + \begin{bmatrix}\underline{\omega}_0^T M_0^{-1}\dfrac{\partial M_0}{\partial\theta_{01}}\\ \underline{\omega}_0^T M_0^{-1}\dfrac{\partial M_0}{\partial\theta_{02}}\\ \underline{\omega}_0^T M_0^{-1}\dfrac{\partial M_0}{\partial\theta_{03}}\end{bmatrix}\left[\frac{\partial \bar{T}}{\partial \underline{\omega}_0}\right] \tag{22}$$

From equation (12),

$$\left[\frac{\partial \bar{T}}{\partial \underline{\omega}_0}\right] = A_1^T\underline{V}_0 + J_0\underline{\omega}_0 + A_3\underline{\omega}_1 + A_6\dot{\underline{q}}_1 \quad . \tag{23}$$

Using the Lagrange's equations with the help of equations (16) and (23)

$$\frac{d}{dt}\left[\frac{\partial T}{\partial \underline{\theta}_0}\right] - \frac{\partial T}{\partial \underline{\theta}_0} = M_0\underline{G}_0 \tag{24}$$

where $\underline{G}_0$ is the net moment about the mass center of the orbiter with respect to the body-fixed frame.

Equation (24) can be simplified by substituting equations (16),(22), and (23) together with the relationship developed in (20) as

$$A_1\dot{\underline{V}}_0 + J_o\dot{\underline{\omega}}_0 + A_3\dot{\underline{\omega}}_1 + A_6\ddot{\underline{q}}_1 = \underline{G}_0 + \underline{N}(\underline{V}_0,\underline{\omega}_0,\underline{\omega}_1,\dot{\underline{q}}_1) \quad . \tag{25}$$

A similar approach is used to obtain the rotational equations of the flexible appendage or the substructure as

$$\underline{\omega}_1 = M_1^T\dot{\underline{\theta}}_1 \tag{26}$$

$$\frac{\partial T}{\partial \underline{\theta}_1} = M_1\left[\frac{\partial \bar{T}}{\partial \underline{\omega}_1}\right] \tag{27}$$

$$\frac{\partial \bar{T}}{\partial \underline{\omega}_1} = J_1\underline{\dot{\omega}}_1 + A_2^T\underline{V}_0 + A_3^T\underline{\dot{\omega}}_0 + A_7\dot{\underline{q}}_1 \tag{28}$$

and the final set of rotational equations are obtained as

$$J_1\dot{\underline{\omega}}_1 + A_2^T\dot{\underline{V}}_0 + A_3^T\dot{\underline{\omega}}_0 + A_7\ddot{\underline{q}}_1 = \underline{N}_1(\underline{V}_0,\underline{\omega}_0,\underline{\omega}_1,\dot{\underline{q}}_1) + \underline{G}_1 \tag{29}$$

where $\underline{G}_1$ is the moment applied at the flexible substructure.

(c) Vibration Equations of the Appendage

Since $\bar{T}$ in equation (16) is given in terms of $\dot{\underline{q}}_1$ which is a vector of generalized velocities,

$$\frac{\partial \bar{T}}{\partial \dot{\underline{q}}_1} = \frac{\partial T}{\partial \dot{\underline{q}}_1}$$

and

$$\left[\frac{\partial T}{\partial \dot{\underline{q}}_1}\right] = A_5^T\underline{V}_0 + A_6^T\underline{\omega}_1 + A_4\dot{\underline{q}}_1 \tag{30}$$

and

$$\left[\frac{\partial U}{\partial \underline{q}_1}\right] = K\underline{q}_1 \quad . \tag{31}$$

Using the Lagrangian Equations (38) and assuming that all control forces for vibration suppression are expressed in terms of generalized forces $\underline{Q}_1$,

$$A_5^T\dot{\underline{V}}_0 + A_6^T\dot{\underline{\omega}}_0 + A_7^T\dot{\underline{\omega}}_1 + A_4\ddot{\underline{q}}_1 = -K\underline{q}_1 + \underline{Q}_1 \quad . \tag{32}$$

Subsystems and State Variable Models

The dynamical subsystems considered for decentralized control are given by equations (18), (25), (29), and (32) which represent four different subsystems in terms of the subsequent control algorithm. The first subsystem describes the translational motion of the space system, whereas the second subsystem describes the rotational motion of the space system. The third subsystem describes the rotational motion of the substructure of the space system and the fourth subsystem describes the elastic motions of the flexible appendage or the substructure. It is important to note that the four subsystems are highly nonlinear and coupled and this coupling can be expressed in terms of interconnecting variables $\underline{z}_i$, where subscript i indicates the specific subsystem. These subsystem dynamics can be represented in terms of the following state variables and equation formulation.

$$\dot{\underline{x}}_i = \underline{f}_i \left[\underline{x}_i(t),\underline{u}_i(t),\underline{z}_i(t),t\right] \qquad i = 1,2,..N \tag{33}$$

here N represents the total number of subsystems.

The Optimal Control Problem

A general problem for the optimal control of interconnected dynamical systems like large flexible spacecrafts can be formulated as

$$\text{Minimize} \quad J(\underline{x}_i,\underline{u}_i,\underline{z}_i) \quad i=1,2,...,N \tag{34}$$
$$w.r.t. \ \underline{u}_i$$

where $\underline{x}_i$ is the n_i dimensional state vector of the ith subsystem, $\underline{u}_i$ is the corresponding m_i dimensional control vector and $\underline{z}_i$ is the r_i dimensional vector of interconnection inputs from the other subsystem. The integer N represents the total number of subsystems and the scalar functional J is defined by

$$J = \sum_{i=1}^{N} \left\{ P_i(\underline{x}_i(t_f)) + \int_{t_o}^{t_f} L_i \left[\underline{x}_i(t),\underline{u}_i(t),\underline{z}_i(t)\right] dt \right\} \tag{35}$$

where $L_i \left[\underline{x}_i(t),\underline{u}_i(t),\underline{z}_i(t)\right]$ is the performance index at time t for i = 1,2,..,N subsystems. The functional J defined in equation (35) is to be minimized subject to the constraints which define the subsystem dynamics, i.e.

$$\dot{\underline{x}}_i = \underline{f}_i \left[\underline{x}_i(t),\underline{u}_i(t),\underline{z}_i(t),t\right], \ t_o \le t \le t_f$$

$$\underline{x}_i(t_o) = \underline{x}_{io} \ , \quad i = 1,2...,N \ . \tag{36}$$

Also, the minimum of J must satisfy the interconnection relationship

$$\sum_{i=1}^{N} \underline{G}_i^*(\underline{x}_i,\underline{z}_j) = \underline{0} \ . \tag{37}$$

The term $\underline{G}_i^*(\underline{x}_i,\underline{z}_j)$ represents the interconnection relationship for ith system in which $\underline{x}_i$ represents the state vector of ith system $\underline{z}_j$ represents the vector of interconnect variables with jth subsystem.

In the subsequent discussion, the open-loop control algorithm is first developed to explain the fundamental ideas behind the closed-loop control algorithm and such has limited practical use.

The Open-loop Hierarchical Control

Using the method of Goal Coordination or infeasible method [17,22], we consider another problem which is obtained by maximizing the dual function $\Phi(\underline{\lambda})$ with respect to $\underline{\lambda}(t)$ ($t_o \le t \le t_f$), where

$$\Phi(\underline{\lambda}) = \underset{\underline{x},\underline{u},\underline{z}}{Min} \left\{ \tilde{J}(\underline{x},\underline{u},\underline{z},\underline{\lambda}) \right\} \tag{38}$$

subject to constraints in equations (36) and (37). Here

$$\underline{x} = \begin{bmatrix} \underline{x}_1 \\ \cdot \\ \cdot \\ \cdot \\ \underline{x}_N \end{bmatrix} \quad \underline{u} = \begin{bmatrix} \underline{u}_1 \\ \cdot \\ \cdot \\ \cdot \\ \underline{u}_N \end{bmatrix} \quad \underline{z} = \begin{bmatrix} \underline{z}_1 \\ \cdot \\ \cdot \\ \cdot \\ \underline{z}_N \end{bmatrix} \tag{39}$$

Also, $\underline{\lambda}$ in equation (108) is a vector of Lagrange multipliers which is given as

$$\underline{\lambda} = \begin{bmatrix} \underline{\lambda}_1 \\ \cdot \\ \cdot \\ \cdot \\ \underline{\lambda}_N \end{bmatrix} . \tag{40}$$

$$\tilde{J} = \sum_{i=1}^{N} \left\{ P_i(\underline{x}_i(t_f)) + \int_{t_o}^{t_f} L_i(\underline{x}_i,\underline{u}_i,\underline{z}_i,t)\, dt + \int_{t_o}^{t_f} \underline{\lambda}^T \underline{G}_i^*(\underline{x}_i,\underline{z}_j,t)\, dt \right\}$$

$$j = 1,2,...,N \tag{41}$$

Rewriting this functional $\tilde{J}$ as

$$\tilde{J} = \sum_{i=1}^{N} J_i$$

$$= \sum_{i=1}^{N} \left\{ P_i(\underline{x}_i(t_f)) + \int_{t_o}^{t_f} \left[L_i(\underline{x}_i,\underline{u}_i,\underline{z}_i,t) + \underline{\lambda}^T \underline{G}_i(\underline{x}_i,\underline{z}_i,t) \right] dt \right\} \tag{42}$$

where,

$$J_i = P_i(\underline{x}_i(t_f)) + \int_{t_o}^{t_f} \left[L_i(\underline{x}_i,\underline{u}_i,\underline{z}_i,t) + \underline{\lambda}^T \underline{G}_i(\underline{x}_i,\underline{z}_i,t) \right] dt \tag{43}$$

and where $\underline{\lambda}^T \underline{G}_i^*(\underline{x}_i,\underline{z}_j,t)$ has been refactored into the form $\underline{\lambda}^T \underline{G}_i(\underline{x}_i,\underline{z}_i,t)$, i.e. into a form separable in the index i.

Thus,

$$\tilde{J} = J + \int_{t_o}^{t_f} \underline{\lambda}^T \sum_{i=1}^{N} \underline{G}_i (\underline{x}_i, \underline{z}_i) \, dt \tag{44}$$

Then by the fundamental theorem of strong Lagrange duality [25]

$$\min_{\underline{u}_i} J = \max_{\underline{\lambda}} \Phi (\underline{\lambda}) , \quad i = 1,2,...,N. \tag{45}$$

Thus an alternative way of optimizing J is to maximize $\Phi (\underline{\lambda})$.

From equation (42), for a given $\underline{\lambda} (t)$, $t_o \le t \le t_f$, the functional $\tilde{J}$ is separable into N independent minimization problems, the ith of which is given by

$$\underset{\underline{x}_i, \underline{u}_i, \underline{z}_i}{Min} \quad J_i = P_i (\underline{x}_i (t_f)) + \int_{t_o}^{t_f} \left[L_i (\underline{x}_i, \underline{u}_i, \underline{z}_i) + \underline{\lambda}^T \underline{G}_i (\underline{x}_i, \underline{z}_i) \right] dt \tag{46}$$

subject to

$$\dot{\underline{x}}_i = \underline{f}_i (\underline{x}_i, \underline{u}_i, \underline{z}_i) , \quad t_o \le t \le t_f$$

$$\underline{x}_i (t_o) = \underline{x}_{io} \tag{47}$$

This leads to a two-level optimization structure where on the first level, for given $\underline{\lambda}$, the N independent minimization problems described in equations (46) and (47) are solved and on the second level, the $\underline{\lambda} (t)$ ($t_o \le t \le t_f$) trajectory is improved by an optimization scheme like the steepest ascent method, i.e. from iteration j to $j+1$

$$\underline{\lambda} (t)^{j+1} = \underline{\lambda} (t)^j + \alpha^j + \underline{d}^j (t) \quad t_o \le t \le t_f \tag{48}$$

where

$$\underline{d}^j = \nabla \Phi(\underline{\lambda}(t)) = \sum_{i=1}^{N} \underline{G}_i (\underline{x}_i, \underline{z}_i) , \tag{49}$$

$\nabla \Phi (\underline{\lambda})$ is the gradient of $\Phi (\underline{\lambda})$, $\alpha_j > 0$ is the step length and $\underline{d}^j$ is the steepest ascent search direction. At the optimum $\underline{d}^j \rightarrow 0$ and the appropriate Lagrange multipier, $\underline{\lambda}$, is the optimum one.

The development of this algorithm depends on the assertion Max $\Phi(\underline{\lambda}) =$ min J and this may not be valid for all nonlinear systems. Consequently, linearization of $\underline{G}_i$, and linearized equations for $\underline{f}_i$ may be required for constraints to be convex and convexity of the constraints is necessary to prove this assertion. Nevertheless, the method is attractive from the standpoint of simplicity and that the dual function is concave for this nonlinear case. This ensures that if the duality assertion is valid, the optimum obtained is the Global Optimum.

On the first level, since equation (46) is to be minimized subject to equation (47), the necessary conditions lead to a two point boundary value problem from which an open loop optimum control could be calculated. However, it is desirable to calculate a closed loop control and for this the quasilinearization approach can be utilized at level one for all subsystems. Thus an iterative scheme can be set up whereby an initial trajectory of $\underline{\lambda} (t)^*$, $t_o \le t \le t_f$ is guessed at level two and provided to level one. At level one the two-point boundary value problems of the subsystems are solved by quasilinearization. The state and control trajectories of all the subsystems obtained at level one are sent to level two. The test for optimality based on equation (49) is conducted at level two and if this is not satisfied, equation (48) is used to obtain the new $\underline{\lambda} (t)$ for the next iteration.

Subsystem Closed Loop Controllers

The closed loop controllers are obtained at the first level by solving the two-point boundary value problems of the subsystems utilizing the quasilinearization procedure. As noted in equations (46) and (47), the first level problem for the ith subsystem is

For given $\underline{\lambda}(t)$, $t_o \le t \le t_f$,

$$\min_{\underline{x}_i, \underline{u}_i, \underline{z}_i} \left\{ P_i \left[\underline{x}_i(t_f) \right] + \int_{t_o}^{t_f} \left[L_i(\underline{x}_i, \underline{u}_i, \underline{z}_i) + \underline{\lambda}^T \underline{G}_i(\underline{x}_i, \underline{z}_i) \right] dt \right\} \tag{46}$$

subject to

$$\dot{\underline{x}}_i = \underline{f}_i (\underline{x}_i, \underline{u}_i, \underline{z}_i) , \quad t_o \le t \le t_f$$

$$\underline{x}_i (t_o) = \underline{x}_{io} \quad . \tag{47}$$

For this problem, the Hamiltonian H_i can be written as

$$H_i = L_i (\underline{x}_i, \underline{u}_i, \underline{z}_i) + \underline{\lambda}^T \underline{G}_i (\underline{x}_i, \underline{z}_i) + \underline{\eta}_i^T \underline{f}_i (\underline{x}_i, \underline{u}_i, \underline{z}_i) \ . \tag{50}$$

For a given $\underline{\lambda}$, the state and costate equations become

$$\dot{\underline{x}}_i (t) = \underline{f}_i (\underline{x}_i, \underline{u}_i, \underline{z}_i) \tag{51}$$

$$\dot{\underline{\eta}}_i (t) = - \frac{\partial H_i}{\partial \underline{x}_i} = - \left[\frac{\partial L_i}{\partial \underline{x}_i} + \frac{\partial \underline{G}_i^T}{\partial \underline{x}_i} \underline{\lambda} + \frac{\partial \underline{f}_i^T}{\partial \underline{x}_i} \underline{\eta}_i \right] \tag{52}$$

with

$$\frac{\partial H_i}{\partial \underline{u}_i} = \underline{0} ; \qquad \frac{\partial H_i}{\partial \underline{z}_i} = \underline{0} \tag{53}$$

It is assumed here that using the equations (52) and (53), it is possible to obtain the control $\underline{u}_i$ and the interconnect variable vector $\underline{z}_i$ which is an explicit function of $\underline{\eta}_i$ and $\underline{x}_i$, i.e.

$$\underline{u}_i = \underline{c}_i (\underline{x}_i, \underline{\eta}_i)$$

$$\underline{z}_i = \underline{d}_i (\underline{x}_i, \underline{\eta}_i) \tag{54}$$

Using these relationships for $\underline{u}_i$ and $\underline{z}_i$ in equations (51) and (52), the following equations are obtained

$$\dot{\underline{x}}_i = \underline{a}_i (\underline{x}_i, \underline{\eta}_i), \quad t_o \le t \le t_f \tag{55}$$

$$\dot{\underline{\eta}}_i = \underline{b}_i (\underline{x}_i, \underline{\eta}_i), \quad t_o \le t \le t_f \tag{56}$$

with the boundary conditions

$$\underline{x}_i (t_o) = \underline{x}_{io} \tag{57}$$

and from the transversality conditions

$$\underline{\eta}_i (t_f) = \frac{\partial P_i \left[\underline{x}_i (t_f) \right]}{\partial \underline{x}_i} . \tag{58}$$

Quasilinearization Procedure

The two-point boundary value problem of ith subsystem is given by equations (55) and (56) subject to boundary conditions of equations (57) and (58). This problem is solved by quasilinearization technique as follows.

Defining $\underline{y} = \begin{bmatrix} \underline{x}_i \\ \underline{\eta}_i \end{bmatrix}$,

equations (55) and (56) can be rewritten as

$$\dot{\underline{y}} (t) = \underline{F} \left[\underline{y} (t) \right] . \tag{59}$$

In the quasilinearization procedure, starting from an initial guessed trajectory for $y = y^j (t)$, successive linearizations are performed of equation (59) in such a way that the final linear equation for y solves equation (59) to an acceptable degree subject to boundary conditions (57) and (58) which could be expressed in a more general form as

$$\underline{y} (t_o)^T A_o = \underline{b}_o^T \tag{60}$$

$$\underline{y} (t_f)^T A_f = \underline{b}_f^T \tag{61}$$

where A_o, A_f are $2n$ x n matrices.

The linearized equation of (59) about a trajectory $\underline{y} = \underline{y}^j (t)$ is obtained by Taylor series expansion as

$$\dot{\underline{y}} = \underline{F} (\underline{y}^j) + J (\underline{y}^j) (\underline{y} - \underline{y}^j) + \underline{\Psi} \tag{62}$$

where $J (y^j)$ is the Jacobian of $\underline{F} \left[\underline{y}(t) \right]$, $t_o \le t \le t_f$, at y^j and $\underline{\Psi}$ represents the contribution of the higher order terms. Neglecting these higher order terms, the following linear equation is obtained

$$\dot{\underline{y}} = \underline{F} (\underline{y}^j) + J (\underline{y}^j) (\underline{y} - \underline{y}^j) . \tag{63}$$

Closed Loop Control

In order to obtain the closed loop control, the solution of the linearized equation (63) can be written as

$$\underline{y}(t_f) = \phi (t_f,t)\underline{y} (t) + \int_t^{t_f} \phi (t_f,\tau) \left[\underline{F} (\tau) - J \, \underline{y} (\tau) \right] d\tau \tag{64}$$

where ϕ is the state transition matrix of the system in equation (63). Rewriting equation (64) in terms of solutions of states and costates and replacing the integral terms by $\underline{p}_i (t)$

$$\begin{bmatrix} \underline{x}_i (t_f) \\ \underline{\eta}_i (t_f) \end{bmatrix} = \begin{bmatrix} \phi_{11} (t_f, t) & \phi_{12} (t_f, t) \\ \phi_{21} (t_f, t) & \phi_{22} (t_f, t) \end{bmatrix} \begin{bmatrix} \underline{x}_i (t) \\ \underline{\eta}_i (t) \end{bmatrix} + \begin{bmatrix} \underline{p}_{i1} (t) \\ \underline{p}_{i2} (t) \end{bmatrix} . \tag{65}$$

From equations (58) and (65)

$$\underline{\eta}_i (t_f) = \frac{\partial P_i}{\partial \underline{x}_i} = \phi_{21} (t_f, t)\underline{x}_i (t_f, t) + \phi_{22} (t_f, t)\underline{\eta}_i (t) + \underline{p}_{i2} (t) . \tag{66}$$

Thus,

$$\underline{\eta}_i (t) = \phi_{22}^{-1} \left[\frac{\partial P_i}{\partial \underline{x}_i} - \phi_{21}\underline{x}_i (t) - p_{i2} (t) \right]$$

$$= \phi_{22}^{-1} \left[\frac{\partial P_i}{\partial \underline{x}_i} - \underline{p}_{i2} (t) \right] - \phi_{22}^{-1} \left[\phi_{21} (t) \right] . \tag{67}$$

It is important to note here that ϕ_{22}^{-1} always exists since it is a principal minor of the state transition matrix.
Substituting equation (67) into equation (54)

$$\underline{u}_i = \underline{c}_i (\underline{x}_i, t) . \tag{68}$$

Example Problem

The foregoing dynamic modeling method and closed-loop control algorithm for decentralized control are applied for large angle rotational slew maneuvers to the NASA Langley Research Center SCOLE system [28,29]. The SCOLE system comprises of the space shuttle attached with a 130 ft. long flexible appendage with a rigid end mass, the reflector which is utilized in measuring the Line Of Sight (LOS) of the configuration during the maneuver [figure 3]. The attachment of the flexible appendage to the rigid shuttle is through a rigid joint and the method developed in this paper is used to obtain the dynamics of the maneuver.

The closed-loop control system is used to perform three-dimension rotational maneuver as well as the vibration suppression of the flexible appendage. The maneuver is developed by using the two-point boundary value problem in terms of the rigid-body slewing and the vibration suppression of the flexible appendage as two separate dynamical subsystems. The hierarchical optimal control scheme in this paper is utilized in order to solve individual boundary-value problem for each of the two subsystems by defining their state variable models and incorporating the coupling variables between the two subsystems in these models. Also, the boundary conditions of the overall system are reworked in terms of boundary conditions of each subsystem. A quadratic performance index is

utilized for the overall system and is subsequently expressed in terms of a sum of two individual performance indices of the subsystems.

The basic algorithm for obtaining an optimal closed-loop state feedback scheme involves using a trajectory in terms of a vector of Lagrange multipliers as an initial guess at level two. This is used at level one in quasilinearization application.

The two-point boundary value problem for each subsystem is solved at level one by using a quasilinearization technique as a trajectory optimization problem. In the quasilinearization procedure, from an initial guessed state trajectory, successive linearizations are performed of state equations in such a way that the final solution of the state trajectory is within an acceptable degree subject to boundary conditions. The state vector definition at this level is an augmented state vector which includes both system states and costates.

These optimum solutions of the system trajectories are utilized at level two to yield the updated trajectory of the vector of lagrange multipliers of the overall system to be used for quasilinearization process at level one. The basic steps of the algorithm are repeated to optimize this second level trajectory with respect to prespecified error criterion to obtain an optimal feedback law.

The plots in fig. 4 show a maneuver of $\theta_1 = 40°$, $\theta_2 = 10°$, and $\theta_3 = 25°$.The plots in fig. 4 (a) show the maneuver angles versus time in seconds. The plots in 4 (b) illustrate the corresponding moment components applied at the tip of the appendage at the shuttle. The torsional deformation for the same time duration is shown in 4 (c). The effect of vibration suppression is calculated in terms of tip deflection in fig. 4 (d). The corresponding control forces applied at the reflector end are shown in figure 4 (e) and since no control force is applied in z direction, only two control force components are shown.

CONCLUSIONS

A method is developed to analyze the dynamics of a flexible multi-body space system where the base platform is undergoing a maneuver and at the same time different articulated flexible payloads deployed from the platform may be undergoing individual large angle maneuvers. The control problem of this system is formulated in terms of a decentralized control algorithm in terms of individual subsystems based on the nonlinear coupled ordinary differential equations. Since the control algorithm can be implemented for individual subsystems at level one after the system trajectory at level two is computed, the control strategy is ideally suited for parallel processing of the algorithm.

REFERENCES

[1] D. J Ness, and R. L. Farrenkopf, "Inductive Methods for Generating the Dynamic Equations of Motion for Multibodied Flexible Systems - Part 1, Unified Approach," Synthesis of Vibrating Systems, ASME, pp 78-91, November 1971.

[2] J. Y. L. Ho, and R. Gluck, "Inductive Methods for Generating the Dynamic Equations of Motion for Multibodied Flexible Systems - Part 2, Perturbation Approach," Synthesis of Vibrating Systems, ASME, pp 92-101, November 1971.

[3] T. A. W. Dwyer, III, "The Control of Angular Momentum for Asymmetric Rigid Bodies," IEEE Trans. Automat. Contr., pp 686-688, June 1982.

[4] D. K. Robertson, "Three-dimensional Vibration Analysis of a Uniform Beam with Offset Inertial Masses at the Ends," NASA TM-86393, September 1985.

[5] T. R. Kane, P. W. Likins, and D. A. Levinson, Spacecraft Dynamics, New York: McGraw-Hill, 1983.

[6] J. Storch, S. Gates and D. O'Connor, "Three Dimensional Motion of a Flexible Beam with Rigid Tip Bodies," Charles Stark Draper Laboratory Report, Interlab Memorandum, October 1985.

[7] Y. P. Kakad, "Slew Maneuver Control of the Spacecraft Control Laboratory Experiment (SCOLE)," Proceedings of ACC Conference, pp 1039-1044 June 1986.

[8] Y. P. Kakad, "Dynamics and Control of Slew Maneuver of Large Flexible Spacecraft," Proceedings of AIAA Guidance, Navigation and Control Conference, pp 629-634, August 1986.

[9] B. Friedland, Control System Design - An Introduction to State-space Methods, New York: McGraw-Hill, 1986.

[10] H. Goldstein, Classical Mechanics, Reading: Addison-Wesley, Second Edition, 1981.

[11] M. Balas, "Feedback Control of Flexible Systems," IEEE Trans. Automat. Contr., pp 673-679, August 1978.

[12] A. E. Bryson, Jr. and Y. C. Ho, Applied Optimal Control - Optimization, Estimation, and Control, New York: John Wiley, revised printing, 1975.

[13] L. Meirovitch, Analytical Methods in Vibrations, New York: The Macmillan Company, 1967.

[14] E. S. Armstrong, "ORACLS - A System for Linear-Quadratic-Gaussian Control Law Design," NASA TP-1106, 1978.

[15] S. Joshi, "SCOLE Equations of Motion-A New Formulation," Proceedings of the 2nd Annual SCOLE Workshop, NASA TM-89048, pp. 14-25, December 1985.

[16] Y. P. Kakad, "Dynamics of Spacecraft Control Laboratory Experiment (SCOLE) Slew Maneuvers," NASA CR-4098, October 1987.

[17] M. G. Singh, Dynamical Hierarchical Control, New York: North-Holland Publishing Company, 1980.

[18] J. L. Junkins, and J. D. Turner, Optimal Spacecraft Rotational Maneuvers, New York: Elsevier Science Publishers B. V., 1986.

[19] P. L. Falb, and J. L. de Jong, Some Successive Approximation Methods in Control and Oscillation Theory, New York: Academic Press, 1969.

[20] H. B. Keller, Numerical Methods for Two-point Boundary-Value Problems, Waltham, Massachsetts: Blaisdal Publishing Company, 1968.

[21] R. E. Bellman, and R. E. Kalaba, Quasilinearization and Nonlinear Boundary-Value Problems, New York: American Elsevier Publishing Company, 1965.

[22] M. G. Singh, and A. Titli, "Closed-loop Hierarchical Control for Non-linear Systems Using Quasilinearization," Automatica, Vol. 11,pp 541-546.

[23] S. M. Roberts, and J. S. Shipman, Two-point Boundary Value Problems: Shooting Methods, New York: American Elsevier Publishing Company, 1972.

[24] M. S. Mahmoud, M. F. Hassan, and M. G. Darwish, Large- Scale Control Systems, New York: Marcel Dekker Inc., 1985.

[25] D. Luenberger, Optimization ny Vector Space Methods, New York: John Wiley, 1969.

[26] D. E. Kirk, Optimal Control Theory - An Introduction, Englewood Cliffs, New Jersey, Prentice-Hall Inc., 1970.

[27] P. W. Likins, "Geometric Stiffness Characteristics of a Rotating Elastic Appendage," International Journal of Solids and Structures, Vol. 10,No. 2, 1974, pp. 161-167.

[28] Y. P. Kakad, "Nonlinear Slew Maneuver Dynamics of Large Flexible Spacecrafts," Advances in Dynamics and Control of Flexible Spacecraft and Space-based Manipulators, ASME DSC-Vol. 20, 1990, pp 19-27.

[29] M. Azam, S. N. Singh, A. Iyer, and Y. P. Kakad, "Detumbling and Reorientation Maneuvers and Stabilization of NASA SCOLE System," IEEE Transactions on Aerospace and Electronic Systems, Vol. 28, No. 1, January 1992, pp. 80-91.

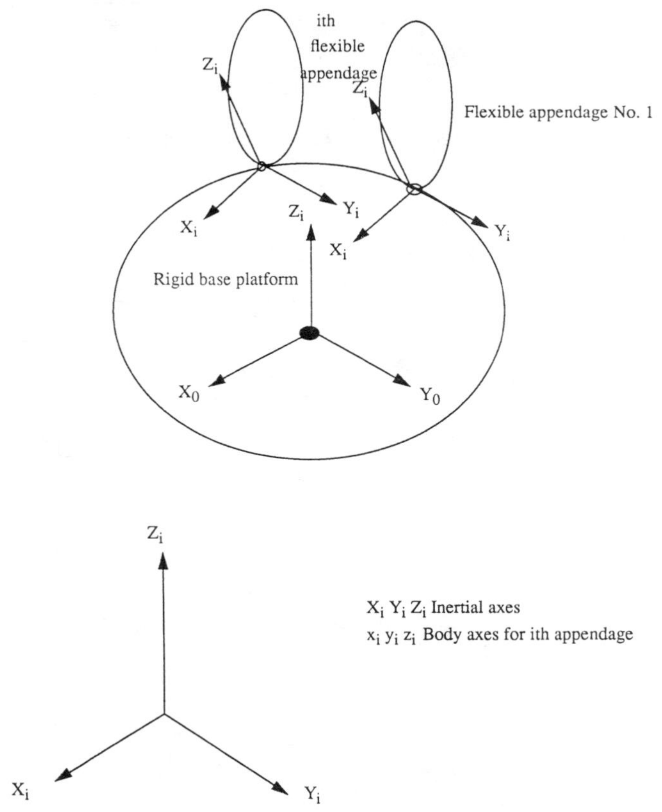

Figure 1. - The Articulated Multibody Flexible Space System

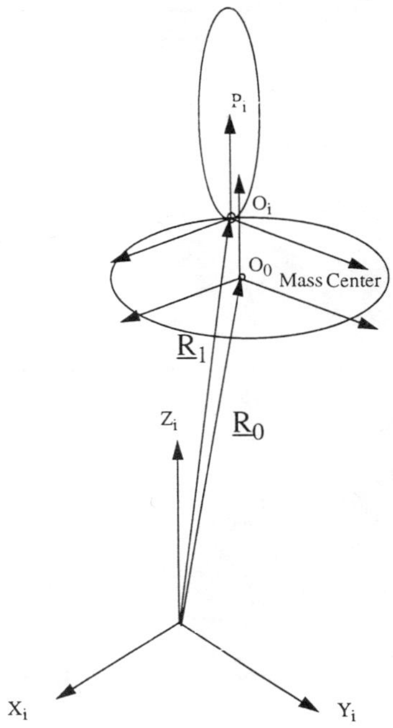

FIGURE 2

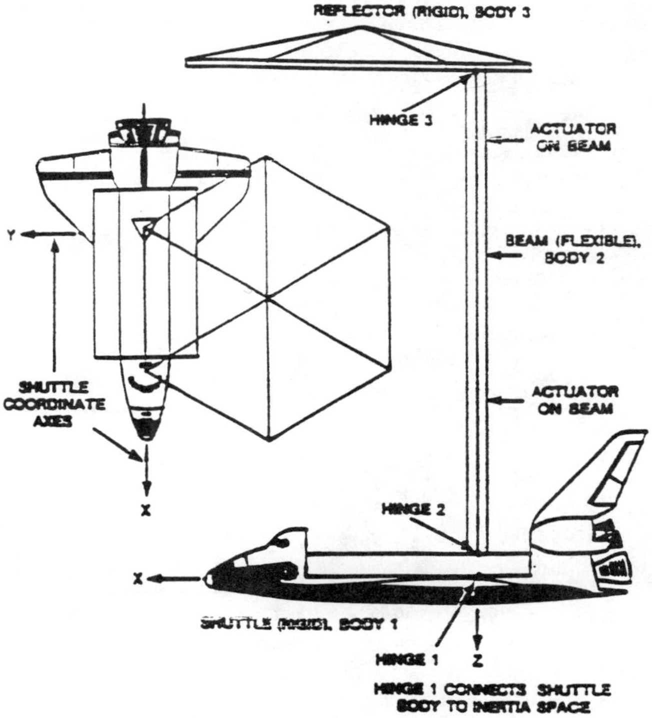

Figure 3. - SCOLE Configuration with Coordinate System and Actuator Control Points

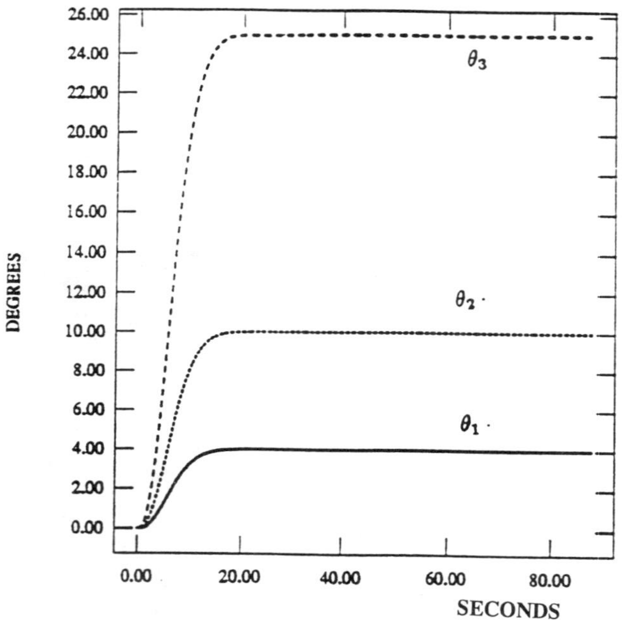

Figure 4(a). - Attitude Angles

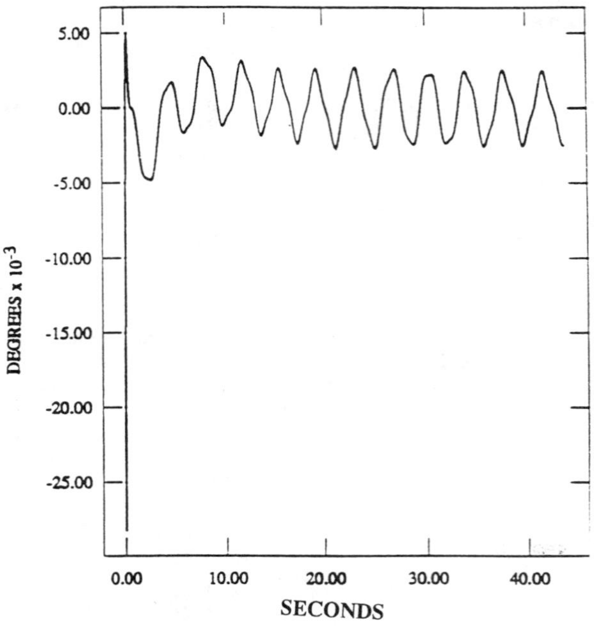

Figure 4(c). - Torsional Deformations

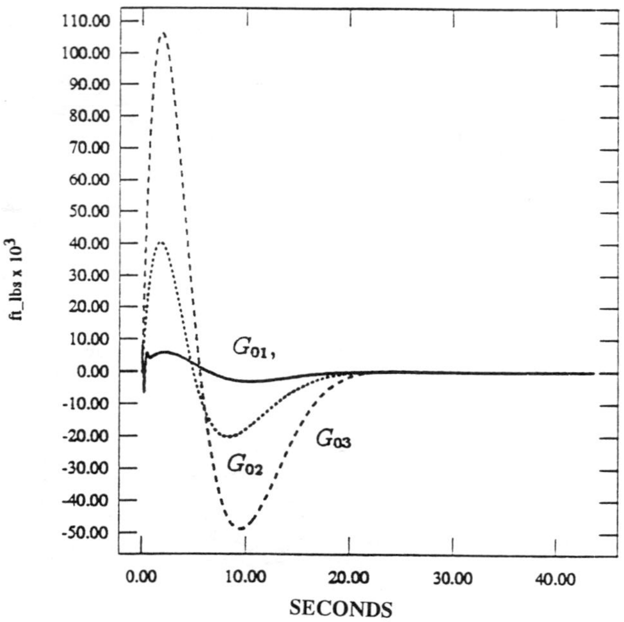

Figure 4(b). - Control Moments G_o

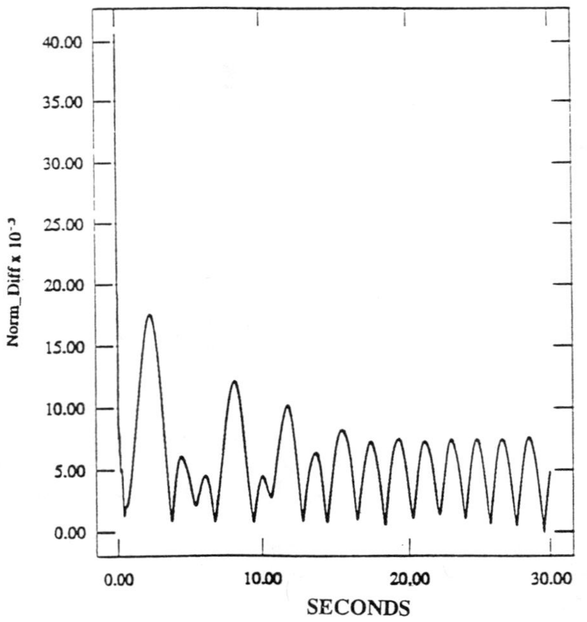

Figure 4(d). - Resultant Tip Deflection

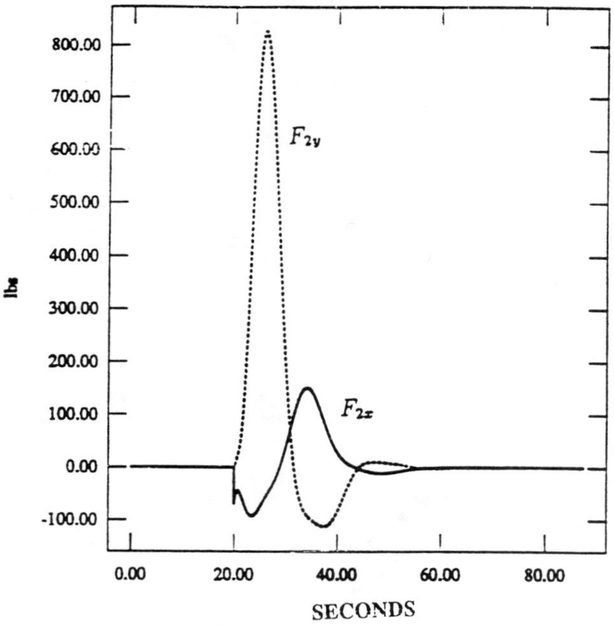

Figure 4(e). - Control Forces

AMD-Vol. 141/DSC-Vol. 37, Dynamics of Flexible Multibody Systems:
Theory and Experiment
ASME 1992

ASYMPTOTIC STABILITY OF DISSIPATIVE COMPENSATORS FOR FLEXIBLE MULTIBODY SYSTEMS IN THE PRESENCE OF ACTUATOR DYNAMICS AND ACTUATOR/SENSOR NONLINEARITIES

A. G. Kelkar and T. E. Alberts
Old Dominion University
Norfolk, Viginia

S. M. Joshi
NASA Langley Research Center
Hampton, Virginia

Abstract

This paper considers the control of a class of nonlinear multibody flexible systems. Global asymptotic stability of such systems under static dissipative compensation is established. Furthermore, the stability is shown to be robust to first-order actuator dynamics, certain actuator and sensor nonlinearities, modeling error, and parametric uncertainty. The results are applicable to a wide class of systems, including flexible space-structures with articulated flexible appendages and terrestrial as well as space-based manipulators. The stability proofs use the Lyapunov and function-space approaches and exploit the inherent passivity-type properties of such systems.

Introduction

Future space missions envisioned will require multibody space systems. Examples of such structures include space platforms with multiple articulated payloads, and space-based manipulators for on-orbit assembly and satellite servicing. The dynamics of such systems are highly nonlinear and complex. Control system design for such multibody flexible systems is a difficult problem because of the large number of significant elastic modes with low inherent damping and the inaccuracies and uncertainties in the mathematical model. The literature contains a number of important stability results for certain subclasses of this problem; e.g., linear flexible structures, nonlinear multibody rigid structures, and most recently multibody flexible structures. The Lyapunov and passivity approaches are used in [1] to demonstrate global asymptotic stability of linear flexible space structures (with no articulated appendages) for a class of dissipative compensators. These include Collocated Attitude Controllers (CAC) and Collocated Damping Enhancement Controllers (CDEC). The stability properties were shown to be robust to first-order actuator dynamics and certain actuator/sensor nonlinearities. Multibody rigid structures comprise another class of systems for which stability results have been advanced. Upon recognition that rigid manipulators belong to the class of natural systems, a number of researchers [2], [3], [4], etc., have established global asymptotic stability of rigid manipulators employing PD control with gravity compensation. Stability of tracking controllers was investigated in [5] and [6] for rigid manipulators. In [7] an extension of the results of [6] to the exponentially stable tracking control for flexible multilink manipulators, local to the desired trajectory, was done. Recently, global asymptotic stability for nonlinear multilink flexible space-structures under static dissipative control law has been established [8]. The stability was also shown to be robust in the presence of a wide range of actuator/sensor nonlinearities. In this paper it is shown that the stability results in [8] are still valid in the presence of actuator dynamics.

Stability With Static Dissipative Control Law

Typically, the mathematical model of flexible multi-body space structures is given by the following dynamical equation of motion:

$$M(p)\ddot{p} + C(p,\dot{p})\dot{p} + D\dot{p} + Kp = B^T u \qquad (1)$$

where $\{p\} = \{\theta^T, q^T\}^T$, θ is the k-vector of measurable coordinates (joint angles in the case of manipulators) and q is the $(n-k)$ vector of the flexural coordinates. $M(p)$ is configuration dependent mass-inertia matrix, $C(p,\dot{p})$ corresponds to Coriolis and centrifugal forces, D is the damping matrix, K is the stiffness matrix, and u is the vector of applied torques. The stiffness and damping matrices K and D have the form,

$$K = \begin{bmatrix} 0_{k\times k} & 0_{k\times(n-k)} \\ 0_{(n-k)\times k} & \overline{K}_{(n-k)\times(n-k)} \end{bmatrix},$$

$$D = \begin{bmatrix} 0_{k\times k} & 0_{k\times(n-k)} \\ 0_{(n-k)\times k} & \overline{D}_{(n-k)\times(n-k)} \end{bmatrix} \tag{2}$$

where $\overline{K}$ and $\overline{D}$ are the flexural stiffness and damping matrices associated with the structural members, and the subscripts indicate sub-matrix dimensions. Since the control inputs considered are all associated with the rigid body coordinates, the input matrix B has the form,

$$B = \begin{bmatrix} I_{k\times k} \\ 0_{(n-k)\times k} \end{bmatrix}.$$

Consider, the static dissipative control law u, given by:

$$u = -G_p y_p - G_r y_r \tag{3}$$

where,

$$y_p = Bp \qquad and \qquad y_r = B\dot{p} \tag{4}$$

y_p and y_r are measured angular position and rate vectors. Then, the following theorem has been proved in [8].

Theorem 1. Suppose G_p and G_r are symmetric and positive definite. Then, the closed-loop system given by equations (1), (3) and (4) is globally asymptotically stable.

Proof.

The details of the proof can be found in [8]. The proof uses the Lyapunov function

$$V = \frac{1}{2}\dot{p}^T M(p)\dot{p} + \frac{1}{2}p^T(K + B^T G_p B)p \tag{5}$$

where $M(p)$ and $(K + B^T G_p B)$ are positive definite symmetric matrices. The proof is based on a very important property of the system, that the matrix $(\frac{1}{2}\dot{M} - C)$ is *skew symmetric*. The time derivative of this Lyapunov function turns out to be negative semidefinite. Then, by Lasalle's theorem it was shown that the closed-loop system was, in fact, globally asymptotically stable.

Robustness to Actuator/Sensor Nonlinearities

Although, as shown in the proof of theorem 1, the static dissipative controller (3) globally asymptotically stabilizes the nonlinear system (1) in the presence of perfect (i.e. linear, instantaneous) actuators and sensors; in practice, these devices have nonlinearities and phase lags. Therefore, for practical applications, the controller (3) should be robust to the nonlinearities and the phase shifts in the actuator/sensor. The following theorem was proved in [8] and it extends the results of [1] to the case of nonlinear flexible multibody systems. That is, the robust stability property of the static dissipative controllers is proved in the presence of a wide class of actuator/sensor nonlinearities. In particular, it is proved that the static dissipative controller preserves global asymptotic stability when actuators have monotonically increasing nonlinearities and sensors have nonlinearities that belong to the $(0, \infty)$ sector. [A function $\psi(\nu)$ is said to belong to the $(0, \infty)$ sector if $\psi(0) = 0$ and $\nu\psi(\nu) > 0$ for $\nu \neq 0$: ψ is said to belong to the $[0, \infty)$ sector if $\nu\psi(\nu) \geq 0$].

In the presence of actuator/sensor nonlinearities, the actual input is given by:

$$u = \psi_a[-G_p\psi_p(y_p) - G_r\psi_r(y_r)] \tag{6}$$

where ψ_a, ψ_p, and ψ_r denote the actuator nonlinearity and the position and rate sensor nonlinearities, respectively. Assuming G_p and G_r are diagonal,

$$u_i = \psi_{ai}[-G_{pi}\psi_{pi}(y_{pi}) - G_{ri}\psi_{ri}(y_{ri})]$$

We assume that ψ_{ai}, ψ_{pi}, and ψ_{ri} $(i = 1, 2, ..., k)$ are continuous single-valued functions: $\mathbf{R} \to \mathbf{R}$. The following theorem was proved [8] to give sufficient conditions for stability.

Theorem 2. Consider closed-loop system given by (1), (3), (4), and (6), where G_p and G_r are diagonal with positive entries. Suppose ψ_{ai}, ψ_{pi}, and ψ_{ri} are single-valued continuous functions, and that, for $i = 1, 2, ..., k$,

(i)$\psi_{ai}(0) = 0$, ψ_{ai} are time invariant and monotonically nondecreasing.

(ii)ψ_{pi}, ψ_{ri}, belong to the $(0, \infty)$ sector and ψ_{pi} are time-invariant.

Then, the closed-loop system is globally asymptotically stable.

Proof.

The proof closely follows [1] and details can be found in [8]. The proof uses Luré-Postnikov type Lyapunov function and the *"skew symmetricity"* of the matrix $(\frac{1}{2}\dot{M} - C)$.

Stability With Dynamic Operator in the Loop

In this section we will prove that the static dissipative control law provides asymptotic stability which is robust to the first order actuator dynamics. Let the actuator be represented by an operator $\mathcal{H}$. In the presence of the actuator dynamics the actual input $u(t)$ is given by:

$$u = \mathcal{H} u_c \tag{7}$$

where u_c is the ideal (desired) input, $\mathcal{H}$ is a nonanticipative, linear or nonlinear, time varying or invariant operator. The control law is now given by:

$$u_c = u_{cp} + u_{cr} \tag{8}$$

where

$$u_{cp} = -G_p y_p = -G_p B p \tag{9}$$

$$u_{cr} = -G_r y_r = -G_r B \dot{p} \tag{10}$$

where u_c represents the command input, u_{cp} and u_{cr} represent command position and rate input, and G_p, G_r are $k \times k$ symmetric positive definite feedback gain matrices.

Suppose $\mathcal{H}$ is a finite-dimensional linear, time-invariant (LTI) operator. [The sensors usually have high bandwidth, and $\mathcal{H}$ then represents the actuators. For the case where $\mathcal{H}, G_p$ and G_r are diagonal, $\mathcal{H}$ represents the combined sensor/actuator dynamics.] We denote $\mathcal{H}g$ by $\mathcal{H}(z_0; g)$ where z_0 is the initial state vector of a minimal realization of $\mathcal{H}$.

Theorem 3: Suppose $\mathcal{H}$ is a non-anticipative, asymptotically stable, observable, LTI operator whose transfer matrix is $H(s) = \epsilon I + \hat{H}(s)$, where $\epsilon > 0$ and $\hat{H}(s)$ is a proper, minimum phase, rational matrix. Under these conditions, the closed-loop system given by Eqs. (1), (4), (5), (6), and (7) is asymptotically stable if

$$\hat{H}(j\omega)(\omega^2 G_r - jG_p\omega) + (\omega^2 G_r + jG_p\omega)\hat{H}^*(j\omega) \geq 0, \quad \forall \ \ real \ \ \omega \tag{11}$$

where $*$ denotes complex conjugate transpose.

Proof.

Define the function

$$V = \frac{1}{2}\dot{p}^T M(p)\dot{p} + \frac{1}{2}q^T \overline{K} q \tag{12}$$

V is clearly positive definite since $M(p)$ and $\overline{K}$ are positive definite symmetric matrices. Taking the time derivative yields,

$$\dot{V} = \dot{p}^T M \ddot{p} + \frac{1}{2}\dot{p}^T \dot{M} \dot{p} + \dot{q}^T \overline{K} q \tag{13}$$

Using (1) in (13), we get,

$$\dot{V} = \dot{p}^T B^T u - \dot{q}^T \overline{D} \dot{q} \tag{14}$$

Substituting (7) and (8) in (14), we obtain,

$$\dot{V} = -\dot{q}^T \overline{D} \dot{q} + y_r^T \mathcal{H} u_c(t) \tag{15}$$

$$\dot{V} = -\dot{q}^T \overline{D} \dot{q} - u_{cr}^T G_r^{-1} \mathcal{H}[z_0; u_c] \tag{16}$$

where $\mathcal{H}$ also depends on its initial state vector z_0. Since $\mathcal{H}$ is linear,

$$\mathcal{H}[z_0; u_c] = h_0(t) + \mathcal{H}[0; u_c] \tag{17}$$

$$\dot{V} = -\dot{q}^T \overline{D} \dot{q} - u_{cr}^T G_r^{-1} h_0(t) - u_{cr}^T G_r^{-1} \mathcal{H}[0; u_c] \tag{18}$$

where $h_0(t)$ is the unforced response of $\mathcal{H}$ due to nonzero initial state. Since $\mathcal{H}$ is strictly stable, $||h_0||$ is finite for any finite z_0. Therefore,

$$\begin{aligned} 0 \leq V(T) = V(0) - &< \dot{q}, \overline{D}\dot{q} >_T - < u_{cr}, G_r^{-1} h_0 >_T \\ &- < u_{cr}, G_r^{-1} \mathcal{H}_p u_{cp} >_T \end{aligned} \tag{19}$$

where $< . >_T$ denotes the truncated inner product for the extended Lebesgue space (see [1]). u_{cp} and u_{cr} are defined in (9) and (10), and

$$\mathcal{H}_p u_{cp} = \mathcal{H}[0; (I + sG_r G_p^{-1}) u_{cp}] \tag{20}$$

In (20), "s" denotes the derivative operator. (s is technically noncausal; however, this difficulty can be overcome by defining the derivative of a truncation at T to be equal to that of a truncated function.) Using Parseval's theorem,

$$\begin{aligned} &<u_{cr}, G_r^{-1} \mathcal{H}_p u_{cp}>_T \\ &= \frac{1}{2\pi}\int_{-\infty}^{\infty} U_{cr_T}^*(j\omega) G_r^{-1} H(j\omega)(I + j\omega G_r G_p^{-1}) \\ &\qquad U_{cp_T}(j\omega) d\omega \\ &= \frac{1}{2\pi}\int_{-\infty}^{\infty} U_{cr_T}^*(j\omega) G_r^{-1} H(j\omega)(\frac{G_p}{j\omega} + G_r) G_r^{-1} \\ &\qquad U_{cr_T}(j\omega) d\omega \\ &= \frac{1}{2\pi}\int_{-\infty}^{\infty} U_{cr_T}^*(j\omega) G_r^{-1} [H(j\omega)(\frac{G_p}{j\omega} + G_r) \\ &\quad + (\frac{G_p}{-j\omega} + G_r) H^*(j\omega)] G_r^{-1} U_{cr_T}(j\omega) d\omega \end{aligned} \tag{21}$$

The matrix in the brackets is positive (from Eq. 11), and we have

$$< u_{cr}, G_r^{-1} \mathcal{H}_p u_{cp} >_T \geq \epsilon_1 ||u_{cr}||_T^2$$

where $\epsilon_1 = \frac{\epsilon}{\lambda_M(G_r)}$. This yields [from(19)]

$$0 \leq V(0) - < \dot{q}, \overline{D}\dot{q} >_T - \epsilon_1 \|u_{cr}\|_T^2 - < u_{cr}, G_r^{-1} h_0 >_T \tag{22}$$

Then, we can write

$$\lambda_m(D)\|\dot{q}\|_T^2 + \epsilon_1 \|u_{cr}\|_T^2 \leq V(0) + \|u_{cr}\| \|G_r^{-1}\|_s \|h_0\| \tag{23}$$

where $\|\cdot\|_s$ denotes the Spectral norm of a matrix. Eq. (23) can be rewritten as

$$\lambda_m(D)\|\dot{q}\|_T^2 + [c_1\|u_{cr}\| - (\frac{c_2}{2c_1})]^2 \leq V(0) + \frac{c_2^2}{4c_1^2} \tag{24}$$

where $c_1 = \sqrt{\epsilon_1}$ and $c_2 = \|h_0\|$. Therefore,

$$\lim_{t\to\infty} \dot{q}(t) = 0 \quad and \lim_{t\to\infty} u_{cr}(t) = 0 \tag{25}$$

$$\Rightarrow \lim_{t\to\infty} \dot{\theta}(t) = 0 \tag{26}$$

Taking the limit of the closed loop equation as $t \to \infty$ we get,

$$Kp = B^T \overline{\mathcal{H}u_{cp}}$$

$$(where \quad \overline{\mathcal{H}u_{cp}} = \lim_{t\to\infty} \mathcal{H}u_c)$$

$$\begin{bmatrix} 0 \\ Kq \end{bmatrix} = \begin{bmatrix} I \\ 0 \end{bmatrix} \overline{\mathcal{H}u_{cp}} \tag{27}$$

$$\Rightarrow q = 0 \quad and \quad \overline{\mathcal{H}u_{cp}} = 0 \Rightarrow \lim_{t\to\infty} \theta(t) = 0 \tag{28}$$

Since $\mathcal{H}$ is observable and minimum phase and its output tends to zero, its state vector tends to zero as $t \to \infty$ and system is globally asymptotically stable.

Remarks

This paper extends the stability properties established in [8] for a flexible multibody systems under static dissipative controller to the case when actuator dynamics are present in the loop. The proof uses Lyapunov's approach and is based on [1]. The Lyapunov function used is an energy-type quadratic function augmented with an appropriate positive definite function to prove global asymptotic stability. In summary, the stability is not only robust to modeling errors, parametric uncertainties, and a wide class of nonlinearities in the actuators/sensors but also to actuator dynamics. This has a significant practical value, since the mathematical models of the system usually have substantial inaccuracies, and the actuation and sensing devices available are not perfect. Future work will also address methods for controller *synthesis* and also dynamic dissipative compensators.

References

1. Joshi, S. M.:*Control of Large Flexible Space Structures.* Berlin Springer-Verlag, 1989 (Vol. 131, Lecture Notes in Control and Information Sciences).

2. Takegaki, M., and Arimoto, S.: A New Feedback Method for Dynamic Control of Manipulators. ASME Journal of Dynamic Systems, Measurement and Control, Vol. 102, June 1981.

3. Koditschek, D. E.: Natural Control of Robot Arms Proc. 1984, I.E.E.E Conference on Decision and Control, Las Vegas, Nevada., pp. 733-735.

4. Arimoto, S., and Miyazaki, F.: Stability and Robustness of PD Feedback Control With Gravity Compensation for Robot Manipulator. ASME Winter Meeting, Anaheim, California, December 1986, pp. 67-72.

5. Wen, J. T., and Bayard, D. S.: A New Class of Control Laws for Robotic Manipulator. Int. Journal of Control, 1988, Vol. 47, No. 5, pp. 1361-1385.

6. Paden, B., and Panja, R.: Globally Asymptotically Stable PD+ Controller for Robot Manipulators. Int. Journal of Control, 1988, Vol. 47, No. 6, pp 1697-1712.

7. Paden, B., Riedle, B., and Bayo, E.: Exponentially Stable Tracking Control for Multi-Joint Flexible-Link Manipulators. Proc. 1990, American Control Conference, San Diego, California, May 23-25, 1990, pp. 680-684.

8. Kelkar, A. G., Joshi, S. M., and Alberts, T. E.: Globally Stabilizing Controllers for Flexible Multibody Systems, to be presented at The 31st IEEE Conference on Decision ans Control, Westin La Paloma, Tucson, Arizona, Dec. 16-18, 1992, paper # S-199.

AMD-Vol. 141/DSC-Vol. 37, Dynamics of Flexible Multibody Systems:
Theory and Experiment
ASME 1992

MODELING AND CONTROL OF A ROTATING FLEXIBLE BEAM ON A TRANSLATABLE BASE

John T. Wen and Michael Repko
Department of Electrical, Computer and Systems Engineering
Rensselaer Polytechnic Institute
Troy, New York

Robert Buche
Department of Spacecraft Engineering
Naval Research Laboratory
Washington, D.C.

Abstract

This paper considers the control of a flexible beam with base rotation torque and translation force input. A nonlinear model is developed by using Hamilton's principle and then linearized by assuming small beam bending displacement. The inherent passivity property in the model is used to construct a simple proportional–derivative type of globally stabilizing control law which requires no model information if the appropriate sensory output is available. Simulation and experimental results confirm the significant improvement of performance with this controller.

1 Introduction

Research in the modeling and control of flexible structures has increased dramatically in the recent years due to the desire for mobility, better energy consumption, and high speed maneuvers, for light–weight structures in the space applications. There are also cases where structural flexibility must be addressed due to the precision requirement of the task, such as positioning in electron microscopes and disk drives, and chip insertion.

The slewing control of a single flexible beam has been used extensively as a simple prototype example to study the control structure interaction issue with the hope that the insight gained in this relatively simple problem can be extrapolated into a more complex domain. Most of the past work on this topic considers a base torque to command the tip of the beam to a desired location, for example, [1, 2, 3, 4, 5], to mention only a few. This problem is inherently difficult as the transfer function from the torque to the arm tip is typically non–minimum–phase. Hence, detail model information is required for the control design and the resulting controller is sensitive with respect to the assumed model parameters.

We observe that when a human being controls a flexible beam (e.g., a fishing rod), the control input is *not* limited just to the wrist torque, but the translation force is used as well. When one is constrained to use only the wrist torque, it becomes more difficult to accurately and rapidly position the tip of the rod as the prediction of the tip motion becomes necessary (for example, before the tip reaches the set point, the torque must be exerted in the opposite direction for deceleration). Another feature of the human control of a flexible beam is that a human being can observe the entire bending deformation of the beam (via vision) and effectively process that information into a corrective force and torque. The control task becomes significantly more difficult if the person is blind–folded or can only observe the tip motion (e.g., through a tip–mounted LED in a dark room). The purpose of this paper is to formalize the above observation by investigating the utilization of the base force input and global sensing (in a loose sense, meaning that the physical sensor is spatially distributed) for vibration suppression. As a concrete example, we consider the control of a flexible beam where the base can both rotate and translate. The more general form of the problem involves a flexible appendage mounted on a rigid body; the flexible appendage is capable of bending and twisting and the rigid body can rotate and translate. This paper considers a special case where the flexible bending and base translation take place in a plane. The hope is to bootstrap from this simpler problem to the better understanding of the general problem. Another possible scenario where base translation and rotation are both available is in the manipulation of flexible objects by using a robot arm.

Interaction between the base force and bending vibration has not been studied as extensively as the torque actuation case. In [6], the effect of bending vibration in a nominally rigid prismatic arm is studied. It is shown that an unguarded bending vibration can lead to instability, and a modified control law is proposed. The energy analysis used there is very closely related to the technique in this paper. Models of a flexible beam involving axial force can be found in [7, 8]. More recently, in the appendix of [9], a nonlinear model of a flexible beam moving in a plane is obtained by using the assumed mode method. The only work that we are aware of that uses base translational motion for vibration suppression appears in [10]. Their model agrees with our linearized model, but their control technique is based on the standard full state feedback pole placement.

In this paper, we extend the modeling approach in [11] to develop a nonlinear dynamic model for a flexible beam moving in a plane. This model is discretized using the natural modes of the fixed base arm to obtain a infinite set of coupled differential equations. It is shown that corresponding to each inputs (torque and a composite force) there is an output such that the input/output pair is passive. In the torque case, the passive output is the hub angular velocity which is a spatially discrete sensor. With hub angular position feedback, the angular velocity is observable, but the Observability Grammian is ill–conditioned due to the weak presence of higher flexible modes in this output. For

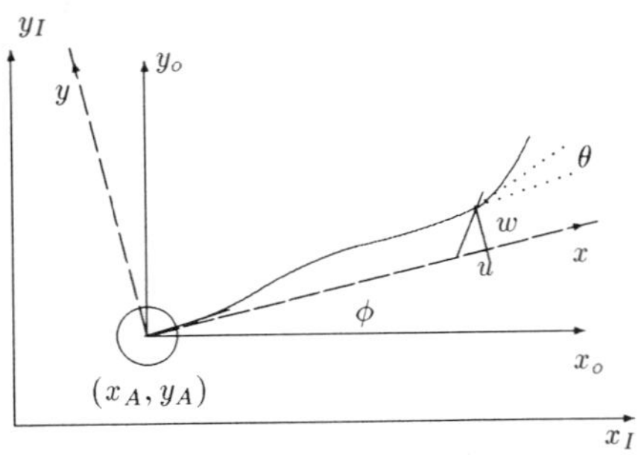

Figure 1: Cantilevered Beam attached to Rotating and Translating Hub

the force input, the passive output can be taken as the rate of change of the spatial integral of the bending displacement which is a spatially distributed output and the corresponding Observability Grammian is better conditioned. By using this passivity property, a simple proportional–derviative (PD) control law, where the derivative portion can be any strictly positive real filter, can be constructed to achieve global asymptotic stability. The advantage of this control law is that explicit model information is not required for stabilization. Any additional model knowledge can be used to improve the closed loop performance. As a validation of the concepts presented in this paper, both simulation and preliminary physical experiments have been performed. The obtained results demonstrate the performance improvement compared to the hub control case. In the experiment, physical sensors consist of four strain gauges and a hub potentiometer. A model–based observer is used to construct the required passive output for the base force feedback. A vision sensor is under development to provide a true model–independent spatially distributed sensory feedback.

The rest of the paper is organized as follows. In section 2, the nonlinear discretized model for a flexible link moving in a plane is obtained. The control design and stability analysis using an energy argument is presented in Section 3. Simulation and experimental results are discussed in Section 4.

The following notations are used throughout the paper:

w	=	bending displacement
u	=	axial displacement
ϕ	=	neutral axis angle
s_ϕ, c_ϕ	=	sine and cosine of ϕ
ρ	=	mass of beam per unit length
L	=	length of beam
M_x, M_y	=	mass of base in x and y directions.

2 Nonlinear Model

Consider a single rotating beam in a plane with the base of the beam free to move in the plane (see Fig. 1). Let the inertial frame in the plane be (x_I, y_I), the translated (but not rotated) frame be (x_o, y_o), and the rotated frame be (x, y). Denote the base of the beam by (x_A, y_A).

An arbitrary point on the beam has the following inertial coordinate:

$$x_I = x_o + x_A \qquad y_I = y_o + y_A. \tag{1}$$

The kinetic energy associated with this unit element is proportional to

$$\frac{1}{2}(\dot{x}_I^2 + \dot{y}_I^2) = \frac{1}{2}(\dot{x}_o^2 + \dot{y}_o^2) + \frac{1}{2}(\dot{x}_A^2 + \dot{y}_A^2) + \dot{x}_A\dot{x}_o + \dot{y}_A\dot{y}_o.$$

Let the total kinetic energy be T and the kinetic energy if (x_o, y_o) frame were inertial be T_o. Then $T = T_o + \Delta T$ with

$$\Delta T = \rho\int_0^L \left(\frac{1}{2}(\dot{x}_A^2 + \dot{y}_A^2) + \dot{x}_A\dot{x}_o + \dot{y}_A\dot{y}_o\right) dx + \frac{1}{2}(M_x\dot{x}_A^2 + M_y\dot{y}_A^2)$$

$$= \Delta T_1 + \Delta T_2$$

where

$$\begin{aligned}\Delta T_1 &= \rho\dot{x}_A\int_0^L ((\dot{u} - w\dot{\phi})c_\phi - (\dot{w} + (u+x)\dot{\phi})s_\phi)\, dx \\ &\quad +\rho\dot{y}_A\int_0^L ((\dot{u} - w\dot{\phi})s_\phi + (\dot{w} + (u+x)\dot{\phi})c_\phi)\, dx \\ &= \rho(\dot{x}_A c_\phi + \dot{y}_A s_\phi)\int_0^L (\dot{u} - w\dot{\phi})\, dx \\ &\quad +\rho(-\dot{x}_A s_\phi + \dot{y}_A c_\phi)\int_0^L (\dot{w} + (u+x)\dot{\phi})\, dx \\ \Delta T_2 &= \frac{1}{2}((M_x + \rho L)\dot{x}_A^2 + (M_y + \rho L)\dot{y}_A^2).\end{aligned}$$

We allow the mass to be different in x and y direction of motion, since in the laboratory experiment, a Cartesian robot is used as the base and the motion in the y direction involves moving the x–axis assembly. Consequently, M_y is much larger than M_x.

The dynamic equation for x_A and y_A can be obtained from the Lagrangian equation:

$$\begin{aligned}\frac{d}{dt}\left(\frac{\partial\Delta T}{\partial\dot{x}_A}\right) - \frac{\partial\Delta T}{\partial x_A} &= F_x \\ \frac{d}{dt}\left(\frac{\partial\Delta T}{\partial\dot{y}_A}\right) - \frac{\partial\Delta T}{\partial y_A} &= F_y.\end{aligned}$$

After some algebra, we obtain:

$$\begin{aligned}M_{T_x}\ddot{x}_A + c_\phi\int_0^L (\ddot{u} - w\ddot{\phi} - 2\dot{w}\dot{\phi} - (u+x)\dot{\phi}^2)\, dx \\ -s_\phi\int_0^L (\ddot{w} + (u+x)\ddot{\phi} + 2\dot{u}\dot{\phi} - w\dot{\phi}^2)\, dx = \rho^{-1}F_x\end{aligned} \tag{2}$$

$$\begin{aligned}M_{T_y}\ddot{y}_A + s_\phi\int_0^L (\ddot{u} - w\ddot{\phi} - 2\dot{w}\dot{\phi} - (u+x)\dot{\phi}^2)\, dx \\ +c_\phi\int_0^L (\ddot{w} + (u+x)\ddot{\phi} + 2\dot{u}\dot{\phi} - w\dot{\phi}^2)\, dx = \rho^{-1}F_y\end{aligned} \tag{3}$$

where $M_{T_x} = \frac{M_x + \rho L}{\rho}$ and $M_{T_y} = \frac{M_y + \rho L}{\rho}$.

Following the same approach as in [11], we discretize the bending and axial coordinates, w and u, and the hub rotation angle ϕ by using the natural modes for the fixed base linear model. The expansion for w and ϕ are given by

$$\phi(t) = \Psi'^T(0)q(t) \tag{4}$$

$$w(t,x) = (\Psi^T(x) - x\Psi'^T(0))q(t). \tag{5}$$

The bending displacement and its derivative, w and w', are assumed to be of order ϵ and *all* time derivatives are assumed to be of order 1. We retain terms up to ϵ^2 for the approximate dynamics. Since the beam is assumed to have no independent axial elongation, u is related to w by (up to the quadratic term)

$$u(t,x) = -\frac{1}{2}\int_0^x w'^2(t,\xi)\, d\xi. \tag{6}$$

With this discretization and approximation, (2) and (3) become (including friction)

$$\begin{aligned}M_{T_x}\ddot{x}_A + \rho^{-1}D_x(\dot{x}_A) - (c_\phi q^T L_1^T + s_\phi L_2^T(q))\ddot{q} \\ +\dot{q}^T(c_\phi C_L - s_\phi C_T)\dot{q} = \rho^{-1}F_x\end{aligned} \tag{7}$$

$$\begin{aligned}M_{T_y}\ddot{y}_A + \rho^{-1}D_y(\dot{y}_A) + (-s_\phi q^T L_1^T + c_\phi L_2^T(q))\ddot{q} \\ +\dot{q}^T(s_\phi C_L + c_\phi C_T)\dot{q} = \rho^{-1}F_y.\end{aligned} \tag{8}$$

where

$$L_1 = P + bd^T - \frac{L^2}{2}bb^T$$

$$
\begin{aligned}
P &= \int_0^L (\Psi'(x) - b)(\Psi'(x) - b)^T (L - x)\, dx. \\
d &= \int_0^L \Psi(x)\, dx \\
b &= \Psi'(0) \\
L_2(q) &= d - \frac{1}{2} b q^T P q \\
C_L(q) &= -P - 2db^T + (\frac{1}{2} q^T P q + \frac{L^2}{2}) bb^T \\
C_T(q) &= -2Pqb^T - (d^T - \frac{L^2}{2} b^T) bb^T
\end{aligned}
$$

The additional terms in the $\ddot{q}$ equation due to the base translational motion can be similarly calculated. After much algebra, the following simple form is obtained:

$$
\begin{aligned}
&M(q)\ddot{q} + D(\dot{q}) - (c_\phi L_1 q + s_\phi L_2(q))\ddot{x}_A \\
&+(-s_\phi L_1 q + c_\phi L_2(q))\ddot{y}_A + C(q,\dot{q})\dot{q} + \Omega^2 q = \rho^{-1} b\tau
\end{aligned}
$$

where M and C are the same as in the fixed base case (see [11] for the exact expression); they have the interpretation of the mass matrix and Coriolis and centrifugal forces expressed in the modal coordinate. M is equal to the identity operator added with terms to the order $\|q\|$. $C(q, \dot{q})$ is linear in $\dot{q}$. Note that $C(q, \dot{q})$ is not unique. M and C are related in an important way: A particular choice of C satisfies the property that $\frac{1}{2}\dot{M} - C$ is skew symmetric. This property follows from the fact that the system is open loop dissipative. $D(\dot{q})$ denotes the damping of the beam.

Combining the above equations, the dynamic equation for a flexible beam on a movable base is

$$
\begin{bmatrix} M(q) & -(c_\phi L_1 q + s_\phi L_2(q)) & (-s_\phi L_1 q + c_\phi L_2(q)) \\ -(c_\phi q^T L_1^T + s_\phi L_2^T(q)) & M_{T_x} & 0 \\ (-s_\phi q^T L_1^T + c_\phi L_2^T(q)) & 0 & M_{T_y} \end{bmatrix} \begin{bmatrix} \ddot{q} \\ \ddot{x}_A \\ \ddot{y}_A \end{bmatrix}
+ \begin{bmatrix} C(q,\dot{q})\dot{q} \\ \dot{q}^T (c_\phi C_L - s_\phi C_T)\dot{q} \\ \dot{q}^T (s_\phi C_L + c_\phi C_T)\dot{q} \end{bmatrix} + \begin{bmatrix} D(\dot{q}) \\ \rho^{-1} D_x(\dot{x}_A) \\ \rho^{-1} D_y(\dot{y}_A) \end{bmatrix} + \begin{bmatrix} \Omega^2 q \\ 0 \\ 0 \end{bmatrix} = \rho^{-1} \begin{bmatrix} b\tau \\ F_x \\ F_y \end{bmatrix} \quad (9)
$$

The augmented mass matrix and Coriolis/centrifugal matrix satisfy the skew symmetric property as in the fixed base case.

To see the influence of the base force on the beam bending vibration, we eliminate $\ddot{x}_A$ and $\ddot{y}_A$ from (9):

$$
\begin{aligned}
&(M(q) - (\frac{c_\phi^2}{M_{T_x}} + \frac{s_\phi^2}{M_{T_y}}) L_1 q q^T L_1^T - (\frac{c_\phi^2}{M_{T_y}} + \frac{s_\phi^2}{M_{T_x}}) L_2(q) L_2^T(q) \\
&+(\frac{1}{M_{T_y}} - \frac{1}{M_{T_x}}) c_\phi s_\phi q^T L_1^T L_2(q))\ddot{q} + D(\dot{q}) \\
&+(c_\phi L_1 q + s_\phi L_2(q)) \frac{D(\dot{x}_A)}{\rho M_{T_x}} - (-s_\phi L_1 q + c_\phi L_2(q)) \frac{D(\dot{y}_A)}{\rho M_{T_y}} + \Omega^2 q \\
&= -\rho^{-1}(\frac{c_\phi}{M_{T_x}} F_x - \frac{s_\phi}{M_{T_y}} F_y) L_1 q - \rho^{-1}(\frac{s_\phi}{M_{T_x}} F_x + \frac{c_\phi}{M_{T_y}} F_y) L_2(q) \\
&+\rho^{-1} b\tau. \qquad (10)
\end{aligned}
$$

Note that the effective mass matrix is positive definite since the mass matrix in (9) is positive definite. If higher order terms in q are ignored by the small bending assumption (including the bilinear terms of $F_x q$ and $F_y q$), then the equation of motion takes on a simpler form ($L_2(q)$ becomes d):

$$
\begin{aligned}
&(I - (\frac{c_\phi^2}{M_{T_y}} + \frac{s_\phi^2}{M_{T_x}}) d^T d)\ddot{q} + D(\dot{q}) \\
&+\frac{d}{\rho}(\frac{s_\phi}{M_{T_x}} D(\dot{x}_A) + \frac{c_\phi}{M_{T_y}} D(\dot{y}_A)) + \Omega^2 q \\
&= \frac{d}{\rho} \frac{s_\phi}{M_{T_x}} F_x + \frac{c_\phi}{M_{T_y}} F_y + \frac{b}{\rho}\tau \qquad (11)
\end{aligned}
$$

The composite force on the right hand side represents an equivalent transversal force (adjusted by the mass) at the base. The longitudal force only has a higher order effect and is ignored. The $\ddot{q}$ term can also be eliminated from the bottom two equations in (9) by using the first equation. This results in (again eliminating higher order terms in q):

$$
\begin{bmatrix} 1 - \frac{s_\phi^2}{M_{T_x}} d^T d & \frac{s_\phi c_\phi}{M_{T_x}} d^T d \\ \frac{s_\phi c_\phi}{M_{T_y}} d^T d & 1 - \frac{c_\phi^2}{M_{T_y}} d^T d \end{bmatrix} \begin{bmatrix} \ddot{x}_A \\ \ddot{y}_A \end{bmatrix}
+ \begin{bmatrix} \frac{D_x(\dot{x}_A)}{\rho M_{T_x}} + \frac{s_\phi d^T D(\dot{q})}{M_{T_x}} \\ \frac{D_y(\dot{y}_A)}{\rho M_{T_y}} - \frac{c_\phi d^T D(\dot{q})}{M_{T_y}} \end{bmatrix} + \begin{bmatrix} \frac{s_\phi d^T \Omega^2 q}{M_{T_x}} \\ \frac{-c_\phi d^T \Omega^2 q}{M_{T_y}} \end{bmatrix}
= \begin{bmatrix} \frac{s_\phi d^T b}{\rho M_{T_x}} \tau \\ -\frac{c_\phi d^T b}{\rho M_{T_y}} \tau \end{bmatrix} + \rho^{-1} \begin{bmatrix} \frac{F_x}{M_{T_x}} \\ \frac{F_y}{M_{T_y}} \end{bmatrix}. \quad (12)
$$

Now we shall make an important simplifying assumption: the effect of the beam motion and hub torque has minimal direct impact on the base translational motion. This assumption is true for our experimental setup since the base is much more massive compared to the beam itself. In the case of a robot arm handling a flexible payload, this assumption also appears to be reasonable. With this assumption, all q and $\dot{q}$ dependent term in the (12) are dropped. Combining with (11), we obtain

$$
\begin{bmatrix} I - (\frac{c_\phi^2}{M_{T_x}} + \frac{s_\phi^2}{M_{T_y}}) dd^T & 0 & 0 \\ 0 & 1 - \frac{s_\phi^2}{M_{T_x}} d^T d & \frac{s_\phi c_\phi}{M_{T_x}} d^T d \\ 0 & \frac{s_\phi c_\phi}{M_{T_y}} d^T d & 1 - \frac{c_\phi^2}{M_{T_y}} d^T d \end{bmatrix} \begin{bmatrix} \ddot{q} \\ \ddot{x}_A \\ \ddot{y}_A \end{bmatrix}
+ \begin{bmatrix} D(\dot{q}) \\ 0 \\ 0 \end{bmatrix} + \begin{bmatrix} \Omega^2 q \\ 0 \\ 0 \end{bmatrix}
= \rho^{-1} \begin{bmatrix} b & \frac{s_\phi}{M_{T_x}} d & -\frac{c_\phi}{M_{T_y}} d \\ 0 & M_{T_x}^{-1} & 0 \\ 0 & 0 & M_{T_y}^{-1} \end{bmatrix} \begin{bmatrix} \tau \\ F_x + D_x(\dot{x}_A) \\ F_y + D_y(\dot{y}_A) \end{bmatrix} \quad (13)
$$

For the simplicity of notation, we rewrite the above equation in the following compact form:

$$
M_e \ddot{z} + D_e \dot{z} + K_e z = B_e u \quad (14)
$$

where z denotes the stacked configuration variable, u denotes the stacked control variable (including the base translational damping). This equation will serve as the basis of our control analysis in the next section. For the stability analysis, we make another simplifying assumption that $\dot{\phi}$ is sufficiently small, that we can treat (14) as a linear equation despite of the trigonometric functions in ϕ.

3 Control Analysis

We only consider the output set point control problem in this paper: *Given the output of interest $y = Cq$ and the desired output y_{des}, find τ and F_T, so that $y \to y_{des}$ and the base returns to its original position, asymptotically.* In the case of flexible beam, y may be tip displacement.

This problem can be decomposed into two parts: 1. Find a steady state configuration to form the error equation, and 2. stabilize the error equation. First we find a steady state solution, q_{des}, such that $Cq_{des} = y_{des}$. Due to the diagonal structure of Ω and the fact the frequency of the rigid body mode is zero, q_{des} can be easily solved:

$$
q_{des} = \begin{bmatrix} \frac{y_{des}}{C_0} & 0 & \ldots & 0 \end{bmatrix}^T \quad (15)
$$

where C_0 is the first component of C represented in the modal coordinate (which is assumed to be non-zero). Let $\Delta q = q - q_{des}$ (and $\Delta z = [\Delta q, 0, 0]^T$), then the error equation is of the following form:

$$
M_e \ddot{z} + D_e \dot{z} + K \Delta z = Bu. \quad (16)
$$

This equation is in the standard second order form and can be easily stabilized with simple proportional–derivative (PD) controllers. The key property to note is that the passivity of the input/output maps from u to $B^T\dot{z}$. We first need to show that this property is preserved under proportional feedback (this actually renders $B^T\dot{z}$ observable, as will be required later). Let

$$u = u_v - K_p\Delta z. \tag{17}$$

To show that the passivity between u and $B^T\dot{z}$ is still maintained, consider the Lyapunov function candidate

$$V = \frac{1}{2}\dot{z}^T M_e \dot{z} + \frac{1}{2}\Delta z^T(K + BK_pB^T)\Delta z. \tag{18}$$

The derivative along the solution yields:

$$\dot{V} = \dot{z}^T(-D_e\dot{z} - Bu_v) \tag{19}$$

Note that $\dot{z}^T D_e \dot{z}$ is negative semidefinite. The velocity feedback, u_v, can be chosen as any strictly positive real function operating on $B^T\dot{z}$, and $B^T\dot{z} \to 0$ as $t \to \infty$ by the Invariance Principle (for an in-depth analysis, see [12]). If the closed loop system is observable with respect to $B^T\dot{z}$, the closed loop error system is asymptotically stable as required. The open loop stiffness matrix for the beam, Ω^2 is diagonal and the only zero diagonal entry is in the leading component. The zero component implies that the open loop system is not observable with respect to $B^T\dot{z}$. With the proportional feedback, the closed loop stiffness matrix is positive definite, and the observability is then obtained. Note that even with hub feedback alone, one can show asymptotical stability since $b^T\dot{q}$ is observable with the hub position feedback. However, the observability Grammian is typically very ill conditioned, due to the "almost" unobservabililty of the higher flexible modes. The advantage provided by the $d^T\dot{q}$ feedback (in $B^T\dot{z}$) is that the higher order modes are typically much more observable. Hence, the gain of the $d^T\dot{q}$ feedback has a larger impact on the overall performance. The overall PD feedback controller, with constant velocity feedback, is summarized below:

$$\tau = -k_p b^T\Delta q - k_v b^T\dot{q} \tag{20}$$

$$\begin{aligned} F_x &= -\frac{k_{p_x}}{M_{T_x}} s_\phi d^T\Delta q - \frac{k_{v_x}}{M_{T_x}} s_\phi d^T\dot{q} - \frac{k_{p_x}}{M_{T_x}}\Delta x_A \\ &\quad - \frac{k_{v_x}}{M_{T_x}}\dot{x}_A - D_x(\dot{x}_A) \end{aligned} \tag{21}$$

$$\begin{aligned} F_y &= \frac{k_{p_y}}{M_{T_y}} c_\phi d^T\Delta q + \frac{k_{v_y}}{M_{T_y}} c_\phi d^T\dot{q} - \frac{k_{p_y}}{M_{T_y}}\Delta y_A \\ &\quad - \frac{k_{v_y}}{M_{T_y}}\dot{y}_A - D_y(\dot{y}_A). \end{aligned} \tag{22}$$

A key assumption of the above analysis is that $B^T\Delta z$ and $B^T\dot{z}$ are available. We now take a closer look at this issue. For simplicity, assume K_p, K_v are both diagonal matrices consisting of diagonal entries, k_p, k_{p_x}, k_{p_y}, and k_v, k_{v_x}, k_{v_y}, respectively. Then the proportional feedback becomes:

$$B^T K_p \Delta z = \begin{bmatrix} k_p b^T\Delta q \\ \frac{k_{p_x} s_\phi}{M_{T_x}} d^T\Delta q + \frac{k_{p_x}}{M_{T_x}} x \\ -\frac{k_{p_y} c_\phi}{M_{T_y}} d^T\Delta q + \frac{k_{p_x}}{M_{T_y}} y \end{bmatrix}. \tag{23}$$

The velocity feedback has a similar expression, except the displacement quantities are replaced by their rates of change. The hub feedback, $b^T\Delta q$ and $b^T\dot{q}$ are the hub angular displacement error and angular velocity. This is typically available. The force feedback consists of two terms, one is proportional to $d^T\Delta q$ and the other to the base translation variable. The second term is reasonable, as the base position can typically be measured. The first term requires a further examination. Choose y as the tip displacement as an example. Then $y(t) = \Psi(L)q(t) = w(t, L) + L\phi(t)$. The steady state solution is $w_{des} = 0$ and $\phi_{des} = y_{des}/L$. Now, $d^T\Delta q$ can be rewritten as:

$$\begin{aligned} d^T\Delta q &= \int_0^L w(t,x)\,dx + \frac{L^2}{2}\Delta\phi(t) \\ &= \int_0^L w(t,x)\,dx + \frac{L^2}{2} b^T\Delta q. \end{aligned} \tag{24}$$

The second term is the same as the hub feedback which is assumed to be available. The first term is a spatially distributed sensor which integrates the total bending displacement over the entire beam. The conclusion of the above analysis suggests the trade–off between the model information versus sensory information. If such global sensor is available (e.g., possibly synthesized via vision), then high performance control (since higher order flexible modes have strong presence in a distributed output) is possible without any explicit model information. Lack of such, model information is required to synthesize the required output, possibly with an observer. In our laboratory experiment, we have begun with the second route by using four strain gauges mounted on the beam. We are also pursuing the first approach by directly mounting a CCD camera overhead. In the above analysis, also note that F_x and F_y both suppress vibration of the beam and also maintain the asymptotic stability of the base position.

4 Simulation and Experimental Results

4.1 Simulation

The analysis presented in the last section is tested on an eight–mode simulation model for a flexible link. The simulation is based on the linearized model.

The relevant parameters are listed below:

$$\begin{aligned} \omega &= [0, 18.7, 45.6, 113.0, 218.4, 359.9, 537.1, 750] \\ b &= [1.44, 3.87, 2.63, 0.98, 0.49, 0.25, 0.17, 0.12] \\ d &= [0.39, 0.02, -0.16, -0.12, -0.08, -0.13, -0.11, -0.09] \\ \Psi(L) &= [1.59, -2.18, 1.57, -1.90, 2.00, -2.03, 2.05, -2.07] \\ M_{T_x} &= M_{T_y} = 50 \qquad \rho = 0.4436 \qquad L = .9. \end{aligned}$$

All dampings are set to zero.

We will compare the open loop damping with the closed loop damping of two cases: hub PD feedback alone and hub PD plus the base force feedback.

To select the hub feedback gains, we target the first two modes: rigid body mode and the first flexible mode. The gains are chosen to critically damp the rigid body mode, and to place the first flexible mode as close to the real axis as possible. The final gains are chosen by root locus; the values are given below:

$$k_p = \frac{\rho\omega_c^2}{b_0^2} = 0.77 \qquad k_v = \frac{2\rho\omega_c}{b_0^2} = 0.81$$

where ω_c is chosen to be 1.9. The force feedback gains are chosen in a similar way, except they target the second and third flexible mode by critically damping the second flexible mode alone. Of course, with all the other modes, critical damping cannot be exactly achieved, so the final gain values are selected based on the root locus method by adjusting the nominal frequency in the critically damped second order system. The selected gain values are:

$$k_{p_x} = k_{p_y} = 493 \qquad k_{v_x} = k_{v_y} = 246.$$

The advantage of the passivity approach in control design is that one does not have to be concern at all about the spillover effect as far as stability is concerned. Gain tuning can then be concentrated on performance improvement.

In Table 1, eigenvalues and damping ratios for the open loop, hub torque PD feedback, and hub torque and base force feedback cases are compared. It is clear that the hub feedback does a good job providing damping for the rigid body and the first flexible modes, but the base force feedback provides additional damping for the second and third flexible modes while further improving the low order modes as well. The performance comparison can be further seen in the bode plots (from hub to tip) shown in Figure 2. A further advantage of the passivity type of controllers is their inherent robustness: the gain margin is infinite (in the positive direction) and the phase margin is at least $\pm 90°$.

mode #	open loop	torque control	torque and force control
1	0	100	100
2	0	100	100
3	0	99.1	74.4
4	0	0.82	17.3
5	0	0.72	2.8
6	0	0.15	0.78
7	0	0.02	1.2
8	0	0.008	0.61
9	0	0.003	0.31

Table 1: Damping Ratio Comparison (all units in %)

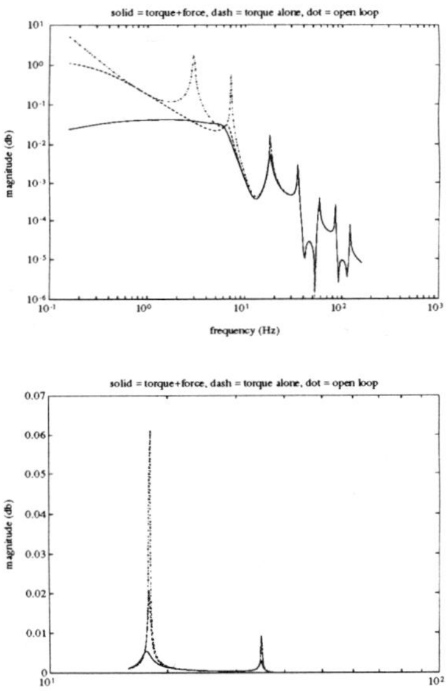

Figure 2: Bode Plots Comparison

4.2 Experimental Results

We have constructed a physical experiment to observe the performance of this controller on an actual apparatus. A GE Cartesian robot has been modified to provide the base planar motion to a rotating beam mounted at the robot wrist. We use three motors, two for the x and y motion causing linear motion via lead screws, and one directly driving the flexible beam. For sensing, we currently have a hub potentiometer and four strain gauges located at 6.5mm, .27m, .37m, and .58m away from the hub. To obtain the required force feedback $d^T \Delta q$ and $d^T \dot{q}$, we use an observer to synthesize estimates of these quantities from the sensory outputs. In order to construct an observer, the model information is required. An extensive identification is conducted through the following steps:

1. The modal frequencies are first identified first by a crude random noise black box transfer function fit, and then refined by a sine sweep close to each of the resonant peaks.

2. The modal dampings are obtained by exciting the beam at each mode and then fit the decay envelop to an exponential.

3. The coefficients in b and d are obtained via a least square fit (assuming that the modal frequencies and dampings are exact).

4. Frictions in the lead screws and the hub assembly are obtained by fitting the physical response to a second order response. Stictions are obtained from the breakaway torque.

The identified results are listed in Table 2.

mode number	frequency (Hz)	damping (%)	b	d
1	3.69	0.29	4.00	0.057
2	8.35	0.26	4.14	-0.16
3	19.31	0.23	2.14	-0.74
4	35.46	0.003	0.95	-0.56

Motor	Static Friction Coeff.
Hub	0.00405 Nm
X	46.1285 N
Y	138.3854 N

Motor	Viscous Friction	Coulomb Friction
Hub	1.6730 $\frac{1}{sec}$	0.0473 $\frac{rad}{sec^2}$
X	6.2021 $\frac{1}{sec}$	0.4683 $\frac{m}{sec^2}$
Y	6.5281 $\frac{1}{sec}$	0.5144 $\frac{m}{sec^2}$

Table 2: Identified Model Parameters

A Kalman filter type observer is then designed by using the identified model. If the observer is an ideal one (i.e., the model is known exactly), then the input/output transfer function is the same as in the exact measurement case (stable observable poles are uncontrollable). There is some robustness margin when the observer is inexact, but a systematic observer design procedure to maximize the robustness margin is yet to be developed. As a preliminary test, a sinusoidal disturbance is fed through the hub torquer, and the open loop response is compared with the closed loop response when the force controller alone is turned on. The responses of strain gauges #1 and #4 are shown in Fig. 3. Similar to the simulation bode plots shown in the last section, a reduction of at least 50% in oscillation is seen in the closed loop case. Another test compares the open loop response with the closed loop responses with the hub torque control alone and with the hub torque combined with base force control. The experimental bode plots in Fig. 4 show much more damped closed loop responses versus the open loop. A more detailed comparison of the two closed loop control cases and their difference (the torque control plot subtracted from the torque plus force control plot) in Fig. 5 shows the enhanced damping provided by the base force control. The damping is not as large as in the simulation due in part to the heavy friction in the x–y linear drives, the increased noise level with three motors running, and the modeling error in the observer design. We are currently incorporating friction compensation, better noise shielding, and improved model identification and direct distributed feedback by using vision sensors. These results will be communicated in the future.

5 Conclusion

In this paper, the flexible beam slewing control problem is investigated when the base of the beam is allowed to both rotate and translate in

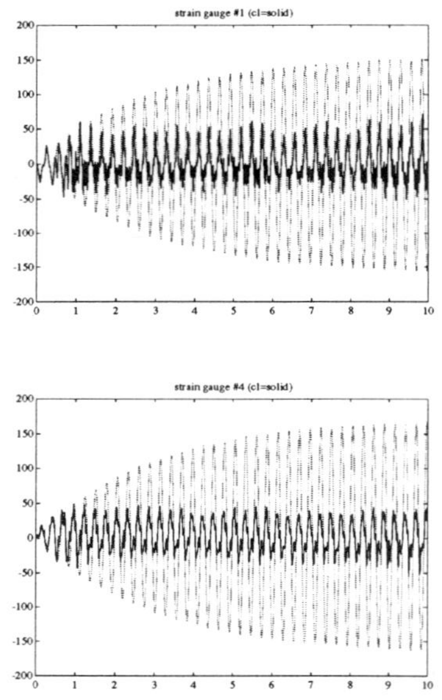

Figure 3: Experimental Strain Gauges Response: Open Loop vs. Closed Loop

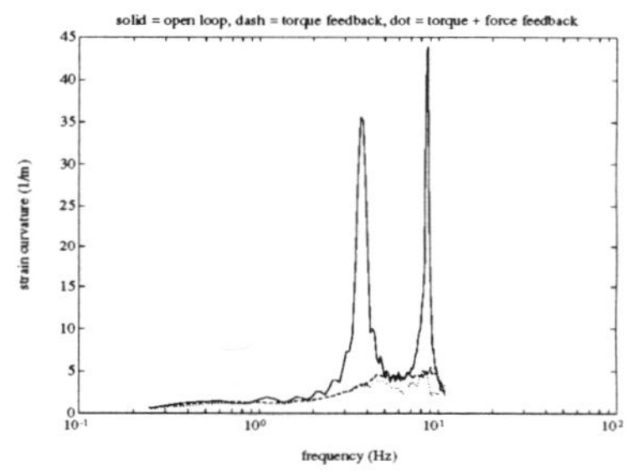

Figure 4: Experimental Hub to Strain Gauge #4 Transfer Function: Open Loop vs. Hub Torque Control vs. Hub Torque and Base Force Control Cases

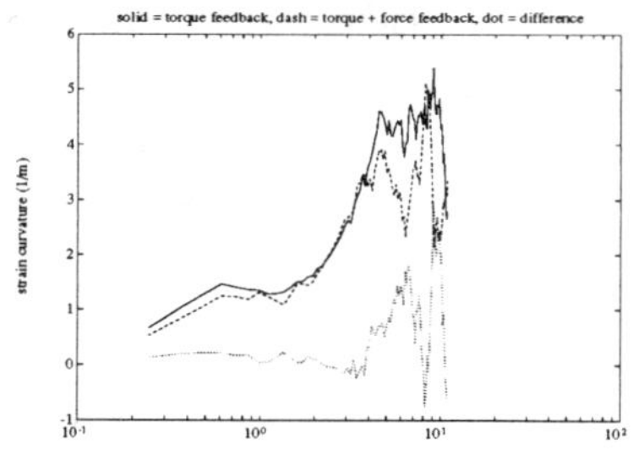

Figure 5: Experimental Hub to Strain Gauge #4 Transfer Function: Hub Torque Control vs. Hub Torque and Base Force Control Cases

a plane. A nonlinear model is obtained by using the Hamilton's Principle. By assuming a massive base, a simple proportional–derivative type control law is obtained by using the inherent passivity property in this system. In the linearized form, measurements of the integral of the bending displacement and velocity over the length of the beam are required, but no other model information is needed for stabilization. Simulation has shown a significant improvement in performance over the torque feedback alone. A laboratory experiment where a flexible beam is mounted at the wrist of a Cartesian robot arm has also been construction to validate this approach experimentally. Preliminary results have shown effective vibration suppression capability by using base translation force alone. We are currently instrumenting a vision sensor to provide a direct measurement of the required spatially distributed output.

Acknowledgment

The work reported in this paper is supported in part by the National Science Foundation under Grant No.MSS–8910437 and MSS–8906219, and in part by the Center for Advanced Technology in Automation and Robotics at Rensselaer Polytechnic Institute, under a grant from the New York State Science and Technology Foundation.

References

[1] R.H. Cannon, Jr. and E. Schmitz. Precise control of flexible manipulators. In M. Brady and R. Paul, editors, *Robotics Research: The First International Symposium*, pages 841–861. MIT Press, Cambridge, MA, 1984.

[2] G. Hastings and W.J. Book. Experiments in the optimal control of a flexible manipulator. In *Proc. 1985 American Control Conference*, pages 728–729, Boston, MA, 1985.

[3] A. De Luca and B. Siciliano. Joint–based control of a nonlinear model of a flexible arm. In *Proc. 1988 American Control Conference*, pages 935–940, Atlanta, GA, 1988.

[4] S. Yurkovich, F. Pacheco, and P. Anthony. On–line frequency domain information for control of a flexible–link robot with varying payload. *IEEE Transactions on Automatic Control*, 34(12):1300–1304, December 1989.

[5] D. Wang and M. Vidyasagar. Modelling and control of flexible beam using the stable factorization approach. In *Proc. 1989 IEEE Robotics and Automation Conference*, pages 1042–1047, Scottsdale, AZ, 1989.

[6] P.K.C. Wang and J.D. Wei. Vibrations in a moving flexible robot arm. *Journal of Sound and Vibration*, 116(1):149–160, 1987.

[7] S. Timoshenko, D.H. Young, and Jr. W. Weaver. *Vibration Problems in Engineering.* John Wiley & Sons, NY, 4th edition, 1974.

[8] L. Meirovitch. *Analytic Methods in Vibration.* The MacMillan Company, New York, 1967.

[9] C.E. Padilla and A.H. von Flotow. Nonlinear strain–displacement relations in the dynamics of a two–link flexible manipulator. SSL Report 6–89, Massachusetts Institute of Technology, May 1989.

[10] A.K. Misra. Vibration control of spacecraft appendages through base motion. In *AIAA/AAS Astrdynamics Conference*, pages 431–437, Portland, OR, April 1990.

[11] F. Wang and J.T. Wen. Nonlinear dynamical model and control for a flexible beam. CIRSSE Report 75, Rensselaer Polytechnic Institute, November 1990.

[12] L. Lanari and J.T. Wen. Asymptotically stabel set point control laws for flexible robots. *to appear in System and Control Letters*, 1992.

AMD-Vol. 141/DSC-Vol. 37, Dynamics of Flexible Multibody Systems:
Theory and Experiment
ASME 1992

VIBRATION CONTROL DURING SLEWING MANEUVERS FOR THE REDUCTION OF LINE-OF-SIGHT ERRORS IN STRUCTURES

Marco D'Amore and Ephrahim Garcia
Department of Mechanical Engineering
Vanderbilt University
Nashville, Tennessee

ABSTRACT

The ASTREX (Advanced Structural Research Experiment) facility located at the Phillips Laboratory, Edwards AFB, is a ground based test facility for the analysis of structural characteristics of a realistic complex structure. The impetus of this paper is to analyze and design control schemes to suppress tip motion of the structure to maintain alignment between the secondary and primary substructures with motivation to enhance line-of-sight (LOS) performance. Various velocity feedback control methods were evaluated based on their ability to control tip deflections of the ASTREX. A reduced order model of the structure was created via the combination of the mass and stiffness information obtained through a finite element model and experimentally determined damping ratios obtained through ERA (Eigensystem Realization Algorithm). A modal structural model and actuator model were then integrated into a single form for use in vibration control. Finally, the effects of slewing on the dynamics of the structure, most notably on the ability to maintain LOS, was considered and the control of these motions was investigated with respect to a quadratic regulator scheme, where an optimal feedback technique was employed.

INTRODUCTION

The inherent flexibility of light weight space structures unfortunately makes them easily susceptible to vibrations caused by various external stimuli. Also, because of various mission requirements, these relatively flimsy space frames with multiple flexible appendages may need to be reoriented into new positions. This may be necessary so that one of these appendages, perhaps an antenna, lens or other optical device can be aimed accurately. The attitude adjustment of this structure, or just that of a single appendage, will induce loads that can cause the structure to vibrate for an unacceptable period of time, making the rapid pointing and repointing virtually impossible. Since the ASTREX is a structural model configured in order to perform tracking studies, the effects of attitude adjustments, i.e. slewing accelerations, on the performance of the structures will need to be investigated. Such a maneuver will induce loads, and hence flexure, on the tripod arms which connect the primary to the secondary. Uncontrolled dynamics caused by these loads will greatly reduce the ability of the structure to maintain signal fidelity. By decreasing the settling time of the structure's response, the structure can be made to retarget more rapidly.

Lim and Horta (1990) investigated the vibration control for the pointing of structures by comparing various techniques such as LVF (local velocity feedback) and the optimal control scheme LQG (Linear Quadratic Gaussian). They developed an LOS criterion to judge the performance of each technique they analyzed. This criterion was based on the pointing geometry of the NASA Langley Evolutionary model. This criteria is the basis for the LQG penalty functions, and hence, used in the formulation of the control gains. Similarly in this paper, suppression of the relative tip motions is used as a performance indicator. These feedback signals are used in the cost function for the optimal control studies. Meirovitch and Quinn (1987) and Meirovitch and Kwak (1988) developed the equations of motion for a structure with a flexible appendage and the later of these papers suggests a possible use of this study as a line-of-sight stabilization of this appendage. Prior to all these aforementioned authors, Hughes (1972) develops similar equations of motion for use in the study of slewing flexible satellites. The significance of their derivation is the development of the equations of motion of a slewing flexible structure while encompassing in these equations the flexibility of the appendage (i.e. antenna or laser) by inclusion of the inertial cross coupling terms. It is the inertia of the structure which induces loads on the appendage that is incorporated into the model via the flexibility of the appendage. In this paper study, a first order approximation to these terms will be derived.

One of the key thrusts behind this study is to develop anticipatory flexure in the structure during slewing maneuvers. That is, to send the slewing command signal not only to the slewing actuators, the gas thrusters, but additionally to the actuators on the structure that control flexure, the torque wheel actuators (see Garcia and Inman 1990). In this way, the structure itself can be made active to anticipate the expected slewing loads, and assume a flexural posture that will aid in the vibration suppression.

SYSTEM DYNAMICS

Actuator Dynamics

In order to accurately determine the performance of the different controller designs, the dynamics of the actuator were modeled and included into the global system model. The actuator used in this model was a torque wheel actuator (TWA). This actuator, mounted on the secondary, can be easily modeled as a motor with a circular disk attached, to provide the necessary inertia to counteract tip deflections. The actuator equations can be expressed in state space form given by

$$\dot{\underline{x}}_a = \begin{bmatrix} -\frac{B}{J} & \frac{K_t}{J} \\ -\frac{K_b}{L} & -\frac{R}{L} \end{bmatrix} \underline{x}_a + \begin{bmatrix} 0 \\ -\frac{1}{L} \end{bmatrix} V_i \tag{1}$$

$$T_m = [\,0 \quad K_t\,]\,\underline{x}_a \tag{2}$$

where the output of this system (Eq. 2) is torque, which is fed back into the structure via the structural input matrix $\underline{u}_s$, (Eq. 4). The actuator state space vector, $\underline{x}_a$, is defined as $[\omega\ i]^T$. This vector consists of the rotational speed and the current of the actuator, respectively. V_i is a function of the sensor signal after a gain has been applied. In Eqs. (1) and (2), the variables B, K_t, K_b, J and R are the damping coefficient, torque constant, back emf constant, inertia and resistance of the actuator, respectively. A complete explanation of these constants can be found in reference [9].

Structural Dynamics

The structural model is used in its modal form because this affords us the ability to integrate experimentally found damping information into the model. A few of the frequencies and corresponding damping ratios are given in Table 1. The modal equation of motion with a multiple input vector is:

$$I\ddot{\underline{q}} + \phi^T D\phi\,\dot{\underline{q}} + \phi^T K\phi\,\underline{q} = \phi^T B_{fy}\,\underline{u}_{sy} + \phi^T B_{fx}\,\underline{u}_{sx} \tag{3}$$

where $\Lambda_K = \phi^T K\phi$ and $\Lambda_D = \phi^T D\phi$. This can be transformed into a state space representation as follows:

$$\dot{\underline{x}}_s = \begin{bmatrix} 0 & I \\ -\Lambda_K & -\Lambda_D \end{bmatrix} \underline{x}_s + \begin{bmatrix} 0 & 0 \\ \phi^T B_{fy} & \phi^T B_{fy} \end{bmatrix} \underline{u}_s \tag{4}$$

where the ϕ (the transformation between physical and modal coordinates) is the mass normalized eigenvector matrix, I is the identity matrix and $\underline{x}^T_s = [\,\underline{q}\ \dot{\underline{q}}\,]^T$. The input vector $\underline{u}_s$ consists of $[\underline{u}_{sy}\ \underline{u}_{sx}]^T$ from Eq. (3). Each one of these vectors comprises the control torques (Eq. 2) distributed over the secondary (Fig. 1). The distribution of these forces over the secondary is accomplished via B_{fy} and B_{fx}, the structural input vectors.

MODE No.	ω Hz	ζ
1	13.2263	.00529
2	13.4066	.00190
3	17.6819	.00880
4	19.2488	.00590
5	19.4623	.00950

Table 1. Table of the first five frequencies and damping ratios.

CONTROLLABILITY AND STABILITY

Typically control laws are based on reduced order models. Experimental implementation of such a controller, may result in stability problems. The closely spaced modes, in particular the first two flexible tripod modes, are spaced such that they display a beating effect which can be seen in the open loop response of the system to a unit impulse in the x and y directions (Fig. 3). It seems that while these modes cannot be classified as repeated, the coupling between these modes is quite evident. It will be shown that the presence of closely spaced frequencies necessitates more than one actuator be employed to effectively control the dynamics.

Placement of sensors and actuators, as well as the type of feedback signal may have a destabilizing effect upon the system. Ideally, to ensure that the system will not be driven unstable by the feedback signal, the sensor and actuator pair is collocated. While precise collocation is possible with piezoelectric actuators (Dosch, et al 1992), care must be taken in applying electromechanical actuators. For these systems, sensors must be physically collocated, and produce a signal which is analytically compatible. The effect of a noncollocated control can potentially destabilize a system by responding with a control force that is in phase with motion of the system, thus adding energy to the system rather than the result intended. The LVF is directly effected by the type of feedback as well as how close to a collocated signal we our able to obtain.

Finally, conditions are considered for ensuring and quantifying the controllability of the system with respect to the each individual mode. The degree of controllability, which can be formulated in several ways, is a vehicle by which we can assess how controllable each mode is with respect to each other. From this information, we can infer the number of actuators warranted for a high degree of authority for the ensuing control law.

The first method discussed was proposed by Hughes and Skelton (1980) for a general second order system. Similarly, Hughes and Skelton [6] published a study under the auspice of including rotational and rigid body modes to model flexible spacecraft dynamics. They propose the development of a controllability norm. Consider the following undamped system.

$$M\ddot{\underline{\xi}} + K\underline{\xi} = B\underline{u} \tag{5}$$

The reason for excluding damping effects in the formulation of the controllability condition is that the dissapative energy effects typically serve as another avenue for coupling in the system and will always inherently increase the controllability. The opposite of this is not valid, and therefore neglecting damping is more conservative design approach.

Making use of the special properties of the mass and stiffness matrices, $M=M^T > 0$ and $K=K^T > 0$. We can formulate the modal matrix ϕ and transform the above equation into

$$I\ddot{\underline{q}} + \Lambda_K \underline{q} = \bar{B}\underline{u} \tag{6}$$

where $I = \phi^T M \phi$, $\Lambda_K = \phi^T K \phi$ and $\bar{B} = \phi^T B$, as previously stated. The system order is N. Transforming this into state space, we have the following matrices

$$\underline{x} = \begin{bmatrix} q \\ \dot{q} \end{bmatrix}, \quad \mathbf{A} = \begin{bmatrix} 0 & I \\ -\Lambda_K & 0 \end{bmatrix}, \quad \mathbf{B} = \begin{bmatrix} 0 \\ \bar{B} \end{bmatrix} \tag{7}$$

where $\mathbf{A}$ is an 2N x 2N matrix. Making use of the typical test for controllability and using the special form of the above matrices, the controllability matrix can be re configured in the following form.

$$C = rank\left[\begin{array}{cc|cc|ccc} 0 & \bar{B} & 0 & & -\Lambda_K\bar{B} & \dots & (-\Lambda_K^{N-1})\bar{B} \\ \hline \bar{B} & 0 & -\Lambda_K\bar{B} & & 0 & \dots & 0 \end{array}\right] = 2N \tag{8}$$

Noticing the repeatability of this matrix, sufficient criteria for controllability can be redefined as

$$C = rank\left[\bar{B} \quad \Lambda_K\bar{B} \quad \Lambda_K^2\bar{B} \quad \dots \quad \Lambda_K^{N-1}\bar{B}\right] = N \tag{9}$$

While equations (8) and (9) make use of the modal form of the system, they only state whether or not the system is or is not controllable. However, it would behoove us to be able to evaluate each mode separately. Thus, if our system was uncontrollable we could deduce from this analysis what mode was the culprit. The results of two methods will be shown. The first is the development of a controllability norm, and the second is the analysis of the singular values of the controllability grammian of a balanced system [12]. These two methods are compared in Table 2.

Comparison of Singular Values and Modal Participation Factors balanced system				
	Single Actuators		Dual Actuators	
ω_i	σ_i	$\lVert b_i \rVert$	σ_i	$\lVert b_i \rVert$
13.2263	0.0016442	0.0240739	0.0406313	0.1966608
13.4066	0.0390196	0.0706435	0.0644075	0.3033061
19.2488	0.0000013	0.0017501	0.0000211	0.0115361
19.4623	0.0018089	0.0299140	0.0006387	0.0310526

* All Values Normalized To One

Table 2. Comparison of methods for relative controllability

Hughes and Skelton make use of the modal form of the system to develope their controllability norm. They distinguish between a system with repeated modes and one which has only distinct modes when formulating the norm. Since our system does not have any repeated modes, the norm for the distinct case is given (Eq. 10).

$$C_n = \lVert \underline{b}_n \rVert > 0, \quad n=1 \dots N \tag{10}$$

where

$$\| b_n \| = \sqrt{b_n^T \, b_n} \tag{11}$$

where this is simply the Euclidean norm of the row vectors of $\bar{B}$, the modal participation factors. The norm in Eq. (10) suggests that the N modes of Λ_K can be ranked according to the size of C_n. This norm provides a relative measure of the controllability of the modes, where the largest C_n is the most controllable.

The above method is compared to a singular value decomposition in Table 2. This norm as well as a singular value decomposition is performed for the two pairs of closely spaced modes at 13 and 19 Hz. The tests are performed for single and dual TWA configuration. For the single actuator case, the TWA produces a torque about the X axis and for the dual actuator case there is an additional TWA producing a moment about the Y axis. From Table 2, it is evident that the value of the norm and the singular value for a particular mode yield consistent results in that they increase at least by an order of magnitude with addition of a second actuator. Intuitively this make sense, since the degree of controllability is ultimately increased with the number of columns in the **B** matrix. Using this information and modeshape analysis, the actuator(s) can be strategically positioned.

VIBRATONAL CONTROL

Before the control of tip motion during a slewing maneuver is pursued, the vibration suppression of the secondary to a unit impulse in the X and Y directions is analyzed. For the orientation of the actuators and the position of the angular velocity sensors refer to Fig. 1. To ensure a collocated feedback, it is assumed that the angular velocity is fed back into a collocated TWA.

The first control method analyzed is a local velocity feedback (LVF) designed via classical methods (i.e. gains chosen by root locus method). As stated earlier the form of the feedback is the angular velocity about point O. The feedback is of the form

$$y = \begin{bmatrix} V_y \\ V_x \end{bmatrix} = K \begin{bmatrix} \omega_y \\ \omega_x \end{bmatrix} \tag{12}$$

where K is the gain matrix and V_x and V_y are the voltage inputs into the actuators. The LVF control design is one of the least complex control schemes to implement both analytically as well as

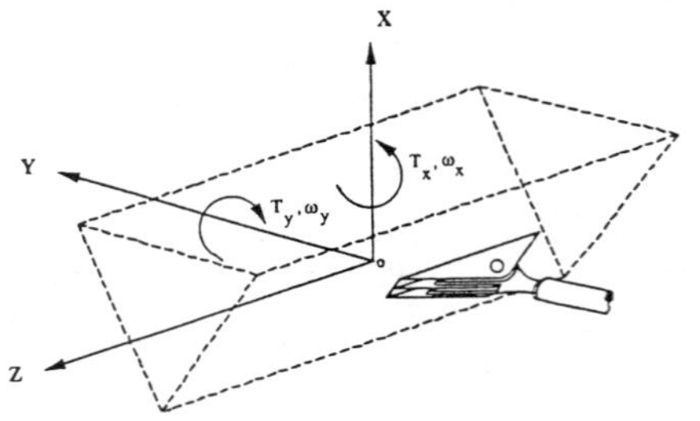

Fig. 1. View of the secondary structure and one of the three tripod legs. View shows the orientation of the coordinate system.

experimentally. In contrast to the simplicity of the LVF method, the linear quadratic regulator (LQR) is a design method which analytically is relatively easy to employ, but experimentally is not implementable without the use of a state estimator. The advent of high powered computers facilitates the analytical implementation of a method such as LQE (Linear Quadratic Estimator), an LQR with a state estimator. This control scenario unlike the LVF method requires full sate feedback which for the purpose of simulations is not unreasonable. However, experimentally this requires the estimation of states which leads to large on line computational burdens for a complex structure. Because of the experimental complexity of the LQE method, the LVF method is often more attractive if satisfactory performance levels are achieved. The LQR design method is an optimal control scheme in which acceptance of a gain matrix is based upon the minimization of a cost function. The cost function chosen for the optimal control analyses in this paper is as follows:

$$\mathbf{J} = \int_{t_1}^{t_2} (y^T Q y + u^T R u)\, dt \tag{13}$$

The reason for adapting this form of **J**, rather than the traditional form, $x^T Q' x$, is that this allows for the direct penalization of the angular rotation and displacement of the secondary. The vector **y** is the same that is used in Eq. (12). This form of **J** is beneficial because it includes the feedback into the cost function. The cost function is transformed into the traditional form by converting the Q in equation (13) into $Q' = C^T Q C$, where C is the output matrix.

The LVF controller substantially reduces the settling time of the response as shown in Fig. 5 to about .5 seconds. The dominant 13 Hz modes were reduced by well over an order of magnitude (Fig. 6). As stated earlier, the simplicity of experimental implementation of the LVF makes it attractive if experimental goals are achievable. To accurately assess the performance of the LVF method it is compared to an LQR analysis. The benefit of an optimal controller is that it will efficiently control the dynamics of the system, while typically not expending any unnecessary power.

Weighting matrices were chosen to closely mimic the time response of the LVF. Fig. 8 shows that like the LVF controller, tip displacements settled in about .5 seconds. Although, the frequency response of the LQR shows a slightly larger reduction in peak frequency (Fig. 9). Since by investigation of the displacement histories both methods were qualitatively performing about the same, it was necessary to see if either method had higher current or speed demands upon the actuators. From Figs. 7 and 10, it is quite evident that the two methods perform similarly. Thus making the complexity of the LQR unnecessary in this case.

OPTIMAL SLEWING CONTROL

Employing a full state feedback scheme, it is proposed to optimize the vibration suppression of the ASTREX during a slewing maneuver about its vertical axis (Fig. 2). This is done by including the inertia terms about the slewing axis as well as the cross coupling flexural terms in the mass matrix. A first order approximation to this system in modal coordinates is:

$$\begin{bmatrix} I_{11} & M_{12}\phi \\ \phi^T M_{21} & I \end{bmatrix} \begin{bmatrix} \ddot{\theta} \\ \ddot{q} \end{bmatrix} + \begin{bmatrix} 0 & 0 \\ 0 & \Lambda_D \end{bmatrix} \begin{bmatrix} \dot{\theta} \\ \dot{q} \end{bmatrix} + \begin{bmatrix} 0 & 0 \\ 0 & \Lambda_K \end{bmatrix} \begin{bmatrix} \theta \\ q \end{bmatrix}$$

$$= \begin{bmatrix} 1 & 0 & 0 \\ 0 & \phi^T B_{fy} & \phi^T B_{fx} \end{bmatrix} \begin{bmatrix} \tau \\ \underline{u}_s \end{bmatrix} \qquad (14)$$

where I_{11} is the moment of inertia of the whole structure about the slewing axis, $\phi^T M_{21}$ and $M_{12}\phi$ are the structural cross coupling terms. The θ and the τ are the rotational angle and the control torque, respectively, for the overall structure about the slewing axis. These cross coupling terms constitute the mass of the appendage being slewed, integrated over the length of the appendage. For a more thorough treatment of these equations see references [2,5,9,10]. Using this model combined with actuator dynamics an optimal study is performed for small slewing angles. For this study a slew angle of 1° was chosen. Small rotational angles were chosen because conceivably spacecraft such as the ASTREX will be aiming at targets at great distances. For instance, a space based antenna communicating with an Earth fixed installation. In such cases the change of attitude by a single degree would mean significant changes on Earth. For the same reasons, tip vibrations, generating LOS errors, would be intolerable. Again the optimal output version of the LQR cost function, Eq. (13), is used, however in this case the output vector **y** not only contains the input voltages to the actuators but also the angle of rotation of the overall structure about the slewing axis. The control vector **u**, in Eq. (13), contains the control voltages for the TWA as well as the control torque for the slewing maneuver.

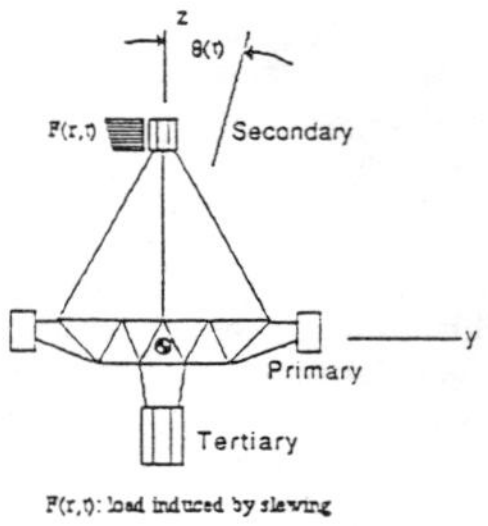

Fig. 2. Schematic of the orientation of the ASTREX about the slewing axis.

The optimal slewing was simulated in such a manner as to track a new reference angle from the initial orientation in Fig. 2. In this case the new reference was a 1° step function. Weighting matrices were chosen to optimize several criteria. First, to minimize the settling time of the tip oscillations. Secondly, to minimize the overshoot to the new reference position while minimizing maneuvering time. Lastly, to minimize the amount of current required by the torque wheel as well as the angular velocity of the rotating masses. To stay within the operating limits of the physical hardware, i.e. limits of the gas-thrusters used in slewing, was precisely why an optimal slewing study was performed.

From Fig. 13a it can be seen that the peak time of the system is about .75 seconds while the system only overshoots by 3.5%. The settling of the tip motion is improved in the X direction to about two seconds whereas in the Y direction the amplitude increase, however oscillations damp out much sooner than in the open loop response of Figs. 11a,c. Also, it is seen in Fig. 11b that the peak time increases to about 1.2 seconds for the open loop maneuver. Comparing the rotational speed of the global structure during the open loop slew (Fig. 11d) to the closed loop (Fig. 13b), the rate for the open loop case is much slower, thus explaining the increased peak time. These results give credence to the benefits of slewing an active structure where the actuators act as an inertial assist for the maneuvering structure.

CONCLUDING REMARKS

The contents of this paper are an attempt to control small amplitude vibrations that will undoubtedly be very significant for satellites that must reorient themselves quickly before transmitting a signal in space. Small attitude changes in space will translate into large errors for a distant target. A small slewing angle was chosen to optimize the tracking of such space structures. Simulations show that vibrations can be suppressed to decrease the settling time of the ASTREX by more than one order of magnitude. Further study will be focussed on the advantages of attitude adjustments with an active structure [2]. This could potentially extend the life of spacecraft by conserving fuel meant for thrusters (where the thrusters are actuators that would only be concerned with the gross motion of the structure, not vibration suppression) by allowing the torque wheels to help in the slewing maneuver where their (TWA) power source would typically be solar. Also it was shown that it is possible to obtain adequate control with a simplistic type controller such as LVF rather than the sophistication necessary for an LQR type controller. This will also be advantageous for spacecraft with weight limits where the need for extravagant computer equipment used for control would be reduced.

ACKNOWLEDGMENT

The authors wish to acknowledge the contributions of the Drs. Nandu Abaankar, Joel Berg and Alok Das of the Phillips Laboratory's Structures and Control Division at Edwards AFB, California, and an AFOSR Research Initiation Program grant #92-32, administered through the Research and Development Laboratory, Incorporated.

REFERENCES

1. Dosch, J., Inman, D.J., and Garcia, E., "A Self-Sensing Piezoelectric Actuator For Collocated Control," Journal of Intelligent Material Systems and Structures, vol. 3, no.1, January, 1992.

2. Garcia, E. and Inman, D.J., "Advantages Of Slewing an Active Structure," Journal of Intelligent Material Systems and Structures, Vol. 1, July 1990, pp. 261 - 272

3. Golub, G. H. and Van Loan, C.F., *Matrix Computations*, Baltimore, Maryland, Johns Hopkins University Press.

4. Hughes, P.C., "Dynamics of Flexible Space Vehicles with Active Attitude Control," Celestial Mechanics, Vol. 9, 1974, pp. 21 - 39

5. Hughes, P.C. and Skelton, R.E., "Controllability and Observability of a Linear Matrix-Second-Order System," Journal of Applied Mechanics, Vol. 47, June 1980, pp. 415 - 420

6. Hughes, P.C. and Skelton, R.E., "Controllability and Observability for Flexible Spacecraft," Journal of Guidance and Control, Vol. 3, No. 5, Sept.- Oct. 1980, pp. 452 - 459

7. Inman, D.J., *Vibration and Control, Measurement, and Stability*, Prentice Hall, Englewood Cliffs, New Jersey, 1989

8. Lim, K.B. and Horta, L.G., "A Line-Of-Sight Performance Criterion for Controller Design of a Proposed Laboratory Model," AIAA-90-1226-CP, pp. 349 - 359

9. Kuo, B.C., *Automatic Control Systems*, Prentice Hall, Englewood Cliffs, New Jersey, 1991

10. Meirovitch, L. and Moon, K.K., "Dynamics and Control of Spacecraft with Retargeting Flexible Antennas," Journal of Guidance, Control and Dynamics, Vol. 13, No. 2, Mar.- Apr. 1990, pp. 241 - 248

11. Meirovitch, L. and Quinn, R.D., "Equations of Motion for Maneuvering Flexible Spacecraft," Journal of Guidance, Control and Dynamics, Vol. 10, No. 5, Sept.- Oct. 1987, pp. 453 - 465

12. Skelton, R.E., *Dynamic Sytsems Control, Linear Systems Analysis and Synthesis*, John Wiley and Sons, Inc., New York, New York, 1988

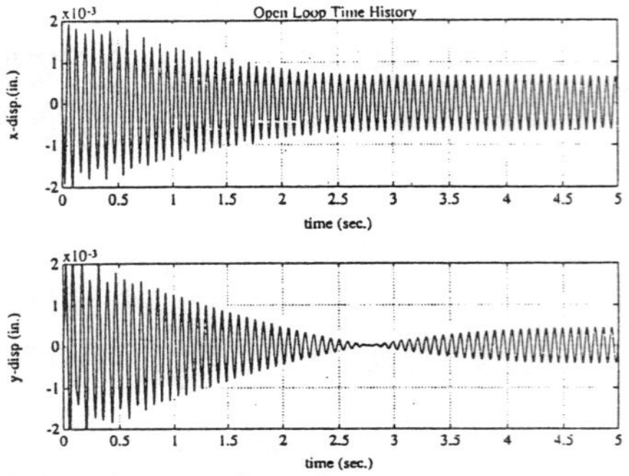

Fig. 3. Open loop time history for an impulse in both x and y directions.

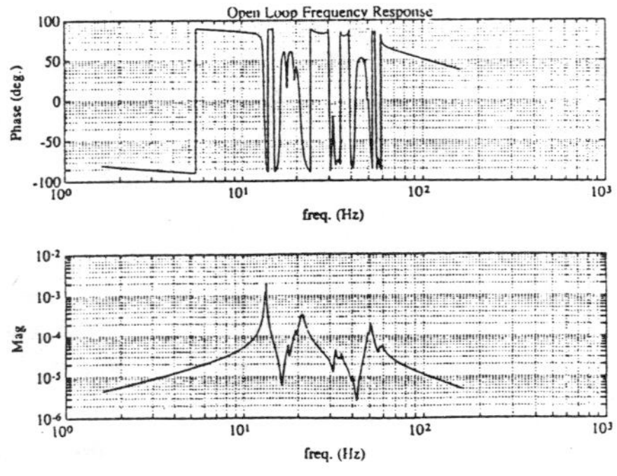

Fig. 4. Open loop frequency response

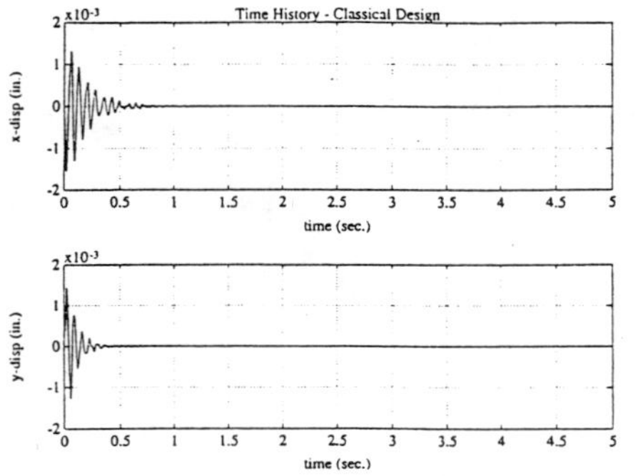

Fig. 5. Closed loop time history for local velocity feedback through the use of classical design methods.

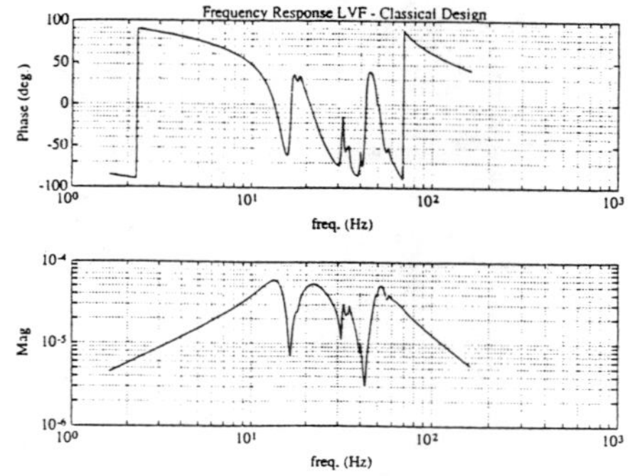

Fig. 6. Closed loop frequency response using classical design methods.

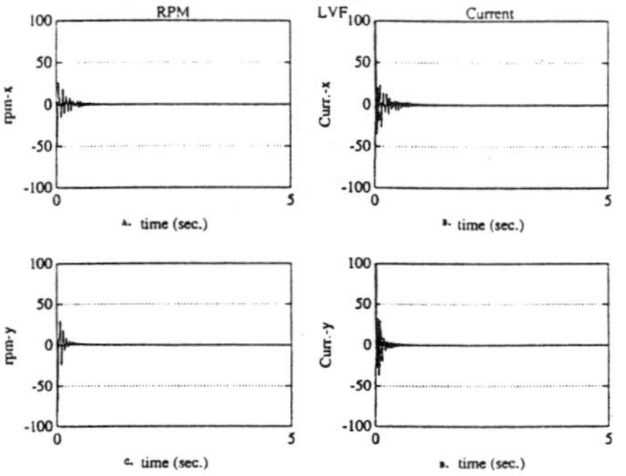

Fig. 7. Closed loop performance using LVF. (a) Ang. Vel. of the TWA the y axis (b) current for the TWA about the y axis (c) Ang. Vel. of the TWA the x axis (d) current for the TWA about the x axis

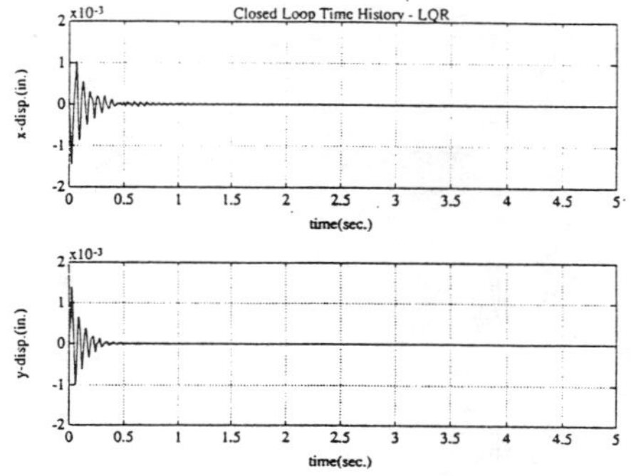

Fig. 8. Closed loop time history for unit impulse in the x and y directions - LQR design.

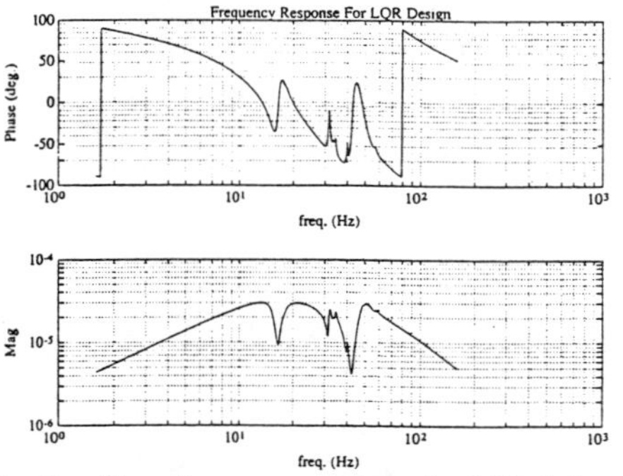

Fig. 9. Closed loop frequency response using LQR design method.

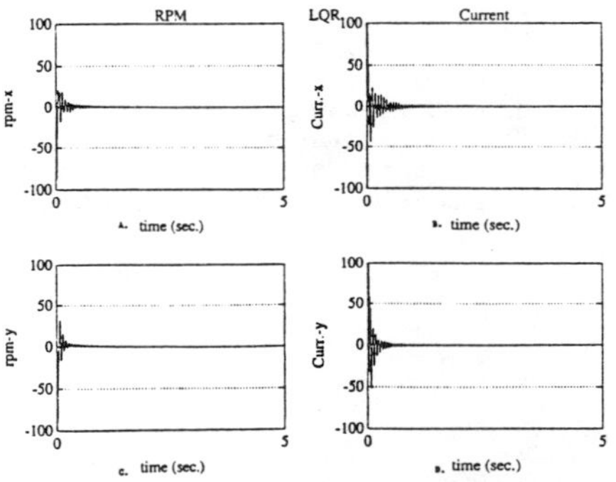

Fig. 10. Closed loop performance using LQR. (a) Ang. Vel. of the TWA the y axis (b) current for the TWA about the y axis (c) Ang. Vel. of the TWA the x axis (d) current for the TWA about the x axis

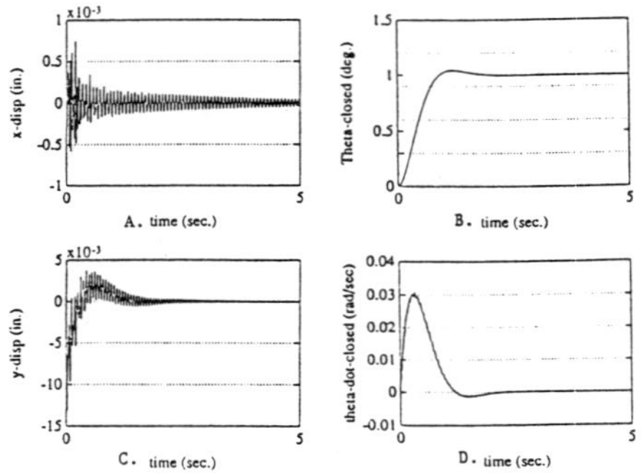

Fig. 11. Open loop slewing simulation. (a) tip displacement in the x directon (b) rotation angle of the global structure (c) tip displacement in the y direction (d) rotational rate of the structure

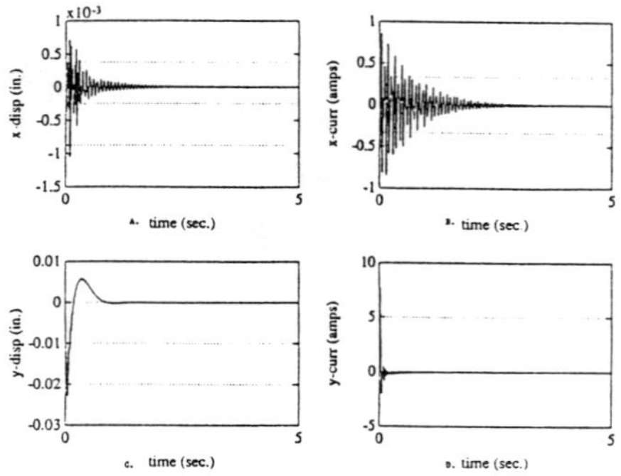

Fig. 12. Closed loop slewing simulation. (a) tip displacement in the x direction (b) current for the TWA in the X (c) tip displacement in the y direction (d) current for the TWA in the Y

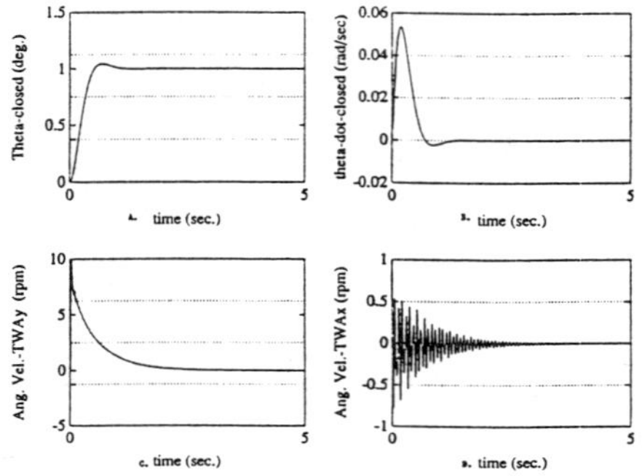

Fig. 13. Closed loop slewing simulation. (a) rotation angle of the global structure (b) rotational rate of the structure (c) Angular vel for the TWA in the about the X (d) Angular vel for the TWA in the about the Y

AMD-Vol. 141/DSC-Vol. 37, Dynamics of Flexible Multibody Systems:
Theory and Experiment
ASME 1992

MSFC MULTIBODY MODELING, VERIFICATION, AND CONTROL LABORATORY

Angelia P. Bukley
Marshall Space Flight Center

Paul M. Christian
Dynacs Engineering, Incorporated

Patrick A. Tobbe
Logicon Control Dynamics

Huntsville, Alabama

INTRODUCTION

Approximate numerical methods are generally employed to solve the nonlinear partial differential equations for flexible multibody dynamics. The TREETOPS multibody modeling code is one such tool. This code uses Kane's equations and the component mode approach for multibody simulation. To date, verification of multibody tools have has been limited to the fixed point case, accomplished by comparing component and system mode results to those of the NASTRAN finite element code. Validation of the modeled nonlinear behavior can not be accomplished in this manner. Hardware experiments highlighting modeling features of interest, such as large angle slewing, are required for such validation. The Multibody Modeling, Verification, and Control (MMVC) Program at Marshall Space Flight Center (MSFC) is focused on the experimental validation of multibody modeling codes and the application of control theory to nonlinear dynamic systems.

The MMVC Program was initiated in November, 1990. The MMVC laboratory is currently under development and will provide a testbed for the execution of experiments designed specifically to validate modeling of complex systems. Modeling features under study are body flexibility, including large motions with small and large deformation; interface degree-of-freedom, including point and line interfaces undergoing translation and rotation; geometric stiffness, including gravity and foreshortening; and constraints, including prescribed motions and closed-tree topologies. The top-level design of a basic set of experiments that emphasize critical modeling features presently included in the TREETOPS simulation has been completed. Beginning with a simple single beam experiment and evolving to multiple beams, joints, and various topologies, the experiments will grow in complexity as each modeling feature is examined. The final experiment will feature a test article traceable to the Advanced X-Ray Astronomical Facility (AXAF). Figure 1 depicts the general methodology of the MMVC validation plan. Experiment hardware has been fabricated, and individual components have been tested. Detailed procedures for system-level experiments are being developed.

Critical to the experiments is the design and development of a test facility. A facility design was chosen such that an existing platform will be modified to accommodate the MMVC experiments. Additional structure will be added to the platform to provide a support base for the test articles and to raise the fundamental frequency of the platform such that it is outside the frequency range of interest for the experiments. The facility design has been finalized, and fabrication should be completed next year. An integral part of the faciltiy is the real-time closed-loop system (RTCS). Its function is to process the sensor inputs, implement the controller, and provide the real-time output signals to the actuators. The RTCS is in place and functionally verified.

As part of the MMVC program, enhancements to the TREETOPS code are planned. The goal is to develop a Government-owned "all-in-one" tool that can be used to develop structural models of multibody systems, perform model order reduction, develop controllers, and assess controller performance in a closed-loop sense via simulation. Currently, the simulation tool is a menu-driven program used to model and analyze flexible multibody structures exhibiting either open- or closed-tree topologies. The menu program provides the means to implement gains for a standard proportional, integral, differential (PID) controller or to include a user-defined controller. The results of this effort will be the enhancement of TREETOPS to include model reduction techniques, thermal effects, optical path analysis capability, expanded controller design capability, and to improve computational efficiency.

The MMVC Program at MSFC will provide experimental validation of multibody simulations and lead to the development of a Government-owned multibody modeling and control system design and analysis tool. The results of the experiments and the enhanced TREETOPS code are and will be publically available upon request to the Government. The following sections contain brief descriptions of the TREETOPS code and planned enhancements, the MMVC experiments and validation plan, the MMVC facility, and highlights of the control design techniques envisioned for use in the closed-loop control experiments.

DESCRIPTION OF THE TREETOPS MODELING TOOL

Introduction

TREETOPS is a time history simulation of the motion of arbitrary complex multibody flexible structures with active control elements.[1] The name TREETOPS, which is not an acronym, refers to the class of structures whose motion can be simulated by the program, those having an open- or a closed-tree topology. The program offers the user an advanced capability for analyzing the dynamics and control-related issues of such structures.

In the simulation, the total structure is considered as an interconnected set of individual bodies, each described by its own modal characteristics with prescribed boundary conditions. An interactive set-up program creates all necessary data files. A linearization option that provides both the simplified model typically used during the initial phases of control system design and the complex model needed for final verification is also available. Thus, TREETOPS can be used throughout the life of a project, and the user is not required to learn a new simulation system as the project progresses.

In addition to multibody simulation, TREETOPS contains subroutines for control system analysis and design. Using this complete capability, the user can create and linearize complex, multibody models, import the plant model into MATLAB, design a feedback compensator in matrix form and export the results back to TREETOPS as a 'matrix controller' for final design verification.

The current version can be configured to execute on most Unix platforms as well as PC class machines. The graphics program, TREEPLOT, is customized for specific monitors and printers and is continuously updated. The PC version of TREEPLOT has yet to be developed; however, TREETOPS is completely compatible with the PC version of MATLAB and this product can be used for obtaining graphical output from TREETOPS.

Planned Enhancements for TREETOPS

A number of enhancements are planned for TREETOPS. Among these enhamcements are order-N formulation for greater computational efficiency, the inclusion of inverse dynamics control and geometric nonlinearities, and an improved graphical user interface (GUI).

The multibody dynamics formulation and corresponding solution algorithm presently employed in TREETOPS is classified as an order-N-cubed approach, where N is the number of degrees of freedom. The dynamic equations of motion are formulated using Kane's Equations. The algorithm currently in use involves a matrix-vector implementation wherein a generalized NxN system mass matrix is formed and inverted to solve for the N degree of freedom accelerations. This proceedure requires N^3 operations. Research in numerical analysis has demonstrated that such problems can be solved using order-N algorithms requiring N operations. These algorithms essentially perform recursive operations to solve the equations of motion wherein the assembly and inversion of a system mass matrix is avoided. For a large system order, order-N techniques result in a substantial savings in computational time.

The increasing demand for high-operating speed, accuracy, and efficiency has led to strict requirements on the design of control systems for space-based manipulators. This requires consideration of a set of highly coupled nonlinear dynamic equations to determine the control torques and forces necessary to produce the desired motion of the manipulator. This also suggests the use of more sophisticated control schemes, such as inverse dynamics controllers. Hence, this feature will be added to TREETOPS. This enhancement is discussed in more detail in a later section.

Another planned enhancement is the inclusion of the effects of geometric nonlinearities. When properly accounted for, these terms will accurately reflect the motion induced change in stiffness of the structure. The current version of TREETOPS uses the assumed modes method to describe the elasticity in the links. The assumption in this method is that the elastic deflection is small and can be obtained as a linear superposition of the modes multiplied by their respective time-dependent amplitudes. These deflections are the axial and transverse elastic displacements, and rotations of a configuration point.

The assumed modes method is perhaps the most suitable method to describe the elasticity in any arbitrarily shaped body. Such a body can be mathematically discretized and its modal frequencies and mode shapes easily obtained using any linear finite element program. An approach is sought to compensate for the change in stiffness created by the use of the linear finite element program. One solution is the retention of the nonlinear part of the strain expression that is omitted in the linear finite element theory.

In the expression for the potential energy due to the nonlinear expression in the strain, the impressed loads (stresses) explicitly appear. Once these loads are specified, a stiffness matrix, called "the geometric stiffness matrix," which is analogous to the linear stiffness matrix, is obtained. This approach will be extended to multibody systems with arbitrarily shaped flexible bodies and included in the analysis code.

A GUI is currently under development. The goals for the GUI development are to increase learning speed and simulation implementation time, reduce errors, and encourage rapid recall for infrequent users. The desktop metaphor, with its windows, icons, and pull down menus, is very popular because it is easy to learn and requires minimal typing skills. The requirement to memorize arcane keyboard commands is also alleviated. The GUI will comprise full screen form using cursor keys and a mouse for movement from field to field. The input options will be designed as a set of icons

TREETOPS currently lacks a unified environment in which to run the constituent programs with transparent data communications. The user must invoke each program at the command line with a problem name. The commands have a three level hierarchy. The user is constrained to sequential movement from a higher level to lower level. In addition the user must remember the exact command for each operation. Thus the user has the burden of with committing the entire command set to memory. With the new GUI, the user will be able to specify a problem name and choose any of the available options, including NASTRAN, TREESET, TREESEL, MATLAB, and others. If the option the user selects requires any interaction, then a form for that interaction is presented on the screen and the users simply provides the required input data. Communication between the different program elements will be through data files, but will be transparent to the user. The GUI will also have an extensive error checking routine executed at all stages of data entry. When an error is detected, the GUI will prompt the user to re-enter the data.

TREETOPS Modeling Features to be Verified via Laboratory Experiments

Several aspects of the flexible multibody modeling problem will be examined in the MMVC program. The primary focus will be on the evaluation of the assumed modes method when applied to multibody systems. In this technique, the structural flexibility of each body is modeled as a linear combination of spatial shape functions and generalized time coordinates. Through proper selection of the component shape functions or Ritz vectors, the system dynamic characteristics may be recovered. Several points will be addressed concerning the selection of the Ritz vectors. First, the type of Ritz vectors that should be used for various classes of multibody systems will be assessed. These vectors may be normal modes, Lanczos modes, block Krylov modes, and shape functions from substructure coupling techniques. Next, the sets of shape functions to be retained for each body will be determined as will the boundary conditions to be used in computing these shape functions. These points will be addressed through a series of increasingly complex experiments to be conducted in the MMVC laboratory. The experiments will be designed such that the flexible effects of the components dominate the time response of the system.

Experiments will also be designed to examine other aspects of multibody systems. Modeling techniques will be evaluated which account for geometric stiffening of systems described through the assumed modes method. These techniques account for changes in structural stiffness induced by motion and gravity. In particular, experiments will be performed to measure the time response of systems undergoing buckling loads and large angular velocities. These results will be compared to analytical predictions which account for the changes in stiffness. Additional studies will be performed to evaluate modeling techniques in the areas of joint friction, joint flexibility, kinematic and closed loop constraints.

Assumed Modes Validation Plan

The MMVC validation plan consists of verification of the assumed modes hypothesis for a multibody structure and will provide insight as to how the multibody structures should be modeled. The current procedure consists of three steps; 1) model development, 2) data collection, 3) post test analysis. The overall plan is illustrated in Figures 2, 3, and 4.

MMVC EXPERIMENTS

The proposed series of experiments for the MMVC program can be classified into three categories: 1) Open-loop topologies, 2) Closed-loop topologies, and 3) Space structures. Each of these categories have specific issues associated with them. For example, the open-loop topologies have one actuator for each joint while the closed topologies have fewer actuators than joints. Furthermore, in closed-loop topologies the component flexible links can be modeled independently, but the system imposes interdependencies between the component modes through closed-loop constraints. Space structures can belong to any of the above categories but elaborate modeling may be required and the control objectives may also differ significantly from those in the first two categories.

A set of experiments has been devised to address the modeling issues identified in the MMVC program. The first group of experiments considers open-loop topologies, the second set is for closed-loops, and the last set focuses on a representative space structure. The experiments are previewed in the following sections and the specific issues of each experiment are addressed. The experiments are ordered according to complexity. Each configuration will be used to address several modeling and dynamics issues and incorporate several control objectives.

Two control objectives will be used in virtually all configurations; pick-and-place control and trajectory control. The objective of pick-and-place is to move from one point to another without regard to the trajectory, while the second approach specifies the trajectory to be followed.

Open Loop Topologies

The experiments designed for this class of problems are composed of single and two link systems connected through active and passive joints to a moving base. The base may be held fixed or actively controlled. The experiment configurations are based on a building block approach using interchangeable components. The designer may select from a wide variety of links with varying dynamic characteristics. There are aluminum and steel beams of varying cross sections and lengths, as well as more complex "geodesic" and "ladder" beams. Each of the beams has been modeled in NASTRAN and its component characteristics documented. There are standard mechanical interfaces to attach the beams to passive and active joints as well as tip masses and counter weights. The active joints are driven by DC torque motors and may be configured for planer or three dimensional experiments. Figures 5, 6, 7, and 8 are typical open loop topology experiments. The objectives of the open loop experiments are:

1) To demonstrate the coupling between rigid body and elatic motion of systems.
2) To address the issue of modal selection and types of shape functions used in the modeling process.
3) To investigate motion induced stiffness changes.

The control objectives are:

1) Pick and place control.
2) Pointing control.
3) Pendulum mode control.

Closed Loop Topologies

Thie class of experiments consists of combinations of rigid and flexible links forming a closed loop mechanism as shown in Figure 9. Typically, the number of active joints in the system is greater than the number of passive joints. These experiments are designed to validate use of kinematic and closed loop constraints equations in multibody codes.

Space Structures

The previous beam experiments were designed to address several aspects of multibody dynamics and control through increasing levels of complexity. The Very Elastic Rotating NASA Experiment (VERNE) will incorporate the experience gained thus far into the modeling and control of a complex spacecraft. VERNE, shown in Figure 10, is composed of a moderately flexible core body, flexible pointing unit, two flexible solar arrays, and a pair of whip antennas with end masses. A rigid beam attaches the core body to the linear motion system of the facility through a ball joint. The experiment will inherently have two pendulum modes, which are rotations about the X and Y axes, and a roll mode about the Z axis. VERNE was designed such that the bending modes of the solar arrays and antenna are highly coupled with the pendulum modes. The pointing unit is connected to the core body through three linear electromechanical actuators, forming a closed loop topology. The pointing unit has a range of motion of ± 30 degrees about the local X and Y axes. The linear actuators can generate a peak force of 200 pounds and have a throw of 18 inches. The pointing resolution of the unit computed from the accuracy of the incremental encoders on the lead screws of the actuators is .002 degrees. The point unit is 2 feet tall and is composed of 3 triangular plates connected by longerons. A generic housing was fabricated with the triangular plates to hold assorted laser or optical sensors.

The flexible solar panels are 8 feet long and 1 foot wide. The panels consist of thin aluminum struts bolted in a truss like fashion. The solar panels have 360 degrees of travel about the X axis and are powered by a direct drive D.C. motor. The drive shafts are instrumented with incremental encoders and tachometers. The encoder resolution is .35 degrees. The peak torque available from the motors is 11 foot-pounds.

The core body is composed of aluminum angle. The whip antenna are rigidly connected to the core body. Three orthogonal reaction wheels are mounted to the core body along the body axes. Each reaction wheel is driven by a D.C. torque motor equipped with a tachometer. The core body is also instrumented with a three axis rate gyro system.

The preliminary system modal characteristics are shown in Table 1. The first two bending modes at .263 and .275 Hertz are torsion modes of the solar panels about the drive shafts. The next mode is a system pendulum mode at .366 Hertz about the Y axis. The bending mode at .484 Hertz is a combination pendulum mode about X and solar panel torsion. These modes may be shifted through the use of counter weights and adjustments to the solar panels and antenna.

Table 1. Preliminary System Modal Characteristics

MODE	Frequency (Hz)	DESCRIPTION
1	0	Rigid Body Rotation About Z
2	.263	Solar Panel Rotation in Phase
3	.275	Solar Panel Rotation Out of Phase
4	.366	Pendulum About Y
5	.484	Pendulum About X / Solar Panel Torsion
6	1.577	Antenna 1st Bending About X in Phase
7	1.640	Antenna 1st Bending About X Our of Phase
8	1.718	Antenna 1st Bending About Z in Phase
9	1.795	Antenna 1st Bending About Z Our of Phase
10	5.164	Solar Panel 1st Bending

VERNE Experiments

The objectives of the experiments proposed for VERNE are divided into dynamics and controls. The objectives of the dynamic open loop tests are:

1) to test the validity of the generalization of modal selection issues from earlier experiments.
2) to study the pendulum modes in a multi-body context.
3) to study motion coupling through various prescribed open loop maneuvers.

The control objectives are:

1) pointing control in the presence of base excitation.
2) pointing control in the presence of solar panel maneuvers.
3) pointing control in the presence of pendulum modes.

Three open loop experiments have been proposed. First, the translational degree of freedom of the linear motion system will be locked and the solar panels will be driven through various slew maneuvers. Next, the solar panels will be held fixed and the system will be driven through base excitation. Finally, the solar panels will again be driven, but this time in the presence of base excitation. The effect of solar array motion and base excitation on the system pendulum modes will be studied using sensor time histories and compared to analytical results.

The controls experiments consist of accurately pointing the lower unit in the presence of solar panel motion and base excitation. The control system designer will have access to line of sight error from a light source on the lower unit illuminating a quad detector on the ground. The designer will also have information from the rate gyros, solar panel drive shaft position and rate, and relative angle between the core body and lower pointing unit. The engineer must design the loops generating torque/force commands for the reaction wheels, solar panel drives, and linear actuators from the feedback of the various sensors.

THE MMVC LABORATORY FACILITY

The MMVC project consist of multibody modeling, verification, and control. Currently dynamic multibody systems with flexible members and large rotations and translations at the joints are modeled using TREETOPS. Information on the flexible modes is input to the code from NASTRAN models of the bodies. There are many open questions as to which modes should be input to TREETOPS - that will be addressed in the modeling experiments. TREETOPS has been widely used for many years, but its results have never been experimentally confirmed. This issue will be addressed in the verification section. Finally, new methods for control of the structures will be investigated in the control section.

Platform and Linear Motion System Design

The MMVC facility will be located in the west high bay area of building 4619 at MSFC. This facility is joined with the Flexible Space Structures (FSS) ground test facilities and is accessed via the control room. The two primary requirements for MMVC facility are experiment work volume and support structure stiffness. The desired work volume is 20' by 20' by 20'. This will allow room for large translations and rotations of the experiments, as well as for larger test articles needed for low frequency modes. The experiment support structure must withstand the static and dynamic loads from the test articles. The structure should also isolate the experiments from unwanted disturbances. Isolation will be accomplished by moving the support structure natural frequencies to a range outside of those under study. Other factors considered in designing the facility were: facility enclosure, power, lighting, ventilation, access, safety, and cost.

Currently, outside of the FSS control room in building 4619, there is a balcony off the third floor in the high bay. Three locations for the facility were considered. First, the experiments could be hung from the existing balcony. Second, the experiments could be enclosed in a standalone structure below the existing balcony on the first floor. Finally, the test articles could be suspended from a fixture above the existing balcony. The last alternative was chosen because of several advantages. The primary advantage is that the real time computer controlling the experiments will be located in the existing FSS control room. Also, test articles will be highly visible from the control room and the current platform or balcony. This location will have a high work volume and require no external lighting or ventilation. The system bending modes computed from finite element analysis are shown in Table 2. These modes were calculated assuming an 800 pound experiment located in the center of the front edge of the new platform. As expected, this is a diving board mode of the new structure at 19.7 Hertz. The frequency is well above those of interest of the experiments.

Table 2. MMVC Facility

Mode	Frequency	Description
1	19.702 Hz	Platform Bending
2	22.381 Hz	Localized Torsion
3	23.540 Hz	Localized Bending
4	25.550 Hz	Localized Torsion
5	27.872 Hz	Localized Bending

A linear motion system will be installed along the front edge of the new balcony. The motion system has a range of travel of 6 feet with a sensor resolution of .003 inches. It is a ball screw system driven by a brushless DC motor with a peak force capability of 430 pounds and can withstand loads well above 800 pounds.

MMVC Real-Time Control System

The hardware chosen for the MMVC Closed-Loop Controller is shown in Figure 11. The user interface is through the Silicon Graphics Personal Iris 4D-25TG console. The real-time functions will be predominantly executed on four Mercury Computer Systems MC860VB-4 single board computers running MC/OS Version 2.0. A SPARC Engine 1E single board computer serves as a host for the MC860VBs. The host interfaces the Mercury boards to a SCSI bus and Ethernet.

The I/O boards consist of a Xycom XVME-203 Counter/Timer Board, a VME Microsystems International VMIVME-2528 128-bit Digital I/O Board, four Datel DVME-611F 14-bit Analog Input Boards, and four VME Microsystems International VMIVME-4100 Analog Output Boards. The MMVC Closed-Loop Controller will be used to provide digital control of the test articles in the MMVC Lab. The controller will be interfaced to the experiment of sensors, compute control outputs, and apply the outputs to the experiment of actuators. The closed-loop control laws will require a large amount of computational power, and must be executed at rates as high as 250 Hz.

MMVC CONTROLLER METHODS

Many control schemes have been evaluated that would not only provide adequate tracking, but also provide vibration suppression. The major problem with these linear design techniques is that the structure (plant) is a highly nonlinear system. Control design studies have showed that a linear controller, designed for the MMVC experiments may result in unstable systems for large-angle slew commands. This is because of the interactions between the control system and the nonlinear centrifugal stiffening, softening, and Coriolis effects. In the following paragraphs are presented three control schemes that may provide acceptable controllability and performance while the system is undergoing these nonlinear interactions.

Inverse Dynamics Controller

One approach to compensate for nonlinear forces is to use a technique referred to as inverse dynamics control.[2] [3] The way the inverse dynamics control law works is illustrated by considering the following equation

$$m (q) \ddot{q} = u (q,\dot{q}) = B (q) r \quad (1)$$

where q is the n-dimensional vector of generalized coordinates, M(q) is the n x n mass matrix, $u(q, \dot{q})$ is the n-dimensional vector including the effect of centripetal, Coriolis, and gravity terms as well as all other stiffness and damping terms, r is the external torque (or force) vector of dimension m, and B(q) is the n x m torque distribution matrix.

The idea of inverse dynamics control is to seek a nonlinear control logic expression

$$r = f (q, \dot{q}) \quad (2)$$

which, when substituted into equation (1), results in a linear closed-loop system. Here, we assume that the state vector, q, is available.

In this paper, we consider the general case where the number of external torques can be less than the number of the generalized coordinates describing the equation of motion (1). Several control logic expressions and their computational steps are developed to apply the inverse dynamics control to this case.

The computational flow for the inverse dynamics terms, M(q), u(q,), and B(q) is shown in Figure 12. TREETOPS subroutine facilities are used to perform this computation.. The state vector, q, is defined to be the set of the hinge angles and translations and the modal coordinates of flex modes. The non-actuator forces, i.e., forces due to gravity, stiffness, damping, etc. are summed with the inertial forces. Also, the torque distribution matrix B(q) is not directly computed.

Model Reference Adaptive Control

Another control design option for the MMVC experiments is a spin-off from the model reference adaptive control (MRAC) methodology referred to as Direct Multivariable Model Reference Adaptive Control (DMMRAC).[4] The primary advantage DMMRAC possesses over conventional MRAC and other control techniques is that it is completely model independent. DMMRAC is a nonlinear adaptive control methodology driven only by the accumulated error between the reference model and plant outputs. The nonlinear part of the filter results from the adapting law being a function of the square of the reference model states. Unlike classical MRAC, DMMRAC does not require any knowledge of the plant. Therefore, the order of the reference model is strictly up to the designer. Conventional MRAC methods require the order of a reference model to be at least equal to that of the plant. This is a major drawback for these other methods because predicting the order of a complex nonlinear plant is essentially impossible. Figure 13 illustrates a simplified block diagram of the DMMRAC controller with plant.

Fuzzy Control

The MMVC team is currently searching for new and innovative control methods for large space structures. Fuzzy logic control holds much promise in this application.[5] [6] [7] Fuzzy logic is a rule-based control methodology based on linguistic phrases and provides control the way a human operator would. It is especially suited for the nonlinear, time varying, and ill-defined systems such as large flexible structures. Another key feature to fuzzy logic is that it is completely model independent. Typical fuzzy rules are of the form:

$$\text{If } X_1 \text{ is } A_{i,1} \text{ and } X_2 \text{ is } A_{i,2} \text{ then U is } B_i \quad (3)$$

where X_1 and X_2 are the inputs to the controller, U is the output, A's and B's are membership functions, and the subscript i denotes the rule number. For example, a rule for line-of-sight error control may state "If the Line Of Sight (LOS) error is negative small and the change in the LOS error is positive big, then torque is positive small". Given input values of X_1 and X_2, the DOF of rule "i" is given by the minimum of the degrees of satisfaction of the individual antecedent clauses i.e.,

$$DOF = \min \{A_{i,1} (X_1), A_{i,2} (X_2),...\} \quad (4)$$

The output value is computed by

$$u = \frac{\sum_{i=1}^{N} (DOF_i) B_i^d}{\sum_{i=1}^{N} (DOF_i)} \quad (5)$$

where u is called the defuzzified value of the membership function B_i and n is the number of rules. The defuzzified value of a membership function is the single value that best represents the linguistic description. For example, assume Figure 14 illustrates the output membership function for some process. If a rule is active for the present conditions such that its output is "increased moderately", the defuzzified value is the centroidal value about the abscissa. In this case the defuzzified value is 3.0.

For control of highly nonlinear, time varying, and hard-to-define dynamics of large flexible structures, fuzzy logic with its model independence properties may prove to be a very practical method of control.

SUMMARY

The MMVC program has been established at MSFC to experimentally validate multibody modeling codes and to improve the computational efficiency of such codes. Experiments have been designed to emphasize modeling features that are to be verified and validated in the effort. A laboratory facility has been designed and is under development. The RTCS is in place and has been functionally verified. Preliminary experiments that do not require the test volume to be provided when construction of the MMVC laboratory is completed are under way. Enhancements to the TREETOPS code are initiated and ongoing. This paper has presented a top-level overview of the MMVC program and its goals and methods.

REFERENCES

1. TREETOPS User's Manual, Dynacs Engineering Company, Inc., Revision 8.
2. H. Baruh and S.S.K. Tadikonda *Issues in the Dynamics and Control of Flexible Robot Manipulators,* AIAA Journal of Guidance, Control, and Dynamics, Vol.12, No.5, Sept.-Oct. 1989.
3. H. Baruh and K. Choe *Sensor Placement in Structural Control,* AIAA Journal of Guidance, Control, and Dynamics, Vol.13, No.3, May-June 1990.
4. Izhak, Bar-Kana, *Direct Multivariable Model Reference Adaptive Control with Applications to Large Structural Systems*, doctoral dissertation, Rensaelar Polytechnic Institute, Troy, NY, 1983.
5. Sugeno, Michio, *An Introductory Survey of Fuzzy Control*, Information Sciences, Vol.36, 1985.
6. Chiv Stephen, *Robustness Analysis of Fuzzy Control Systems with Application to Aircraft Roll Control*, AIAA, J. Guidance and Control, June 1991.
7. Zadeh, Lotfi A., *Making Computers Think Like People*, IEEE Spectrum, August 1984.

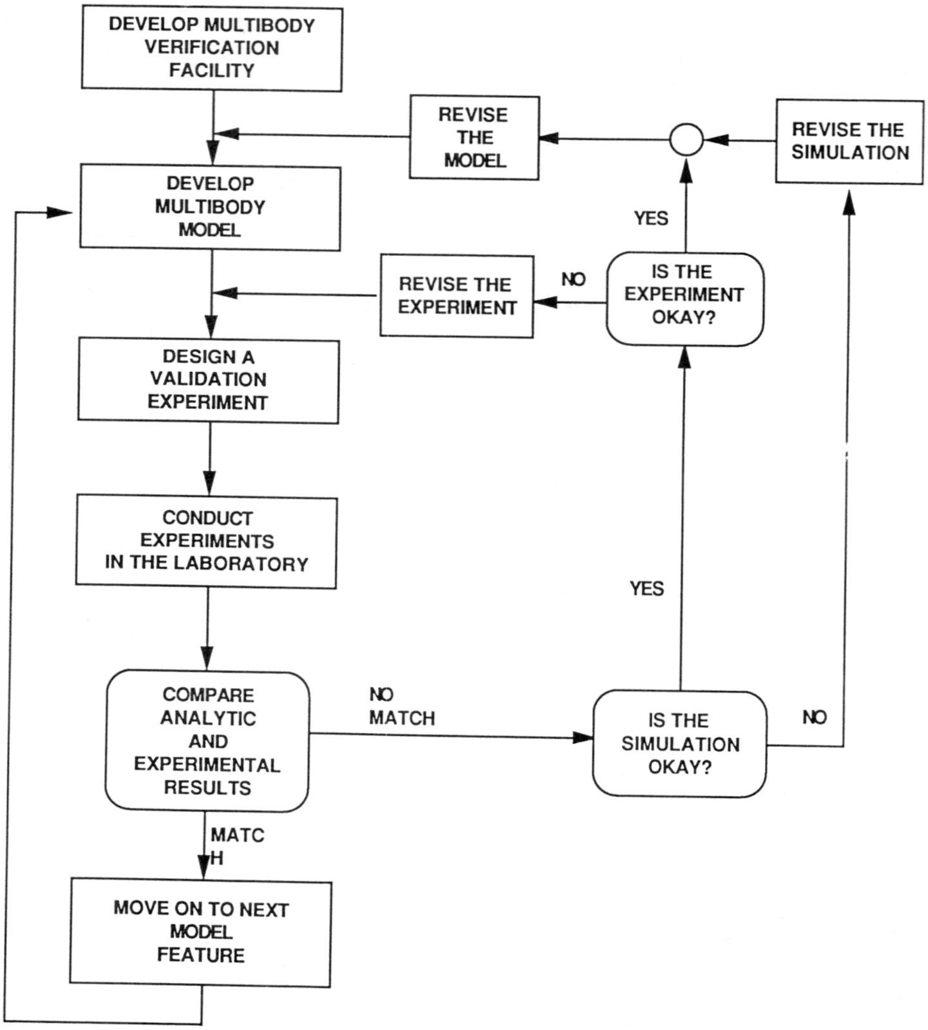

Figure 1

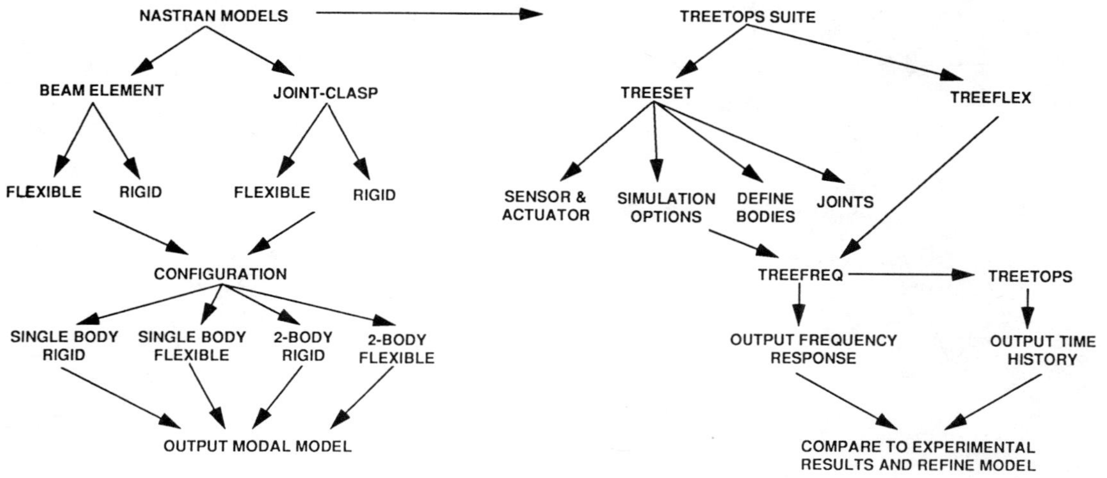

Figure 2

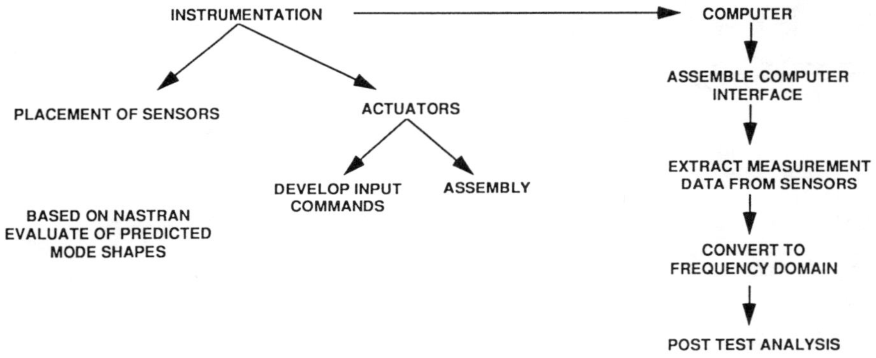

Figure 3

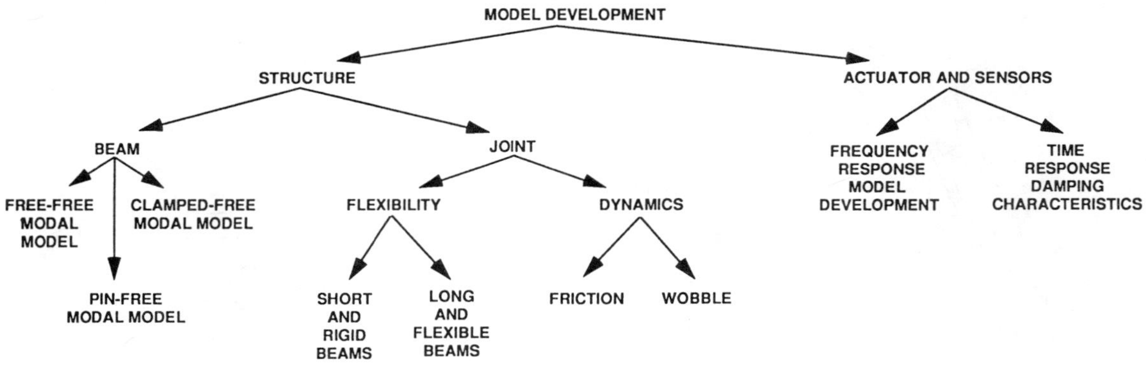

Figure 4

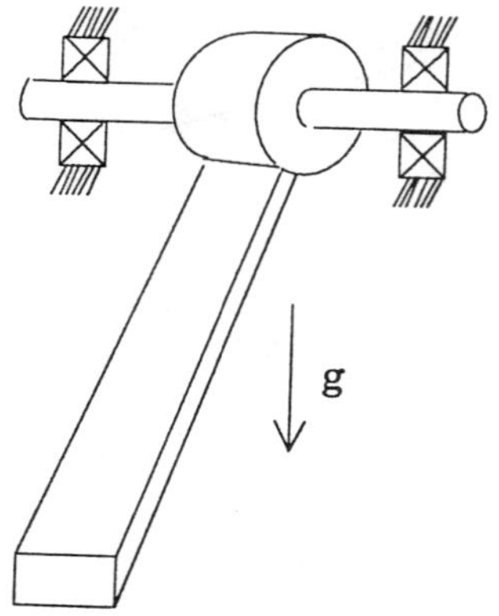

Figure 5

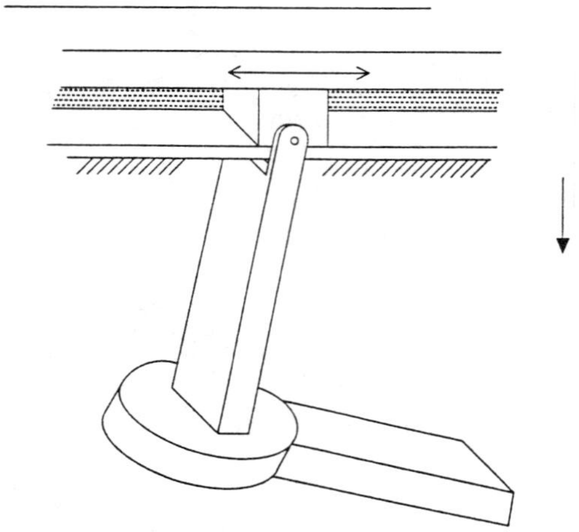

Figure 8

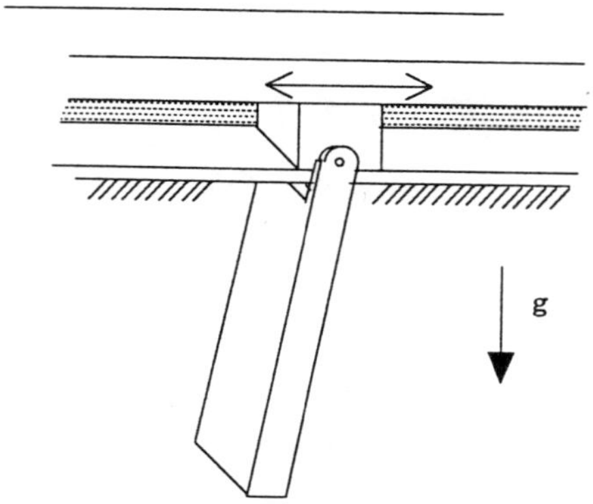

Figure 6

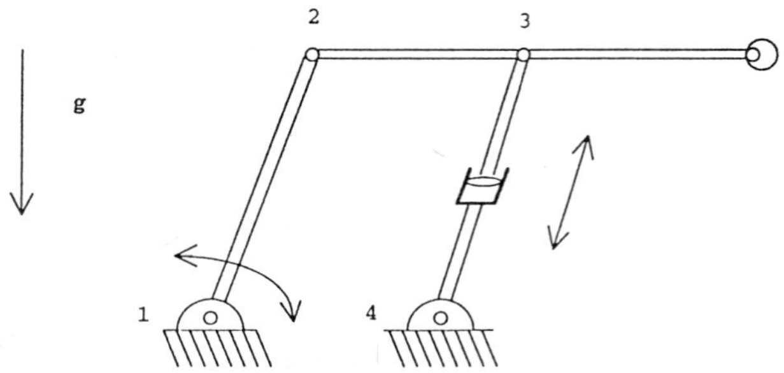

Figure 9

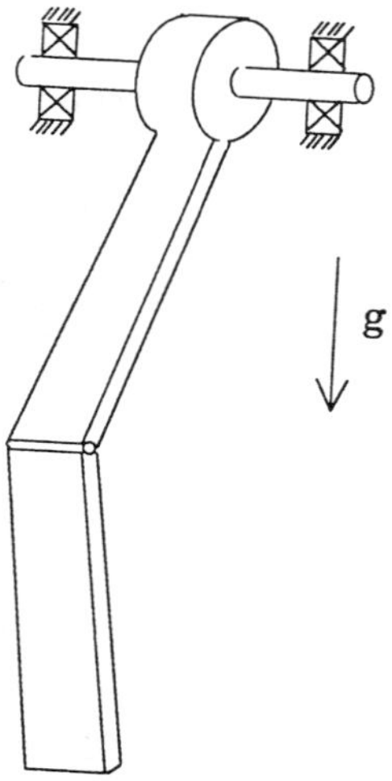

Figure 7

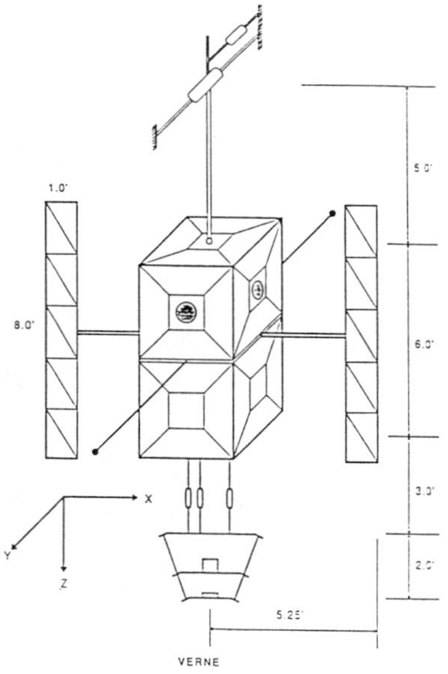

Figure 10

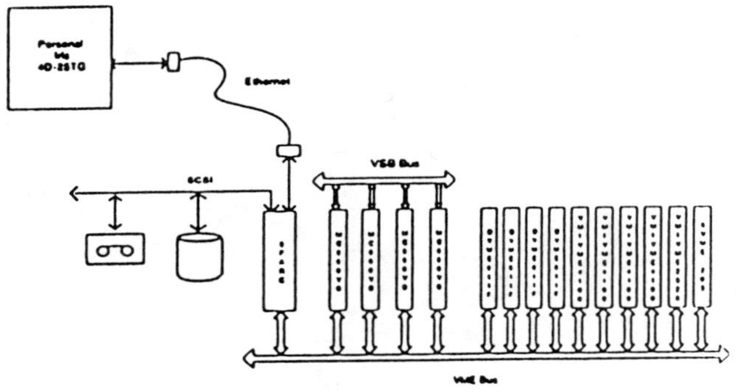

Figure 11

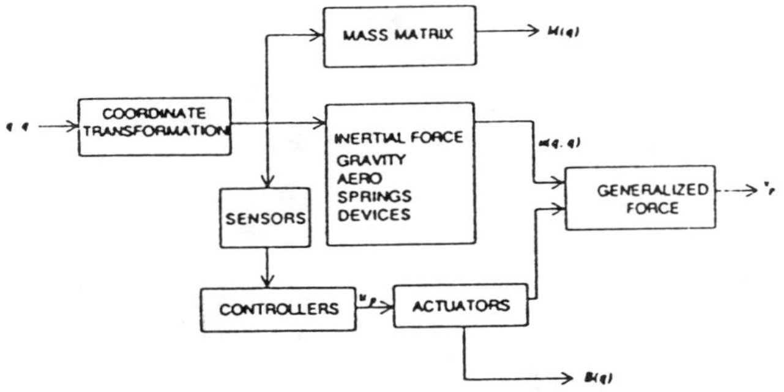

Figure 12

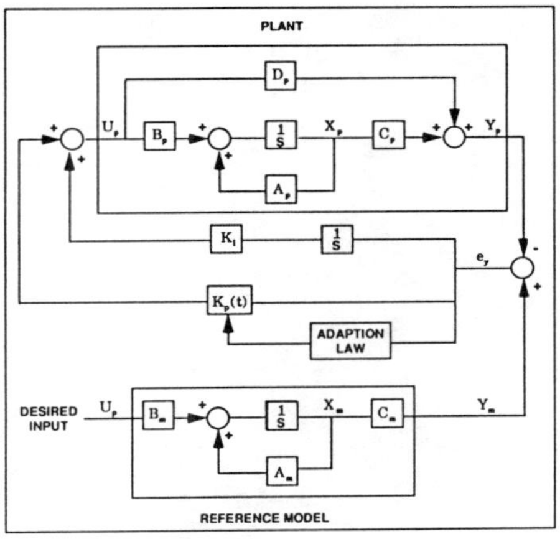

Figure 13

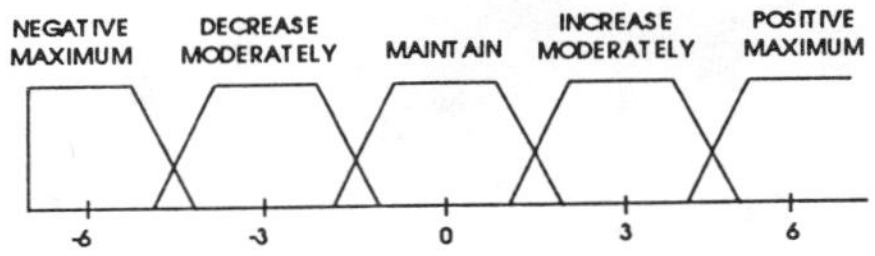

OUTPUT MEMBERSHIP FUNCTION

Figure 14

AMD-Vol. 141/DSC-Vol. 37, Dynamics of Flexible Multibody Systems:
Theory and Experiment
ASME 1992

EXPERIMENTAL VERIFICATION OF COMPONENT MODE TECHNIQUES FOR A FLEXIBLE MULTIBODY SYSTEM

S. C. Sinha, J. W. Benner, and G. J. Wiens
Department of Mechanical Engineering
Auburn University
Auburn, Alabama

SUMMARY

The Component Mode Synthesis technique is applied to a large, three link, flexible structure. This technique is used to obtain an approximation of the natural frequencies and mode shapes (i.e. eigenvalues and eigenvectors) of the structure in varying configurations. The mode functions are selected such that the method can be made computationally as simple as possible, without compromising accuracy. In this light, a simple power series is selected. In order to verify the theoretical results obtained from the component mode technique, experimental modal tests were conducted at NASA's LSS GTF (Large Space Structures Ground Test Facility) located at MSFC (Marshall Space Flight Center). The experimental data is also compared with the results obtained from a dynamic software simulation package, entitled TREETOPS, which is also based on the component mode procedure.

It is found that the experimental rersults are in excellent agreement with those obtained by the use of component mode techniques for the aforementioned structures. The average percent change in the frequencies is observed to be 6.09%, while the average mode shape correspondence is noted as 94.5%.

1. INTRODUCTION

In recent years, the aerospace industry has been involved in the extensive use of multibodied structures. This is evidenced by many examples, such as, the space station, the space shuttle, space robots, and the list goes on. With the incorporation of composite materials, the ability to achieve a high payload to weight ratio has been facilitated. Also, scientists and engineers have been asking large lightweight robotic manipulators to perform precise, high speed maneuvers. In consequence, these lightweight structures exhibit high flexibilities, and due to the stringent requirements demanded of them, the study of their vibration characteristics has become of paramount importance.

The complexity of the LSS (Large space structures) is such that it has become increasingly difficult to accurately model their dynamic behavior. Therefore, high speed computers have been heavily relied upon for simulation and analysis. The modeling techniques utilized in the software packages are predominantly based on the finite element method. As the size of the system increases, the model becomes too complicated and simulation time is undesirably high. Therefore, the component mode techniques are being turned to as an alternate approach in simplifying the system model.

As industry is turning to the component mode techniques, the need to verify this procedure is becoming increasingly important. Despite the generous amounts of analytical research being conducted, there is a lack of experimental data available for verification. Therefore, NASA's MSFC has embarked on creating a national testbed for such structures. This testbed will be used to explore new issues associated with large flexible structures and also provide research data for which one use is the experimental verification of component mode techniques.

The component mode technique is based on simplifying the system model by reducing the system into separate components or substructures. This reduces the system of partial differential equations of motion, which sometimes can not be solved, down to a group of component equations that have known solutions. The knowledge of the component behavior, along with how they are constrained by the system configuration, can then be used to assemble a feasible mathematical model for the entire system. This is significant when one considers that many times the major components of large systems are being built by separate contractors and analysis of the entire assembled system is often times not feasible. Therefore, the component mode techniques allow for much of the analysis to be done on the component level.

Many methods have been developed on the basis of the component mode approach. Hurty, who is considered the pioneer in this field, developed the Component Mode Synthesis technique [1,2]. The Craig-Bampton method differs only slightly from Hurty's and the two give the same numerical results [3]. Gladwell also developed a variant of Hurty's method which has come to be called the Branch Mode Analysis [4]. Furthermore, Benfield and Hruda developed the Component Mode Substitution technique [5]. Dowell has also developed a technique where he incorporates the use of Lagrange multipliers, in a systematic approach, to permit for various system constraints [6,7,8,9,10,11,12]. In general, these methods tend to differ only in the selection of the various component modes. In this paper Hurty's Method is utilized. For complete reference of this method see [1,2,13,14].

2. COMPONENT MODE SYNTHESIS FOR A THREE LINK SYSTEM

The various component mode techniques tend to differ, in general, only in the selection of the component mode functions, all of which can be determined by solving some type of component eigenvalue problem. For further reference one may look into Benfield and Hruda [15], Rubin [16], and Craig and Bampton [3]. In this investigation, we will be considering a slightly new approach. The component mode techniques, as stated earlier, were introduced to simplify the computation problems associated with other methods, and still produce a reliable approximation. With this premise in mind, we suggest the use of simple polynomials as alternative mode functions. Hurty has previously suggested this [1], along with Meirovitch and Hale [14].

The first criterion placed on these polynomials, is that they qualify as admissible functions. In general, the admissible functions must be selected such that displacements and forces exist at the internal boundaries of the system. The other criteria are that the admissible functions selected form a complete set, and they should be linearly independent. A reasonable check to see if the frequencies converge will provide assurance that a given set of mode functions forms a complete set. In considering the problem associated with early truncation of the mode function set, Macneal [17] and Rubin [16] proposed a technique to include the effect of residual modes not retained [13,14].

We will now consider a three link flexible system as shown in Figure 1, and examine the undamped, free vibration problem. We will denote Figure 1 as Configuration 1. The equation of motion for an arbitrary substructure, is given as

$$[M_s]\{\ddot{\zeta}_s(t)\}+[K_s]\{\zeta_s(t)\}=\{0\}. \tag{1}$$

where $[M_s]$ and $[K_s]$ are the substructure mass and stiffness matrices respectively, $\{\zeta_s(t)\}$ is the time dependent generalized coordinate, and the $\dot{}$ represents the derivative with respect to time ([ ] denotes a matrix and { } represents a column vector). We shall restrict our analysis to motions of small amplitude so that the free vibrations in longitudinal, transverse, and twisting motion are decoupled. Also, we shall assume infinite axial rigidity of all three members, which is later supported by experimentation. That is, the system considered in Figure 1 has two adjustable joints (top revolute joint at $x_1 = 0$ and bottom revolute joint between link 1 and the composite body formed by links 2 and 3). It should be noted that links 2 and 3 are rigidly clamped together at $x_2 = L_2$, $x_3 = L_3$. The other joints are locked in various configurations; hence a quasi-dynamic analysis is presented here. In addition, we are only concerned with vibration in the plane of the paper as shown by Figure 1.

Next, we will choose the transverse displacement vector for component 1 to be

$$w_1(x_1,t)=\left(\frac{x_1}{L_1}\right)^2\zeta_1+\left(\frac{x_1}{L_1}\right)^3\zeta_2+\dots+\left(\frac{x_1}{L_1}\right)^n\zeta_{n-1} \tag{2}$$

where ζ is taken to be a function of time although not explicitly shown, and n is equal to the highest power of the last spatial function. The displacement vector chosen for component 1 satisfies all geometric and force conditions at the external and internal boundaries of component 1, whereby the interface of the components is designated the internal boundary. These conditions are

$$\begin{gathered} w_1(0)=0,\ w_1'(0)=0,\ w_1''(0)\neq 0,\ w_1'''(0)\neq 0 \\ w_1(L_1)\neq 0,\ w_1'(L_1)\neq 0,\ w_1''(L_1)\neq 0,\ w_1'''(L_1)\neq 0\ . \end{gathered} \tag{3}$$

where the $'$ represents the derivative with respect to the spatial coordinate. Note that shear forces and moments are allowed at the interface.

We choose the displacement vectors for component 2

$$\begin{aligned} w_2(x_2,t)=1(x_2)\zeta_n+\left(\frac{x_2}{L_2}\right)\zeta_{n+1}+\left(\frac{x_2}{L_2}\right)^4\zeta_{n+2}+ \\ \dots+\left(\frac{x_2}{L_2}\right)^m\zeta_{m+1} \end{aligned} \tag{4}$$

$$u_2(x_2,t)=1(x_2)\zeta_{m+2} \tag{5}$$

where $u_2(x_2,t)$ is the axial displacement of component 2, and m is the highest power of the last spatial function. Here again, these displacement vectors were chosen so that they satisfy all geometric and force conditions at the boundaries of component 2. These conditions are as follows:

$$\begin{gathered} w_2(0)\neq 0,\ w_2'(0)\neq 0,\ w_2''(0)=0,\ w_2'''(0)=0 \\ w_2(L_2)\neq 0,\ w_2'(L_2)\neq 0,\ w_2''(L_2)\neq 0,\ w_2'''(L_2)\neq 0\ . \end{gathered} \tag{6}$$

In equations (2), (4), and (5), rigid body translation and rigid body rotation modes are represented by 1(x) and (x/L) respectively, whereas the remaining functions represent deformation modes. And similarly for component 3,

$$\begin{aligned} w_3(x_3,t)=1(x_3)\zeta_{m+3}+\left(\frac{x_3}{L_3}\right)\zeta_{m+4}+\left(\frac{x_3}{L_3}\right)^4\zeta_{m+5}+ \\ \dots+\left(\frac{x_3}{L_3}\right)^m\zeta_{2m+1} \end{aligned} \tag{7}$$

$$u_3(x_3,t)=1(x_3)\zeta_{2m+2} \tag{8}$$

where these displacement vectors satisfy similar boundary conditions as in equation (6).

Now we compute the generalized mass matrix for each component which is derived as follows. Consider the displacement vector for a given beam

$$w_s(x_s,t)=\sum_{i=1}^{N}\phi_i(x_s)\zeta_i(t) \tag{9}$$

where ϕ is the spatial function and N is the desired number of mode functions. The velocity is

$$v_s(x_s)=\sum_{i=1}^{N}\phi_i(x_s)\dot{\zeta}_i(t) \qquad (10)$$

and therefore the kinetic energy becomes

$$T_s=\frac{1}{2}\sum_{i=1}^{N}\sum_{j=1}^{N}\dot{\zeta}_i\dot{\zeta}_j\int\phi_i(x_s)\phi_j(x_s)\,dm \qquad (11)$$

or

$$T_s=\frac{1}{2}\sum_{i=1}^{N}\sum_{j=1}^{N}m_{ij}\dot{\zeta}_i\dot{\zeta}_j \qquad (12)$$

where the generalized mass is defined to be

$$m_{ij}=\int\phi_i(x_s)\phi_j(x_s)\,dm \qquad (13)$$

and the integration is carried out over the entire system. Therefore applying equation (13) to component 1 we obtain

$$m_{ij}=\int_0^{L_1}\rho_1(x_1)\phi_i(x_1)\phi_j(x_1)\,dx_1 \qquad (14)$$

where ρ_1 is the mass density per unit length of component 1. For example, when n=3 in equation (2), we obtain

$$m_{11}=\int_0^{L_1}\rho_1\phi_1\phi_1dx_1=\int_0^{L_1}\rho_1\left(\frac{x_1}{L_1}\right)^4dx_1=\left(\frac{1}{5}\right)\rho_1L_1 \qquad (15)$$

$$m_{12}=m_{21}=\int_0^{L_1}\rho_1\phi_1\phi_2dx_1=\int_0^{L_1}\rho_1\left(\frac{x_1}{L_1}\right)dx_1=\left(\frac{1}{6}\right)\rho_1L_1 \qquad (16)$$

$$m_{22}=\int_0^{L_1}\rho_1\phi_2\phi_2dx_1=\int_0^{L_1}\rho_1\left(\frac{x_1}{L_1}\right)^6dx_1=\left(\frac{1}{7}\right)\rho_1L_1 \qquad (17)$$

substituting in the appropriate values (Table 1)

$$[M_1]=\rho_1L_1\begin{bmatrix}\frac{1}{5} & \frac{1}{6}\\ \frac{1}{6} & \frac{1}{7}\end{bmatrix}. \qquad (18)$$

In like manner, the generalized mass matrices for component 2 and 3 are computed for m=6. Noting that there is no coupling between the lateral and longitudinal displacements we obtain,

$$[M_2]=\rho_2L_2\begin{bmatrix}1 & \frac{1}{2} & \frac{1}{5} & \frac{1}{6} & \frac{1}{7} & 0\\ \frac{1}{2} & \frac{1}{3} & \frac{1}{6} & \frac{1}{7} & \frac{1}{8} & 0\\ \frac{1}{5} & \frac{1}{6} & \frac{1}{9} & \frac{1}{10} & \frac{1}{11} & 0\\ \frac{1}{6} & \frac{1}{7} & \frac{1}{10} & \frac{1}{11} & \frac{1}{12} & 0\\ \frac{1}{7} & \frac{1}{8} & \frac{1}{11} & \frac{1}{12} & \frac{1}{13} & 0\\ 0 & 0 & 0 & 0 & 0 & 1\end{bmatrix}. \qquad (19)$$

Changing subscripts 2 to 3, one obtains the definition for $[M_3]$. We arrange matrices, $[M_1]$, $[M_2]$, and $[M_3]$, into the system block diagonal matrix as

$$[M^d]=\begin{bmatrix}[M_1] & & \\ & [M_2] & \\ & & [M_3]\end{bmatrix}. \qquad (20)$$

We form the generalized stiffness matrix by expressing the flexural potential energy for an Euler-Bernoulli beam in bending.

$$U_s=\frac{1}{2}\int E_sI_s\left(\frac{dw_s}{dx_s}\right)^2dx_s \qquad (21)$$

Substituting for w_s, as in equation (9), we get

$$U_s=\frac{1}{2}\sum_i\sum_j\zeta_i\zeta_j\int E_sI_s\phi_i''\phi_j''dx_s \qquad (22)$$

or

$$U_s=\frac{1}{2}\sum_i\sum_j k_{ij}\zeta_i\zeta_j \qquad (23)$$

where the generalized stiffness is defined to be

$$k_{ij}=\int_0^{L_s}E_sI_s\phi_i''\phi_j''dx_s \qquad (24)$$

Therefore, applying this equation to component 1

$$k_{11}=E_1I_1\int_0^{L_1}\phi_1''\phi_1''dx_1=E_1I_1\int_0^{L_1}\left(\frac{2}{L_1^2}\right)dx_1=4\frac{E_1I_1}{L_1^3} \qquad (25)$$

$$k_{12}=k_{21}=E_1I_1\int_0^{L_1}\left(\frac{2}{L_1^2}\right)\left(\frac{6x_1}{L_1^3}\right)dx_1=6\frac{E_1I_1}{L_1^3} \qquad (26)$$

$$k_{22}=12\frac{E_1I_1}{L_1^3} \qquad (27)$$

which results in

$$[K_1] = \frac{E_1 I_1}{L_1^3}\begin{bmatrix} 4 & 6 \\ 6 & 12 \end{bmatrix} . \tag{28}$$

Similarly, for component 2 we arrive at

$$[K_2] = \frac{E_2 I_2}{L_2^3}\begin{bmatrix} 0 & 0 & 0 & 0 & 0 & 0 \\ 0 & 0 & 0 & 0 & 0 & 0 \\ 0 & 0 & 28.8 & 40 & 51.43 & 0 \\ 0 & 0 & 40 & 57.143 & 75 & 0 \\ 0 & 0 & 51.43 & 75 & 100 & 0 \\ 0 & 0 & 0 & 0 & 0 & 0 \end{bmatrix} \tag{29}$$

Again, changing subscripts 2 to 3, $[K_3]$ is similarly defined for component 3. Furthermore, assembling these matrices into the block diagonal stiffness matrix we obtain

$$[K^d] = \begin{bmatrix} [K_1] & & \\ & [K_2] & \\ & & [K_3] \end{bmatrix} . \tag{30}$$

Since we have formed the mass and stiffness matrix for the decoupled system, the equations must be coupled by a simple coordinate transformation. The transformation matrix $[\beta]$ is a transformation from the dependent generalized coordinate $\{\zeta(t)\}$, into a set of independent generalized coordinates $\{q(t)\}$

$$\{\zeta(t)\} = [\beta]\{q(t)\} . \tag{31}$$

This matrix, $[\beta]$, is formed directly from the equations of constraint for the system, which arise due to the force and displacement compatibility requirements at the component interfaces. These constraint equations may be written in local generalized coordinates $\{\zeta(t)\}$. Upon forming all the constraint equations for each interface, we obtain a set of linear constraint equations which can be written as

$$[A]\{\zeta(t)\} = \{0\} \tag{32}$$

where [A] is a (c x M) matrix of constant coefficients. Note that c is the total number of constraint equations, and M is the dimension of $\{\zeta(t)\}$. Next, the matrix may be partitioned as such

$$[A] = [A_1 \,\vdots\, A_2] \tag{33}$$

where $[A_1]$ is a square matrix of order (c x c). Furthermore, equation (33) may now be written in the following form

$$[A_1]\{\zeta(t)\}_d + [A_2]\{q(t)\} = \{0\} \tag{34}$$

where $\{q(t)\}$ is a subset of $\{\zeta(t)\}$ which is chosen to include the independent variables, and $\{\zeta(t)\}_d$ is chosen to include the dependent variables. This is achieved by choosing $[A_1]$ to be nonsingular. Next, equation (34) can be rewritten into the following form

$$\{\zeta(t)\}_d = -[A_1]^{-1}[A_2]\{q(t)\} . \tag{35}$$

Finally, the complete vector $\{\zeta(t)\}$ can be expressed in terms of the independent subset $\{q(t)\}$, by supplying the identity matrix where needed,

$$\{\zeta(t)\} = \left[\frac{[I]}{-[A_1]^{-1}[A_2]}\right]\{q(t)\} . \tag{36}$$

Therefore,

$$[\beta] = \left[\frac{[I]}{-[A_1]^{-1}[A_2]}\right] \tag{37}$$

which forms the coordinate transformation matrix $[\beta]$ [15].

Consider the case where the three link structure is in the configuration shown in Figure 1. In general, the displacement equations at the junction are

$$\begin{aligned} w_1(L_1) + u_2(L_2) &= 0 \\ w_1(L_1) - u_3(L_3) &= 0 \end{aligned} \tag{38}$$

$$\begin{aligned} w_2(L_2) &= 0 \\ w_3(L_3) &= 0 \end{aligned} \tag{39}$$

$$\begin{aligned} w_1'(L_1) - w_2'(L_2) &= 0 \\ w_1'(L_1) - w_3'(L_3) &= 0 . \end{aligned} \tag{40}$$

Equations (39) arise due to the assumption of infinite axial rigidity. Equations (40) enforce that the component rotations due to bending at the junction are equal. However, the links are modeled as Euler-Bernoulli beams and therefore, in the absence of shear deflection this rotation angle reduces to the slope. The force equilibrium equation is

$$M_1(L_1) + M_2(L_2) + M_3(L_3) = 0 \tag{41}$$

where M's are the moments. This equation is further simplified to

$$E_1 I_1 w_1''(L_1) + E_2 I_2 w_2''(L_2) + E_3 I_3 w_3''(L_3) = 0 . \tag{42}$$

In general, there would also be equations that would enforce the shear and axial force at the junction, but again for our assumptions these are neglected. These equations are now evaluated and expressed in terms of $\{\zeta(t)\}$ as in equation (32), where [A] is equal to the following constant matrix.

$$\begin{bmatrix} 0 & 0 & 1 & 1 & 1 & 1 & 1 & 0 & 0 & 0 & 0 & 0 & 0 & 0 \\ 1 & 1 & 0 & 0 & 0 & 0 & 0 & 1 & 0 & 0 & 0 & 0 & 0 & 0 \\ \frac{2}{L_1} & \frac{3}{L_1} & 0 & \frac{-1}{L_2} & \frac{-4}{L_2} & \frac{-5}{L_2} & \frac{-6}{L_2} & 0 & 0 & 0 & 0 & 0 & 0 & 0 \\ \left(\frac{2I_1}{I_2 L_1^2}\right) & \left(\frac{6I_1}{I_2 L_1^2}\right) & 0 & 0 & \frac{12}{L_2^2} & \frac{20}{L_2^2} & \frac{30}{L_2^2} & 0 & 0 & 0 & \frac{12}{L_3^2} & \frac{20}{L_3^2} & \frac{30}{L_3^2} & 0 \\ 0 & 0 & 0 & 0 & 0 & 0 & 0 & 0 & 1 & 1 & 1 & 1 & 1 & 0 \\ 1 & 1 & 0 & 0 & 0 & 0 & 0 & 0 & 0 & 0 & 0 & 0 & 0 & -1 \\ \frac{2}{L_1} & \frac{3}{L_1} & 0 & 0 & 0 & 0 & 0 & 0 & 0 & \frac{-1}{L_3} & \frac{-4}{L_3} & \frac{-5}{L_3} & \frac{-6}{L_3} & 0 \end{bmatrix} \tag{43}$$

Because there are 14 generalized coordinates and 7 constraints this means that there are 14-7=7 redundant coordinates. The [A] matrix is now partitioned as indicated equation (33). Let

$$\zeta_1=q_1,\ \zeta_{12}=q_{12},\ \zeta_2=q_2,\ \zeta_{13}=q_{13},\quad \zeta_3=q_3,\ \zeta_{14}=q_{14},\ \zeta_4=q_4 \tag{44}$$

where these are chosen to ensure that $[A_1]$ is nonsingular. Therefore, $[A_1]$ is formed by taking the first four and last three columns of [A], and $[A_2]$ is formed by taking columns five through eleven of [A]. The transformation matrix $[\beta]$ can now be formed from equation (37), and is shown in the APPENDIX.

Finally, the system matrices are formed by the transformation defined below

$$[M]=[\beta]^T\,[M^d][\beta]\ \ and\ \ [K]=[\beta]^T\,[K^d][\beta] \tag{45}$$

and are listed in the APPENDIX. Once the system matrices have been constructed, the natural frequencies can be obtained from the resulting eigenvalue problem, where the first four natural frequencies are

$$\begin{aligned}\omega_{n1}&=2.5390\ Hz\\ \omega_{n2}&=3.3339\ Hz\\ \omega_{n3}&=3.9432\ Hz\\ \omega_{n4}&=19.6606\ Hz\end{aligned} \tag{46}$$

To examine the possibility of premature truncation of the mode functions in the component displacement vectors, we will consider adding more terms in equations (2), (4) and (7) to test for convergence of the natural frequencies. Table 2 summarizes the results. At this stage, the mode shapes can be determined as described in the following. The eigenvalue problem can be solved for the independent coordinate vector $\{q(t)\}$ with respect to an arbitrary reference. Subsequently, $\{q(t)\}$ can be transformed into the dependent coordinate vector $\{\zeta(t)\}$, which can be used to find the respective component displacement vectors defined by equations (2), (4), (5), (7), and (8). Therefore, these component modes can be assembled to obtain the system natural modes.

In a similar manner, the structure is analyzed in two more configurations shown in Figures 6 and 11. The convergence of the natural frequencies are summarized in the Tables 3 and 4, while the mode shapes are located in Figures (2-5), (7-10) and (12-15).

3. EXPERIMENTAL VERIFICATION

As was stated earlier, the experimental procedure was conducted at NASA Marshall Space Flight Center and signified the advent for the Multibody Modeling Verification and Control (MMVC) project.

The test article was designed to simulate a large flexible multibody system. One of the design criteria was that the structure should exhibit dense mode concentration at low frequencies. The closely packed modes were desirable because this typifies the dynamic behavior of multibodied, flexible structures in space environments. It was also designed so that the maximum deflections would fall within the linear deflection range and that the vibration was restricted to be in the plane shown by Figure 1. The resulting dimensions for each component can be found in, Table 1. Essentially, the design process yielded a linear, time invariant, observable structure, so that analysis technique could be applied for the study.

After the structure was assembled, the external conditions were examined. The external joint that suppc structure was designed to facilitate the interchanging of the system components. Therefore, this fixed boundary condition was validated by testing the dynamic response of an isolated cantilever beam, which indeed closely resembled an actual cantilever beam response. Therefore, this joint closely approximated a fixed boundary. Also, each component was tested to show that the axial vibration for each respective component was negligible, and thus, the assumption of infinite axial rigidity was justified.

Each component was discretized into nine measurement response and force input points. These points were chosen to accurately excite and observe all modes of interest within the desired frequency range. The excitation was achieved by the direct impact hammer technique. This technique utilized the use of a load cell at the tip of the hammer to measure the force input. The tip of the hammer was selected to exhibit good frequency content throughout the frequency range of interest.

The piezoelectric accelerometer, or the response transducer, was located at the end of component 3. It was mounted on the structure with a type of strong adhesive, and therefore, accurately described the response of the structure. The structure was impacted by the hammer at all the points on each component, while frequency response data was collected. The accelerometer did not mass load the structure, as was evidenced by doubling the weight of the accelerometer and noticing that there were no shifts in the frequency response function.

The hardware used in this experiment is documented in, Table 5. The listed spectrum analyzer had all necessary analytical capabilities to produce and animate the system mode shapes. A single degree of freedom curve fit algorithm was employed, and the mode shapes were normalized with respect to the maximum degree of freedom in the structure.

The structure was tested in nine different configurations. It was speculated that gravity might have an effect on highly flexible, possibly nonlinear, structures. It was discovered that gravity had negligible effect on the frequencies and mode shapes of this structure, and therefore, the original nine configurations reduced to three unique configurations. These configurations are depicted in Figures 1, 6, and 11. The experimental natural frequencies of the structure for all three configurations are summarized in the Tables 6-8. The experimental system mode shapes corresponding to these configurations are shown in Figures (2-5), (7-10) and (12-15).

4. CORRELATION OF THEORETICAL AND EXPERIMENTAL RESULTS

A comparison of analytical and experimental values of frequencies are sumarized in Tables 9-11. In addition to the results from sections 2. and 3., the results as computed by TREETOPS are also presented. As stated before, TREETOPS is a dynamic simulation software package that is based on the component mode technique. The TREETOPS results are based on the first five cantilever beam modes selected as component modes. This flexible data was generated by another software package called TREEBEAMS.

The percent change between all of the results, is shown in, Tables 12-14. The percent change was calculated relative to the results located on the right side of the forward slash, as listed in the column headings of each table.

The theoretical and experimental mode shapes are compared by the Modal Assurance Criterion (MAC). This condition number ranges from 0 to 1 where a 1 signifies perfect correspondence of the two modes. The MAC is defined as

$$0 \leq \frac{(v^T u)^2}{(v^T v)(u^T u)} \leq 1 \qquad (47)$$

where v and u are the theoretical and experimental mode shapes, respectively. Tables 15-17 summarize the information for the given configurations.

5. DISCUSSION AND CONCLUSIONS

For the first time, an experiment is conducted in order to test the use of component mode synthesis techniques in determining natural frequencies and modes of vibration for a large, flexible, three-link structure. Standard modal analysis techniques were used in analyzing the structure. This structure is representative of the types of large structures employed in space applications today, and therefore, the experimental results are valuable in providing confidence in the analysis procedures used on such structures.

The component mode technique eliminates the difficulties involved in choosing the correct system eigenfunctions, a problem that the Rayleigh-Ritz method as been unable to overcome. It is shown that simpler polynomials prove to yield accurate enough results for engineering analysis. Although component eigenfunctions can also produce accurate results, they tend to increase the computational demands. Also, this approach eliminates the need for component testing if a suitable admissible function is known that can accurately describe the dynamic behavior of the component.

The theoretical results in this study show that the use of integral powers of (x/L) is an adequate approximation of the modes of vibration for Euler-Bernoulli beams in a chain-like structure. These polynomials prove to be suitable admissible functions for the structure, as it is analyzed in three different configurations. The experimental evidence not only verified the use of these simple mode functions, but also validated the dynamic software simulation package, TREETOPS, which is based on the component mode technique.

The results show that the average frequency difference between the experimental and the theoretical results was 6.09%. Yet this number can be misleading considering that seventy-five percent of the frequency results had an average percent difference of 2.22%. Also, the experimental and theoretical mode shapes, on the average, had 94.5% correspondence.

A major concern is how to accurately control large, flexible, multibody structures to perform precise maneuvers. In the future, as in the spirit of this research, decentralized control, or control on the component level should be investigated. Also, the nonlinear behavior of such structures undergoing large displacements and exhibiting large deflections will need to be addressed. Here again, the experimental evidence will prove to be a necessity in the exploration of these areas.

ACKNOWLEDGEMENTS

The authors would like to thank NASA for the funding of this research under Grant Numbers NGT 50567 and NAG8-123(16). Thanks are also due to Dr. Henry B. Waites, who, in large part, was responsible for the supervision of the experiment, and Allen Patterson and David Smart for help in the coordination of the experiment.

REFERENCES

1. W. C. Hurty, "Vibrations of Structural Systems by Component Synthesis," Journal of the Engineering Mechanics Division, Proceedings of ASCE, August 1960, pp. 51-69.

2. W. T. Thomson, Theory of Vibration with Applications, Prentice Hall, 1988, pp. 246-251.

3. R. R. Craig Jr. and M. C. C. Bampton, "Coupling of Substructures for Dynamic Analyses," AIAA Journal, Volume 6, Number 7, July 1968, pp. 1313-1319.

4. G. M. L. Gladwell, "Branch Mode Analysis of Vibrating Systems," Journal of Sound and Vibration, Volume 1, January 1964, pp. 41-59.

5. W. A. Benfield and R. F. Hruda, "Vibration Analysis of Substructures by Component Mode Substitution," AIAA Journal, Volume 9, Number 7, July 1971, pp. 1255-1261.

6. E. H. Dowell, "On the Modal Approach to Structural Modification," Journal of the American Helicopter Society, February 1984, pp. 75-77.

7. E. H. Dowell, "Free Vibrations of an Arbitrary Structure in Terms of Component Modes," Journal of Applied Mechanics, September 1972, pp. 727-732.

8. E. H. Dowell and L. R. Klein, "Analysis of Modal Damping by Component Modes Method Using Lagrange Multipliers," Journal of Applied Mechanics, June 1974, pp.527-528.

9. E. H. Dowell, "On Some General Properties of Combined Dynamical Systems," Journal of Applied Mechanics, March 1979, Volume 46, pp. 206-209.

10. E. H. Dowell, "Bounds on Damping by a Component Modes Method Using Lagrange Multipliers," Journal of Applied Mechanics, March 1980, Volume 47, pp. 211-213.

11. E. H. Dowell, "Component Mode Analysis of Nonlinear and Nonconservative Systems," Journal of Applied Mechanics, March 1980, Volume 47, pp. 172-176.

12. E. H. Dowell, "Component Mode Analysis of Nonlinear, Nonconservative Systems," Journal of Applied Mechanics, March 1983, Volume 50, pp. 204- 209.

13. L. Meirovitch, Computational Methods in Structural Dynamics, Sijthoff and Noordhoff, 1980, pp.286-298.

14. L. Meirovitch and A. L. Hale, "On the Substructure Synthesis Method," An International Conference on Recent Advances in Structural Dynamics, Southampton England, July 1980, pp. 7-11.

15. W. C. Hurty, "Dynamic Analysis of Structural Systems Using Component Modes," AIAA Journal, Volume 3, Number 4, April 1965, pp. 678-684.

16. S. Rubin, "Improved Component Mode Representation for Structural Dynamic Analysis," AIAA Journal, Volume 13, Number 8, August 1975, pp. 995-1006.

17. R. H. Macneal, "A Hybrid Method of Component Mode Synthesis," Computers and Structures Journal, Volume 1, 1971, pp. 581-601.

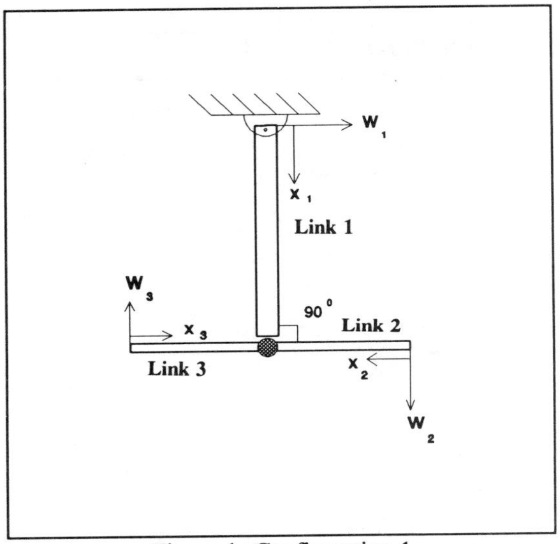

Figure 1. Configuration 1

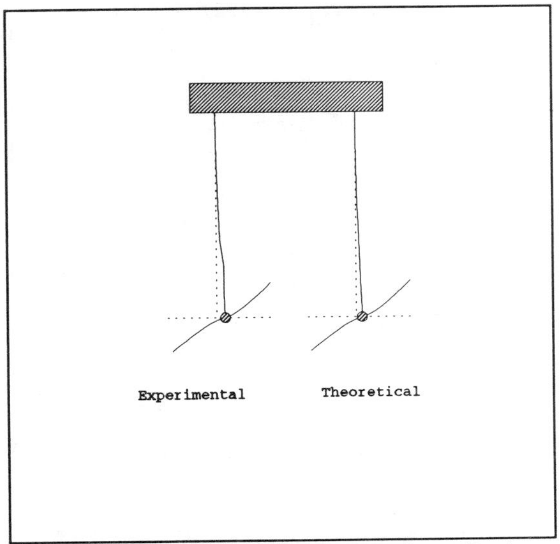

Figure 2. System Modes 1 for Configuration 1

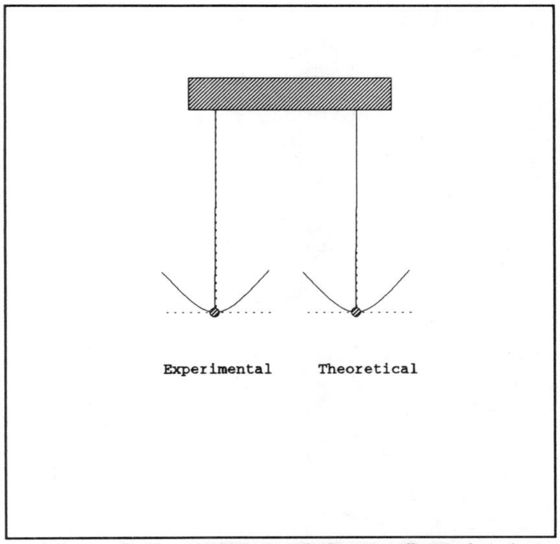

Figure 3. System Modes 2 for Configuration 1

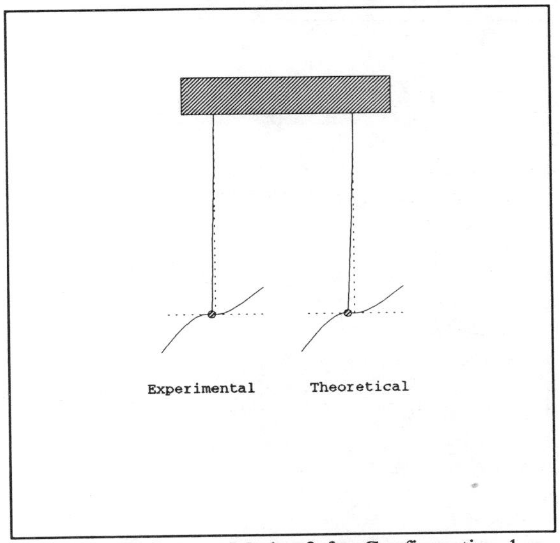

Figure 4. System Modes 3 for Configuration 1

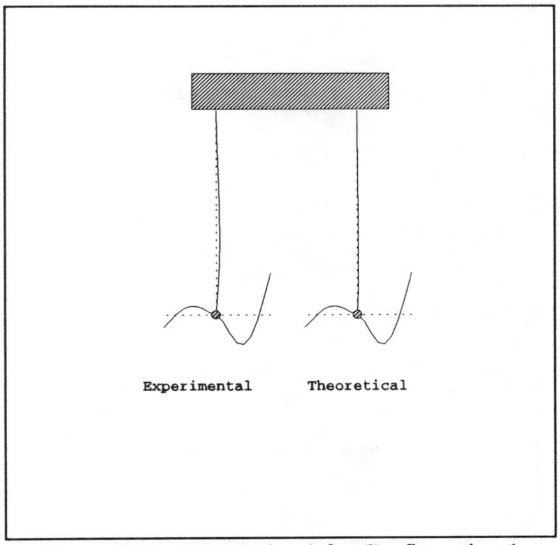

Figure 5. System Modes 4 for Configuration 1

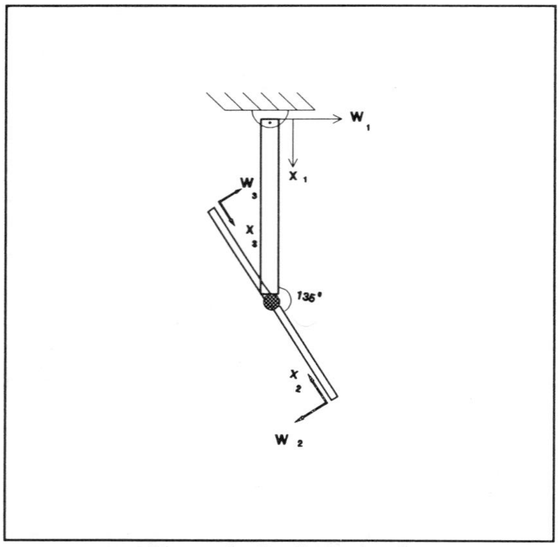

Figure 6. Configuration 2

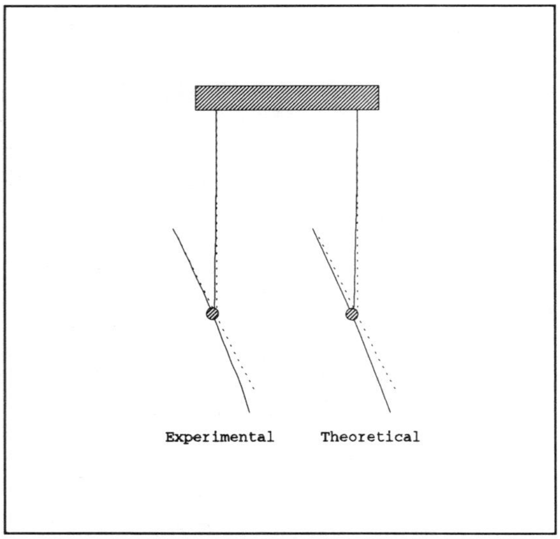

Figure 7. System Modes 1 for Configuration 2

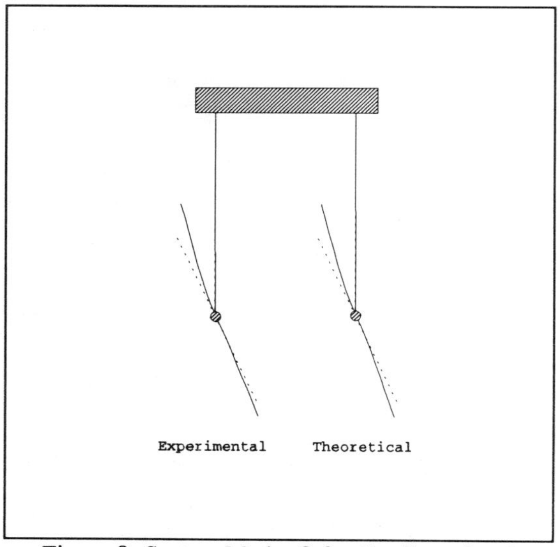

Figure 8. System Modes 2 for Configuration 2

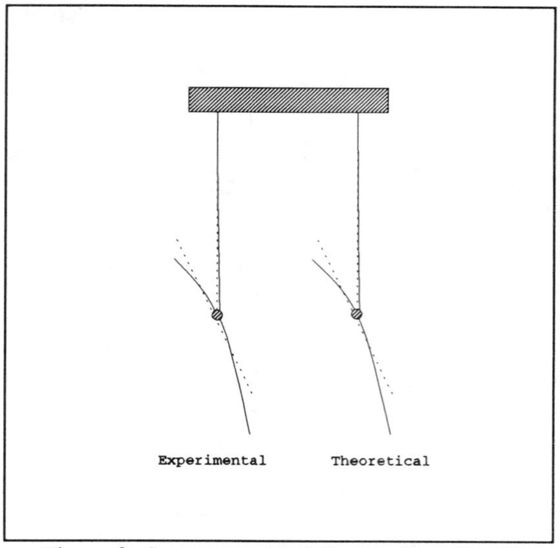

Figure 9. System Modes 3 for Configuration 2

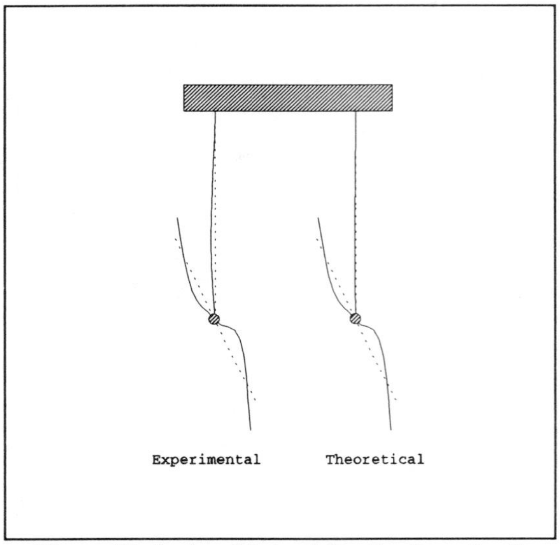

Figure 10. System Modes 4 for Configuration 2

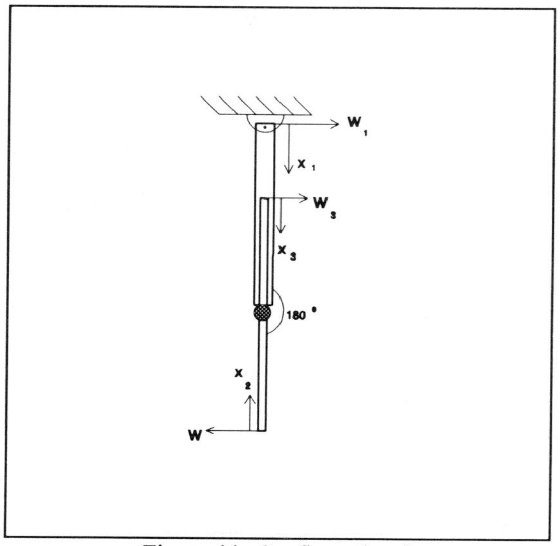

Figure 11. Configuration 3

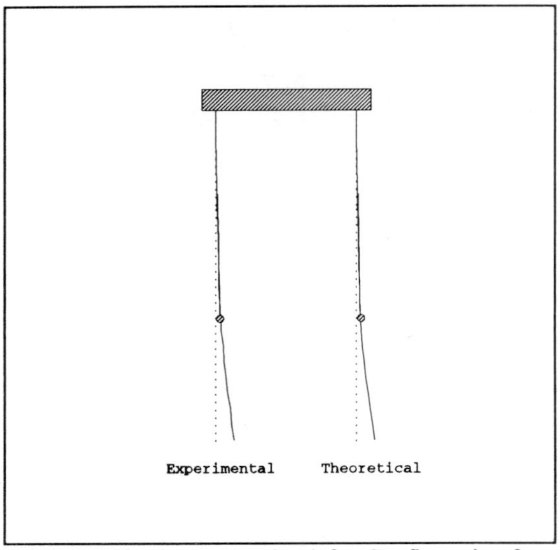

Figure 12. System Modes 1 for Configuration 3

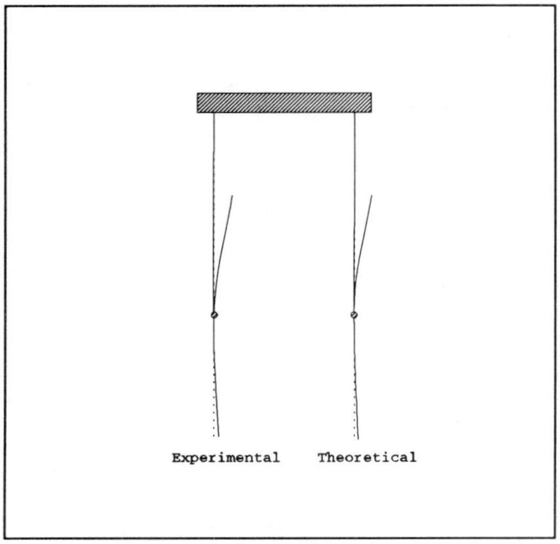

Figure 13. System Modes 2 for Configuration 3

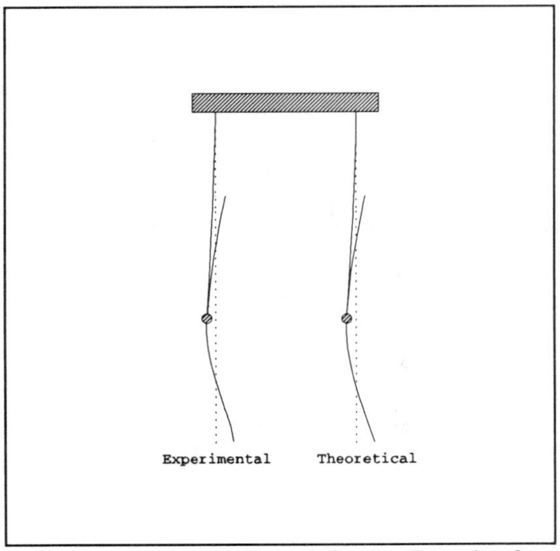

Figure 14. System Modes 3 for Configuration 3

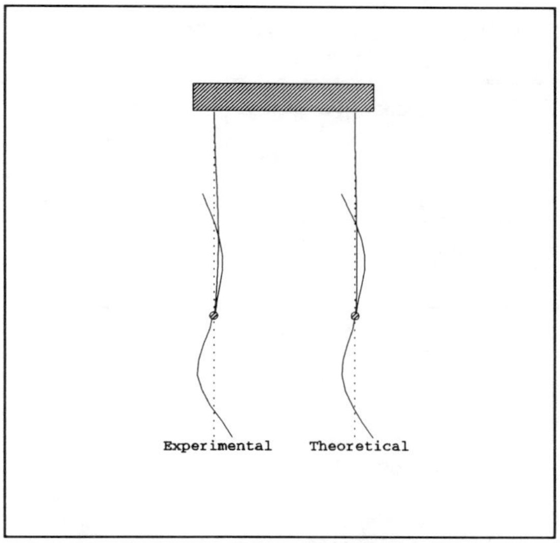

Figure 15. System Modes 4 for Configuration 3

TABLE 1. BASIC DATA *

	Component 1 (Box Beam)	Component 2 (Solid Beam)	Component 3 (Solid Beam)
Material	Aluminum	Aluminum	Aluminum
E	68.95E+9	68.95E+9	68.95E+9
Volume Density	2768	2768	2768
Base	0.0635	0.0335	0.0335
Height	0.0318	0.009	0.009
Thickness	0.003175	______	______
Length	2.44	1.50	1.45
Moment of Inertia	9.13E-8	2.04E-9	2.04E-9
Area	5.64E-4	3.02E-4	3.02E-4

* All in S.I base units (Kg,m)

TABLE 2. CONVERGENCE OF FREQUENCIES FOR CONFIGURATION 1

Frequency (Hertz) n=3 m=6	Frequency (Hertz) n=4 m=7	Frequency (Hertz) n=5 m=8	Frequency (Hertz) n=6 m=9	Frequency (Hertz) n=7 m=10
ω_{n1}=2.5390	2.5385	2.5385	2.5385	2.5385
ω_{n2}=3.3339	3.3338	3.3338	3.3338	3.3338
ω_{n3}=3.9432	3.9399	3.9398	3.9398	3.9398
ω_{n4}=19.6606	19.5271	19.5220	19.5340	19.5246

TABLE 3. CONVERGENCE OF FREQUENCIES FOR CONFIGURATION 2

Frequency (Hertz) n=3 m=6	Frequency (Hertz) n=4 m=7	Frequency (Hertz) n=5 m=8	Frequency (Hertz) n=6 m=9	Frequency (Hertz) n=7 m=10
ω_{n1}=1.6049	1.6048	1.6048	1.6048	1.6048
ω_{n2}=3.3070	3.3069	3.3069	3.3069	3.3069
ω_{n3}=4.1550	4.1537	4.1537	4.1538	4.1537
ω_{n4}=19.8378	19.5957	19.5894	19.6003	19.5914

TABLE 4. CONVERGENCE OF FREQUENCIES FOR CONFIGURATION 3

Frequency (Hertz) n=3 m=6	Frequency (Hertz) n=4 m=7	Frequency (Hertz) n=5 m=8	Frequency (Hertz) n=6 m=9	Frequency (Hertz) n=7 m=10
ω_{n1}=2.2716	2.2713	2.2713	2.2713	2.2713
ω_{n2}=3.3541	3.3540	3.3540	3.3541	3.3540
ω_{n3}=5.8058	5.7812	5.7801	5.7803	5.7801
ω_{n4}=19.9180	19.6036	19.5963	19.6069	19.5981

TABLE 5. HARDWARE DOCUMENTATION

Amplifier/Power Unit	PCB Model # 483A07
Accelerometer	PCB Model # 303A
Load Cell	PCB Model # 208A03
Analyzer	HP 5423

TABLE 6. EXPERIMENTAL FREQUENCIES FOR CONFIGURATION 1

Mode Number	Frequency (Hz)
1	ω_n=2.25
2	ω_n=3.37
3	ω_n=3.75
4	ω_n=19.83

TABLE 7. EXPERIMENTAL FREQUENCIES FOR CONFIGURATION 2

Mode Number	Frequency (Hz)
1	ω_n=2.12
2	ω_n=3.37
3	ω_n=4.12
4	ω_n=19.79

TABLE 8. EXPERIMENTAL FREQUENCIES FOR CONFIGURATION 3

Mode Number	Frequency (Hz)
1	ω_n=2.12
2	ω_n=3.37
3	ω_n=5.00
4	ω_n=19.8

TABLE 9. CORRELATION OF FREQUENCIES FOR CONFIGURATION 1

	Hurty's Method Frequency (Hz)	Experimental Frequency (Hz)	TREETOPS Frequency (Hz)
ω_{n1}	2.54	2.25	2.55
ω_{n2}	3.33	3.37	3.33
ω_{n3}	3.94	3.75	3.96
ω_{n4}	19.52	19.83	19.65

TABLE 10. CORRELATION OF FREQUENCIES FOR CONFIGURATION 2

	Hurty's Method Frequency (Hz)	Experimental Frequency (Hz)	TREETOPS Frequency (Hz)
ω_{n1}	1.60	2.12	2.37
ω_{n2}	3.31	3.37	3.40
ω_{n3}	4.15	4.12	4.72
ω_{n4}	19.59	19.79	19.68

TABLE 11. CORRELATION OF FREQUENCIES FOR CONFIGURATION 3

	Hurty's Method Frequency (Hz)	Experimental Frequency (Hz)	TREETOPS Frequency (Hz)
ω_{n1}	2.27	2.12	2.28
ω_{n2}	3.35	3.37	3.38
ω_{n3}	5.78	5.00	5.78
ω_{n4}	19.60	19.8	19.74

TABLE 12. PERCENT CHANGE IN FREQUENCIES FOR CONFIGURATION 1

	% Change Hurty/Experiment	% Change Hurty/TREETOPS	% Change TREETOPS/Experiment
ω_{n1}	12.89 %	0.39 %	13.33 %
ω_{n2}	1.19 %	0.00 %	1.19 %
ω_{n3}	5.07 %	0.51 %	5.60 %
ω_{n4}	1.56 %	0.66 %	0.91 %

TABLE 13. PERCENT CHANGE IN FREQUENCIES FOR CONFIGURATION 2

	% Change Hurty/Experiment	% Change Hurty/TREETOPS	% Change TREETOPS/Experiment
ω_{n1}	24.53 %	32.49 %	11.79 %
ω_{n2}	1.78 %	2.65 %	0.89 %
ω_{n3}	0.73 %	12.08 %	14.56
ω_{n4}	1.01 %	0.46 %	0.56 %

TABLE 14. PERCENT CHANGE IN FREQUENCIES FOR CONFIGURATION 3

	% Change Hurty/Experiment	% Change Hurty/TREETOPS	% Change TREETOPS/Experiment
ω_{n1}	7.08 %	0.44 %	7.55 %
ω_{n2}	0.59 %	0.89 %	0.30 %
ω_{n3}	15.60 %	0.00 %	15.60 %
ω_{n4}	1.01 %	0.71 %	0.30 %

TABLE 15. MAC FOR CONFIGURATION 1

MODE	MAC
1	0.982
2	0.981
3	0.926
4	0.929

TABLE 16. MAC FOR CONFIGURATION 2

MODE	MAC
1	0.909
2	0.834
3	0.907
4	0.954

TABLE 17. MAC FOR CONFIGURATION 3

Mode	MAC
1	0.992
2	0.994
3	0.970
4	0.957

APPENDIX

$$
\begin{bmatrix}
1.5245 & 2.5409 & 3.8114 & -2.7119 & 4.0787 & 2.7192 & 0.2719 \\
-1.5245 & -2.5409 & -3.8114 & 1.7119 & -4.0787 & -2.7192 & -0.2719 \\
3.9372 & 5.5620 & 7.3430 & 0.1771 & 2.5074 & 1.6716 & 0.1672 \\
-4.9372 & -6.5620 & -8.3430 & -0.1771 & -2.5074 & -1.6716 & -0.1672 \\
1 & 0 & 0 & 0 & 0 & 0 & 0 \\
0 & 1 & 0 & 0 & 0 & 0 & 0 \\
0 & 0 & 1 & 0 & 0 & 0 & 0 \\
0 & 0 & 0 & 1 & 0 & 0 & 0 \\
0 & 0 & 0 & 0 & 1 & 0 & 0 \\
0 & 0 & 0 & 0 & 0 & 1 & 0 \\
0 & 0 & 0 & 0 & 0 & 0 & 1 \\
0.9060 & 1.5100 & 2.2649 & 0.1712 & -3.5762 & -3.3841 & -1.8384 \\
-0.9060 & -1.5100 & -2.2649 & -0.1712 & 2.5762 & 2.3841 & 0.8384 \\
0 & 0 & 0 & -1 & 0 & 0 & 0
\end{bmatrix}
$$

Matrix 1. Transformation Matrix β for Configuration 1

$$
\begin{bmatrix}
5.3875 & 7.8156 & 10.5016 & -0.0262 & 3.9472 & 2.6300 & 0.2628 \\
 & 11.3488 & 15.2603 & -0.1001 & 5.7802 & 3.8510 & 0.3847 \\
 & & 20.5318 & -0.2096 & 7.8281 & 5.2151 & 0.5209 \\
 & \textit{SYMMETRIC} & & 3.8721 & -0.5829 & -0.3888 & -0.0389 \\
 & & & & 4.0576 & 2.4250 & 0.2266 \\
 & & & & & 1.5527 & 0.1495 \\
 & & & & & & 0.0147
\end{bmatrix}
$$

Matrix 2. System Mass Matrix for Configuration 1

$$
\begin{bmatrix}
0.5500 & 0.8833 & 1.2892 & -0.3152 & 1.0756 & 0.7220 & 0.0737 \\
 & 1.4324 & 2.1039 & -0.5253 & 1.7926 & 1.2034 & 0.1228 \\
 & & 3.1038 & -0.7879 & 2.6890 & 1.8051 & 0.1842 \\
 & \textit{SYMMETRIC} & & 0.3856 & -0.8573 & -0.5706 & -0.0568 \\
 & & & & 2.9415 & 1.9798 & 0.1970 \\
 & & & & & 1.3398 & 0.1333 \\
 & & & & & & 0.0137
\end{bmatrix}
$$

Matrix 3. System Stiffness Matrix for Configuration 1

AMD-Vol. 141/DSC-Vol. 37, Dynamics of Flexible Multibody Systems:
Theory and Experiment
ASME 1992

TABULATED MODE CALCULATIONS FOR CHAINED FLEXIBLE MULTIBODY SYSTEMS

Hanching Wang
Department of Electrical Engineering
University of California, Los Angeles
Los Angeles, California

Abstract

A new tabulated approach to find modes of homogeneous (or homogenized) chained flexible multibody systems is proposed. The n-body system is modeled as n-wave equations coupled through boundary conditions. Each wave equation is represented in its generic basis for convenience. The (6x6) Timoshenko-type wave equation with lumped masses at tips (simple beam) is used throughout this paper for examples. The boundary conditions can also take account of lumped masses such as actuators and sensors.

A characteristic matrix (CM) is defined such that, when the determinant is set to zero, the roots are the mode frequencies. The table of simple beam's CM's, with clamped-clamped (C-C), free-free (F-F), clamped-free (C-F) and free-clamped (F-C) boundary conditions, respectively, are derived using techniques of functional analysis. The derivation reveals that there is a rule to write down these matrices by inspection. This table is subsequently called library and will be searched by a table look-up procedure in software implementation.

The CM of multibody systems is derived mathematically to show that it is a binary composition of the CM's of simple beams. The CM of n-body system then can be obtained easily in a tabulated way or programmed using binary tree data structure and table look-up scheme. This greatly simplifies the derivation of CM for multibody systems.

The major contribution of this paper is the novel that the CM of multibody systems is composed from CM's of simple beams. Numerical example is carried out for the Evolutionary Model to show its feasibility and simplicity.

1 Generalized Timoshenko Beams

The kinetic energy and potential energy of a Timoshenko beam are

$$KE = \frac{1}{2}\int \dot{X}^T M \dot{X}\ dx$$

and

$$PE = \frac{1}{2}\int \epsilon^T C \epsilon \; dx$$

respectively, where M and C have the general pattern as

$$M = \begin{bmatrix} m_{11} & \cdot & \cdot & \cdot & \cdot & \cdot \\ & m_{22} & \cdot & \cdot & \cdot & \cdot \\ & & m_{33} & \cdot & \cdot & \cdot \\ & & & m_{44} & \cdot & \cdot \\ s & y & m & & m_{55} & m_{56} \\ & & & & m_{56} & m_{66} \end{bmatrix}$$

$$\begin{aligned} C &= \begin{bmatrix} c_{11} & c_{12} & c_{13} & \cdot & \cdot & \cdot \\ & c_{22} & c_{23} & \cdot & \cdot & \cdot \\ & & c_{33} & \cdot & \cdot & \cdot \\ & & & c_{44} & c_{45} & c_{46} \\ s & y & m & & c_{55} & c_{56} \\ & & & & & c_{66} \end{bmatrix} \\ &= \begin{bmatrix} C_1 & 0_{3\times3} \\ 0_{3\times3} & C_2 \end{bmatrix} \quad \text{and} \\ X &= [u \; v \; w \; \phi_1 \; \phi_2 \; \phi_3]^T \\ \epsilon &= X' + \begin{bmatrix} 0_{3\times3} & L_1 \\ 0_{3\times3} & 0_{3\times3} \end{bmatrix} X \\ L_1 &= \begin{bmatrix} 0 & 0 & 0 \\ 0 & 0 & -1 \\ 0 & 1 & 0 \end{bmatrix} \end{aligned}$$

The dynamic equation of Timoshenko beam is

$$M\ddot{X} - CX'' - A_1 X' + A_0 X = 0$$

$$A_0 = \begin{bmatrix} 0_{3\times3} & 0_{3\times3} \\ 0_{3\times3} & L_1^T C_1 L_1 \end{bmatrix}$$

$$L_1^T C_1 L_1 = \begin{bmatrix} 0 & 0 & 0 \\ 0 & c_{33} & -c_{23} \\ 0 & -c_{23} & c_{22} \end{bmatrix}$$

$$A_1 = L - L^T \; , \quad L = \begin{bmatrix} 0_{3\times3} & C_1 L_1 \\ 0_{3\times3} & 0_{3\times3} \end{bmatrix}$$

$$C_1 L_1 = \begin{bmatrix} 0 & c_{13} & -c_{12} \\ 0 & c_{23} & -c_{22} \\ 0 & c_{33} & -c_{32} \end{bmatrix}$$

The boundary conditions are:
(a) for geometric B.C.

$$X_b \;=\; \text{specified function of } t$$

(b) for force B.C.

$$M_b \ddot{X}_b - C \; \Delta X_b' - L \; \Delta X_b = F_b$$

where $\quad \Delta X_b = X(b^+) - X(b^-)$
M_b is the lumped boundary mass matrix, and F_b the concentrated force. See [1] for a detailed derivation and a concrete example.

2 Dynamic Equations of Multibody Systems

Here we consider a flexible structure which consists of n-homogeneous Timoshenko beams in a chained configuration.

Let E_i be a matrix such that $E_i(1)$, $E_i(2)$, and $E_i(3)$ are the basis for the coordinates of the i th beam. The coordinates are principal coordinates (of cross-sectional inertia) with $E_i(1)$ along the generic axis.

For each beam the Timoshenko beam equation is

$$M_i\ddot{X}_i - C_i X_i'' - A_{1,i} X_i' + A_{0,i} X_i = 0 \qquad i = 1\;(1)\;n$$

where X_i is the vector X for i th beam measured w.r.t. the equilibrium position, observed from an inertia frame and represented in E_i coordinates.

Let $E_i {}^i\tilde{T}^j = E_j$, where ${}^i\tilde{T}^j$ is the representation of E_j basis w.r.t. E_i basis, ie.,

$${}^i\tilde{T}^j(k,l) = [E_i(k), E_j(l)]$$

Note that$({}^i\tilde{T}^j)^{-1} = ({}^i\tilde{T}^j)^T = {}^j\tilde{T}^i$.

The boundary conditions can be obtained from section (1) for each beam, except at the joints which need further consideration.

1. The geometric condition at (stationary) joints :
If the j th beam has a joint with the i th beam, then the rigid-body motion at each beam should be equal when represented in the same coordinates, i.e.,

$$\begin{bmatrix} u \\ v \\ w \\ \frac{1}{2}\nabla \otimes \begin{pmatrix} U \\ V \\ W \end{pmatrix} \end{bmatrix}_i = {}^iT^j \begin{bmatrix} u \\ v \\ w \\ \frac{1}{2}\nabla \otimes \begin{pmatrix} U \\ V \\ W \end{pmatrix} \end{bmatrix}_j$$

$${}^iT^j = \begin{bmatrix} {}^i\tilde{T}^j & 0 \\ 0 & {}^i\tilde{T}^j \end{bmatrix}$$

The above equation can be well approximated by

$$X_i = {}^iT^j X_j$$

2. The force condition at the joints :
For the joint of the i th and j th beam, we have

$$\begin{aligned} M_{ij}\ddot{X}_i &- C_i\,\Delta X_i' - L_i\,\Delta X_i \\ &- {}^iT^j[C_j\,\Delta X_j' + L_j\,\Delta X_j] = F_{ij} \end{aligned}$$

where M_{ij} is the lumped mass matrix at the joint and F_{ij} is the external force.

3 Example: Two-Beam Structure

The formulation procedures above are better demonstrated by an example. Consider two beams with the same M and C matrices in the principal coordinates (of cross-sectional inertia) as shown in figure (1). Assume there is no lumped mass at location s for simplicity, and only one actuator at location ℓ. Let

$$\{\tilde{T}(i,j)\} = \{[x_i, y_j]\} \qquad i,j = 1(1)3$$

then

$$[y_1\ y_2\ y_3] = [x_1 x_2 x_3]\tilde{T}$$

and

$$X = TY\ , \qquad T = \begin{bmatrix} \tilde{T} & 0 \\ 0 & \tilde{T} \end{bmatrix}$$

The dynamic equations are

$$M\ddot{X} - CX'' - A_1X' + A_0X = 0, \quad (\bullet)' = \frac{\partial(\bullet)}{\partial x_1}$$

$$M\ddot{Y} - CY'' - A_1Y' + A_0Y = 0, \quad (\bullet)' = \frac{\partial(\bullet)}{\partial y_1}$$

BC's:

$$\begin{aligned} -CX'(0) - LX(0) &= 0 \\ M_b\ddot{Y}(\ell) + CY'(\ell) + LY(\ell) &= F_b \end{aligned}$$

at the stationary joint :

$$X(L) = TY(0)$$

$$\begin{aligned} 0 &= -C\Delta X'(L) - L\Delta X(L) \\ &\quad + T[-C\Delta Y'(0) - L\Delta Y(0)] \\ &= CX'(L) + LX(L) - T[CY'(0) + LY(0)] \end{aligned}$$

The advantage of the above formulation is that both beam equations have the same parameter matrices for saving solution effort. The transformation T is entered at the joint boundary conditions only. An alternative formulation can be developed using a similarity transformation.

4 Modes of Simple Beams

The approximation of mode calculations using Transfer Matrix Method can be traced back to late 1950s [2], see pages 204-225 on [3] for a historical review. An extension of this transfer matrix concept, based on Functional Analysis, has been developed recently. For all the details one should refer to [4, 5]. This paper here will concentrate on the mode calculations based on Functional Analysis for chained flexible multibody systems.

Consider a simple Timoshenko beam on a bounded domain described by

$$M\ddot{X} - CX'' - A_1X' + A_0X = 0$$

The modes are the solution of

$$CX'' + A_1X' + (\omega^2M - A_0)X = 0$$

with conditions at boundary points.

The above problem can be cast in state-space form with

$$A(\omega) = \begin{pmatrix} 0 & I \\ -C^{-1}(\omega^2M - A_0) & -C^{-1}A_1 \end{pmatrix}$$

and

$$\begin{bmatrix} X(s) \\ X'(s) \end{bmatrix} = e^{As} \begin{bmatrix} X(0) \\ X'(0) \end{bmatrix}$$

where e^{As} is the spatial transition matrix. Let

$$P = \begin{bmatrix} P_{11} & P_{12} \\ P_{21} & P_{22} \end{bmatrix} = e^{A\ell}$$

where ℓ is the length of the beam.

We can solve mode frequencies in terms of P. Here we summarize some of the results.

1. Clamped-clamped beam(C-C):

   $$X(0) = 0 = X(\ell)$$

   characteristic matrix (CM) $= P_{12}(\omega)C^{-1}$
   characteristic function(CF)

   $$CF = Det[P_{12}(\omega)]$$

   mode frequencies =

   $$\{\omega > 0 \quad | \quad Det[P_{12}(\omega)] = 0\}$$

   The characteristic matrix is unique up to nonsingular-constant-matrix transformation. The characteristic function is unique up to constant multiplication. The characteristic matrices chosen here are in the most convenient forms as will become clear later.

2. Free-free beam(F-F):

   $$-CX'(0) - LX(0) = 0 = CX'(\ell) + LX(\ell)$$

   characteristic matrix(CM)

   $$CM = (LP_{11}+CP_{21})-(LP_{12}+CP_{22})C^{-1}L$$

   mode frequencies =
   $\{\omega \geq 0 \mid Det\,[CM] = 0\}$

3. Clamped-free beam(C-F):

   $$X(0) = 0 = CX'(\ell) + LX(\ell)$$

   characteristic matrix$=(LP_{12}+CP_{22})C^{-1}$

4. Free-clamped beam(F-C):

   $$-CX'(0) - LX(0) = 0 = X(\ell)$$

   characteristic matrix $= P_{11} - P_{12}C^{-1}L$

It is a generic property that all characteristic matrices should contain the matrix P_{12} . Case 3 and case 4 have the same mode frequencies. Case 1 and case 2 have the same mode frequencies except that case 2 includes $\omega = 0$ as rigid-body modes.

By noting that $e^{A\ell}$ is analytic in ω, we have

$$Det[C^{-1}LP_{12}+P_{22}] = \alpha Det[P_{11}-P_{12}C^{-1}L]$$

and

$$Det[(C^{-1}LP_{11} + P_{21})-(C^{-1}LP_{12} + P_{22})C^{-1}L] = \beta\omega^{18}Det[P_{12}]$$

where the ω^{18}, after canceling out the w in $Det[P_{12}]$, provides the six rigid-body modes.

5 Modes of Beams with Lumped Masses

Let's consider a Timoshenko beam with lumped mass at point s as shown in figure

(2). The boundary conditions at s are

$$\begin{bmatrix} X(s^+) \\ X'(s^+) \end{bmatrix} = \begin{bmatrix} I & 0 \\ -\omega^2 C^{-1} M_s & I \end{bmatrix} = \begin{bmatrix} X(s^-) \\ X'(s^-) \end{bmatrix}$$

Let

$$\mathcal{L} = \begin{bmatrix} \mathcal{L}_{11} & \mathcal{L}_{12} \\ \mathcal{L}_{21} & \mathcal{L}_{22} \end{bmatrix} = e^{As}$$

$$R = \begin{bmatrix} R_{11} & R_{12} \\ R_{21} & R_{22} \end{bmatrix} = e^{A(\ell - s)}$$

and $L_s = L + \omega^2 M_s$

We summarize the results below with the following notation adopted.

C: clamped

F: free

RB: right hand part of the beam, from s to ℓ with M_s lumped at s.

LB: left hand part of the beam, from 0 to s.

B: whole beam without lumped mass

5. Clamped-clamped beam with lumped mass:
characteristic matrix

$$\begin{aligned} &= (R_{11} - R_{12} C^{-1} L_s) \mathcal{L}_{12} C^{-1} \\ &+ R_{12} C^{-1} (L\mathcal{L}_{12} + C\mathcal{L}_{22}) C^{-1} \end{aligned}$$

From case 1 , 3 and 4, the characteristic matrix is that of
$[F - C \quad RB][C - C \quad LB]$
$+[C - C \quad RB][C - F \quad LB]$
It makes no difference whether M_s goes with RB or LB or divides between them. However, we will make it a convention to let the lumped mass go with RB.

6. Free-free beam with lumped mass:
characteristic matrix =
$[(LR_{11} + CR_{21}) - (LR_{12} + CR_{22}) C^{-1} L_s]$
$[\mathcal{L}_{11} - \mathcal{L}_{12} C^{-1} L]$
$+[LR_{12} + CR_{22}] C^{-1}$
$[(L\mathcal{L}_{11} + C\mathcal{L}_{21}) - (L\mathcal{L}_{12} + C\mathcal{L}_{22}) C^{-1} L]$

$$\begin{aligned} CM &= [F - F \quad RB][F - C \quad LB] \\ &+ [C - F \quad RB][F - F \quad LB] \end{aligned}$$

7. Free-clamped beam with lumped mass:
characteristic matrix
$= [F - C \quad RB][F - C \quad LB]$
$+[C - C \quad RB][F - F \quad LB]$

8. Clamped-free beam with lumped mass:
characteristic matrix
$= [F - F \quad RB][C - C \quad LB]$
$+[C - F \quad RB]C - F \quad LB]$
The asymptotic modes can be obtained by applying the generic property of spatial transition matrix, that

$$P_{11} \sim \omega P_{12} \sim \frac{1}{\omega} P_{21} \sim P_{22}$$

are of the same order of magnitude in ω.
For example, in this case:
For $\omega \rightarrow 0$,
characteristic matrix $\rightarrow [C - F \quad B]$
For $\omega \rightarrow \infty$, characteristic matrix $\rightarrow$ $[C - F \quad RB][C - C \quad LB]$

The interesting part is that case 5, case 6, case 7 and case 8 can be obtained easily by inspection, as will be shown in the next section. Those terms which contains point s as a free point correspond to lower mode frequencies, while those terms which contains point s as a clamped point correspond to higher mode frequencies.

6 Modes of Multi-Beam Systems

We will derive a procedure to obtain the characteristic matrix of a multi-beam structure by inspection. The following are concluded from the previous two sections.

For a beam with L_L and L_R at the left end and right end respectively, a rule may be stated to derive the characteristic matrix. L after P_{ij} should be substituted by L_L to read $P_{ij}C^{-1}L_L$; and asymptotically $[P_{ij}C^{-1}]M_L$. L before P_{ij} should be substituted by L_R to read $C^{-1}L_RP_{ij}$; and asymptotically $C^{-1}M_R[P_{ij}]$.

We tabulate the characteristic matrices of simple beams for easy reference.

C-C	$P_{12}C^{-1}$
F-F	$(L_RP_{11} + CP_{21}) - (L_RP_{12} + CP_{22})C^{-1}L_L$
C-F	$(L_RP_{12} + CP_{22})C^{-1}$
F-C	$P_{11} - P_{12}C^{-1}L_L$

For derivation of characteristic matrices and demonstration of lumped effects on mode frequencies, we consider the same structure in figure (1) with three actuators.

6.1 Characteristic Matrices

From section (3) ,

$$CX'' + A_1X' + (\omega^2M_0 - A_0)X = 0$$

$$CY'' + A_1Y' + (\omega^2M_0 - A_0)Y = 0$$

$$CX'(0) + L_0X(0) = 0\ , \quad L_0 = L + \omega^2M_0$$

$$CY'(\ell_2) + L_2Y(\ell_2) = 0\ , \quad L_2 = L - \omega^2M_2$$

Joint:

$$\begin{bmatrix} Y(0) \\ Y'(0) \end{bmatrix} = \begin{bmatrix} T^T & 0 \\ C^{-1}(T^TL - LT^T) & C^{-1}T^TC \end{bmatrix} \begin{bmatrix} X(\ell) \\ X'(\ell) \end{bmatrix}$$

Point s:

$$\begin{bmatrix} X(s^+) \\ X'(s^+) \end{bmatrix} = \begin{bmatrix} I & 0 \\ -\omega^2C^{-1}M_s & I \end{bmatrix} \begin{bmatrix} X(s^-) \\ X'(s^-) \end{bmatrix}$$

Letting $Q = e^{A\ell_2}$, $\quad R = e^{A(\ell_1 - s)}$ and $\mathcal{L} = e^{As}$, we have

$$\begin{aligned} 0 &= [L_2 \quad C] \begin{bmatrix} Y(\ell_2) \\ Y'(\ell_2) \end{bmatrix} \\ &= [L_2 \quad C]Q \begin{bmatrix} T^T & 0 \\ C^{-1}(T^TL - LT^T) & C^{-1}T^TC \end{bmatrix} \\ &\quad R \begin{bmatrix} I & 0 \\ -\omega^2C^{-1}M_s & I \end{bmatrix} \mathcal{L} \begin{bmatrix} X(0) \\ X'(0) \end{bmatrix} \\ &= \begin{bmatrix} [(L_2Q_{11} + CQ_{21}) - & (L_2Q_{12} + CQ_{22}) \\ (L_2Q_{12} + CQ_{22})C^{-1}L]T^T & C^{-1}T^TC \\ +(L_2Q_{12} + CQ_{22})C^{-1}T^TL & \end{bmatrix} \\ &\quad R \begin{bmatrix} I & 0 \\ -C^{-1}(L_s - L) & I \end{bmatrix} \mathcal{L} \begin{bmatrix} X(0) \\ X'(0) \end{bmatrix} \\ &= \begin{bmatrix} [F-F \quad AB]T^TR_{11} & [F-F \quad AB]T^TR_{12} \\ +[C-F \quad AB]T^T & +[C-F \quad AB]T^T \\ (LR_{11} + CR_{21}) & (LR_{12} + CR_{22}) \end{bmatrix} \\ &\quad \begin{bmatrix} I & 0 \\ -C^{-1}(L_s - L) & I \end{bmatrix} \mathcal{L} \begin{bmatrix} X(0) \\ X'(0) \end{bmatrix} \end{aligned}$$

AB : attached beam

$$\begin{aligned} &= \begin{bmatrix} [F-F \quad AB]T^T & [F-F \quad AB] \\ (R_{11} - R_{12}C^{-1}(L_s - L)) & T^TR_{12} \\ +[C-F \quad AB]T^T & +[C-F \quad AB] \\ [LR_{11} + CR_{21} - & T^T(LR_{12} + CR_{22}) \\ (LR_{12} + CR_{22})C^{-1}(L_s - L)] & \end{bmatrix} \\ &\quad \begin{bmatrix} \mathcal{L}_{11} - \mathcal{L}_{12}C^{-1}L_0 \\ \mathcal{L}_{21} - \mathcal{L}_{22}C^{-1}L_0 \end{bmatrix} X(0) \end{aligned}$$

$$\begin{aligned} CM &= [F-F \quad AB]T^T[F-C \quad RB][F-C \quad LB] \\ &\quad + [C-F \quad AB]T^T \\ &\quad [F-F \quad RB][F-C \quad LB] \end{aligned}$$

$$+ \quad [F-F \quad AB]T^T R_{12} C^{-1} L \quad [\mathcal{L}_{11} - \mathcal{L}_{12} C^{-1} L_0]$$

$$+ \quad [F-F \quad AB]T^T R_{12} \quad [\mathcal{L}_{21} - \mathcal{L}_{22} C^{-1} L_0]$$

$$+ \quad [C-F \quad AB]T^T (LR_{12} + CR_{22}) \quad C^{-1} L [\mathcal{L}_{11} - \mathcal{L}_{12} C^{-1} L_0]$$

$$+ \quad [C-F \quad AB]T^T (LR_{12} + CR_{22}) \quad [\mathcal{L}_{21} - \mathcal{L}_{22} C^{-1} L_0]$$

$$= \quad [F-F \quad AB]T^T [F-C \quad RB] \quad [F-C \quad LB]$$

$$+ \quad [F-F \quad AB]T^T [C-C \quad RB] \quad [F-F \quad LB]$$

$$+ \quad [C-F \quad AB]T^T [F-F \quad RB] \quad [F-C \quad LB]$$

$$+ \quad [C-F \quad AB]T^T [C-F \quad RB] \quad [F-F \quad LB]$$

6.2 Tabulated Modes Calculations

We will use the same example to demonstrate the general procedures to write down the characteristic matrix by inspection.

1. Set exterior boundary condition: points 0 and ℓ_2 free

AB	T^T	RB	LB
-F			F-
-F			F-
-F			F-
-F			F-

} 4 terms

Total terms $= 2^n$, n : number of interior boundary points.

2. Set point ℓ_1 free

AB	T^T	RB	LB
F-F			F-
F-F			F-
-F		-F	F-
-F		-F	F-

(a) Set point s free

AB	T^T	RB	LB
F-F			F-
F-F			F-
-F		F-F	F-
-F		-F	F-F

(b) Set point s clamped

AB	T^T	RB	LB
F-F			F-
F-F			F-
-F		F-F	F-C
-F		C-F	F-F

3. Set point ℓ_1 clamped, and do (a) and (b) as in (2)

AB	T^T	RB	LB
F-F		F-C	F-C
F-F		C-C	F-F
C-F		F-F	F-C
C-F		C-F	F-F

For $\omega \to 0$, L_0, L_s, L_2, $\to L$ (all lumped masses neglected)

1. Set exterior points: removing lumped masses

AB	T^T	B=LB+RB
-F		F-
-F		F-

2. Set interior points (which have no lumped masses)

AB	T^T	B=LB+RB
F-F		F-C
C-F		F-F

For $\omega \longrightarrow \infty$, points 0, s and $\ell_2 \longrightarrow$ clamped

1. Set exterior points: setting all points with lumped masses to be clamped

M_2	AB	T^T	RB	M_s	LB	M_0
	-C		C-		C-C	
	-C		C-		C-C	

2. Set interior points (which have no lumped masses)

M_2	AB	T^T	RB	M_s	LB	M_0
	F-C		C-C		C-C	
	C-C		C-F		C-C	

Note that the rule to add M_0, M_s, M_2 is the same as that for L_0, L_s, L_2. Also, there is a hidden identity transformation between RB and LB.

The asymptotic modes are very sensitive to small perturbation. The accuracy of asymptotic modes depends on the ratio of lumped mass to total beam mass. Caution is advised when dealing with asymptotic modes, because a more complete and complicated model is indispensable in order to get accurate high-frequency modes.

7 Numerical Method

Suppose we want to calculate the flexible modes of the Evolutionary Model as shown in figure (3). Let the lumped mass remain with the left end of the sub-beam on its right as we have already made it to be a convention in section (5). Follow the procedures outlined in section (6.2) to tabulate the characteristic matrix which has $2^4 = 16$ terms.

B5	B4	B3	B2	B1
F-F	F-C	F-C	F-C	F-C
F-F	F-C	F-C	C-C	F-F
F-F	F-C	C-C	F-F	F-C
F-F	F-C	C-C	C-F	F-F
F-F	C-C	F-F	F-C	F-C
F-F	C-C	F-F	C-C	F-F
F-F	C-C	C-F	F-F	F-C
F-F	C-C	C-F	C-F	F-F
C-F	F-F	F-C	F-C	F-C
C-F	F-F	F-C	C-C	F-F
C-F	F-F	C-C	F-F	F-C
C-F	F-F	C-C	C-F	F-F
C-F	C-F	F-F	F-C	F-C
C-F	C-F	F-F	C-C	F-F
C-F	C-F	C-F	F-F	F-C
C-F	C-F	C-F	C-F	F-F

characteristic matrix(CM) =

$$[F-F \quad B5] \left\{ \begin{array}{l} [F-C \quad B4] \left\{ \begin{array}{l} [F-C \quad B3]\{\cdots\} \\ +[C-C \quad B3]\{\cdots\} \end{array} \right\} \\ +[C-C \quad B4]\{\cdots\} \end{array} \right\} +[C-F \quad B5]\{\cdots\}$$

$$f(\omega) = Det[CM]$$

$$\begin{aligned} \text{Let } P(s) &= e^{As} \\ P1 &= e^{125A} = P(125) \\ P2 &= e^{30A} = P(30) \\ P3 &= e^{170A} = P(170) \\ P4 &= e^{230A} = P(230) \end{aligned}$$

$$P5 = e^{65A} = P(65)$$

$$\begin{aligned} \text{Let } L_{c1} &= L + \omega^2 M_{c1} \\ L_{c2} &= L_{c1} \\ L_{s1} &= L + \omega^2 M_{s1} \\ L_{s2} &= L_{s1} \\ L_t &= L + \omega^2 M_t \\ L_a &= L - \omega^2 M_a \end{aligned}$$

The elements in the characteristic matrix can be found from the library in section (6) with suitable substitution.

	B1	B2	B3	B4	B5
L_L	L_{c1}	L_{s1}	L_t	L_{c2}	L_{s2}
L_R	L	L	L	L	L_a
P	P1	P2	P3	P4	P5

Numerical Algorithm:

0. Input M,C,A_i, M_i. Form $A(\omega)$.

1. Set up beam library.

2. Form CM of the problem of interest. CM is automatically generated, given the exterior points, interior points and substitution table.

3. Newton-Raphson scheme

a. For each ω_i, $\omega_{i+1} = \omega_i + \Delta\omega$ $\Delta\omega$: frequency increasement
calculate $P = e^{A(\omega_i)s}$

b. Calculate $f(\omega_i) = Det[CM]$

c. If a root in the vicinity is detected, use N-F scheme to find ω^j, where ω^j is the jth mode.

Advantages:

1. We can have as many modes as we want without any reformulation (FEM needs to regrid).
2. Computation burden does not increase for higher modes calculation (FEM needs to increase the dimension of matrix eigenvalue problem). In fact, the higher modes are much easier to detect since they are well-separated. Note that scaling of mode frequency is needed to preserve accuracy in higher modes calculation [6].

8 Acknowledgement

Grateful acknowledgement to Professor A. V. Balakrishnan for discussion. Research supported in part by NASA Langley FRC.

References

[1] A. V. Balakrishnan. Compensator design for stability enhancement with colocated controllers. *IEEE Trans. on Automatic Control*, Vol. 36(No. 9):pp. 994–1007, September 1991.

[2] W. T. Thomson. Matrix solution for the vibration of nonuniform beams. *J. Appl. Mech.*, Vol. 17(No. 3):pp. 337–339, 1950.

[3] W. C. Hurty and M. F. Rubinstein. *Dynamics of Structures*. Prentice-Hall, Inc., Englewood Cliffs, N.J., 1964.

[4] A. V. Balakrishnan. Mode calculation for the anisotropic Timoshenko model of EPS truss. unpublished internal report, May 1990.

[5] A. V. Balakrishnan. Modes of interconnected lattice trusses using continuum models, Part (I). NASA Contract Report 189568, December 1991.

[6] C. B. Moler and C. F. Van Loan. Nineteen dubious ways to compute the exponential of a matrix. *SIAM Rev.*, Vol. 20:pp. 801–836, 1978.

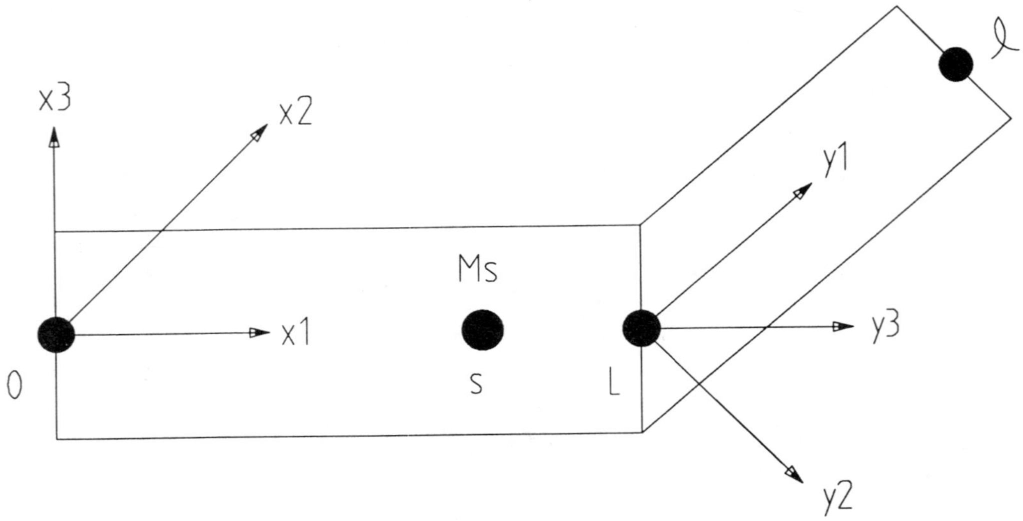

Figure 1: Two Beam Structure

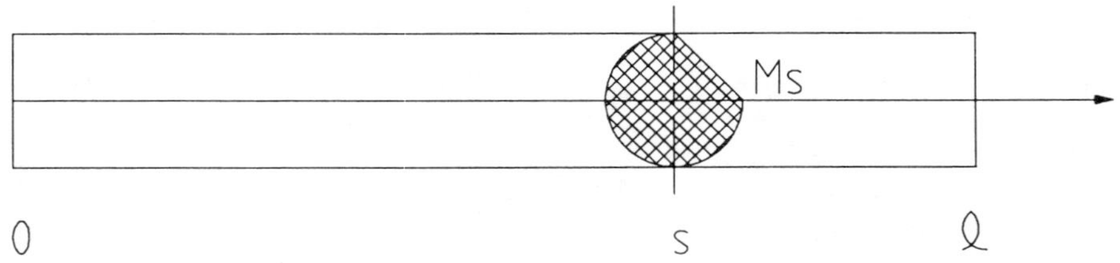

Figure 2 : Beam with Lumped Mass

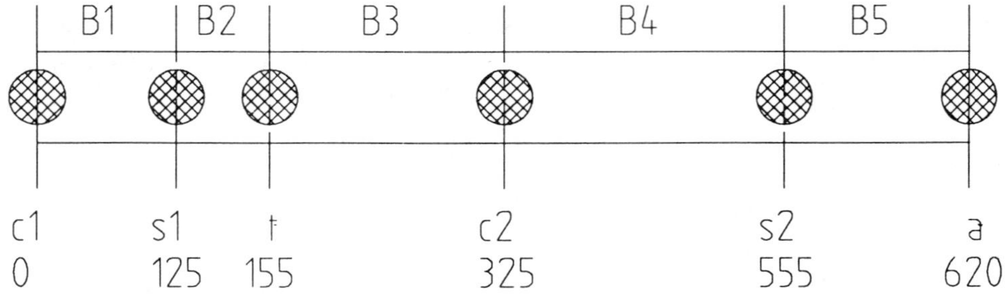

Figure 3 : Evolutionary Model

AMD-Vol. 141/DSC-Vol. 37, Dynamics of Flexible Multibody Systems:
Theory and Experiment
ASME 1992

FINITE ELEMENT ANALYSIS OF STRUCTURAL SYSTEMS USING SYMBOLIC PROCESS

Junghsen Lieh and Srinivas Tummarakota
Department of Mechanical and Materials Engineering
Wright State University
Dayton, Ohio

Abstract

The understanding of mechanics is important to the development of better design methodologies and control algorithms for structural multibody systems. The formulation of a complex mathematical model using manual approach is a very difficult task and prone to errors. It is nearly impossible when a system contains a large number of degrees of freedom. For nonlinear and/or time-varying systems, numerical formulation can only provide very limited information about physical insight. Symbolic equations of motion help understand the system behavior thus will assist in the design process and control synthesis. In this paper, it is intended to explore the use of symbolic method to generate finite element equations of motion for structural multibody systems. Lagrange's equation is expended to facilitate symbolic programming. Base motion effect is incorporated in the formulation allowing simulation of a structural system on a moving base or on a curved path.

Introduction

There has been an increasing need for developing computer-aided formalism for structural multibody systems. Formulation of dynamic equations usually does not present theoretical difficulties, but can be cumbersome and prone to errors. For a complex system, hand derivation is very difficult because it involves tremendous algebraic manipulation. The development of a computer-aided formalism requires it be simple, accurate and computationally attractive.

The efficiency of a multibody formalism varies depending on how the constraints are treated, the methods used and the coordinates chosen [1-9]. The constraints are generally from interaction of bodies in the form of kinematic loops, varying system topology or nonholonomic configurations. The use of Newton-Euler method, Lagrange's equations, Hamilton's principle or virtual work principle affects the formulation efficiency. Different coordinate systems, for example, Cartesian coordinates, generalized coordinates, moving coordinates and natural coordinates [10] yield different efficiency as well. Multibody formalisms that do not include base motion effect can produce erroneous results under certain conditions [11-12].

Design and analysis of flexible spacecraft and mechanisms, and control of structures were discussed in numerous publications [13-17]. As the materials become lighter and operating speeds become higher, rigid body analysis is no longer valid. Recently, increasing research in the dynamics and control of flexible manipulators and space structures has renewed the interest in these areas [18-29].

The goal of a computer-aided approach is to minimize the analyst's burden, to avoid errors, and to retain higher-order elastic modes and nonlinearities in the design process. The objective of order reduction is to express the equations of motion in terms of a minimum set of independent coordinates so that the redundant coordinates and singularity problems can be avoided. One may generate a very large set of algebraic differential equations (ADE) with sparse matrices and treat Lagrange multipliers as additional unknowns. It is possible to minimize the number of equations using generalized coordinate partitioning [30]. A functional with fictitious energies and a penalty factor as part of the Lagrangian may also be used [31]. A new separated-form virtual work formalism was recently evolved by Lieh [32] which separates formulation of each physical term.

Numerical method repeats formulation for every time step and each new set of parameters which leads to inefficiency in analysis of a multibody system. Symbolic method only requires to formulate once for a specific model. It provides explicit equations which can facilitate parameter identification. However, so far there were very few articles related to symbolic formulation of finite element equations of motion [33-35].

The symbolic method is used to generate the finite element equations of motion for structural multibody systems. In the current study, only planar case is considered. Both lateral and longitudinal deformations are included. Nonlinear strain is used in the element formulation.

Symbolic Finite Element Formulation for Structural Multibody Systems

The dynamics of a structural system may be described by partial differential equations with an infinite number of degrees of freedom or may be discretized with a finite number of generalized coordinates. The choice between these two approaches must be made with careful considerations, such as, the purpose of analysis and design, the requirement in accuracy, and the computational cost. Discretization of a continuous system can be carried out using the lumped-mass method [36], finite element method [37-38] or assumed-mode method.

The nodes for lumped-mass method are placed at the geometric centers of lumps by assuming the entire lump has the same properties. Assumed-mode method, which is essentially a truncated distributed method, uses several lower modes to approximate the transient response. In general, the mode shapes are transcendental functions comprising an infinite number of natural frequencies and normal modes that are influenced by boundary conditions, material density and geometry. Assumed-mode method may lead to a small number of degrees of freedom without significant loss of accuracy [39].

In using finite element method, a deformable body is divided into several small elements and each is considered as an elastic member. The elements are interconnected at certain points known as nodes or joints. Elastic deformation (transverse, torsional or longitudinal) of each element is expressed in terms of a series of geometrically dependent trial or shape functions and time dependent generalized coordinates.

Research on finite element analysis of structural systems appears in numerous articles. Finite element equations of motions were often derived from Lagrangian method [40-43], variational principle [44], Hamilton's principle [45], Galerkin method and Ritz method [46-48]. Other methods include the component mode synthesis technique and virtual work principle can also be found in the literature [49-51]. In this paper, Lagrangian method is pursued and modified so that it can be suitable for symbolic programming. Many symbolic languages including FORMAC [52], LISP [53], MACSYMA [54], MAPLE [55], muMATH [56], MATHEMATICA [57] and REDUCE [58] may be used for this purpose. MAPLE is adopted in this paper.

Coordinate Transformation

Figure 1 shows the nodes, elements, coordinate systems and displacement vectors of a structural system undergoing two dimensional deformations (lateral, angular and longitudinal). $\underline{R}_{ie}$ and $\underline{R}_i$ represent the absolute position vectors of node **i** and an arbitrary point on element **i**, respectively. $\underline{e}_{ri}$ and $\underline{e}_{\theta i}$ are the unit vectors along the tangential and transverse directions at node **i**.

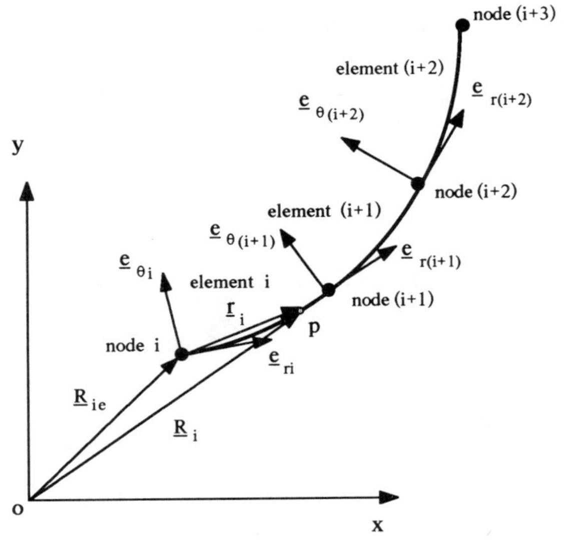

Figure 1. Definition of coordinate systems for finite element formulation.

The field variables at node **i** are represented by u_{ie}, θ_{ie} and v_{ie} as lateral, angular and longitudinal deformations, respectively. For convenience, these field variables are replaced by q_{i1}, q_{i2} and q_{i3}. The local deformations at any arbitrary point p on element **i** are defined as u_i and v_i in lateral and longitudinal directions, respectively (see Figure 2.). The transformation from $\mathbf{i}^{th}$ elemental coordinates to $(\mathbf{i+1})^{th}$ elemental coordinates (see Figure 3) is through a rotational matrix $\underline{T}_{(i+1),\,i}$ which is defined as

$$\begin{Bmatrix} \underline{e}_{r(i+1)} \\ \underline{e}_{\theta(i+1)} \\ \underline{e}_{k} \end{Bmatrix} = \underline{T}_{(i+1),\,i} \begin{Bmatrix} \underline{e}_{ri} \\ \underline{e}_{\theta i} \\ \underline{e}_{\kappa} \end{Bmatrix} \tag{1}$$

where

$$\underline{T}_{(i+1),\,i} = \begin{bmatrix} \cos(q_{(i+1),2}) & \sin(q_{(i+1),2}) & 0 \\ -\sin(q_{(i+1),2}) & \cos(q_{(i+1),2}) & 0 \\ 0 & 0 & 1 \end{bmatrix} \tag{2}$$

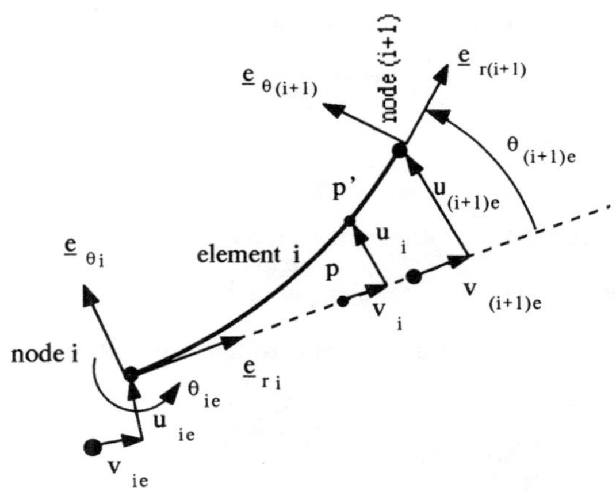

Figure 2. Deformation of element **i**.

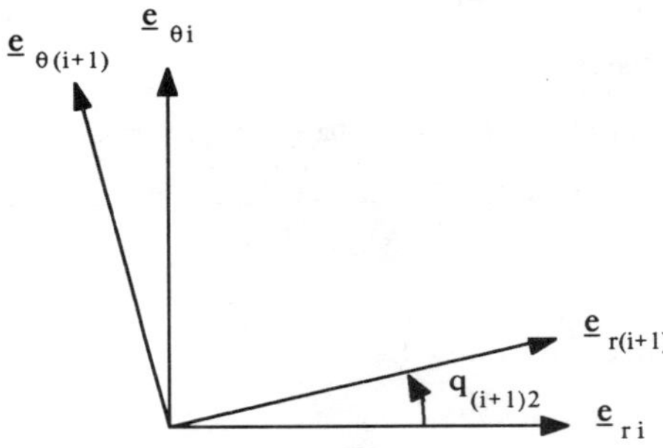

Figure 3. Transformation of coordinate systems for element **i**.

Local position vector at point p' (after deformation) with respect to node **i** is defined in terms of coordinate system of node **i** as

$${}^{i}\underline{r}_i = \begin{Bmatrix} r + v_i \\ u_i \\ 0 \end{Bmatrix} \tag{3}$$

At the elemental boundary, i.e., $r = L_i$ where L_i is the elemental length

$${}^{i}\underline{r}_{ie} = \begin{Bmatrix} L_i + q_{(i+1)3} \\ q_{(i+1)1} \\ 0 \end{Bmatrix} \tag{4}$$

The absolute position vector at point p' in terms of ith coordinate system becomes

$$ {}^{i}\underline{R}_i = {}^{i}\underline{R}_{ie} + {}^{i}\underline{r}_i \quad (5) $$

${}^{i}\underline{R}_{ie}$ is the absolute displacement of node **i** expressed in terms of ith local coordinate system and can be obtained through a series of coordinate transformations from previous nodal displacements and coordinates

$$ {}^{i}\underline{R}_{ie} = \prod_{n=o}^{i-1} \underline{T}_{n+1,n}\, {}^{o}\underline{r}_{oe} + \prod_{n=1}^{i-1} \underline{T}_{n+1,n}\, {}^{1}\underline{r}_{1e} + \prod_{n=2}^{i-1} \underline{T}_{n+1,n}\, {}^{2}\underline{r}_{2e} + \prod_{n=3}^{i-1} \underline{T}_{n+1,n}\, {}^{3}\underline{r}_{3e} + \cdots\cdots $$

$$ = \sum_{g=o}^{i-1} \left\{ \prod_{n=g}^{i-1} \underline{T}_{n+1,n}\, {}^{g}\underline{r}_{ge} \right\} \quad (6) $$

where ${}^{o}\underline{r}_{oe}$ is the displacement of base with respect to global coordinate system.

The absolute angular displacement of ith coordinate system is a summation of base angular motion, θ_o, and angular displacement at each node, i.e.,

$$ \theta_i = \theta_o + \sum_{g=1}^{i} q_{g2} \quad (7) $$

Please note $q_{g2} = \theta_{ge} = \dfrac{\partial u_g(r=0,t)}{\partial r}$ is the angular deformation at node **g**.

Elemental Mode Shapes

The lateral deformation of a beam corresponds to a fourth-order partial derivative of position r. The local deformation may be expressed as a function of mode shapes and generalized coordinates, i.e.,

$$ u_i(r,t) = \Phi_{u1}(r)\, u_{ie}(t) + \Phi_{\theta 1}(r)\, \theta_{ie}(t) + \Phi_{u2}(r)\, u_{(i+1)e}(t) + \Phi_{\theta 2}(r)\, \theta_{(i+1)e}(t) \quad (8) $$

where $\Phi_{u1}(r)$, $\Phi_{\theta 1}(r)$, $\Phi_{u2}(r)$, and $\Phi_{\theta 2}(r)$ are the shape functions. For convenience, $u_i(r, t)$ is rewritten as

$$ u_i(r,t) = \Phi_{u1}(r)\, q_{i1}(t) + \Phi_{\theta 1}(r)\, q_{i2}(t) + \Phi_{u2}(r)\, q_{(i+1)1}(t) + \Phi_{\theta 2}(r)\, q_{(i+1)2}(t) \quad (9) $$

where q_{i1} and $q_{(i+1)1}$ are the generalized coordinates corresponding to the lateral deformations, and q_{i2} and $q_{(i+1)2}$ are the generalized coordinates corresponding to the angular deformations at nodes **i** and **(i+1)**, respectively.

The elemental mode shapes for lateral deformation are often defined as cubic polynomials in terms of r such as

$$ \Phi(r) = c_o + c_1 r + c_2 r^2 + c_3 r^3 \quad (10) $$

By applying boundary conditions of element **i**, the following mode shapes are obtained

$$ \Phi_{u1}(r) = 1 - 3\frac{r^2}{L_i^2} + 2\frac{r^3}{L_i^3} \quad (11) $$

$$ \Phi_{\theta 1}(r) = r - 2\frac{r^2}{L_i} + \frac{r^3}{L_i^2} \quad (12) $$

$$ \Phi_{u2}(r) = 3\frac{r^2}{L_i^2} - 2\frac{r^3}{L_i^3} \quad (13) $$

$$ \Phi_{\theta 2}(r) = -\frac{r^2}{L_i} + \frac{r^3}{L_i^2} \quad (14) $$

The local longitudinal deformation is a second-order partial derivative of r and may be expressed as

$$ v_i(r,t) = \Phi_{v1}(r)\, q_{i3}(t) + \Phi_{v2}(r)\, q_{(i+1)3}(t) \quad (15) $$

where q_{i3} and $q_{(i+1)3}$ are the generalized coordinates corresponding to the longitudinal deformation at nodes **i** and **(i+1)**, respectively. Φ_{v1} and Φ_{v2} are the associated mode shapes. The mode shapes for longitudinal deformation can be obtained by using the same procedure mentioned above for lateral deformation.

$$ \Phi_{v1}(r) = 1 - \frac{r}{L_i} \quad (16) $$

$$ \Phi_{v2}(r) = \frac{r}{L_i} \quad (17) $$

Kinetic and Potential Energies

Absolute velocity of p' at element **i** is defined as

$$ {}^{i}\underline{V}_i = {}^{i}\dot{\underline{R}}_i = ({}^{i}\dot{\underline{R}}_i)_{rel} + \tilde{\underline{\Omega}}_i\, {}^{i}\underline{R}_i = \frac{\partial {}^{i}\underline{R}_i}{\partial \underline{q}} \dot{\underline{q}} + \frac{\partial {}^{i}\underline{R}_i}{\partial t} + \tilde{\underline{\Omega}}_i\, {}^{i}\underline{R}_i \quad (18) $$

where $\underline{q}$ is the vector of generalized coordinates and

$$\underline{\tilde{\Omega}}_i = \begin{bmatrix} 0 & -\dot{\theta}_i & 0 \\ \dot{\theta}_i & 0 & 0 \\ 0 & 0 & 0 \end{bmatrix} \tag{19}$$

The kinetic energy of element **i** is

$$KE_i = \frac{1}{2}\int_0^{L_i} \rho_i A_i \, {}^i\underline{V}_i^T \, {}^i\underline{V}_i \, dr \tag{20}$$

where ρ_i and A_i are the density and area of cross section of element **i**. Potential energy of element **i** contains the nonlinear strain energies due to lateral deformation and total stretch and is expressed as

$$PE_i = \frac{1}{2}\int_0^{L_i} \left\{ E_i I_{zi} u_{i,rr}^2 + E_i A_i \left(v_{i,r}^2 + v_{i,r} u_{i,r}^2 + \frac{1}{4} u_{i,r}^4 \right) \right\} dr \tag{21}$$

where E_i and I_{zi} are the Young's modulus and area moment of inertia, respectively.

Lagrange's Equations of motion

The elemental equations of motion may be set up individually and assembled later to form system equations of motion. In this paper, system equations of motion are obtained directly. The first step to formulate Lagrange's equations of motion is to obtain Lagrangian by subtracting potential energy from kinetic energy, i.e.,

$$L(\underline{q}, \dot{\underline{q}}, t) = \sum_{i=1}^{N_e} KE_i - \sum_{i=1}^{N_e} PE_i \tag{22}$$

where N_e is the number of elements in the system. The general form of Lagrange's equations of motion is written as

$$\frac{d}{dt}\left(\frac{\partial L}{\partial \dot{q}_{ij}}\right) - \frac{\partial L}{\partial q_{ij}} + \frac{\partial D}{\partial \dot{q}_{ij}} = Q_{ij} + \sum_{k=1}^{n_k} \lambda_k a_{kij} \, , \tag{23}$$

for $i = 1$ to N_n, $j = 1$ to N_d .

subject to constraints

$$\sum_{i=1}^{N_n} \sum_{j=1}^{N_d} a_{kij} \dot{q}_{ij} + a_{kt} = 0 \tag{24}$$

where N_n is the number of nodes, N_d is the number of degrees of freedom of each node ($N_d = 3$ for planar motion and 6 for three-dimensional motion), D is the dissipative function, Q_{ij} is the generalized force, λ_k is Lagrange multiplier corresponding to $\mathbf{k}^{th}$ constraint. Since Lagrangian L is a function of $\dot{\underline{q}}$, $\underline{q}$ and t, Eqn (23) can be expanded in the following form to facilitate symbolic programming, i.e.,

$$\frac{\partial}{\partial \dot{\underline{q}}}\left(\frac{\partial L}{\partial \dot{q}_{ij}}\right)\ddot{\underline{q}} = -\frac{\partial}{\partial \underline{q}}\left(\frac{\partial L}{\partial \dot{q}_{ij}}\right)\dot{\underline{q}} - \frac{\partial}{\partial t}\left(\frac{\partial L}{\partial \dot{q}_{ij}}\right) + \frac{\partial L}{\partial q_{ij}} - \frac{\partial D}{\partial \dot{q}_{ij}} + Q_{ij} + \sum_{k=1}^{n_k} \lambda_k a_{kij} \tag{25}$$

or

$$\sum_{a=1}^{N_n} \sum_{b=1}^{N_d} \frac{\partial}{\partial \dot{q}_{ab}}\left(\frac{\partial L}{\partial \dot{q}_{ij}}\right)\ddot{q}_{ab} = f_{ij}(\dot{\underline{q}}, \underline{q}, t) \tag{26}$$

where n_k is the number of constraints and

$$f_{ij}(\dot{\underline{q}}, \underline{q}, t) = -\sum_{a=1}^{N_n} \sum_{b=1}^{N_d} \frac{\partial}{\partial q_{ab}}\left(\frac{\partial L}{\partial \dot{q}_{ij}}\right)\dot{q}_{ab} - \frac{\partial}{\partial t}\left(\frac{\partial L}{\partial \dot{q}_{ij}}\right) + \frac{\partial L}{\partial q_{ij}} - \frac{\partial D}{\partial \dot{q}_{ij}} + Q_{ij} + \sum_{k=1}^{n_k} \lambda_k a_{kij} \tag{27}$$

To facilitate symbolic programming, new coordinates $x_m = q_{ij}$ are introduced, where $m = (i-1)*N_d + j$. Eqns (26 & 27) become

$$\underline{M}(\underline{x}, t)\,\ddot{\underline{x}} = \underline{f}(\dot{\underline{x}}, \underline{x}, t) \tag{28}$$

The entries of inertia matrix $\underline{M}$ and vector $\underline{f}$ may be expressed as follows:

$$M_{mn} = \frac{\partial^2 L}{\partial \dot{x}_n \partial \dot{x}_m} \, , \tag{29}$$

$$f_m = -\sum_{n=1}^{N_n*N_d} \frac{\partial}{\partial x_n}\left(\frac{\partial L}{\partial \dot{x}_m}\right)\dot{x}_n - \frac{\partial}{\partial t}\left(\frac{\partial L}{\partial \dot{x}_m}\right) + \frac{\partial L}{\partial x_m} - \frac{\partial D}{\partial \dot{x}_m} + Q_m + \sum_{k=1}^{n_k} \lambda_k a_{km} \, , \tag{30}$$

for $m = 1$ to $N_n * N_d$.

Based on the above procedure, a symbolic program is developed. The symbolic program generates finite element equations of motion in FORTRAN as the form of Eqn (28). Given boundary conditions, the known coordinates may be eliminated. Some of the MAPLE's symbolic commands such as **int** (integration), **diff** (differentiation), **multiply** (matrix/vector multiplication), **add** (matrix/vector addition), **array** (matrix/vector declaration), **sum** (summation), **fortran** (converting equations into FORTRAN format) etc. provide a very easy way to develop the program. The output FORTRAN equations of motion are ready to combine numerical differential equation solver for simulation and control synthesis. Figure 4 shows the flow chart of symbolic formulation of finite element equations of motion.

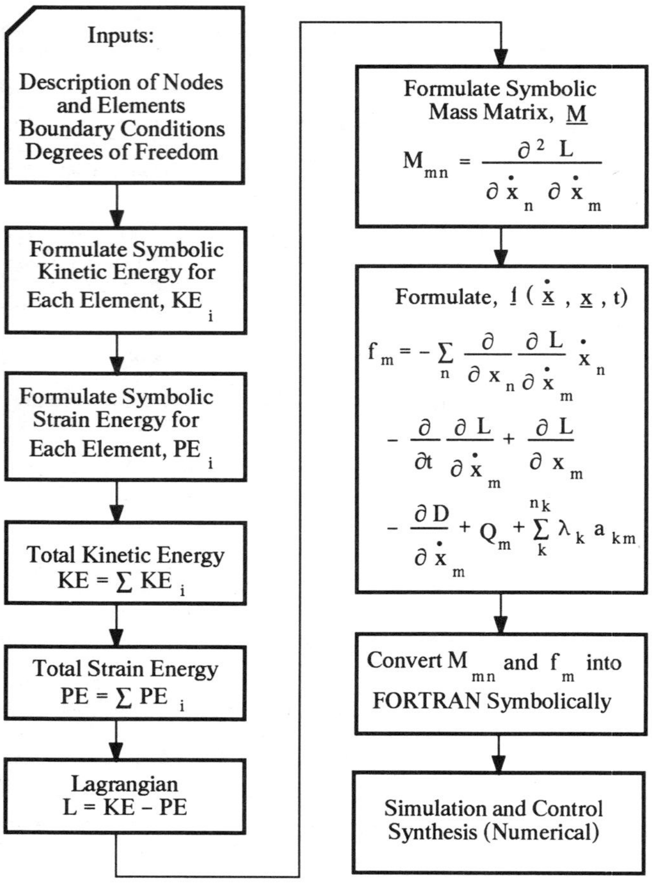

Figure 4. Flow chart of symbolic finite element equation formulation.

Example

The example used here is a uniform cantilever beam on a rotating base as shown in Figure 5. The base of the beam is rotating with an angular velocity $\theta_o(t)$. SYMFEM generates symbolic finite element equations in FORTRAN. The equations have been compared with cumbersome equations carefully derived by hand.

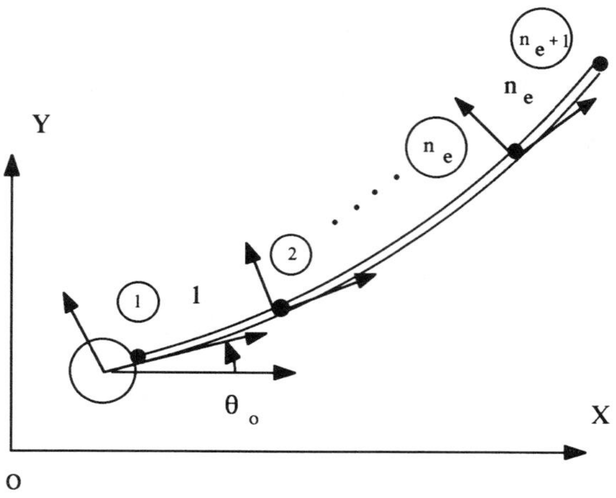

Figure 5. A cantilever beam on a rotating base.

Conclusion

This paper presents the use of symbolic method in formulating finite element equations of motion. The formulation differs from conventional finite element approach that only considers linear inertia, stiffness and generalized force terms. The method used here includes coupling terms that may have a significant influence on deformable behavior. Finite element equations of motion of a cantilever beam are generated symbolically and compared with hand-derived equations. Numerical results will be presented in the conference. Future work will be extended to include 3D deformation. Comparison with other methods such as the virtual work principle will also be addressed.

References

1. Magnus, K. Edited, Dynamics of Multibody Systems, IUTAM Symp., Munich, Germany, Aug. 29 - Sept. 3, Springer-Verlag, 1977.

2. Wittenburg, J., Dynamics of Systems of Rigid Bodies, Stuttgart, Germany, 1977.

3. Luh, J. Y. S. et al., "On-Line Computational Scheme for Mechanical Manipulators," J. Dynamic Systems, Measurement and Control, Vol. 102, 1980, pp. 69-76.

4. Kane, T. R. and Levinson, D. A., "Multibody Dynamics," J. Applied Mechanics, Vol. 50, 1983, pp. 1071-1078.

5. Book, W. J., "Recursive Lagrangian Dynamics of Flexible Manipulator Arms," Int. J. Robotics Research, Vol. 3, 1984, pp. 87-101.

6. Haug, E. J. Edited, Computer Aided Analysis and Optimization of Mechanical Systems Dynamics, NATO ASI Series F: Computer and Systems Sciences, Vol. 9, Springer-Verlag, 1984.

7. Bianchi, G. and Schiehlen, W. Edited, Dynamics of Multibody Systems, IUTAM/IFToMM Symp., Udine, Italy, Springer-Verlag, 1985.

8. Roberson, R. E. and Schwertassek, R., Dynamics of Multibody Systems, Springer-Verlag, 1988.

9. Shabana, A. A., Dynamics of Multibody Systems, John Wiley & Sons, 1989.

10. Garcia de Jalon, J., Unda and J., Avello, A., "Natural Coordinates for the Computer Analysis of Multibody Systems," Computer Methods in Applied Mechanics and Engineering, Vol. 56, 1986, pp. 309-327.

11. Kane, T. R., Ryan, R. R. and Banerjee, A. K., "Dynamics of a Cantilever Beam Attached to a Moving Base," J. Guidance and Control, Vol. 10, 1987, pp. 139-151.

12. Hanagud, S. and Sarkar, S., "Problem of the Dynamics of a Cantilever Beam Attached to a Moving Base," J. Guidance and Control, Vol. 12, 1989, pp. 438-441.

13. Winfrey, R. C., "Dynamic Analysis of Elastic Link Mechanisms by Reduction of Coordinates," J. Engineering for Industry, 1972, pp. 577-582.

14. Sadler, J. P. and Sandor, G. N., "Nonlinear Vibration Analysis of Elastic Four-Bar Linkages," J. Engineering for Industry, 1974, pp. 411-419.

15. Likins, P.W., "Analytical Dynamics and Nonrigid Spacecraft Simulation," NASA Technical Report 32-1593, 1974.

16. Dubowsky, S. and Maatuk, J., "The Dynamic Analysis of Elastic Spatial Mechanisms," I. Mech. E., 1975, pp. 927-932.

17. Book, W. J., Maizza-Neto, O. and Whitney, D. E., "Feedback Control of Two Beam, Two Joint Systems with Distributed Flexibility," J. Dynamics Systems, Measurement and Control, 1975, pp. 424-431.

18. Cannon, R. H. and Schmitz, E., "Initial Experimental on the End-Point Control of a Flexible One-Link Robot," Int. J. Robotics Research, Vol. 3, 1984, pp. 62-75.

19. Usoro, P. B., Nadira, R. and Mahil, S. S., "A Finite Element/Lagrange Approach to Modeling Lightweight Flexible Manipulators," J. Dynamic Systems, Measurement and Control, Vol. 108, 1986, pp. 198-205.

20. Hastings G. G. and Book, W. J., "A Linear Dynamic Model for Flexible Robotic Manipulators," IEEE Control Systems Magazine, 1987, pp. 61-64.

21. Low, K. H. and Vidyasagar, M., "A Lagrangian Formulation of the Dynamic Model for Flexible Manipulator Systems," J. Dynamic Systems, Measurement and Control, Vol. 110, 1988, pp. 175-181.

22. Huang, H. and Lee, C. S. G., "Generation of Newton-Euler Formulation of Dynamic Equations to Nonrigid Manipulators," J. Dynamic Systems, Measurement and Control, Vol. 110, 1988, pp. 308-315.

23. Yang, G.-B. and Donath, M., "Dynamic Model of a Two-Link Robot Manipulator with Both Structural and Joint Flexibility," Symp. on Robotics, ASME Winter Annual Meeting, 1988, pp. 37-44.

24. Naganathan, G. and Soni, A. H., "Nonlinear Modeling of Kinematic and Flexible Effects in Manipulator Design," J. Mechanisms, Transmissions, and Automation in Design, Vol. 110, 1988, pp. 243-254.

25. Changizi, K. and Shabana, A. A., "A Recursive Formulation for the Dynamic Analysis of Open Loop Deformable Multibody Systems," J. Applied Mechanics, 1988, pp. 687-693.

26. Modi, V.J. and Chan, J. K., "Performance of an Orbiting Flexible Mobile Manipulator," Proc. Flexible Mechanism, Robotics, and Robot Trajectories, ASME Design Technical Conf., DE-Vol. 24, 1990, pp. 375-384.

27. Kakizaki, T., Deck, J. F. and Dubowsky, S., "Modeling the Spatial Dynamics of Robotic Manipulators with Flexible Links and Joint Clearances," ASME Design Technical Conf., Proc. Flexible Mechanism, Robotics, and Robot Trajectories, DE-Vol. 24, 1990, pp. 343-350.

28. Simo, J. C. and Vu-Quoc, L., "On the Dynamics of Flexible Beams Under Large Overall Motions - The Plane Case: Part I & Part II," J. Applied Mechanics, Vol. 53, 1986, pp. 849-863.

29. Oh and Vanderploeg, "Modeling of a Two Arm Flexible Robot in Gravity," 1991 ASME Design Automation Conf., DE-V. 32-2, 523-530.

30. Wehage, R. A. and Haug, E. J., "Generalized Coordinate Partitioning for Dimension Reduction in Analysis of Constrained Dynamic Systems," J. Mechanical Design, Vol. 104, pp. 247-255, 1982.

31. Bayo, E. and Serna, M. A., "Penalty Formulations for the Dynamic Analysis of Elastic Mechanisms," J. Mechanisms, Transmissions, and Automation in Design, Vol. 111, 1989, pp. 312-326.

32. Lieh, J., "An Alternative Method to Formulate Closed-Form Dynamics for Elastic Mechanical Systems Using Symbolic Process," to appear in the Mechanics of Structures and Machines, Vol. 20, No. 2, June 1992.

33. Wang, P.S., "FINGER: A Symbolic System for Automatic Generation of Numerical Programs for Finite Element Analysis," Journal of Symbolic Computation, Vol. 2, 1986, pp. 305-316.

34. Barbier, C. et al., "Automatic Generation of Shape Functions for Finite Element Analysis Using Reduce," University of Newcastle, United Kingdom, 1990.

35. Sharma, N. and Wang, P. S., "Generating Finite Element Programs for Shared Memory Multiprocessors," Symbolic Computations and Their Impact on Mechanics, PVP-Vol. 205, 1990 ASME WInter Annual Meeting, pp. 63-79.

36. Sadler, J. P., "On the Analytical Lumped-Mass Model of an Elastic Four-Bar Mechanism," J. Engineering for Industry, 1975, pp. 561-565.

37. Zienkiewicz, O. C., The Finite Element Method, New York: McGraw-Hill, 1977.

38. Turcic, D. A. and Midha, A., "Dynamic Analysis of Elastic Mechanism Systems, Part I: Applications," J. Dynamic Systems, Measurement and Control, Vol. 106, 1983, pp. 249-254.

39. Hastings, G. G. and Book, W. J., "Verification of a Linear Dynamic Model for Flexible Robotic Manipulators," Proc. 1986 IEEE Int. Conf. on Robotics and Automation, pp. 1024-1029.

40. Midha, A., Erdman, A. G. and Frohrib, D. A., "Finite Element Approach to Mathematical Modeling of High-Speed Elastic Linkages," Mechanisms and Machine Theory, Vol. 13, 1978, pp. 603-618.

41. Sunada, W. and Dubowsky, S., "The Applications of Finite Element Methods to the Dynamic Analysis of Flexible Spatial and Co-Planar Linkage Systems," J. Mechanical Design, Vol. 103, 1981, pp. 643-651.

42. Nagaranjan, S. and Turcic, D, A., "Lagrangian Formulation of the Equations of Motion for Elastic Mechanisms with Mutual Dependence Between Rigid Body and Elastic Motions: Part I & II," J. Dynamic Systems, Measurement and Control, Vol. 112, 1990, pp. 203-224.

43. Weeën, F. V. D., "A Finite Element Approach to Three-Dimensional Kineto-Elastodynamics," Mechanisms and Machine Theory, Vol. 23, 1988, pp. 491-500.

44. Thompson, B. S. and Sung, C. K., "A Variational Formulation for the Nonlinear Finite Element Analysis of Flexible Linkages: Theory, Implementation, and Experimental Results," J. Mechanisms, Transmissions, and Automation in Design, Vol. 106, 1984, pp. 482-488.

45. Cavin III, R. K. and Dusto, A. R., "Hamilton's Principle: Finite-Element Methods and Flexible Body Dynamics," AIAA Journal, Vol. 15, 1977, pp. 1684-1690.

46. Stasa, F., Applied Finite Element Analysis for Engineers, CBS Pub., 1985.

47. Strang, G. and Fix G. J., An Analysis of the Finite Element Method, Englewood Cliffs, N.J.: Prentice-Hall.

48. Ortega, J. M. and Poole, W. G., Numerical Methods for Differential Equations, New York: John Wiley & Sons, 1984.

49. Smet, Liefooghe, Sas and Snoeys, "Dynamics Analysis of Flexible Structures Using Component Mode Synthesis," J. Applied Mechanics, Vol. 56, 1989, pp. 874-880.

50. Liou, F. W. and Erdman, A. G., "Analysis of a High-Speed Flexible Four-Bar Linkage: Part I-Formulation and Solution," J. Vibration, Acoustics, Stress, and Reliability in Design, Vol. 111, 1989, pp. 35-41.

51. Jonker, B., "A Finite Element Dynamic Analysis of Spatial Mechanisms with Flexible Links," Computer Methods in Applied Mechanics and Engineering, Vol. 76, 1989, pp. 17-40.

52. Bahr, K., SHARE FORMAC/FORMAC73, SHARE Program Library Agency, Triangle University Computation Center, Research Triangle Park, N.C., Program 360D-03.3.013, 1975.

53. Berk, A. A., LISP: The Language of Artificial Intelligence, Van Nostrand Reihold, 1985.

54. Rand, R. H., Computer Algebra in Applied Mathematics: An Introduction to MACSYMA, Pitman Advanced Pub., 1984.

55. Char, B. W., Geddes, K. O., Gonnet, G. H. and Watt, S. M.,, MAPLE User's Guide, 5th Ed., Waterloo, Canada: WATCOM Pub. 1988.

56. Wooff, C. and Hodgkinson, muMATH: A Microcomputer Algebra System, Academic Press, London, 1987.

57. Wolfram, S., Mathematica™: A System for Doing Mathematics by Computer, Reading, MA.: Addison-Wesley, 1988.

58. Rayna, G., REDUCE: Software for Algebraic Computation, Springer-Verlag, 1987.

AMD-Vol. 141/DSC-Vol. 37, Dynamics of Flexible Multibody Systems:
Theory and Experiment
ASME 1992

GEOMETRICALLY NONLINEAR COUPLING BETWEEN AXIAL AND FLEXURAL MODES OF DEFORMATION OF MULTIBODY SYSTEMS

J. Mayo and J. Domínguez
Department of Mechanical Engineering
University of Seville
Seville, Spain

Abstract

Linear elastic theory is not valid for some applications. A nonlinear elastic theory that accounts for small strains, stresses under the elastic limit of the material, but that allows for large relative displacements between points within bodies must be used. A formulation is presented that accounts for the geometrically nonlinear effects by retaining the terms in the strain-displacement relationship that couple the axial and flexural deformations. Two different approaches are presented. A assumed-modes method is developed for fully physical understanding of the geometrically nonlinear effects. The second approach is a finite-element one. The method is easily implemented in a standard flexible multibody dynamics code. However, some difficulties might be overcome when using component-mode synthesis to reduce the number of coordinates. A comparative study between the formulation developed in this work and other nonlinear formulations is presented and numerical results for a slider crank mechanism are shown.

1 Introduction

Linear elastic theory assumes that both displacement and strain in flexible components are infinitesimally small and the material is linearly elastic. There exists, however, a number of practical applications in which these assumptions are no longer valid. For instance, these assumptions are not applicable in the study of stability of elastic components. They are not valid when large elastic deflections exist, although strainsremain small, as in the case of components that are large in size. Futhermore, they shouldn't be applied to the case of bending deflection (even small) that is influenced by axial forces as in the case of high-speed rotating components such as helicopters or turbine blades. In these applications the nonlinear elastic theory should be used. The nonlinear elastic theory assumes that strains are small and the elastic limit of the material is not exceeded, but large relative displacement is permitted between points within bodies. This is achieved by retaining the terms in the strain-displacement equations that couple the axial and flexural deformations. *Geometrically nonlinear analysis* is the name which is often used in the literature for those techniques which employ higher-order theories.

Many efforts have been devoted to the geometrically nonlinear analysis of mechanical systems. There are several works which don't consider the axial deformations. They dealt with particular applications and showed that the first axial natural frequency is much higher than the first bending natural frequency. Therefore, they consider that the strain is due exclusively to the nonlinear term in the strain-displacement relationship. Viscomi and Ayre [1] used this method to study a slider-crank mechanism. Sadler and Sandor [2] obtained the response for a four-bar linkage. Yigit et al. [3] studied a radially rotating beam attached to a rigid body. They considered in the analysis the work done by the axial forces arising from centrifugal effects and called it centrifugal stiffening effect.

Turcic and Midha [4] , Bakr and Shabana [5] and Liou and Erdman [6] use a method very similar to the geometric stiffness matrix approach used in structural dynamics. It is perhaps the most common method found in the literature as it is very easily implemented in a finite-element code. It accounts for the effect of the axial force on the transversal displacement, but it doesn't allow for the shortening due to bending. In this analysis the longitudinal and transversal shape functions are decoupled.

The methodology by Wu and Haug [7] is an extension of the work by Song [8]. Components are divided into substructures, on each of which the theory of linear elasticity relative to a body reference frame is adequate to describe deformation and its coupling with system motion. Compatibility conditions are derived and imposed as constraints equations at boundary points between substructures.

The work of Kane et al. [9] presents a completely nonlinear formulation. The equations of motion are derived taking into account all the nonlinear terms that arise from considering the nonlinear contribution of the strain-displacement relationship. The longitudinal and transverse deformations aren't independent as considered in the linear theory, because this would be in direct conflict with the strain-displacement relationship, i.e., it would be in conflict with the physical fact that every transverse displacement of a point on the neutral axis of a beam gives rise to an axial displacement. The particular case of a cantilever beam attached to a moving base is studied.

In this paper a completely nonlinear formulation has been developed. The denomination of partially nonlinear formulation is used for that technique which accounts for the stiffening effect due to traction or the softening effect due to compresion, but neglects the foreshortening effect. The influence of this shortening due to the transverse deflection, in dynamics of multibody systems, is not only important in the longitudinal displacement but also affect the transverse deflection because of the coupling between longitudinal and transverse deformations through the finite rotation of the body. The approach developed in this paper follows the work done by Kane et al. [9], but its objective is to derive a formulation easily implemented in a standard finite-element multibody code. In order to achieve this goal, the equations are written in terms of the longitudinal and transverse deformations measured from the undeformed state. Kane et al. [9] used as longitudinal coordinate the one along the deformed neutral axis. So, in this formulation additional stiffness terms appear while in formulation presented in [9] additional inertia terms appeared.

2 Completely Nonlinear Formulation

Large deflections of components lead to nonlinearities in the stiffness terms of the equations of motion when these are expressed using the coordinates measured from the undeformed state. Two different methods have been approached. First, a classical method or assumed-mode method has been used in order to get a fully physical understanding of the phenomenon. As this method is not easily generalized in the case of deformable bodies with complex geometrical shapes, the completely nonlinear formulation has been derived using a finite-element approach. In what follows a two-dimensional beam formulation is assumed just for simplicity reasons, but the formulation is easily generalized to any other planar or spacial component.

In calculating the strain energy U for a beam in 2-D, the contributions from the shearing strains will be neglected. Thus, only the normal strains will be included, leading to the following expression

$$U = \int_V [\int_0^{\varepsilon_{xx}} (\sigma_x d\varepsilon_{xx})] dV \tag{1}$$

These strains, for large deflections on a beam but still small deformations can be written as

$$\begin{aligned} \varepsilon_{xx} &= \varepsilon^l_{xx} + \varepsilon^{nl}_{xx} \\ &= \frac{\partial u}{\partial x} + \frac{1}{2}\left(\frac{\partial v}{\partial x}\right)^2 \end{aligned} \tag{2}$$

The second term is neglected when the linear theory is assumed, but it will be shown later in this paper how important it can be for some applications.

Assuming that plane sections remain plane, the strain of a point out of the neutral axis can be written as [10]

$$\varepsilon_{xx} = \frac{\partial u_0}{\partial x} - \frac{\partial^2 v_0}{\partial x^2} y + \frac{1}{2}\left(\frac{\partial v_0}{\partial x}\right)^2 \tag{3}$$

where y is measured from the neutral axis of the beam and u_0 and v_0 denote the u and v displacements at $y = 0$.

Restricting the attention to the case of isotropic linear material,

$$\sigma_x = E\varepsilon_{xx} \tag{4}$$

where E is the Young's modulus of the material.

Introducing Eq. 4 into Eq. 1 and integrating, it is obtained that

$$U = \frac{1}{2}\int_0^L E(x)A(x)\varepsilon_{xx}^2 dx \tag{5}$$

in which A is the cross sectional area of the beam and L is the length of the beam.

Now, substituting Eq. 3 into Eq. 5 and neglecting the terms that have the form $\int y dA$ as it equals zero because y is measured from the neutral axis, one obtains that

$$\begin{aligned} U = & \frac{1}{2}\int_0^L EA\left(\frac{\partial u}{\partial x}\right)^2 dx + \frac{1}{2}\int_0^L EI\left(\frac{\partial^2 v}{\partial x^2}\right)^2 dx \\ + & \frac{1}{2}\int_0^L EA\frac{\partial u}{\partial x}\left(\frac{\partial v}{\partial x}\right)^2 dx + \frac{1}{2}\int_0^L \frac{EA}{4}\left(\frac{\partial v}{\partial x}\right)^4 dx \end{aligned} \tag{6}$$

where $I = \int_A y^2 dA$ is the second moment of area.

Note that the two last integrals of Eq. 6 are the contribution arising from considering the nonlinear strain components. They will be responsible for the coupling between axial and transverse deformations.

2.1 Assumed-mode Method

2.1.1 Formulation

The axial and transverse displacements are not independent as in the linear or partially nonlinear theory as they are related through the strain-displacement relationship (see Eq. 2). From this equation it can be seen that to be consistent the coupling between the longitudinal and transverse displacements must be considered.

It should be noticed at this stage that the shape functions are required to satisfy just geometric boundary conditions, but not natural ones, i.e., those arising from forces equilibrium [11]. As the Eq. 2, which couples the axial and transverse deflections, implies natural boundaries conditions, there is no need for the assumed displacements to satisfy them, and therefore, they can be independent, but the selection of the shape functions should be made trying to represent the real deformation as close as possible. In this approach, however, the coupling will be considered, but it won't in the finite-element approach.

When considering linear or partially nonlinear theories, the axial and transverse modes of deformation are decoupled. This is so because expression $\varepsilon_{xx} = \partial u/\partial x$ is considered, instead of the general expression of Eq. 2.

But when the completely nonlinear formulation is used it should be noticed that the longitudinal displacements arise from two sources, both from axial forces and from bending deflections.

The methodology employed to obtain the shape functions is discussed in [12]. The deformation shape is assumed for the transversal and longitudinal deflections but measured the last one in the deformed position. The reason for doing this is that these two deformations are independent.

$$\frac{s(\xi,t)}{L} = \boldsymbol{\varphi}^T(\xi)\boldsymbol{f}(t) \tag{7}$$

$$\frac{v(\xi,t)}{L} = \boldsymbol{\psi}^T(\xi)\boldsymbol{g}(t) \tag{8}$$

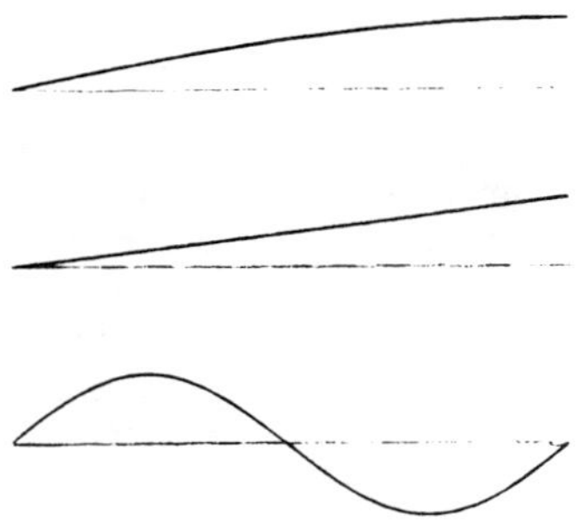

Figure 1: Axial shape functions for the assumed-modes approach

Then, using Eq. 2, it is easy to find the shape functions for the longitudinal deformation in the undeformed state

$$\frac{u(\xi,t)}{L} = [\boldsymbol{\varphi}^T(\xi) - \boldsymbol{\varphi}^T(0)]\boldsymbol{f}(t) - \frac{1}{2}\boldsymbol{g}^T \int_0^1 \boldsymbol{\psi}'\boldsymbol{\psi}'^T d\xi \boldsymbol{g} \tag{9}$$

In Figure 1 the combination of shape functions that will represent the longitudinal displacement in the case of a simply supported beam is shown, when only the first transverse mode is important. For this particular case:

$$\frac{u}{L} = \frac{2}{\pi}\sin\left(\frac{\pi}{2}\xi\right) f - \left(\frac{\pi}{2}\right)^2 \xi g^2 - \left(\frac{\pi}{2}\right)^2 \frac{\sin(2\pi\xi)}{2\pi} g^2 \tag{10}$$

$$\frac{v}{L} = \sin(\pi\xi)g \tag{11}$$

The first term in Eq. 10 is the same used in linear and partially nonlinear theories and it is multiplied by the time-dependent longitudinal coordinate. The two last term are the contribution from bending deflection and therefore are multiplied by the square of the time-dependent transversal coordinate.

Introducing the assumed modes shown in Eqs. 8 and 9 in the strain energy (see Eq. 6), this becomes exactly the same as in the linear theory, while the kinetic energy becomes more complicated. Therefore, by using this methodology in deriving the shape functions, the same approach found in [9] is reached.

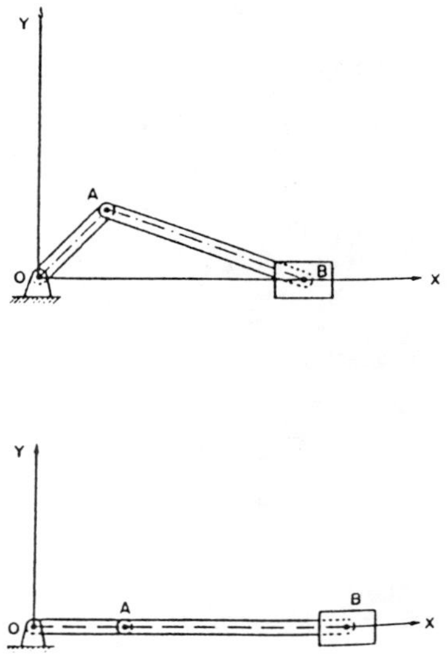

Figure 2: (a) Slider-crank mechanism. (b) Initial position

Different applications will need different shape functions. So, if both ends are fixed the first mode plot in Figure 1 won't have any contribution. On the other hand, if higher transverse modes are to contribute to the response of the system, then higher axial modes should be introduced in the analysis to be consistent with the formulation developed above.

2.1.2 Numerical Results

In order to examine the effects of including geometric nonlinearities on the dynamic response of multibody systems, the responses of the slider-crank mechanism shown in Figure 2 are presented in this section. Gravity is not considered in the analysis. The initial position for the mechanism corresponds to a zero crank angle and the initial deformations and velocities are assumed to be zero.

Figures 3–4 show the dimensionless vertical component of the midpoint displacement of the connecting rod versus the crank angle obtained through three different formulations: A linear formulation, which doesn't include any geometric nonlinearities in the equations of motion, a partially nonlinear formulation, where the geometric nonlinearities are included using the geometric stiffness matrix, and the completely nonlinear formulation developed in this work.

Data for the example are listed in Table 1, where a is the crank length, b is the connecting rod length, D is the diameter

Table 1: Slider crank mechanisms—dimensional and inertial properties

	a (m)	b (m)	D (m)	ω (rad/s)	E (Pa)	ρ (kg/m^3)	M_{sb}/M_{cr}
1	0.15	0.3	0.006	150	0.2×10^{12}	7870	0.5
2	0.15	0.3	0.006	150	0.2×10^{12}	7870	3

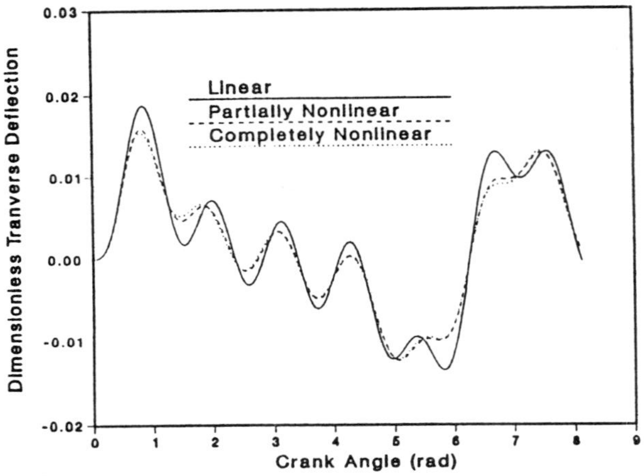

Figure 3: Dimensionless transverse deflection for $M_{sb}/M_{cr} = 0.5$

of the cross section of the beam, ω is the crank angular speed, E is the Young's modulus, ρ is the density of the material and M_{sb}/M_{cr} is the ratio of the mass of the slider block to the mass of the connecting rod.

Figure 3 shows the solution found for the first example using the three formulations. It can be seen that the two nonlinear responses are very similar and they also agree with the results presented in [5], where linear and partially nonlinear formulations were compared.

If axial force is increased by increasing the mass of the slider block, but always under the lowest Euler buckling load, the geometric nonlinearity effect become more important. This can be seen in the second example. Results are shown in Figure 4. It is clear from this figure that the elastic nonlinearities have significant effect on the behavior of the solution.

Figures 5–6 show the dimensionless longitudinal component of displacement at point B, for the last example data. Results obtained with the two nonlinear formulations are very different.

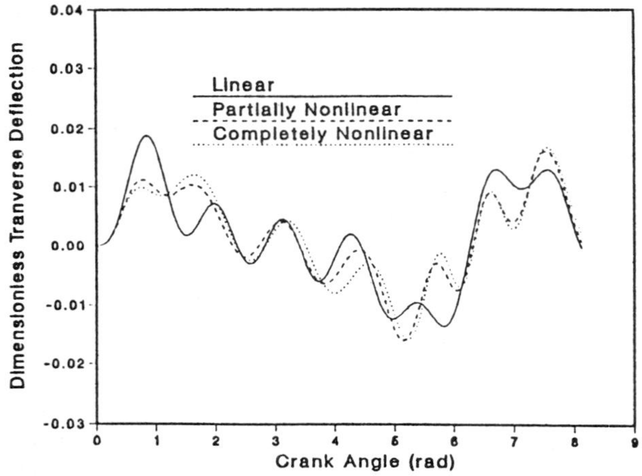

Figure 4: Dimensionless transverse deflection for $M_{sb}/M_{cr} = 3$

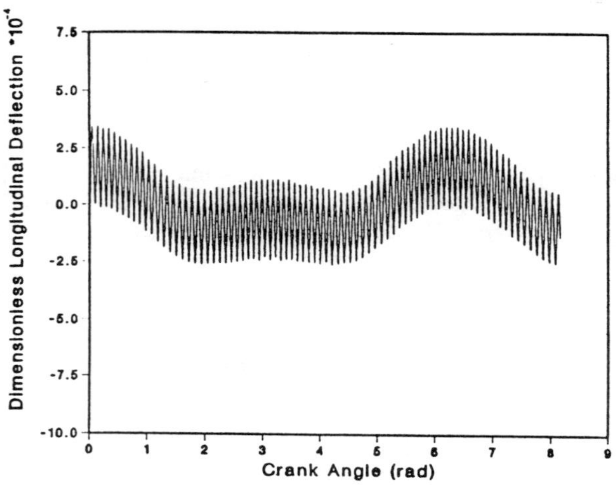

Figure 5: Dimensionless longitudinal deflection for $M_{sb}/M_{cr} = 3$ using the partially nonlinear formulation

The longitudinal response for the linear formulation and for the partially nonlinear formulation are identical, this is so because the partially nonlinear formulation is neglecting the foreshortening effect. The completely nonlinear formulation allows for this effect as can be seen in Figure 6. It should be noticed that the higher peaks correspond to the large transversal deflections shown in Figure 4. They are always negative because of the shortening effect due to transverse displacement. It should also be noticed that the highest peak for the longitudinal deflection for the completely nonlinear formulation is of the order of four times the highest peak for the partially nonlinear formulation. It has been shown, therefore, the importance of the component of the movement of point B due to the connecting rod transverse displacement.

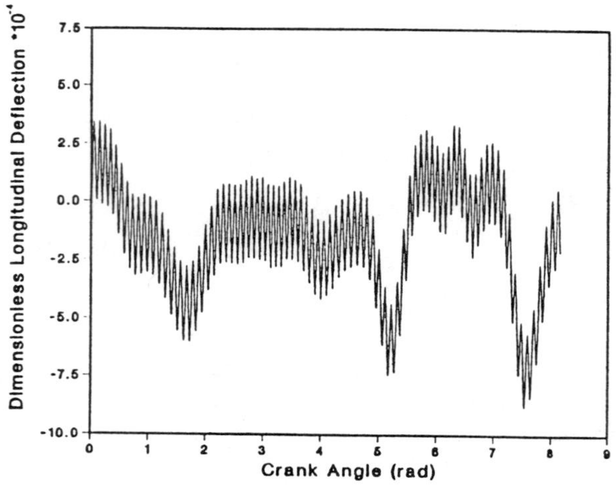

Figure 6: Dimensionless longitudinal deflection for $M_{sb}/M_{cr} = 3$ using the completely nonlinear formulation

2.2 Finite Element Method

2.2.1 Formulation

The finite element formulation employed in this work can be found in Ref. [13]. The finite element configuration is defined in a fixed global coordinate system. The general displacement of the finite element in a deformable body in the multibody system can be described by using a coupled set of body reference coordinates and element nodal elastic coordinates. In order to define a unique displacement field, the rigid body modes of the element shape functions have to be eliminated by using a set of reference conditions. These reference conditions must be consistent with the kinematic constraints imposed on the multibody system.

The total Lagrangian formulation leads to a nonlinear mass matrix which can be expressed in terms of a set of invariants that are dependent on the assumed displacement field of the finite element. They can be calculated just once at the beggining of the analysis.

By using the assumed displacements fields,

$$\frac{u(x,t)}{L} = \boldsymbol{\varphi}^T\left(\frac{x}{L}\right)\boldsymbol{f}(t) \tag{12}$$

$$\frac{v(x,t)}{L} = \boldsymbol{\psi}^T\left(\frac{x}{L}\right)\boldsymbol{g}(t) \tag{13}$$

the differential equations of motion of the system are derived using Lagrange's equations. This leads to

$$\boldsymbol{M}^i\ddot{\boldsymbol{q}}^i + \boldsymbol{K}^i\boldsymbol{q}^i = \boldsymbol{Q}_e^i + \boldsymbol{Q}_v^i + \boldsymbol{Q}_{ff}^i - \boldsymbol{C}_{\boldsymbol{q}^i}^T\boldsymbol{\lambda} \tag{14}$$

where $\boldsymbol{M}^i$ and $\boldsymbol{K}^i$ are, respectively, the symmetric mass and linear stiffness matrices of body i, $\boldsymbol{Q}_e^i$ is the vector of externally applied forces, $\boldsymbol{C}_{\boldsymbol{q}^i}$ is the constraint Jacobian matrix, $\boldsymbol{\lambda}$ is the vector of Lagrange multipliers, $\boldsymbol{Q}_v^i$ is the quadratic velocity vector that arises from differentiating the kinetic energy with respect to time and with respect to the generalized coordinates of body i, and $\boldsymbol{Q}_{ff}^i$ is the vector of nonlinear stiffness terms that arises from differentiating the non-quadratic strain energy with respect to the generalized coordinates of body i. Calculating the derivative of the total strain energy with respect to the time-dependent generalized coordinates, one obtains,

$$\left(\frac{\partial U}{\partial \boldsymbol{f}}\right)^T = L\int_0^1 EA\boldsymbol{\varphi}'\boldsymbol{\varphi}'^T d\xi \boldsymbol{f} + L\int_0^1 \frac{EA}{2}\boldsymbol{\varphi}'\boldsymbol{g}^T\boldsymbol{\psi}'\boldsymbol{\psi}'^T\boldsymbol{g}\,d\xi \tag{15}$$

$$\left(\frac{\partial U}{\partial \boldsymbol{g}}\right)^T = \frac{1}{L}\int_0^1 EI\boldsymbol{\psi}''\boldsymbol{\psi}''^T d\xi \boldsymbol{g} + L\int_0^1 EA\boldsymbol{\varphi}'^T\boldsymbol{f}\boldsymbol{\psi}'\boldsymbol{\psi}'^T d\xi \boldsymbol{g} + L\int_0^1 \frac{EA}{2}\boldsymbol{\psi}'\boldsymbol{\psi}'^T\boldsymbol{g}\boldsymbol{g}^T\boldsymbol{\psi}'\boldsymbol{\psi}'^T\boldsymbol{g}\,d\xi$$

The first vectorial equation correspond to the axial stiffness, while the second is for the transversal stiffness. It should be noticed that the first terms in each equation are the elements of the linear stiffness matrix, while the other three are coming from the nonlinear strain components, and, therefore, are the components of $\boldsymbol{Q}_{ff}^i$.

In order to reduce the number of coordinates, component mode synthesis techniques are employed. Using these techniques together with the completely nonlinear formulation is not straight foward as it should be discussed in the following section for the case of a slider crank mechanism.

2.2.2 Numerical Results

The numerical results that will be presented in this section have been obtained using the program DAMS (Dynamic Analysis of Multibody Systems) developed in the Department of Mechanical Engineering of the University of Illinois at Chicago.

The first slider-crank mechanism analyzed with the classical method will be studied here. The flexible connecting rod will be discretized using 2-D beam elements. The procedure of the numerical analysis of the mechanism can be divided into two distinct steps.

First, a finite element preprocessor is used to generate the linear stiffness matrix and the invariants of the mass matrix. The output of this finite element preprocessor also defines the mode shapes based on a set of boundary conditions selected for the analysis.

The connecting rod will be divided into four elements and simply supported ends reference conditions are selected ($u_A = v_A = 0, v_B = 0$).

The output results from the preprocessor finite-element simulation are introduced into the main processor dynamic analysis computer program, together with the rigid body parameters as well as the entire multibody system configuration. The nonlinear stiffness terms must be updated along the analysis. The transverse deflection of the connecting rod is used to study the response of the system as it was done in the classical approach.

The numerical integration has been performed with the following parameters: an output step size of $6 \times 10^{-5}s$, a tolerance parameter for the Newton-Raphson algorithm of 1×10^{-3}, and an absolute and relative errors of the numerical integration of values 1×10^{-3}.

The number of elastic coordinates are reduced using component mode techniques. Agrawal and Shabana [14] showed that two modes are sufficient to accurately represent the solution in the linear case. In fact, the major contribution arises from the fundamental mode. Bakr and Shabana [5] studied the partially nonlinear formulation and showed that at least six modes should be introduced in the analysis because the sixth mode is the first axial mode, therefore, using less than six modes yields a zero geometric stiffness matrix. To the authors' knowledge, a similar analysis hasn't been done before for the case of completely nonlinear formulation, so a first approach to the problem was performed using a six modes solution. The results are shown in Figure 7. The results for the linear and partially nonlinear cases agrees perfectly with those obtained in [14] and [5]. The solution for the completely nonlinear formulation shows high differences from the other two formulations. It is also very different from the one obtained using classical methods.

In using twelve modes instead of six, the linear and partially nonlinear solutions remain the same, as it should be expected, but the completely nonlinear solution changes dramatically, as shown in Figure 8. A comparison between the six and twelve modes completely nonlinear tranverse responses is presented in Figure 9. The higher frequency and the smaller amplitude of the six modes solution indicates a more rigid response. The differences in the response arise from the contribution of the

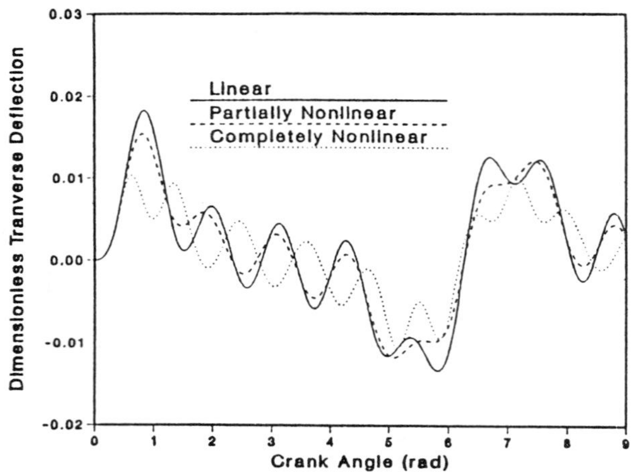

Figure 7: Six modes solution using finite-element approach

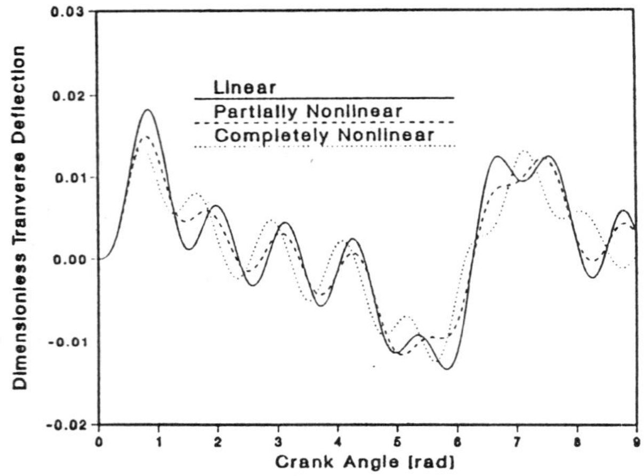

Figure 8: Twelve modes solution using finite-element approach

eleventh and twelfth modes. While when using six modes only the first axial mode was included, when using twelve modes four axial modes are used.

Several important conclusions can be drawn from these results.

> For the partially nonlinear formulation the contribution of only the first axial mode is important, as the solution doesn't change when using six or twelve modes of deformation. This is so because for the only nonlinear term in the partially nonlinear formulation what it is important is the axial displacement due just to the axial forces.

> For the completely nonlinear formulation, there exists a very *high coupling between axial and transverse deformations. The transverse response changes completely by adding higher frequency axial modes.* For the completely nonlinear formulation adding higher transverse modes doesn't improve the solution. The solution is improved by adding higher axial modes

> Not all the axial modes contribute to the solution. It could be shown that the second axial mode for the simply supported reference does not significantly contribute to the solution.

Figures 10–12 compares the responses obtained using the assumed-mode method and that obtained using the finite-element approach. They show that the linear and partially nonlinear responses agree but there exist some discrepancies for the completely nonlinear formulation. These differences are due to the fact that different axial modes are being used in the analysis. A plot of the axial modes shapes introduced in each analysis are shown in Figures 1 and 13. In the case studied in this work the longitudinal displacement is not constrained, therefore the foreshortening effect leads to a longitudinal displacement and not to an axial force. The longitudinal shape functions must be able to represent the shortening due to transverse deflection. Because of the high axial stiffness, a small error in the longitudinal displacement could yield a high error in the axial force and therefore, a wrong transversal response. When using linear functions to rep-

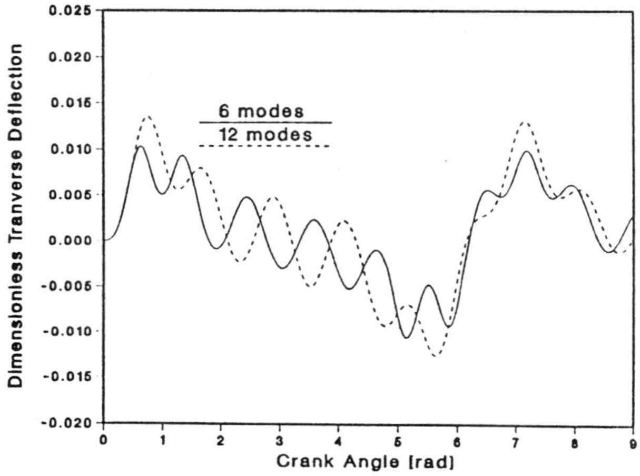

Figure 9: Comparison of the completely nonlinear response using 6 and 12 modes

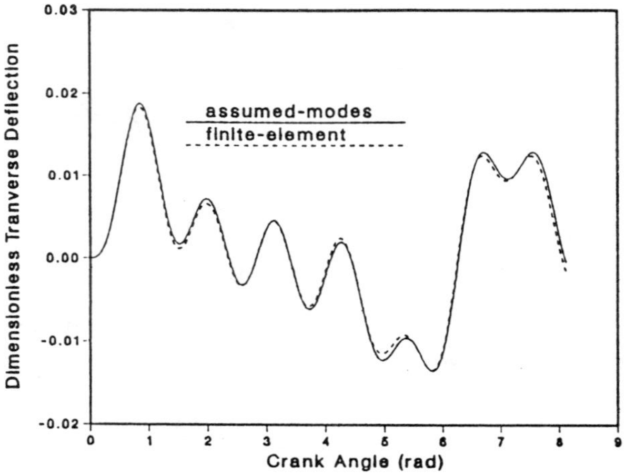

Figure 10: Comparison of the linear responses using classical and finite-element methods

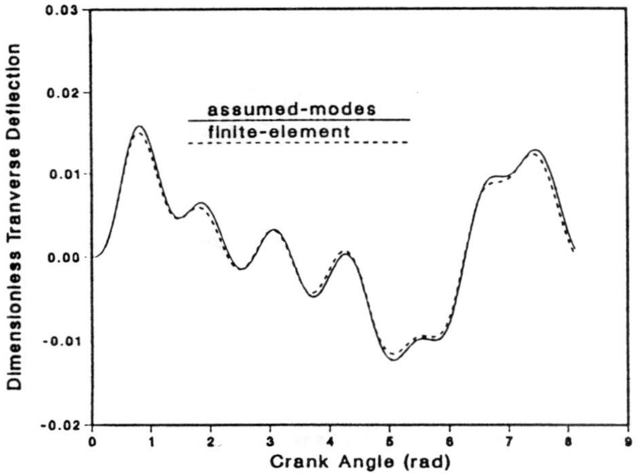

Figure 11: Comparison of the partially nonlinear responses using classical and finite-element methods

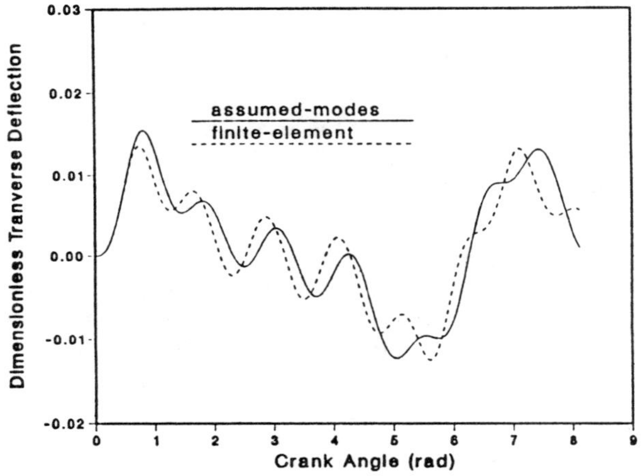

Figure 12: Comparison of the completely nonlinear responses using classical and finite-element methods

resent the longitudinal displacement, a high number of elements must be used in order to obtain an accurate transversal response.

Nonlinear terms become important when high axial forces are present in the analysis and/or if the components undergo large transverse deformations. The first applies to both partially and completely nonlinear formulations but the second only applies to completely nonlinear as the partially nonlinear neglects the shortening due to bending deflection of the beam. By changing the angular velocity of the crank shaft, the transverse deformations of the connecting rod can be controled as they are due mainly to the inertia forces. When the crank angular velocity is decreased to $75rad/s$, half the value used in the above figures, the transverse deformations are reduced considerably and therefore, a perfect agreement is found using the linear and nonlinear formulations. On the other hand, if the angular velocity of the crank is increased to $175rad/s$, the transverse deformations increase significantly. This value of the angular velocity is quite

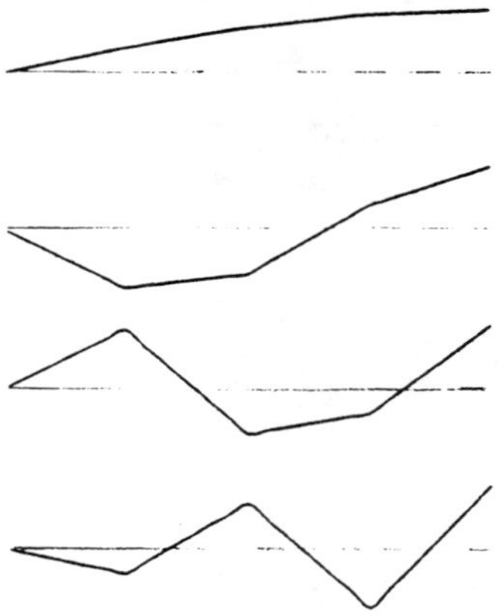

Figure 13: Axial modes for the simply-supported reference using a four-element discretization

close to the resonant frequency. Therefore, a small increment of the velocity leads to a high increment of the deformations. These results are shown in Figures 14 and 15.

From Figure 15, it can be observed that the discrepancies among the three analyse are higher than in the case of $w = 150rad/s$ as shown in Figure 8. It should be noticed that increasing the angular velocity of the crank not only increase the transverse deformations but also the axial forces, therefore the discrepancies between the linear and the partially nonlinear analyse also increase. Figure 16 shows the dimensionless longitudinal deflection for the case of $w = 175rad/s$ using the completely nonlinear formulation. It should be noticed the increasing importance of the contribution of the nonlinear term.

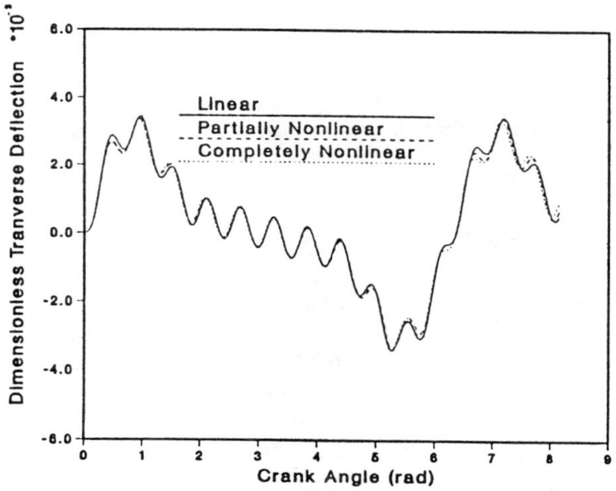

Figure 14: Dimensionless transverse deflection with w=75 rad/s

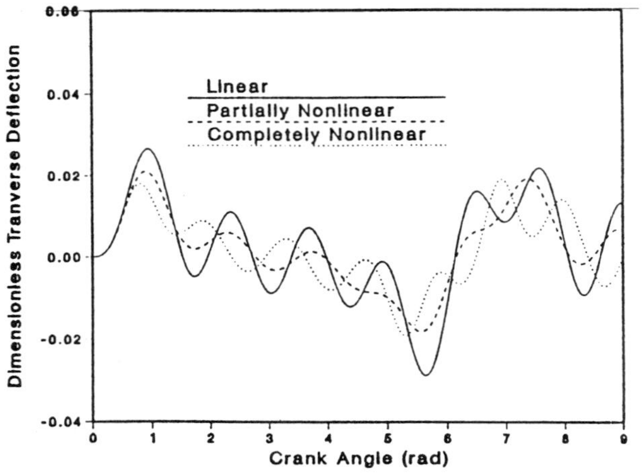

Figure 15: Dimensionless transverse deflection with w=175 rad/s

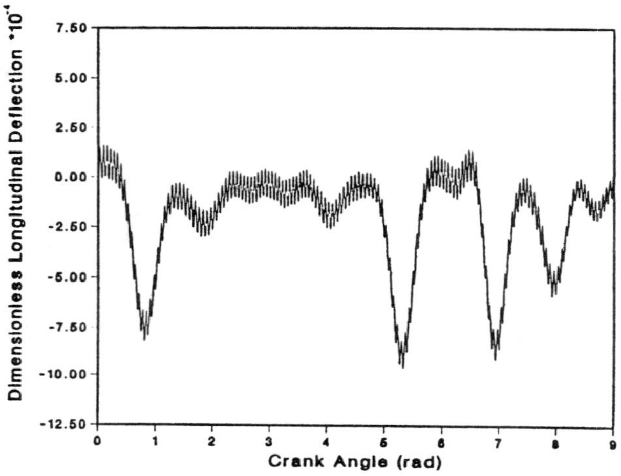

Figure 16: Dimensionless longitudinal deflection with w=175 rad/s

It is concluded that the two following facts validate the completely nonlinear formulation developed in this work.

Small transverse deformations and small axial forces lead to similar responses using the linear and nonlinear theories (see Figure 14).

Large transverse deformations yield very different results for the two nonlinear deformations. If small axial forces are present the partially nonlinear response will be similar to the linear solution, while if high axial forces exist the three formulations will lead to different solutions (see Figure 15).

3 Conclusions

A completely nonlinear formulation for the dynamics of multibody systems easily implemented in a finite-element code has been developed. Numerical results for a slider-crank mechanism with a flexible connecting rod have been analyzed yielding the main following conclusions:

In axial vibrations, the nonlinear higher-order stiffness terms may produce a superior contribution to the response than the linear terms.

The effect of tne nonlinear terms on transverse vibrations may be important when either significant axial forces or large transverse vibrations are present.

For the completely nonlinear formulation there exists a very high coupling between axial and transverse deformations. Adding higher transverse modes doesn't improve the solution, but the transverse response changes completely by adding higher frequency axial modes.

The solution shows a strong dependence not only of the number of axial modes introduced in the analysis but also of their shapes.

The major drawback with this formulation is the fact that in order to represent the axial deformation high-frequencies axial modes are needed. This can be both time-consuming and yield a stiff system of equations of motion.

Acknowledgement

The authors are very grateful to the University of Illinois at Chicago where all the finite element numerical results have been obtained using the the computer program DAMS and to Dr. A.A. Shabana for his invaluable advice and support.

References

[1] Viscomi, B.V. and Ayre, R.S., 'Nonlinear dynamic response of elastic slider-crank mechanism', *Journal of Engineering for Industry*, 1971, 251–262.

[2] Sadler, J.P. and Sandor G.N., 'Nonlinear vibration analysis of elastic four-bar linkages', *Journal of Engineering for Industry*, 1974, 411–419.

[3] Yigit, A., Scott, R.A. and Ulsoy A.G., 'Flexural motion of a radially rotating beam attached to a rigid body', *Journal of Sound & Vibration* **121(2)**, 1988, 201–210.

[4] Turcic, D.A. and Midha A., 'Dynamic analysis of elastic mechanism systems. Part I: Applications', *Journal of Dynamic Systems, Measurement, and Control* **106**, 1984, 249–254.

[5] Bakr, E.M. and Shabana, A.A., 'Geometrically nonlinear analysis of multibody systems', *Computers & Structures* **23(6)**, 1986, 739–751.

[6] Liou, F.W. and Erdman, A.G., 'Analysis of a high-speed flexible four-bar linkage: Part I — Formulation and solution', *Journal of Vibration, Acoustics, Stress, and Reliability in Design* **111**, 1989, 35–47.

[7] Wu, S.C. and Haug, E.J., 'Geometric non-linear substructuring for dynamics of flexible mechanical systems', *International Journal for Numerical Methods in Engineering* **26**, 1988, 2211–2226.

[8] Song, J.O., *Dynamic Analysis of Flexible Mechanisms*, PhD Dissertation. The University of Iowa, 1979.

[9] Kane, T.R., Ryan, R.R. and Banerjee A.K., 'Dynamics of a cantilever beam attached to a moving base', *Journal of Guidance and Control* **10(2)**, 1987, 13

[10] Przemieniecki, J.S., 'Nonlinear structural analysis', *Theory of Matrix Structural Analysis*, McGraw Hill, 1968, 383–409.

[11] Meirovitch, L., *Analytical Methods in Vibration*, F. Landis (Ed.), The MacMillan Company, New York, 1967, 308–313.

[12] Mayo, J.M., Dominguez, J. and Garcia-Lomas J., 'Continuous modelling of flexible mechanisms: Geometrically nonlinear analysis', *Proceedings of the Eighth World Congress on Theory of Machines and Mechanisms* **6**, 1991, (to be published).

[13] Shabana, A.A., *Dynamics of Multibody Systems*, Wiley, New York, 1989.

[14] Agrawal, O.P. and Shabana, A.A., 'Dynamic analysis of multibody systems using component modes', *Computers & Structures* **21(6)**, 1985, 1303–1312.

AMD-Vol. 141/DSC-Vol. 37, Dynamics of Flexible Multibody Systems:
Theory and Experiment
ASME 1992

MODELING AND CONTROL OF FLEXIBLE MANIPULATORS: PART I – DYNAMIC ANALYSIS AND CHARACTERIZATION

Zhijie Xia and Chia-Hsiang Menq
Department of Mechanical Engineering
Ohio State University
Columbus, Ohio

ABSTRACT

The objective of this paper is to develop a methodology that can be used to accurately predict the dynamic response of a flexible manipulator. Based on the finite element model, a model reduction technique is proposed in this paper. In the proposed approach, the elastic mode shape functions are determined by taking into account of both the rigid motion and the elastic deformation as well as the coupling effect among individual links. Since the rigid motion has strong influence on the mode shape functions, it is found that the mode shape vectors are functions of link configurations. In this paper, using the mode shape functions of the manipulator at different link configurations along with an interpolation approach, an explicit expression of the mode shape functions in terms of the manipulator's configuration can be synthesized. In this model reduction technique, only the dominate modes of the system remain, and the time variant nature of their mode shape functions are reserved. Consequently, the required computation for dynamic analysis is greatly reduced while the accuracy of the system response is maintained. A flexible two-link manipulator is used to illustrate the proposed approach. Using the results of the flexible two-link manipulator, the accuracy and application range of the various assumptions, that were previously used for modeling the dynamics of a flexible manipulator, are examined and the dynamic response of flexible manipulators is characterized.

1. Introduction

In this paper, based on the finite element model, the dynamic analysis and characterization of flexible multi-link manipulators is attempted. There are three main difficulties in the dynamic analysis of a flexible manipulator: (1) the system parameters are the function of the system configuration and thus are time varying; (2) the system dynamic equations are strong nonlinear and the rigid and elastic motions are coupled together; (3) the finite element model consists of high frequency elastic modes. While these high frequency elastic modes do not significantly contribute to the system response, the computational demands required by them are so heavy that the numerical solution of the system dynamic equations is very difficult, if not impossible. For these reasons, most of the previous work, which used the finite element approach and considered the mutual dependence between the rigid and the elastic motions, failed to give numerical results (Nagarajan and Turcic, 1990a, Nagarajan and Turcic, 1990b). Consequently, neither the accuracy and application range of the various assumptions, that were previously used for modeling the dynamics of a flexible manipulator, be examined, nor the dynamic response of flexible manipulators be characterized.

To avoid these problems in solving the system dynamic equations, two approaches have been reported in the literature. The first one (Schmitz, 1986, Jonker, 1990) is to linearize the nonlinear dynamic equations of the system and therefore the nonlinear coupling can be avoided. The shortcoming of this approach is that the linearized dynamic equations may not be able to accurately represent the actual dynamics of the system. The second approach (Oakley and Cannon, 1988, Oakley and Cannon, 1989) is to use the assumed mode method instead of the finite element method in deriving the system dynamic equations. Although this approach can avoid the problem induced by the high frequency elastic modes in the finite element model, the modeling flexibility is limited and the determination of the assumed mode shapes is subjective. Moreover, since the rigid motion has strong influence on the mode shape functions, it is evident that for high speed applications, the mode shape vectors are not constant and should be functions of link configurations.

The objective of this paper is to develop a methodology that can be used to accurately predict the dynamic response of a flexible manipulator based on the finite element model. A model reduction technique is proposed. In the proposed approach, the elastic mode shape functions are determined by taking into account of both the rigid motion and the elastic deformation as well as the coupling effect among individual links. Since the rigid motion has strong influence on the mode shape functions, it is found that the mode shape vectors are functions of link configurations. In this paper, using the mode shape functions of the manipulator at different link configurations along with an interpolation approach, an explicit expression of the mode shape functions in terms of the manipulator's configuration can be synthesized. In this model reduction technique, only the dominate modes of the system remain, and their time variant nature of the mode shape functions are reserved. Consequently, the required computation for dynamic analysis is greatly reduced while the accuracy of the system response is maintained. A flexible two-link manipulator is used to illustrate the proposed approach. Using the results of the flexible two-link manipulator, the accuracy and application range of the various assumptions, that were previously used for modeling the dynamics of a flexible manipulator, are examined and the dynamic response of flexible manipulators is characterized.

2. Dynamic Equations for a Flexible Two-link Manipulator

A flexible two-link manipulator is used in this paper to illustrate the proposed approach for the dynamic analysis of flexible manipulators. The schematic diagram of the flexible two-link manipulator is depicted in Figure 1 and the geometric dimensions of the two links are shown in Figure 2 and Figure 3. The material property of the system is given in Table 1. In view of the fact that the motion of the flexible two-link manipulator under study is on the same plane, the elastic deformation perpendicular to that plane can be neglected. The flexible manipulator is driven by two actuators on the hubs of individual links and no other external forces acting on it.

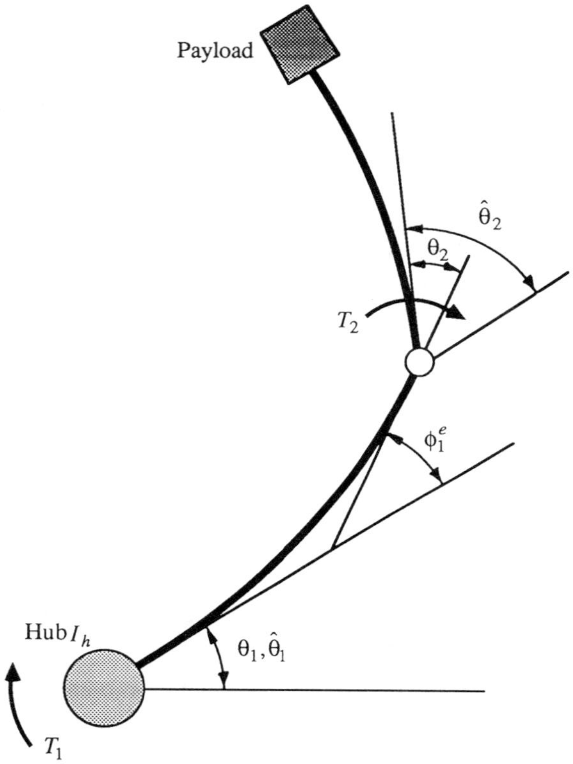

Figure 1 Schematic Diagram of a Flexible Two-Link Manipulator

Table 1 Material Property of the Flexible Links

Material	6061-T6 Aluminum Alloy
Young's Modules	10.3×10^6 psi
Unit Weight	0.098 lb/inch^3

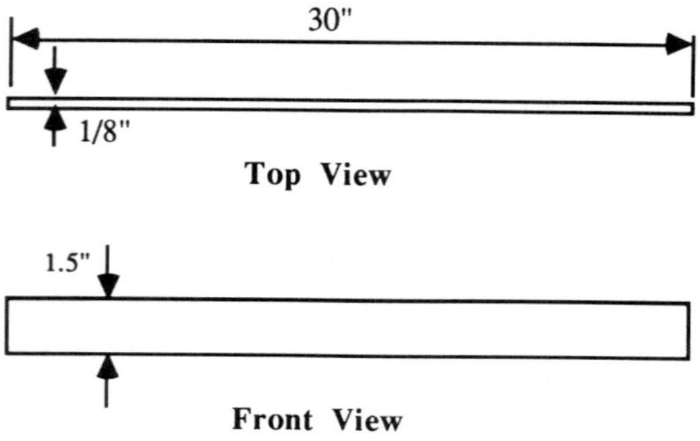

Figure 2 Geometric Dimension of Link 1

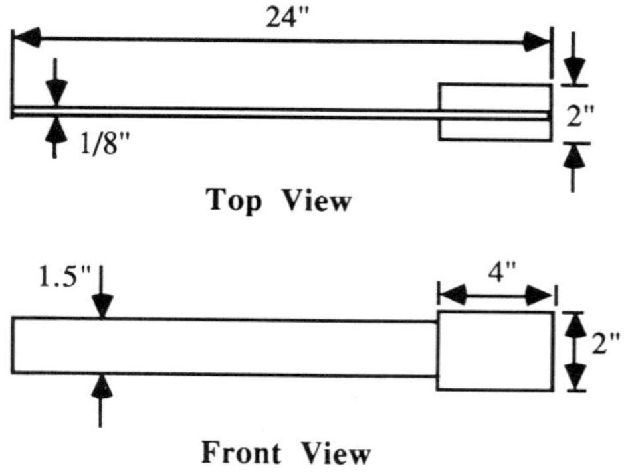

Figure 3 Geometric Dimension of Link 2

Based on the link model as shown in Figures 2 and 3, a beam element with six degrees of freedom is chosen here. The elastic coordinates of the e-th element of the i-th link are expressed as follows,

$$\{^i u\}_e = (u_1 \quad v_1 \quad \phi_1 \quad u_2 \quad v_2 \quad \phi_2)^T \tag{1}$$

where u_1 and u_2 are the respective elastic deformations along the longitudinal direction of the two nodes, v_1 and v_2 are the respective elastic deformations in the vertical direction, and ϕ_1 and ϕ_2 are the respective slopes of the elastic deformation in the vertical direction. The element shape function N is chosen as,

$$N = \begin{bmatrix} N_{u1} & 0 & 0 & N_{u2} & 0 & 0 \\ 0 & N_{v1} & N_{v2} & 0 & N_{v3} & N_{v4} \end{bmatrix} \tag{2}$$

where,

$$N_{u1} = 1 - \frac{\xi}{l_n} \qquad N_{u2} = \frac{\xi}{l_n}$$

$$N_{v1} = 1 - \frac{3\xi^2}{l_n^2} + \frac{2\xi^3}{l_n^3} \qquad N_{v2} = \xi - \frac{2\xi^2}{l_n} + \frac{\xi^3}{l_n^2}$$

$$N_{v3} = \frac{3\xi^2}{l_n^2} - \frac{2\xi^3}{l_n^3} \qquad N_{v4} = -\frac{\xi^2}{l_n} + \frac{\xi^3}{l_n^2}$$

and ξ is the local coordinate of the element, l_n is the length of the element.

For the flexible two-link manipulator as shown in Figure 1, each link is mounted to its hub. Due to the fact that the dimension of the cross section of the hub is much greater than that of the flexible link, the deformation of the hub is very small and can be considered as rigid. The deformation of the flexible link at the nodes being attached to the hub can also be neglected. Therefore, the boundary conditions of the flexible two-link manipulator will be,

$$u_0{}^i = 0 \qquad i = 1,2 \tag{3}$$

$$v_0{}^i = 0 \qquad i = 1,2 \tag{4}$$

$$\phi_0{}^i = 0 \qquad i = 1,2 \tag{5}$$

where the superscript i denotes the i-th link of the manipulator and the subscript 0 denotes the respective nodes of the two links being attached to the hubs. By using the equations of motion for a general flexible manipulator (Xia, 1992) and the given boundary conditions, the equations of motion for a flexible two link manipulator can be obtained. The equations of motion for coordinate angles can be written as,

$$\begin{aligned}&\left\{M_\theta^{11}\ddot{\hat\theta}_1 + M_\theta^{21}\ddot{\hat\theta}_2 + [M_{ru}^1]\{\ddot q\}\right\} + c_1\dot{\hat\theta}_1 + \left\{(I_{rr}^{121} + I_{rr}^{211})\dot{\hat\theta}_1\dot{\hat\theta}_2\right.\\ &\left.+ I_{rr}^{221}\dot{\hat\theta}_2{}^2\right\} + \left\{[I_{\theta u}^{12}]\{q\}\dot{\hat\theta}_1\dot{\hat\theta}_2 + [I_{\theta u}^{221}]\{q\}\dot{\hat\theta}_2{}^2 + [I_{\theta uD}^{11}]\{\dot q\}\dot{\hat\theta}_1\right.\\ &\left.+ [I_{\theta uD}^{21}]\{\dot q\}\dot{\hat\theta}_2\right\} - \left\{[G_u^1]\{q\} + G_\theta^1\right\} = T_1\end{aligned} \tag{6}$$

$$\begin{aligned}&\left\{M_\theta^{12}\ddot{\hat\theta}_1 + M_\theta^{22}\ddot{\hat\theta}_2 + [M_{ru}^2]\{\ddot q\}\right\} + \left\{c_2\dot{\hat\theta}_2 - c_2\dot\phi_1^e\right\} + I_{rr}^{112}\dot{\hat\theta}_1{}^2\\ &+ \left\{[I_{\theta u}^{112}]\{q\}\dot{\hat\theta}_1{}^2 + [I_{\theta uD}^{12}]\{\dot q\}\dot{\hat\theta}_2 + [I_{\theta uD}^{22}]\{\dot q\}\dot{\hat\theta}_2\right\}\\ &- \left\{[G_u^2]\{q\} + G_\theta^2\right\} = T_2\end{aligned} \tag{7}$$

where $\hat\theta_1$ and $\hat\theta_2$ are the coordinate angles of link 1 and link 2 respectively, and,

$$\hat\theta_1 = \theta_1 \tag{8}$$

$$\hat\theta_2 = \theta_2 + \phi_1^e \tag{9}$$

where θ_1 and θ_2 are the encoder readings, ϕ_1^e is the slope of the elastic deformation at the tip of the first link, and $\{q\}$ is the system's elastic coordinates. In equations (6) and (7), the three terms in the first brace are the inertial force. The term $c_1\dot{\hat\theta}_1$ of equation (6) and the two terms in the second brace of equation (7) are the damping force. The two terms in the second brace of equation (6) and the term $I_{rr}^{112}\dot{\hat\theta}_1{}^2$ of equation (7) are the force induced by the coupling of the rigid motion, in which $(I_{rr}^{121} + I_{rr}^{211})\dot{\hat\theta}_1\dot{\hat\theta}_2$ is the Coriolis force and $I_{rr}^{221}\dot{\hat\theta}_2{}^2$ and $I_{rr}^{112}\dot{\hat\theta}_1{}^2$ are the centrifugal force. The four terms in the third brace of equation (6) and the three terms in the third brace of equation (7) are the forces induced by the coupling between the rigid motion and the elastic motion. The two terms in the fourth brace of equations (6) and (7) are the effects of gravity force on the rigid motion.

The equations of motion for elastic freedoms can be expressed as,

$$\begin{aligned}&\left\{\{M_{\theta\theta}^1\}\ddot{\hat\theta}_1 + \{M_{\theta\theta}^2\}\ddot{\hat\theta}_2 + [M]\{\ddot q\}\right\} + \left\{[C]\{\dot q\} - c_2(\dot{\hat\theta}_2 - \dot\phi_1^e)[B_1^\phi]^T\right\}\\ &+ [K]\{q\} + \left\{\{I_\theta^{11}\}\dot{\hat\theta}_1{}^2 + (\{I_\theta^{12}\} + \{I_\theta^{21}\})\dot{\hat\theta}_1\dot{\hat\theta}_2 + \{I_\theta^{22}\}\dot{\hat\theta}_2{}^2\right\} + \left\{\left([I_{u\theta}^{11}]\dot{\hat\theta}_1{}^2\right.\right.\\ &\left.\left.+ ([I_{u\theta}^{12}] + [I_{u\theta}^{21}])\dot{\hat\theta}_1\dot{\hat\theta}_2 + [I_{u\theta}^{22}]\dot{\hat\theta}_2{}^2 + [M_{u\theta}^1]\ddot{\hat\theta}_1 + [M_{u\theta}^2]\ddot{\hat\theta}_2\right)\{q\}\right\}\\ &+ \left\{[I_{uD\theta D}^1]\dot{\hat\theta}_1\{\dot q\} + [I_{uD\theta D}^2]\dot{\hat\theta}_2\{\dot q\}\right\} - \{G\} = \{F^e\}\end{aligned} \tag{10}$$

In equation (10), the three terms in the first brace are the inertial force, the two terms in the second brace are the damping force, the term $[K]\{q\}$ is the elastic force induced by the structural stiffness, the three terms in the third brace are the force induced by the coupling of the rigid motions, the five terms in the fourth brace are the elastic force induced by the rigid motion, the two terms in the fifth brace are the force induced by the coupling of the rigid motion and the elastic deformation rate, the matrix $\{G\}$ is the effects of the gravity force on the elastic motion, and the matrix $\{F^e\}$ is the projection of the external force on the flexible freedoms.

For the applications of a flexible manipulator considered in this paper, the flexible link will rotate at an angular rate, which is much less than its first elastic frequency. For these applications (Cyril etc., 1989), the axial stiffening effect can be neglected and,

$$\begin{aligned}&\left([I_{u\theta}^{11}]\dot{\hat\theta}_1{}^2 + ([I_{u\theta}^{12}] + [I_{u\theta}^{21}])\dot{\hat\theta}_1\dot{\hat\theta}_2 + [I_{u\theta}^{22}]\dot{\hat\theta}_2{}^2 + [M_{u\theta}^1]\ddot{\hat\theta}_1\right.\\ &\left.+ [M_{u\theta}^2]\ddot{\hat\theta}_2\right)\{q\} << [K]\{q\}\end{aligned} \tag{11}$$

i.e., the elastic force induced by the rigid motion is much less than the elastic force induced by the structural stiffness. Therefore the terms in the third brace of equation (10) can be neglected and the equation of motion for the elastic freedoms as shown in equation (10) can be rewritten as,

$$\begin{aligned}&\left\{\{M_{\theta\theta}^1\}\ddot{\hat\theta}_1 + \{M_{\theta\theta}^2\}\ddot{\hat\theta}_2 + [M]\{\ddot q\}\right\} + \left\{[C]\{\dot q\} - c_2(\dot{\hat\theta}_2 - \dot\phi_1^e)[B_1^\phi]^T\right\}\\ &+ [K]\{q\} + \left\{\{I_\theta^{11}\}\dot{\hat\theta}_1{}^2 + (\{I_\theta^{12}\} + \{I_\theta^{21}\})\dot{\hat\theta}_1\dot{\hat\theta}_2 + \{I_\theta^{22}\}\dot{\hat\theta}_2{}^2\right\}\\ &+ \left\{[I_{uD\theta D}^1]\dot{\hat\theta}_1\{\dot q\} + [I_{uD\theta D}^2]\dot{\hat\theta}_2\{\dot q\}\right\} - \{G\} = \{F^e\}\end{aligned} \tag{12}$$

Equations (6), (7) and (12) can be combined together and the equation of motion of the manipulator can be expressed in a matrix form as follows,

$$[FM]\ddot X + [FC]\dot X + [FK]X + NL(\dot{\hat\theta}_1, \dot{\hat\theta}_2, \{q\}, \{\dot q\}) - \{FG\} = \{F\} \tag{13}$$

where $X = (\hat\theta_1 \quad \hat\theta_2 \quad \{q\})^T$,

$$[FM] = \begin{bmatrix} M_\theta^{11} & M_\theta^{21} & [M_{ru}^1] \\ M_\theta^{12} & M_\theta^{22} & [M_{ru}^2] \\ \{M_{\theta\theta}^1\} & \{M_{\theta\theta}^{22}\} & [M] \end{bmatrix}$$

$$[FC] = \begin{bmatrix} c_1 & 0 & 0_{1n} \\ 0 & c_2 & [c_\phi] \\ 0_{n1} & -c_2[B_1^\phi]^T & [C] + c_2[B_1^\phi]^T[B_1^\phi] \end{bmatrix}$$

where $[c_\phi] = [0 \ \ldots \ -1 \ 0 \ \ldots \ 0]$, n is the number of the elastic freedoms of the system after applying the boundary conditions.

$$[FK] = \begin{bmatrix} 0 & 0 & 0_{1n} \\ 0 & 0 & 0_{1n} \\ 0_{n1} & 0_{n1} & [K] \end{bmatrix}$$

$$NL(\dot{\hat\theta}_1, \dot{\hat\theta}_2, \{q\}, \{\dot q\}) = \begin{Bmatrix} NL(1) \\ NL(2) \\ NL(3) \end{Bmatrix}$$

$$\begin{aligned}NL(1) = &\left[(I_{rr}^{121} + I_{rr}^{211}) + [I_{\theta u}^{12}]\{q\}\right]\dot{\hat\theta}_1\dot{\hat\theta}_2 + \left[I_{rr}^{221} + [I_{\theta u}^{221}]\{q\}\right]\dot{\hat\theta}_2{}^2\\ &+ \left[[I_{\theta uD}^{11}]\dot{\hat\theta}_1 + [I_{\theta uD}^{21}]\dot{\hat\theta}_2\right]\{\dot q\}\end{aligned}$$

$$NL(2)=\left[I_{rr}^{112}+[I_{\theta u}^{112}]\{q\}\right]\dot{\hat{\theta}}_1^{\,2}+\left[[I_{\theta uD}^{12}]\dot{\hat{\theta}}_1+[I_{\theta uD}^{22}]\dot{\hat{\theta}}_2\right]\{\dot{q}\}$$

$$NL(3)=\{I_{\theta}^{11}\}\dot{\hat{\theta}}_1^{\,2}+(\{I_{\theta}^{12}\}+\{I_{\theta}^{21}\})\dot{\hat{\theta}}_1\dot{\hat{\theta}}_2+\{I_{\theta}^{22}\}\dot{\hat{\theta}}_2^{\,2}+\left[[I_{uD\theta D}^{1}]\dot{\hat{\theta}}_1+[I_{uD\theta D}^{2}]\dot{\hat{\theta}}_2\right]\{\dot{q}\}$$

$$\{FG\}=\begin{Bmatrix}G_\theta^1+[G_u^1]\{q\}\\ G_\theta^2+[G_u^{22}]\{q\}\\ \{G\}\end{Bmatrix}\qquad \{F\}=\begin{Bmatrix}T_1\\ T_2\\ \{F^e\}\end{Bmatrix}$$

Equation (13) represents the dynamics of a flexible two-link manipulator. This system will be used in this paper for the dynamic analysis. The dynamic response of the flexible two-link manipulator can be examined by solving equation (13). Due to their nonlinear, time-varying characteristics and the high frequency elastic modes in the finite element model, it is difficult to solve this dynamic equation directly (Nagarajan and Turcic, 1990b). Therefore, model reduction process must be performed to simplify the system model.

3. Model Reduction

Due to its distributive nature, a flexible manipulator has infinite number of elastic modes. When the finite element method is used to model the system, the number of elastic modes that can be estimated will then depend on the number of elements being used. When increasing the number of elements, not only more elastic modes can be obtained but also more accurate estimation of the dominant modes of the system can be achieved. However, when including high frequency modes of the system into the system model, computational burden makes the dynamic simulation become very difficult if not impossible (Weaver and Johnson, 1987). In the usual dynamic simulation of flexible structures, if the contribution of high frequency elastic modes to the solution is insignificant, these elastic modes can be eliminated by diagonalizing the dynamic equation of the system (Weaver and Johnson, 1987). This approach was successfully employed in the model reduction of a flexible one-link manipulator (Menq and Xia, 1991). However, this model reduction method can not be applied directly to a flexible multi-link manipulator because of the time varying nature of the system.

In order to reduce the amount of computation in solving the system dynamic equation as shown in equation (13), a model reduction technique based on the modal analysis is employed here. For the purpose of model reduction, the system variable X is expressed as,

$$X=[P]\{\eta\} \tag{14}$$

where $[P]$ is the mode shape function matrix and $\{\eta\}$ is the principal mode coordinates. For the two link flexible manipulators, the first two principle modes will be the rigid modes and the other modes will be the elastic modes. Therefore, the principle modes $\{\eta\}$ can be divided into the rigid modes and the elastic modes.

$$\{\eta\}=\begin{Bmatrix}\begin{Bmatrix}\eta_1\\ \eta_2\end{Bmatrix}\\ \{z_e\}\end{Bmatrix} \tag{15}$$

where η_1 and η_2 are the two rigid modes, which correspond to the rigid motions of the two links, and $\{z_e\}$ are the elastic modes, which can be separated into two parts as,

$$\{z_e\}=\begin{Bmatrix}\{z\}\\ \{z_h\}\end{Bmatrix} \tag{16}$$

where $\{z\}$ is the dominant elastic modes and $\{z_h\}$ is the high frequency elastic modes. Therefore, the system principal modes can be partitioned as,

$$\{\eta\}=\begin{Bmatrix}\begin{Bmatrix}\eta_1\\ \eta_2\end{Bmatrix}\\ \{z\}\\ \{z_h\}\end{Bmatrix} \tag{17}$$

Corresponding to the partition of the system principal modes as shown in equation (17), the mode shape function matrix $[P]$ can also be partitioned as,

$$[P]=\left[[P_r]\quad [P_e]\quad [P_e^h]\right] \tag{18}$$

where $[P_r]$ is the eigenvectors corresponding to the rigid modes, $[P_e]$ is the eigenvectors corresponding the the dominant elastic modes, which is called the dominant elastic mode shape function matrix, and $[P_e^h]$ is the eigenvectors corresponding to the high frequency elastic modes. The matrix $[P_r]$ can be partitioned corresponding to the coordinate angles and the elastic freedoms as,

$$[P_r]=\begin{bmatrix}I_{22}\\ 0_{n2}\end{bmatrix} \tag{19}$$

where I_{22} is a 2×2 identity matrix, 0_{n2} is a $n\times2$ null matrix. Similarly, the dominant elastic mode shape function matrix $[P_e]$ can also be partitioned as,

$$[P_e]=\begin{bmatrix}P_{re}\\ P_{ee}\end{bmatrix} \tag{20}$$

where $[P_{re}]$ is the first two rows of matrix $[P_e]$ and represents the effects of the elastic modes on the motion of the coordinate angles and $[P_{ee}]$ represents the effects of the elastic modes on the elastic motion.

In order to reduce the amount of computation in the system dynamic simulation, high frequency elastic modes $\{z_h\}$, which contribute little to the system response while demanding large amount of computation, can be eliminated. By eliminating the high frequency elastic modes $\{z_h\}$, the system variable X can be expressed in terms of the rigid modes and the dominant elastic modes as,

$$X=\begin{Bmatrix}\hat{\theta}_1\\ \hat{\theta}_2\\ \{q\}\end{Bmatrix}=[PP]\begin{Bmatrix}\begin{Bmatrix}\eta_1\\ \eta_2\end{Bmatrix}\\ \{z\}\end{Bmatrix} \tag{21}$$

where

$$[PP]=\left[[P_r]\quad [P_e]\right]=\begin{bmatrix}I_{22} & [P_{re}]\\ 0_{n2} & [P_{ee}]\end{bmatrix}$$

It can be seen from equation (21) that the coordinate angles $\hat{\theta}_1$ and $\hat{\theta}_2$ will be the function of both the rigid modes η_1 and η_2 and the elastic modes $\{z\}$, and the elastic coordinate $\{q\}$ will be the function of the elastic modes $\{z\}$ only. From equation (21), the coordinate angles can be expressed in terms of the rigid modes and the dominant elastic modes as,

$$\begin{Bmatrix}\hat{\theta}_1\\ \hat{\theta}_2\end{Bmatrix}=\begin{Bmatrix}\eta_1\\ \eta_2\end{Bmatrix}+\begin{bmatrix}P_{re}^1\\ P_{re}^2\end{bmatrix}\{z\} \tag{22}$$

where P_{re}^1 is the first row of matrix $[P_{re}]$, P_{re}^2 is the second row of matrix $[P_{re}]$. Similarly, the elastic freedoms $\{q\}$ can be expressed in terms of the dominant elastic modes as,

$$\{q\}=[P_{ee}]\{z\} \tag{23}$$

Substituting equations (21), (22) and (23) into equation (13) and pre-multiply equation (13) with $[PP]^T$, equation (13) becomes,

$$[RM]\ddot{Y}+[RC]\dot{Y}+[RK]Y+RN(\dot{\eta}_1,\dot{\eta}_2,\{z\},\{\dot{z}\})-\{RG\}=\{RF\} \tag{24}$$

where $Y=(\eta_1 \quad \eta_2 \quad \{z\})^T$, $[RM]$ is the mass matrix, $[RC]$ is the damping matrix, $[RK]$ is the stiffness matrix, $RN(\dot{\eta}_1,\dot{\eta}_2,\{z\},\{\dot{z}\})$ is the nonlinear term induced by the coupling of the rigid motion and the elastic motion, and $\{RG\}$ is the gravity force, $\{RF\}$ is the external force.

$$[RM]=[PP]^T[FM][PP]$$

$$[RC]=[PP]^T[FC][PP]+2[PP]^T[FM]\frac{d}{dt}[PP]$$

$$[RK]=[PP]^T[FK][PP]+[PP]^T[FC]\frac{d}{dt}[PP]+[PP]^T[FM]\frac{d^2}{dt^2}[PP]$$

$$RN(\dot{\eta}_1,\dot{\eta}_2,\{z\},\{\dot{z}\})=[PP]^T\begin{Bmatrix} NL_1 \\ NL_2 \\ \{NL\}_u \end{Bmatrix}$$

$$\begin{aligned} NL_1 = &\left[(I_{rr}^{121}+I_{rr}^{211})+[I_{\theta u}^{12}]\{q\}\right](\dot{\eta}_1+P_{re}^1\{\dot{z}\}+\dot{P}_{re}^1\{z\})(\dot{\eta}_2 \\ &+P_{re}^2\{\dot{z}\}+\dot{P}_{re}^2\{z\})+\left[I_{rr}^{221}+[I_{\theta u}^{221}][P_{ee}]\{z\}\right](\dot{\eta}_2+P_{re}^2\{\dot{z}\} \\ &+\dot{P}_{re}^2\{z\})^2+\Big[[I_{\theta uD}^{11}](\dot{\eta}_1+P_{re}^1\{\dot{z}\}+\dot{P}_{re}^1\{z\}) \\ &+[I_{\theta uD}^{21}](\dot{\eta}_2+P_{re}^2\{\dot{z}\}+\dot{P}_{re}^2\{z\})\Big]([P_{ee}]\{\dot{z}\}+[\dot{P}_{ee}]\{z\}) \end{aligned}$$

$$\begin{aligned} NL_2 = &\left[I_{rr}^{112}+[I_{\theta u}^{112}][P_{ee}]\{z\}\right](\dot{\eta}_1+P_{re}^1\{\dot{z}\}+\dot{P}_{re}^1\{z\})^2 \\ &+\Big[[I_{\theta uD}^{12}](\dot{\eta}_1+P_{re}^1\{\dot{z}\}+\dot{P}_{re}^1\{z\})+[I_{\theta uD}^{22}](\dot{\eta}_2+P_{re}^2\{\dot{z}\} \\ &+\dot{P}_{re}^2\{z\})\Big]([P_{ee}]\{\dot{z}\}+[\dot{P}_{ee}]\{z\}) \end{aligned}$$

$$\begin{aligned} \{NL\}_u = &\{I_\theta^{11}\}(\dot{\eta}_1+P_{re}^1\{\dot{z}\}+\dot{P}_{re}^1\{z\})^2+(\{I_\theta^{12}\}+\{I_\theta^{21}\})(\dot{\eta}_1 \\ &+P_{re}^1\{\dot{z}\}+\dot{P}_{re}^1\{z\})(\dot{\eta}_2+P_{re}^2\{\dot{z}\}+\dot{P}_{re}^2\{z\})+\{I_\theta^{22}\}(\dot{\eta}_2 \\ &+P_{re}^2\{\dot{z}\}+\dot{P}_{re}^2\{z\})^2+\Big[[I_{uD\theta D}^{1}](\dot{\eta}_1+P_{re}^1\{\dot{z}\}+\dot{P}_{re}^1\{z\}) \\ &+[I_{uD\theta D}^{2}](\dot{\eta}_2+P_{re}^2\{\dot{z}\}+\dot{P}_{re}^2\{z\})\Big]([P_{ee}]\{\dot{z}\}+[\dot{P}_{ee}]\{z\}) \end{aligned}$$

$$\{RG\}=[PP]^T\begin{Bmatrix} [G_\theta^1]+G_u^1][P_{ee}]\{z\} \\ [G_\theta^2]+[G_u^{22}][P_{ee}]\{z\} \\ \{G\} \end{Bmatrix} \qquad \{RF\}=[PP]^T\{F\}$$

By eliminating the high frequency elastic modes, the dimension of system dynamic equation is reduced, and the time step in the direct time integration approach can be greatly increased. Therefore, the computational time can be drastically reduced. The number of dominant elastic modes can be chosen based on the speed of the system response and the desired accuracy. Once the solution of equation (24) is obtained, the actual response of the system can be synthesized using equation (21).

4. Mode Shape Function

It can be seen from the previous section that in order to reduce the system model, the dominant elastic mode shape function matrix $[P_e]$ must first be constructed. One of the commonly used methods is to assume that the mode shape function of the elastic deformation for a flexible manipulator is constant as that of a flexible beam in the structural dynamics. The mode shape function matrix $[P_e]$ can be constructed by using the mode shape functions of a flexible beam with the cantilever or simple support boundary conditions. This approach basically assumes that the elastic mode shape function will not be affected by the rigid motion or the configuration of the manipulator and the elastic deformation of different links is decoupled. However, in the real situation, the rigid motion will change the boundary conditions of the elastic motion of a flexible link (Oakley and Cannon, 1988) and the elastic motion of different links will couple together. Moreover, the link configuration will have notable impact on the elastic mode shape function. In this paper, the dominant elastic mode shape function matrix is determined by considering the effects of the rigid motion, the system configuration, and the coupling of the elastic deformation of the two links. In other words, the mode shape function matrix $[P]$ should be obtained using the system mass matrix $[FM]$ and the stiffness matrix $[FK]$, both of which include the rigid and elastic freedoms as well as their coupling. Once the $[P]$ matrix is derived, the dominant elastic mode shape function matrix $[P_e]$ can be obtained. However, since the system mass matrix $[FM]$ is a function of the manipulator's configuration, a direct solution of the eigenvalue problem in usual modal analysis is not feasible. In the two-link flexible manipulator as shown in Figure 1, it is found that matrix $[FM]$ is a function of $\hat{\theta}_2$, which is the coordinate angle of the second link. Therefore, it can be expected that the mode shape function matrices $[P]$ and $[P_e]$ are also functions of $\hat{\theta}_2$. For a specified coordinate angle $\hat{\theta}_2$, the mode shape function matrix can be determined as in a usual generalized eigenvalue problem.

$$\{[FK]-[FM](\omega_i^2)\}[P_i]=0 \tag{25}$$

where ω_i^2 and $[P_i]$ are the i-th eigenvalue and eigenvector of the system respectively. Using this approach, the mode shape functions of the two-link manipulator at different configurations can be determined. Figure 4 shows the frequencies of the first four elastic

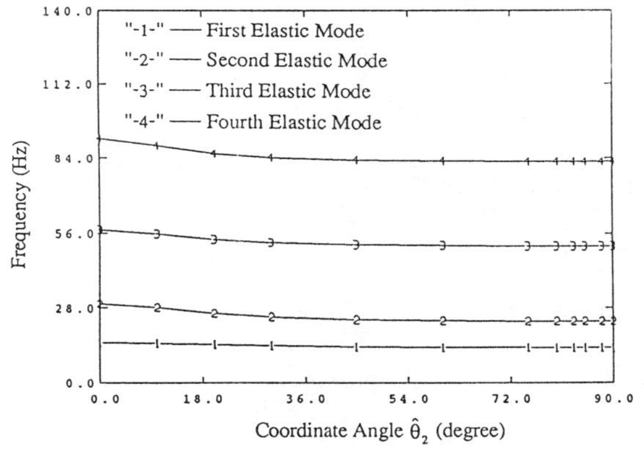

Figure 4 Frequencies of the Elastic Modes

modes. Figures 5, 6, 7 and 8 show the mode shape functions of the manipulator for the first four elastic modes where the configurations are specified as $\hat{\theta}_2=0^0$, 20^0, 45^0, 90^0. From these results, it is evident that neither the frequencies nor the mode shape functions are constant. They are functions of the link configuration, i.e., the coordinate angle $\hat{\theta}_2$. The elastic motions of the two links couple with each other except for the case of $\hat{\theta}_2=90^0$. It can also be seen from the figures that the first and the third elastic modes are the dominant elastic modes for link 1 and the second and the fourth elastic modes are the dominant elastic modes for link 2. At the configuration of $\hat{\theta}_2=90^0$, the elastic motion of the two links are

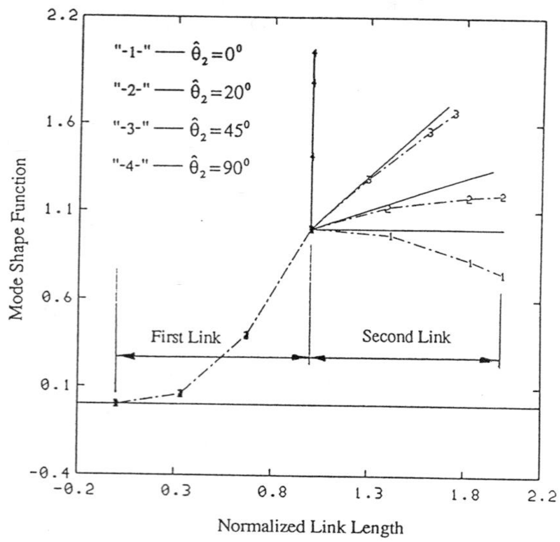

Figure 5 Mode Shape Function of the First Elastic Mode

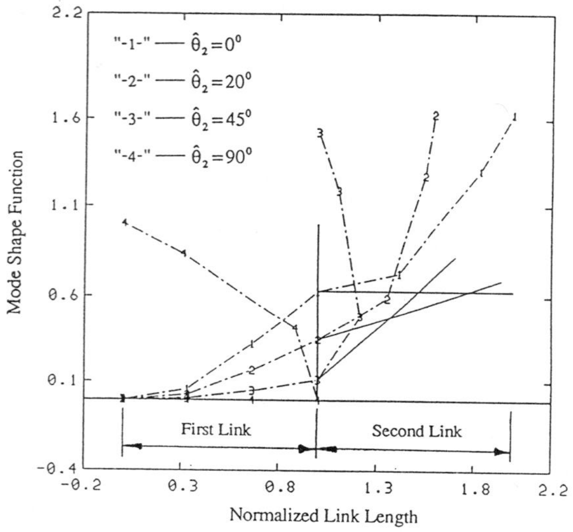

Figure 6 Mode Shape Function of the Second Elastic Mode

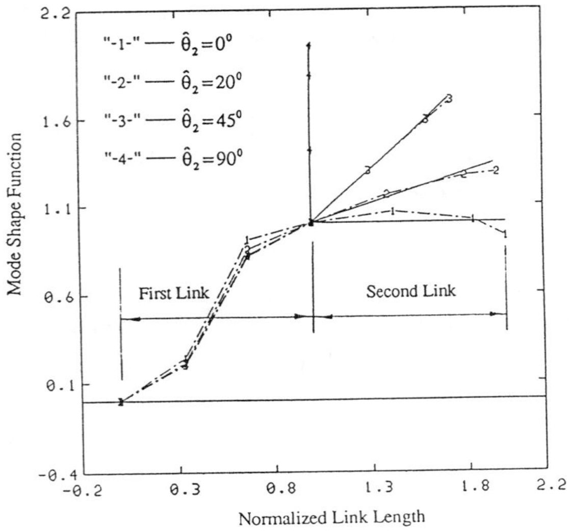

Figure 7 Mode Shape Function of the Third Elastic Mode

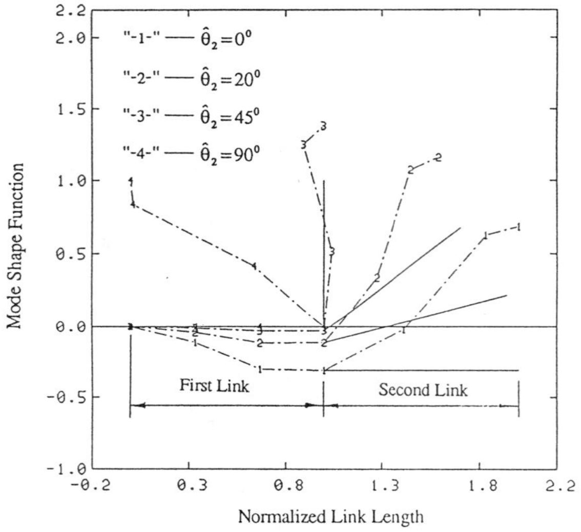

Figure 8 Mode Shape Function of the Fourth Elastic Mode

decoupled and the contribution of the elastic motion at this configuration will come from the dominant elastic modes of each link.

Due to the fact that it is not computationally feasible to calculate the eigenvector matrix for each configuration during the computer simulation, it is desired to express the dominant elastic mode shape function matrix $[P_e]$ in terms of the link coordinate angle $\hat{\theta}_2$. Since such an explicit function can not be obtained theoretically, numerical interpolation method is employed.

In order to obtain the function $[P_e(\hat{\theta}_2)]$, the elastic mode shape function of each link are normalized with their respective tip nodes. For each elastic mode, the mode shape function is expressed in terms of two angular ratio functions, which represent the effects of the elastic modes on the motion of the two coordinate angles, the tip ratio function, which represents the ratio of the two links' mode shape function at the tip nodes, and the normalized mode shape functions of the two links, which represent the mode shape of each individual link. For the dominant elastic mode of link 1 such as the first and the third elastic modes, the elastic mode shape function $\{P_e^j\}$, j=1, 3, ... is expressed as,

$$\{P_e^j\} = \begin{Bmatrix} f_1^j(\hat{\theta}_2) \\ f_2^j(\hat{\theta}_2) \\ \{P_e^j(\hat{\theta}_2)\}_1^n \\ \{P_e^j(\hat{\theta}_2)\}_2^n f_t^j(\hat{\theta}_2) \end{Bmatrix} \tag{26}$$

where $f_1^j(\hat{\theta}_2)$ is the angular ratio function for the coordinate angle $\hat{\theta}_1$, $f_2^j(\hat{\theta}_2)$ is the angular ratio function for the coordinate angle $\hat{\theta}_2$, $\{P_e^j(\hat{\theta}_2)\}_1^n$ is the normalized j-th elastic mode shape function of link 1, $\{P_e^j(\hat{\theta}_2)\}_2^n$ is the normalized j-th elastic mode shape function of link 2, and $f_t^j(\hat{\theta}_2)$ is the tip ratio function of the j-th elastic mode.

The angular ratio function $f_1^j(\hat{\theta}_2)$ is defined as,

$$f_1^j(\hat{\theta}_2)=\frac{P_e(1,j)}{(P_e^j)_1^t} \quad (27)$$

where $(P_e^j)_1^t$ is the value of the mode shape function at the tip node of link 1. The angular ratio function $f_2^j(\hat{\theta}_2)$ is defined as,

$$f_2^j(\hat{\theta}_2)=\frac{P_e(2,j)}{(P_e^j)_1^t} \quad (28)$$

The tip ratio function $f_t^j(\hat{\theta}_2)$ is defined as,

$$f_t^j(\hat{\theta}_2)=\frac{(P_e^j)_2^t}{(P_e^j)_1^t} \quad (29)$$

where $(P_e^j)_2^t$ is the value of the mode shape function at the tip node of link 2. For the dominant elastic mode of link 2, such as the second and the fourth elastic modes, the elastic mode shape function $\{P_e^j\}$, j=2, 4, ... is expressed as,

$$\{P_e^j\} = \begin{Bmatrix} f_1^j(\hat{\theta}_2) \\ f_2^j(\hat{\theta}_2) \\ \{P_e^j(\hat{\theta}_2)\}_1^n f_t^j(\hat{\theta}_2) \\ \{P_e^j(\hat{\theta}_2)\}_2^n \end{Bmatrix} \quad (30)$$

where the angular ratio function $f_1^j(\hat{\theta}_2)$ is defined as,

$$f_1^j(\hat{\theta}_2)=\frac{P_e(1,j)}{(P_e^j)_2^t} \quad (31)$$

the angular ratio function $f_2^j(\hat{\theta}_2)$ is defined as,

$$f_2^j(\hat{\theta}_2)=\frac{P_e(2,j)}{(P_e^j)_2^t} \quad (32)$$

and the tip ratio function $f_t^j(\hat{\theta}_2)$ is defined as,

$$f_t^j(\hat{\theta}_2)=\frac{(P_e^j)_1^t}{(P_e^j)_2^t} \quad (33)$$

In order to perform the numerical interpolation, the dominant elastic mode shape function $[P_e(\hat{\theta}_2)]$ is calculated for different link configurations, i.e., for different values of the link coordinate angle $\hat{\theta}_2$. From the numerical results, it is found that the function $[P_e(\hat{\theta}_2)]$ will be symmetrical to the configuration in which two links are coaxial. It is also found from the numerical results that the elastic motion of two links is decoupled in the configuration in which two links are perpendicular and the mode shape function of the elastic motion $[P_{ee}]$ is also asymmetrical to this configuration. By using the symmetricity of $[P_e(\hat{\theta}_2)]$, it is only necessary to construct the mode shape function $[P_{re}]$ in the first and the second quadrants of $\hat{\theta}_2$ and to construct the mode shape function $[P_{ee}]$ in the first quadrant of $\hat{\theta}_2$. In this paper, the numerical data of the mode shape function at every ten degree are used to construct the mode shape function matrix. With the data of the dominant elastic mode shape function matrix at different configurations available, the numerical interpolation is divided into two steps. The first step is to obtain the angular ratio function $f_1^j(\hat{\theta}_2)$ and $f_2^j(\hat{\theta}_2)$ and tip ratio function $f_t^j(\hat{\theta}_2)$ of each dominant elastic mode. For the flexible two link manipulator under study, the angular ratio functions in the first and the second quadrants of $\hat{\theta}_2$ and the tip ratio functions in the first quadrant of $\hat{\theta}_2$ of the first dominant elastic mode are shown in Figure 9 to Figure 11. From Figure 9 to Figure 11, it can be seen that the tip ratio functions approach zero as $\hat{\theta}_2$ approaches 90^0, which is corresponding to the fact that the elastic motions of the two links are decoupled at the configuration of $\hat{\theta}_2=90^0$. To fit the angular and the tip ratio functions, cubic spline functions are used. The angular ratio functions and the tip ratio functions obtained through the numerical interpolation are compared with the real angular and tip ratio functions and are shown in Figure 9 to Figure 11. It can be seen from the figures that these fitted functions agree well with the real angular and tip ratio functions.

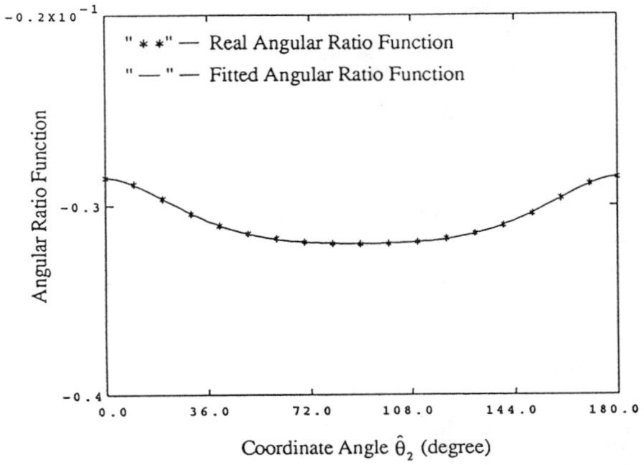

Figure 9 Angular Ratio Function for $\hat{\theta}_1$ of the First Elastic Mode

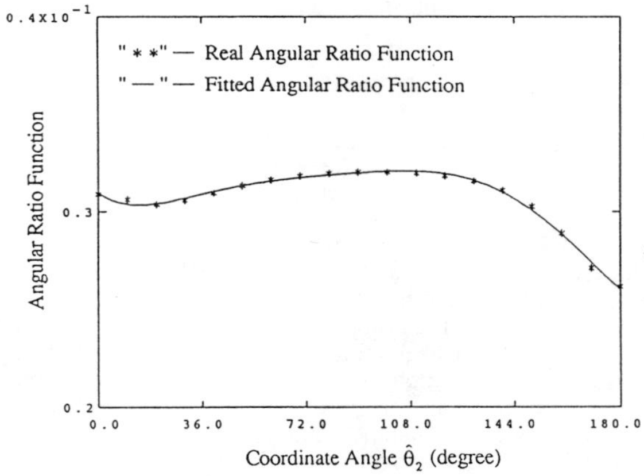

Figure 10 Angular Ratio Function for $\hat{\theta}_2$ of the First Elastic Mode

The second step of the numerical interpolation is to obtain the explicit expression of the normalized mode shape function $\{P_e^j(\hat{\theta}_2)\}_1^n$ and $\{P_e^j(\hat{\theta}_2)\}_2^n$, both of which are the function of both the coordinate angle $\hat{\theta}_2$ and the link coordinate. Similar approach as in obtaining the angular and tip ratio functions can be used to obtain the normalized mode shape function $\{P_e^j(\hat{\theta}_2)\}_1^n$ and $\{P_e^j(\hat{\theta}_2)\}_2^n$. Once the

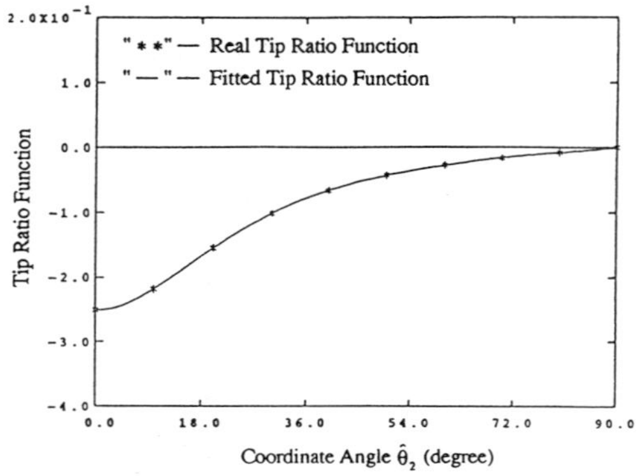

Figure 11 Tip Ratio Function of the First Elastic Mode

explicit expression for the angular ratio function $f_1^j(\hat{\theta}_2)$ and $f_2^j(\hat{\theta}_2)$, the tip ratio function $f_t^j(\hat{\theta}_2)$, and the normalized mode shape function $\{P_e^j(\hat{\theta}_2)\}_1^n$ and $\{P_e^j(\hat{\theta}_2)\}_2^n$ as a function of the coordinate angle $\hat{\theta}_2$ for each dominant elastic mode are obtained, the dominant elastic mode shape function matrix $[P_e(\hat{\theta}_2)]$ can be constructed and the system finite element model as shown in equation (13) can be reduced to the system model as shown in equation (24). With the reduced system model, system dynamic simulation can be performed, and the system dynamic responses can be predicted.

5. Modeling Accuracy and Simulation Results

To show the effectiveness of the proposed model reduction technique and to examine the various assumptions in modeling flexible manipulators, simulation results of a flexible two link manipulator are presented here. Both links are initially fully extended with $\theta_1(0) = \theta_2(0) = 0$ and then moved under the action of the gravity force. No other external forces are applied here. The payload at the end-point of the flexible manipulator is equal to 0.9 lb.

5.1 Effects of Model Reduction

In the dynamic simulation, each flexible link as shown in Figure 2 and Figure 3 is modeled by three elements. In the finite element model, the highest frequency of the elastic modes is found to be 2.44×10^4 Hz as $\hat{\theta}_2(0)=0$. To ensure the stability of the numerical integration, the step size should be very small. For each time step, several matrices need to be constructed and the inverse mass matrix need to be obtained. The computational burden for the simulation of such a system is so heavy that it is almost impossible with the current available computer facilities. With the model reduction technique presented in the previous section, the high frequency elastic modes are eliminated. Therefore the numerical integration interval can be greatly increased, and large amount of computations can be reduced. Although equation (24) is still nonlinear and has strong coupling, its numerical solution can now be obtained.

To examine the modeling accuracy of the reduced system model, the simulation results for the first four elastic modes are shown in Figures 12 and 13. As mentioned in the previous section, the first and the third elastic modes are the dominant elastic modes for the first link and the second and the fourth elastic modes are the dominant elastic modes for the second link. As can be seen from Figures 12 and 13, the second dominant elastic mode for each link is much smaller than the first one as expected. Depending on the desired accuracy and the frequency content of the excitation signal, the number of required dominant elastic modes can be chosen. If more elastic modes are used, better numerical accuracy will be achieved, however, more computation is needed. For computational

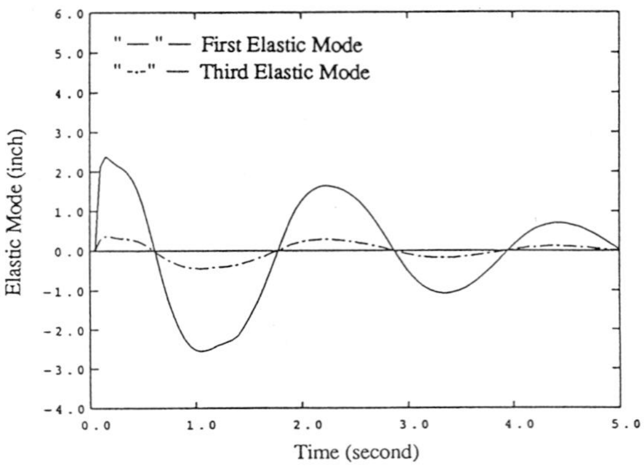

Figure 12 The Dominant Elastic Modes of Link 1

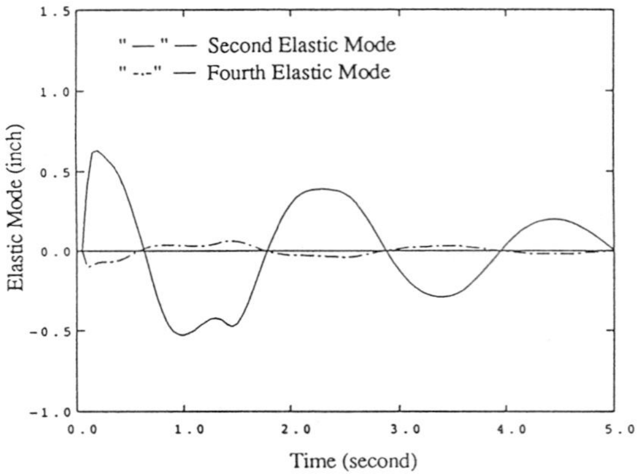

Figure 13 The Dominant Elastic Modes of Link 2

convenience, the system model with the first two elastic modes, i.e., one dominant elastic mode for each link, are used in the subsequent simulation of this paper.

5.2 Effects of the Mode Shape Function

In the assumed mode approach, the elastic deformation of each link is expressed in terms of the dominant elastic modes. The mode shape function is assumed to be constant and can be obtained using a single flexible beam with either the cantilever or simple support boundary conditions. Such an assumption neglects both the effects of the rigid motion on the mode shape functions and the coupling of the elastic deformation of the individual elastic links. To examine the consequence of this assumption, the simulation result using constant mode shape function is compared with that using the approach proposed in this paper. In the assumed mode approach, if one dominant elastic mode is used for each link, the elastic deformations for both links are expressed as,

$$\{q_1\}=\{P_1\}z_1 \tag{34}$$

$$\{q_2\}=\{P_2\}z_2 \tag{35}$$

where $\{q_1\}$ and $\{q_2\}$ are the elastic coordinates of the first link and the second link respectively, z_1 and z_2 are the respective dominant elastic modes, and $\{P_1\}$ and $\{P_2\}$ are the respective mode shape functions. The mode shape functions $\{P_1\}$ and $\{P_2\}$ are constructed as that the two individual links are cantilever beams. Due to the fact that the second link has a payload attached to its tip, a finite element model is first constructed, and its cantilever beam mode shape functions are obtained from the mass and the stiffness matrix of the finite element model. From equations (34) and (35), the system variable X is expressed as,

$$X = \begin{Bmatrix} \hat{\theta}_1 \\ \hat{\theta}_2 \\ \{q\} \end{Bmatrix} = [PP_a] \begin{Bmatrix} \hat{\theta}_1 \\ \hat{\theta}_2 \\ z_1 \\ z_2 \end{Bmatrix} \tag{36}$$

The mode shape function matrix $[PP_a]$ is constructed as,

$$[PP_a] = \begin{bmatrix} I_{22} & 0 & 0 \\ 0_{n_1 2} & \{P_1\} & 0_{n_1} \\ 0_{n_2 2} & 0_{n_2} & \{P_2\} \end{bmatrix} \tag{37}$$

where n_1 is the number of elastic coordinates of link 1 and n_2 is the number of the elastic coordinates of link 2. The mode shape functions in the assumed mode approach as shown in equation (37) will form a constant mode shape function matrix. The predicted tip deformations of the two links using the assumed mode approach are compared with those obtained using the proposed approach. The results of the tip elastic deformation for the two individual links are shown in Figures 14 and 15, respectively. It can be seen from the figures that when constant mode shape matrix is used, a 40% error will exist in the prediction of the elastic deformation.

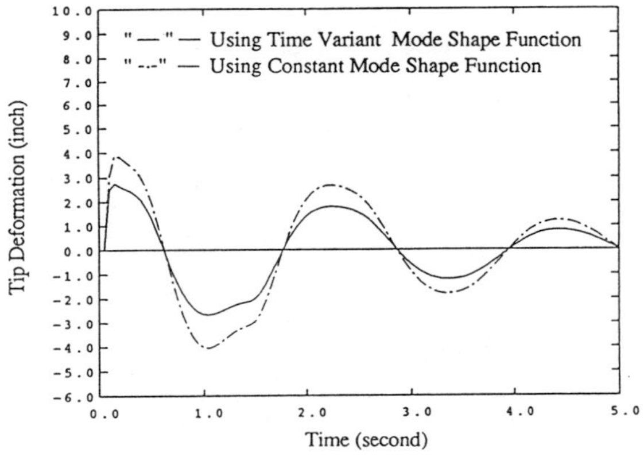

Figure 14 The Tip Deformation of Link 1

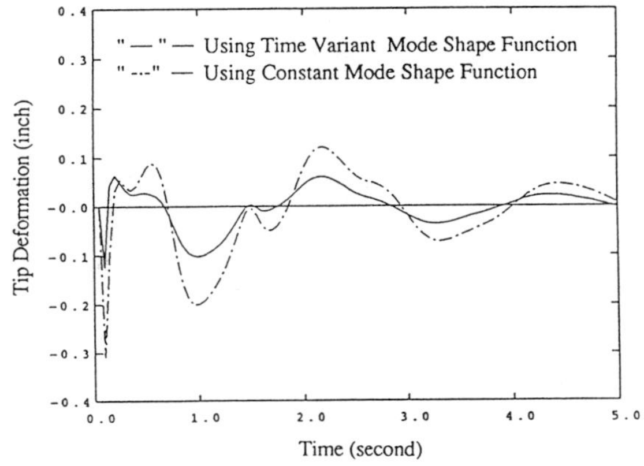

Figure 15 The Tip Deformation of Link 2

5.3 Effects of the Transformation Matrix

The system model used this work includes the effects of the elastic deformation on the kinematic transformation matrix. In the previous approach, such effects are often neglected, and the transformation matrix is approximated as that of a rigid manipulator. To examine the consequence of this approximation, simulation results of the tip deformation are shown in Figures 16 and 17. It

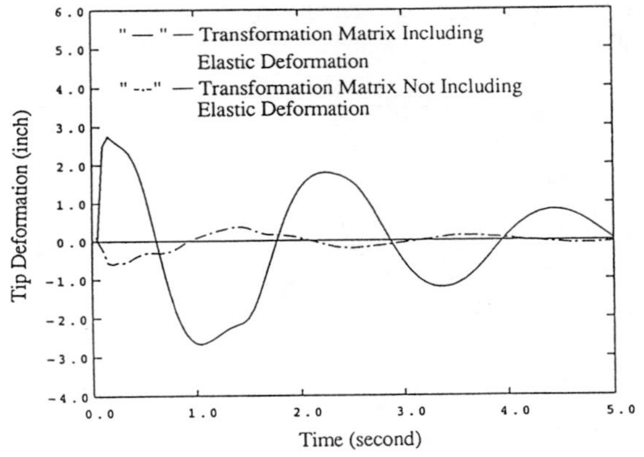

Figure 16 The Tip Deformation of Link 1

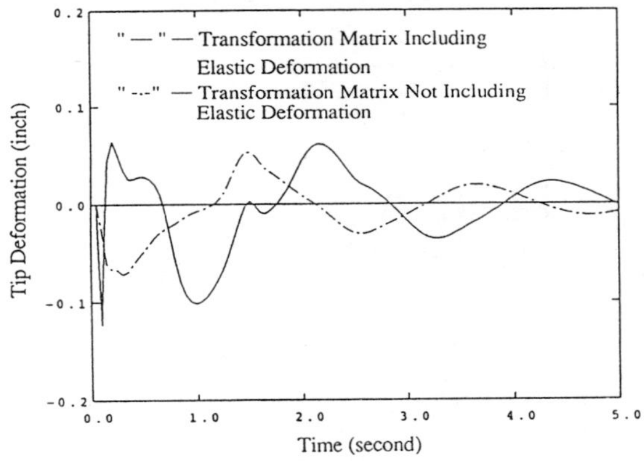

Figure 17 The Tip Deformation of Link 2

can be seen from the figures that when the effects of the elastic motion on the kinematic transformation matrix is neglected, the predicted elastic deformation has a notable discrepancy from that predicted by the system model that includes the elastic deformation in the transformation matrix. This is particularly evident for the tip deformation of link one. Therefore, the effects of the elastic deformation on the kinematic transformation matrix are significant in predicting the end point deformation and should not be neglected.

In various applications, the end-point position of a flexible manipulator is of main concern. For the system model used in this research, the end-point position can be directly calculated as,

$$\{r\} = \left[{}^{0}T_i\right]\{{}^{i}r\} \tag{38}$$

where $\{r\}$ is the position vector of point P in the global coordinate, $\{{}^{i}r\}$ is the position vector of point P in the link coordinate, $\left[{}^{0}T_i\right]$ is the transformation matrix between the link coordinate and the global coordinate. For previous models, if the joint angles and the tip deformations of each link are available, the end-point position for a flexible two-link manipulator can be approximately calculated as,

$$X_e = a_1 \cos\theta_1 - v_1^e \sin\theta_1 + a_2 \cos(\theta_1+\theta_2) - v_2^e \sin(\theta_1+\theta_2) \qquad (39)$$

$$Y_e = a_1 \sin\theta_1 + v_1^e \cos\theta_1 + a_2 \sin(\theta_1+\theta_2) + v_2^e \cos(\theta_1+\theta_2) \qquad (40)$$

With the end-point position available, the end-point position errors can be calculated. Figures 18 and 19 show the end-point position errors in both X and Y coordinates, respectively, for the previous models in which the elastic deformation in the coordinate transformation matrix is neglected. It can be seen from the figures that the end-point error will be in the range of several inches. It is expected that this modeling error will increase as the speed of the system increases.

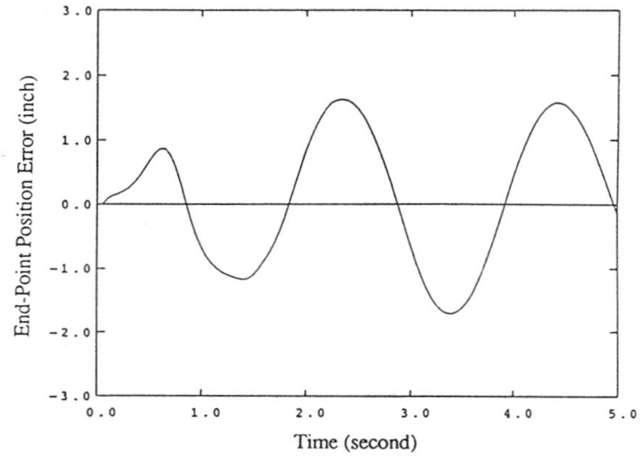

Figure 18 The End-Point Position Error (X-Direction)

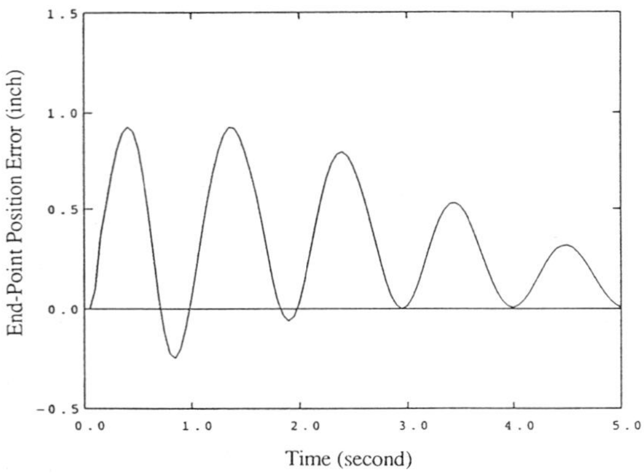

Figure 19 The End-Point Position Error (Y-Direction)

6. Conclusion

A method based on the finite element approach is developed in this paper for the dynamic analysis of flexible multi-link manipulators. A model reduction technique is proposed and employed to reduce the order of the finite element model. In the proposed approach, the elastic mode shape function matrix is obtained by considering both the effects of the rigid motion and the configuration of the manipulator and the coupling among the elastic deformations of individual links. An explicit expression of the reduced order elastic mode shape matrix as a function of the link coordinate angle is synthesized through an interpolation approach. By applying this model reduction technique, the high frequency elastic modes, which have little contribution to the system response while demanding large amount of computation, are eliminated. With the reduced system model, the amount of computation is greatly reduced and it becomes possible to obtain the numerical solution of the system equation of motion.

A flexible two-link manipulator is employed to illustrate the proposed approach. The simulation results for the manipulator acting by the gravity force are presented. The effects of model reduction, the mode shape function, and the transformation matrix are studied to examine the accuracy and application range of the various assumptions, that were previously used for modeling the dynamics of a flexible manipulator. From the simulation results, notable error was found in the prediction of the the system's elastic motion if constant mode shape matrix is assumed. It is also shown from the simulation result that if neglecting the elastic deformation in the kinematic transformation, large discrepancy will exist in the prediction of the tip elastic deformation, and the end-point position errors will be in the range of several inches for the case under study.

References

Cyril, X, Angeles, J., and Misra, A.K., 1989, "Flexible-Link Robotic Manipulator Dynamics," Proceeding of 1989 American Control Conference, pp. 2346-2351.

Jonker, B., 1990, "A Finite Element Dynamic Analysis of Flexible Manipulators," The International Journal of Robotics Research, vol. 9, pp. 59-74.

Menq, C.-H. and Xia, Z., "Modeling and Tracking Control of A Flexible One-Link Manipulator," submitted to the ASME Journal of Dynamic Systems, Measurement and Control.

Nagarajan, S. and Turcic, D.A., 1990a, "Lagrangian Formulation of the Equations of Motion for Elastic Mechanisms with Mutual Dependence Between Rigid Body and Elastic Motions. Part I: Element Level Equations," ASME, Journal of Dynamic Systems, Measurement, and Control, vol. 112, pp. 203-214.

Nagarajan, S. and Turcic, D.A., 1990b, "Lagrangian formulation of the Equations of Motion for Elastic Mechanisms with Mutual Dependence Between Rigid Body and Elastic Motions. Part II: System Equations," ASME, Journal of Dynamic Systems, Measurement, and Control, vol. 112, pp. 215-224.

Oakley, C.M. and Cannon, R.H., Jr., 1988, "Initial Experiments on the Control of a Two-link Manipulator with a very Flexible forearm," Proceeding of 1988 American Control Conference, pp. 996-1002.

Oakley, C.M. and Cannon, R.H., Jr., 1989, "End-point Control of a Two-link Manipulator with a very Flexible Forarm: Issues and Experiments," Proceeding of 1989 American Control Conference, pp. 1381-1388.

Schmitz, E., 1986, "Dynamics and Control of a Planar Manipulator with Elastic Links," Proceeding of 25th Conference of Decision and Control, pp. 1135-1139.

Weaver, W. Jr., Johnson, P.R., 1987 , "Structural Dynamics by Finite Elements," Prentice-Hall, Inc., New Jersey.

Xia, Z., 1992, "Modeling and Control of Flexible Manipulators," Ph.D. dissertation, The Ohio State University.

AMD-Vol. 141/DSC-Vol. 37, Dynamics of Flexible Multibody Systems:
Theory and Experiment
ASME 1992

MODELING AND CONTROL OF FLEXIBLE MANIPULATORS: PART II – END-POINT TRACKING CONTROL

Zhijie Xia and Chia-Hsiang Menq
Department of Mechanical Engineering
Ohio State University
Columbus, Ohio

ABSTRACT

In this paper, an elastic-deformation estimator is proposed for real time end-point tracking control of a flexible two-link manipulator. Due to the non-colocated characteristics of the system, the inverse model (from end-point motion to control torques) is divided into two subsystems, namely, the stable subsystem and the unstable one, corresponding to the causal part and the noncausal part of the system's elastic motion, respectively. A digital filter is formulated to replace the unstable subsystem so as to estimate the noncausal part of the elastic motion associated with a specified end-point motion. For the design of the filter, the frequency response ratio between the filter and the unstable subsystem is used as the criterion, the objective of which is to have the frequency response ratio have zero phase shift as well as unity gain within a specified frequency range. It is shown that due to the noncausal characteristics of the unstable subsystem, preview information of the input trajectory is required for implementing the proposed digital filter, and the accuracy of the estimation increases as the preview steps increases. Based on the stable subsystem and the proposed digital filter, a time-varying estimator is designed to estimate the elastic motion of the system when the end-point motion is specified. A command feedforward controller is then used to calculate the required control torques based on the estimated elastic deformation and the desired end-point motion. The computed torques along with a feedback controller then form a control scheme making the flexible two-link manipulator become capable of precision end-point tracking. Simulation results are presented to show the performance of the proposed end-point tracking scheme as well as the elastic-deformation estimator.

1. Introduction

In the part I of this research, a methodology is developed for the dynamic analysis of flexible manipulators. In part II, the end-point tracking control of flexible manipulators will be investigated. The main difficulties in the end-point tracking control of a flexible multi-link manipulator are: (1) the system is nonlinear, and there are different types of strong nonlinear terms in the system's dynamic equations; (2) the parameters of the dynamic equations vary with system configuration, and therefore the system is time varying; (3) for end-point tracking control, the control inputs are the torques acting on the hub of each individual link, and the control objective is the motion of the end-point of the manipulator, therefore the control system is non-colocated. Due to these three main difficulties, the control of a flexible multi-link manipulator poses a great challenge to control engineers.

The control of flexible manipulators has been an active research subject in recent years. Several control schemes for the end-point tracking of a flexible one-link manipulator have been developed, and good tracking performance has been achieved for both the computer simulation and experimental results (Kwon and Book, 1990, Menq and Xia, 1992). The point-to-point feedback regulatory control of a flexible multi-link manipulator has been studied based on the linearized system model (Oakley and Cannon, 1988, Yurkovich etc., 1990, Domroese and Hardt, 1990). For the end-point tracking control, Oakley and Cannon (1989) designed a linear quadratic regulator based on the linearized system model, and no feedforward compensation is used. Singh and Schy (1986) proposed a joint space closed-loop control for flexible manipulators based on nonlinear inversion and modal damping, which requires the availability of all the state variables for feedback control. Bayo etc. (1989) transfers the system dynamic equations from the time domain to the frequency domain, and the desired control torque is obtained through an iterative process. Asada etc.(1990) formulates the inverse problem in the time domain. The control torque is obtained by superposing the rigid torque with the compensating torque.

The objective of this paper is to develop a control scheme for the end-point tracking control of a flexible two-link manipulator. A feedforward tracking control scheme is presented based on the reduced order inverse model of the system. The inherent instability of the system's inverse model is characterized. Due to the non-colocated characteristics of the system, the inverse model will be divided into two subsystems, namely, the stable subsystem and the unstable one. A digital filter is formulated to replace the unstable subsystem in estimating the noncausal part of the elastic motion. Based on the stable subsystem and the proposed digital filter, a time-varying estimator is designed to estimate the elastic motion of the system when the end-point motion is specified. A command feedforward controller is then used to calculate the required control torques based on the estimated elastic deformation and the desired end-point motion. The computed torques along with a feedback controller then form a control scheme that enables the flexible two-link manipulator of precision end-point tracking. Simulation results are presented to show the performance of the proposed end-point tracking control scheme as well as the elastic-deformation estimator.

2. System Modeling

The system under study is a flexible two-link manipulator, as shown in Figure 1, moving on the horizontal plane. By using the system modeling and model reduction technique presented in Part I, and choosing two dominant elastic modes, in addition to the two rigid modes, the system dynamic equation can be obtained as follows,

$$[RM]\ddot{Y}+[RC]\dot{Y}+[RK]Y+RN(\dot{\eta}_1,\dot{\eta}_2,\{z\},\{\dot{z}\})=\{RF\} \tag{1}$$

where $Y=[(\eta_1,\eta_2),(z_1,z_2)]^T$, in which η_1 and η_2 are the two rigid modes of the system, $\{z\}^T=(z_1,z_2)$ are the two dominant elastic modes, $[RM]$ is the mass matrix, $[RC]$ is the damping matrix, $[RK]$ is the stiffness matrix, $RN(\dot{\eta}_1,\dot{\eta}_2,\{z\},\{\dot{z}\})$ is the nonlinear force, which represents the coupling among the rigid and elastic motions, and $\{RF\}$ is the external force. In this paper, the hub torques are the only external forces. If this is the case, $\{RF\}$ can be expressed as,

$$\{RF\}=\begin{Bmatrix} I_{22} \\ [B] \end{Bmatrix}\begin{Bmatrix} T_1 \\ T_2 \end{Bmatrix} \tag{2}$$

where T_1 and T_2 are the respective torques acting on the hubs of the first and the second links. The matrix $[B]$ represents the projection of the hub torques on the system dominant elastic modes. I_{22} is the 2×2 identity matrix. The detail description of each individual term of equations (1) and (2) can be referred to Xia (1992).

For end-point tracking, the control objective is to command the end-point of a flexible manipulator moving along a specified trajectory. The control inputs are the torques acting on the hub of each individual link. In order to design a controller to achieve the control objective, which is specified in the Cartesian space in this paper, the control specifications need to be transferred to the joint space.

For a flexible manipulator, the end-point motion includes the contribution of both the elastic deformation and the joint motion. Therefore, the end-point position of a flexible manipulator can not be directly transferred to the joint angular positions as in the case of rigid manipulators. In order to establish the relationship between the Cartesian space and the joint space, an imaginary manipulator, which consists of a set of virtual links (Asada et al , 1990), is introduced here. The virtual link, which is rigid and straight, shares the same end-point of the corresponding flexible link. The joint angle of the two adjacent virtual links is defined as the elastic joint angle. The imaginary manipulator and the elastic joint angles for the flexible two-link manipulator under study are defined in Figure 1.

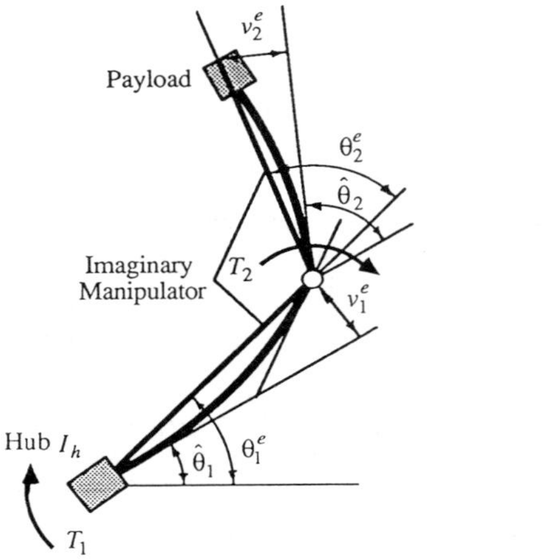

Figure 1 Schematic Diagram of a Flexible Two-Link Manipulator and Its Imaginary Manipulator

Using the imaginary manipulator, the specification of the end-point position of the flexible manipulator can be easily transferred to that of the elastic joint angles by the inverse kinematics of the rigid manipulators. For the flexible two-link manipulator, once the trajectory of the end-point $x_d^e(t)$ and $y_d^e(t)$ is specified, the corresponding elastic joint angles $\theta_{1d}^e(t)$ and $\theta_{2d}^e(t)$ of the first and the second link can be obtained as,

$$\theta_{2d}^e(t)=A\tan 2(\frac{\sqrt{1-a^2}}{a}) \tag{3}$$

$$\theta_{1d}^e(t)=A\tan 2\left(\frac{y_d^e(t)}{x_d^e(t)}\right)-A\tan 2\left(\frac{a_2\sin\theta_{2d}^e(t)}{a_1+a_2\cos\theta_{2d}^e(t)}\right) \tag{4}$$

where a_1 and a_2 are the length of the two flexible links, respectively, and,

$$a=\frac{[x_d^e(t)^2+y_d^e(t)^2]-(a_1^{\ 2}+a_2^{\ 2})}{2a_1a_2} \tag{5}$$

From Figure 1, the elastic joint angles can be expressed in terms of the joint coordinate angles $\hat{\theta}_1$ and $\hat{\theta}_2$ and the tip elastic deformations v_1^e and v_2^e.

$$\begin{Bmatrix} \theta_1^e \\ \theta_2^e \end{Bmatrix}=\begin{Bmatrix} \hat{\theta}_1 \\ \hat{\theta}_2 \end{Bmatrix}+\begin{bmatrix} \frac{1}{a_1} & 0 \\ -\frac{1}{a_1} & \frac{1}{a_2} \end{bmatrix}\begin{Bmatrix} v_1^e \\ v_2^e \end{Bmatrix} \tag{6}$$

Therefore, from Part I, the relationship between the elastic joint angles and the system variable Y can be expressed as,

$$\begin{Bmatrix} \theta_1^e \\ \theta_2^e \end{Bmatrix}=CY \tag{7}$$

where C is the output matrix, and

$$C=\begin{bmatrix} I_{22} & [S] \end{bmatrix}$$

$$[S]=\begin{bmatrix} PP(1,3) & PP(1,4) \\ PP(2,3) & PP(2,4) \end{bmatrix}+\begin{bmatrix} \frac{1}{a_1} & 0 \\ -\frac{1}{a_1} & \frac{1}{a_2} \end{bmatrix}\times\begin{bmatrix} PP(n_1+1,3) & PP(n_1+1,4) \\ PP(n_1+n_2+1,3) & PP(n_1+n_2+1,4) \end{bmatrix}$$

in which $[PP]$ is the system dominant mode shape function matrix, n_1 is the number of the elastic freedoms of link 1, and n_2 is the number of the elastic freedoms of link 2.

Equations (1), (2) and (7) represent the dynamics of the flexible two link manipulator with the hub torques as the control inputs and the elastic joint angles as the output variables. For the given hub torques, the elastic joint angles can be predicted from equations (1), (2), and (7). Since a given end-point trajectory can be transferred to the trajectories of the elastic joint angles in the joint space through equations (3) and (4), the control objective for the end-point tracking can be equivalently specified as to command the elastic joint angles moving along the desired trajectories. Therefore, the focus of the end-point tracking control will be on the design of a feedforward controller that provides the desired hub torques making the elastic joint angles, which can be predicted from equations (1), (2), and (7), follow the respective desired trajectories, which can be obtained from equations (3) and (4) for the specified end-point trajectory.

3. Computation of Control Torques

In order to develop a control law that achieves the control objective, the system dynamic equations should be transferred to the representation in terms of the elastic joint angles and the dominant elastic modes. From equation (7), the relationship between the elastic joint angles and the system rigid as well as elastic modes can be expressed as,

$$\begin{Bmatrix} \theta_1^e \\ \theta_2^e \end{Bmatrix} = \begin{Bmatrix} \eta_1 \\ \eta_1 \end{Bmatrix} + [S]\begin{Bmatrix} z_1 \\ z_2 \end{Bmatrix} \tag{8}$$

Using this transformation formula, the dynamic equation of the system can be rewritten in terms of the elastic joint angles and the dominant elastic modes as,

$$[MM]\ddot{X} + [CC]\dot{X} + [KK]X + NN(\dot{\theta}_1^e, \dot{\theta}_2^e, \{z\}, \{\dot{z}\}) = \{FF\} \tag{9}$$

where $X = [(\theta_1^e, \theta_2^e), (z_1, z_2)]^T$, which can be related to Y as follows,

$$X = [Q]^{-1}Y \tag{10}$$

in which,

$$[Q] = \begin{bmatrix} I_{22} & -[S] \\ 0_{22} & I_{22} \end{bmatrix}$$

where 0_{22} is the 2×2 null matrix and therefore,

$$[MM] = [RM][Q] = \begin{bmatrix} [M_{rr}] & [M_{ru}] \\ [M_{ur}] & [M_{uu}] \end{bmatrix}$$

$$[CC] = [RC][Q] = \begin{bmatrix} [C_{rr}] & [C_{ru}] \\ [C_{ur}] & [C_{uu}] \end{bmatrix}$$

$$[KK] = [RK][Q] = \begin{bmatrix} 0_{22} & 0_{22} \\ 0_{22} & [K_{uu}] \end{bmatrix}$$

$$NN(\dot{\theta}_1^e, \dot{\theta}_2^e, \{z\}, \{\dot{z}\}) = RN(\dot{\eta}_1, \dot{\eta}_2, \{z\}, \{\dot{z}\}) = \begin{Bmatrix} \begin{Bmatrix} N_1 \\ N_2 \end{Bmatrix} \\ \begin{Bmatrix} N_3 \\ N_4 \end{Bmatrix} \end{Bmatrix}$$

$$\{FF\} = \{RF\}$$

It is worth noting that the time derivative of matrix $[Q]$ is neglected in equation (9). This assumption is valid because the change of matrix $[S]$ with link configuration is slow, and its time derivative can be neglected.

The system dynamic equation can be further separated into two parts,

$$[M_{rr}]\begin{Bmatrix} \ddot{\theta}_1^e \\ \ddot{\theta}_2^e \end{Bmatrix} + [M_{ru}]\begin{Bmatrix} \ddot{z}_1 \\ \ddot{z}_2 \end{Bmatrix} + [C_{rr}]\begin{Bmatrix} \dot{\theta}_1^e \\ \dot{\theta}_2^e \end{Bmatrix} + [C_{ru}]\begin{Bmatrix} \dot{z}_1 \\ \dot{z}_2 \end{Bmatrix} + \begin{Bmatrix} N_1 \\ N_2 \end{Bmatrix} = \begin{Bmatrix} T_1 \\ T_2 \end{Bmatrix} \tag{11}$$

and

$$[M_{ur}]\begin{Bmatrix} \ddot{\theta}_1^e \\ \ddot{\theta}_2^e \end{Bmatrix} + [M_{uu}]\begin{Bmatrix} \ddot{z}_1 \\ \ddot{z}_2 \end{Bmatrix} + [C_{ur}]\begin{Bmatrix} \dot{\theta}_1^e \\ \dot{\theta}_2^e \end{Bmatrix} + [C_{uu}]\begin{Bmatrix} \dot{z}_1 \\ \dot{z}_2 \end{Bmatrix}$$
$$+ [K_{uu}]\begin{Bmatrix} z_1 \\ z_2 \end{Bmatrix} + \begin{Bmatrix} N_3 \\ N_4 \end{Bmatrix} = [B]\begin{Bmatrix} T_1 \\ T_2 \end{Bmatrix} \tag{12}$$

It can be seen from equation (11) that if both the elastic joint motion and the elastic deformation are available, the desired hub torques T_1 and T_2 for achieving the specified end-point trajectory can be calculated.

4. Estimation of Elastic Deformation

Once the desired end-point motion is specified, the desired elastic joint motion can be obtained from the inverse kinematics of the imaginary manipulator. However, the elastic deformation of the flexible links remains unknown. Therefore, in order to achieve the control objective, the elastic deformation of the system must be estimated based on the given end-point motion. As a matter of fact, estimating the elastic deformation for a given end-point motion is the major difficulty for the tracking control of flexible manipulators.

To estimate the elastic deformation of the system, equation (11) can be rewritten as,

$$\begin{Bmatrix} T_1 \\ T_2 \end{Bmatrix} = [M_{rr}]\begin{Bmatrix} \ddot{\theta}_1^e \\ \ddot{\theta}_2^e \end{Bmatrix} + [M_{ru}]\begin{Bmatrix} \ddot{z}_1 \\ \ddot{z}_2 \end{Bmatrix} + [C_{rr}]\begin{Bmatrix} \dot{\theta}_1^e \\ \dot{\theta}_2^e \end{Bmatrix} + [C_{ru}]\begin{Bmatrix} \dot{z}_1 \\ \dot{z}_2 \end{Bmatrix} + \begin{Bmatrix} N_1 \\ N_2 \end{Bmatrix} \tag{13}$$

Substituting equation (13) into equation (12), it becomes,

$$[M_z]\begin{Bmatrix} \ddot{z}_1 \\ \ddot{z}_2 \end{Bmatrix} + [C_z]\begin{Bmatrix} \dot{z}_1 \\ \dot{z}_2 \end{Bmatrix} + [K_z]\begin{Bmatrix} z_1 \\ z_2 \end{Bmatrix} = \{U_z\} \tag{14}$$

where

$$[M_z] = [M_{uu}] - [B][M_{ru}]$$

$$[C_z] = [C_{uu}] - [B][C_{ru}]$$

$$[K_z] = [K_{uu}]$$

$$\{U_z\} = \{[B][M_{rr}] - [M_{ur}]\}\begin{Bmatrix} \ddot{\theta}_1^e \\ \ddot{\theta}_2^e \end{Bmatrix} + \{[B][C_{rr}] - [C_{ur}]\}\begin{Bmatrix} \dot{\theta}_1^e \\ \dot{\theta}_2^e \end{Bmatrix}$$
$$+ [B]\begin{Bmatrix} N_1 \\ N_2 \end{Bmatrix} - \begin{Bmatrix} N_3 \\ N_4 \end{Bmatrix}$$

4.1 Configuration-Dependent Eigenvalues

Equation (14) represents the dynamics of the elastic motion of the system, which can be represented in the state space form as,

$$\dot{Z} = A_z Z + B_z U_z \tag{15}$$

where $Z = [(z_1, z_2), (\dot{z}_1, \dot{z}_2)]^T$, and

$$A_z = \begin{bmatrix} 0_{22} & I_{22} \\ -[M_z]^{-1}[K_z] & -[M_z]^{-1}[C_z] \end{bmatrix} \qquad B_z = \begin{Bmatrix} 0_{22} \\ [M_z]^{-1} \end{Bmatrix}$$

Ideally, equation (15) should be used to estimate the link deflection for a specified end-point motion. However, this dynamic equation is unstable. In other words, positive eigenvalues of matrix A_z exist. As a matter of fact, this instability problem is due to the non-colocated nature of the end-point control of a flexible manipulator. The non-colocated control scheme results in the nonminimum phase zeros of the system transfer function from the control inputs to the system outputs. When the inverse model of such a system is formulated, instability problem will be introduced by these nonminimum phase zeros. This problem is well known for the case of flexible one-link manipulators (Cannon and Schmitz, 1984), and several approaches have been proposed to counteract this problem for the end-point tracking control (Kwon and Book, 1990, Menq and Xia, 1992). This problem also exists for the end-point control of multi-link flexible manipulators. In equation (13), the control torques are calculated based on the desired end-point motions, which is in fact the inverse model of the flexible two-link manipulator. This inverse model of the non-colocated system will be unstable. Such an instability problem appears in estimating the elastic motion based on the desired end-point motion and is represented by the positive eigenvalues of matrix A_z. For the case of flexible multi-link manipulators, in addition to instability, the system is also configuration-dependent. For the two-link flexible manipulator having the geometry shown in Figures 2 and 3 of Part I and the material properties shown in Table 1 of Part I, four

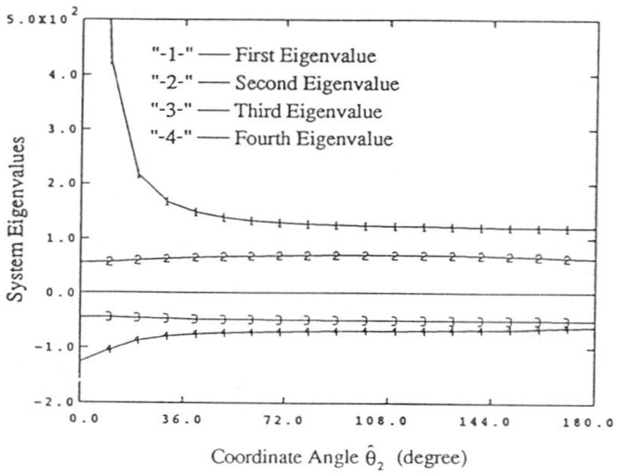

Figure 2 The Eigenvalues of the System

eigenvalues are calculated and shown in Figure 2. It can be seen from the figure that the system has two positive as well as two negative eigenvalues and all of them are functions of system configurations.

4.2 Stable Subsystem and Unstable Subsystem

In order to obtain the solution of the system that has positive eigenvalues, Kwon and Book (1990) proposed an approach, in which the system is separated into two subsystems, namely, the unstable subsystem and the stable one. Such an approach was applied to the end-point tracking control of a flexible one link manipulator. In their approach, the noncausal part of the elastic motion is obtained by integrating the equations of the unstable subsystem from the end to the beginning of the specified time range. The result was proven to be very good although the integration must be performed before actual control taking place.

In this paper, real time control of the end-point tracking for a flexible two-link manipulator is attempted. By diagonalizing equation (15), the system can be separated into two subsystems.

$$\begin{Bmatrix} \dot{Z}_u \\ \dot{Z}_s \end{Bmatrix} = \begin{bmatrix} A_z^u & 0_{22} \\ 0_{22} & A_z^s \end{bmatrix} \begin{Bmatrix} Z_u \\ Z_s \end{Bmatrix} + \begin{Bmatrix} B_z^u \\ B_z^s \end{Bmatrix} \{U_z\} \qquad (16)$$

and

$$Z = [\Gamma] \begin{Bmatrix} Z_u \\ Z_s \end{Bmatrix} \qquad (17)$$

where $[\Gamma]$ is the eigenvector matrix of A_z, and

$$A_z^u = \begin{bmatrix} \lambda_1^u & 0 \\ 0 & \lambda_2^u \end{bmatrix} \quad A_z^s = \begin{bmatrix} \lambda_1^s & 0 \\ 0 & \lambda_2^s \end{bmatrix} \quad \begin{Bmatrix} B_z^u \\ B_z^s \end{Bmatrix} = [\Gamma]^{-1}[B_z]$$

in which λ_1^u and λ_2^u are the two positive eigenvalues of the system, and λ_1^s and λ_2^s are the two negative eigenvalues. From equation (16), we have,

$$\dot{Z}_u = A_z^u Z_u + B_z^u \{U_z\} \qquad (18)$$

and

$$\dot{Z}_s = A_z^s Z_s + B_z^s \{U_z\} \qquad (19)$$

Due to the fact that the flexible manipulator is controlled by a digital computer, the information for the desired torques and the estimated elastic motion is only needed at each sampling instant. In order to improve the efficiency of the control algorithm, the differential equations can be approximated by difference equations (Nicosia, etc. 1990). For a linear time invariant or a slowly varying system, such an approximation can be best achieved by the z-transform. For a time varying system whose system parameters change rapidly, such an approximation can be more accurately performed with special techniques (Young, 1979, Schoukens, 1990). For the system under study, the eigenvalues of the system are shown in Figure 2. It can be seen from the figure that for a small sampling time, the eigenvalues can be assumed to be constant within the sampling interval. Therefore, z-transform is used here and the two differential equations (18) and (19) can be approximated by two difference equations as,

$$Z_u(k+1) = F_z^u Z_u(k) + G_z^u \{U_z(k)\} \qquad (20)$$

and

$$Z_s(k+1) = F_z^s Z_s(k) + G_z^s \{U_z(k)\} \qquad (21)$$

in which,

$$F_z^u(k) = \exp\left[A_z^u(kT_s)T_s\right] = \begin{bmatrix} z_1^u & 0 \\ 0 & z_2^u \end{bmatrix}$$

$$F_z^s(k) = \exp\left[A_z^s(kT_s)T_s\right] = \begin{bmatrix} z_1^s & 0 \\ 0 & z_2^s \end{bmatrix}$$

$$G_z^u(k) = \int_0^{T_s} \exp\left[A_z^u(kT_s)(T_s - \tau)\right] B_z^u(\tau) d\tau$$

$$G_z^s(k) = \int_0^{T_s} \exp\left[A_z^s(kT_s)(T_s - \tau)\right] B_z^s(\tau) d\tau$$

where T_s is the sampling time, z_1^u and z_2^u are the two unstable eigenvalues of the discrete system, and z_1^s and z_2^s are the two stable eigenvalues. As mentioned above, the flexible two link manipulator is a time varying system, and matrices A_z and B_z are functions of the link configurations. Therefore, matrices $[\Gamma]$, F_z^u, F_z^s, G_z^u, and G_z^s will also be the functions of the link configurations. For the two-link flexible manipulator as shown in Figure 1, matrices $[\Gamma]$, F_z^u, F_z^s, G_z^u, and G_z^s are the functions of coordinate angle $\hat{\theta}_2$. Due to the fact that it is not computationally feasible to calculate the eigenvector matrix and the z-transform in each sampling interval, it is desired to express matrices $[\Gamma]$, F_z^u, F_z^s, G_z^u, and G_z^s in terms of the link coordinate angle $\hat{\theta}_2$. Since such an explicit function can not be obtained theoretically, numerical interpolation method is employed.

In order to perform the numerical interpolation, matrices $[\Gamma]$, F_z^u, F_z^s, G_z^u, and G_z^s are calculated for different link configurations, i.e., for different values of the link coordinate angle $\hat{\theta}_2$. In this paper, the numerical data is obtained at every ten degree of the link coordinate angle $\hat{\theta}_2$. From the numerical results, it is found that matrices $[\Gamma]$, F_z^u, F_z^s, G_z^u, and G_z^s are symmetrical to the configuration in which the two links are coaxial. By using the symmetrical property and the numerical data at different configurations, the numerical interpolation is performed by fitting the entries of matrices $[\Gamma]$, F_z^u, F_z^s, G_z^u, and G_z^s with cubic splines in the first and the second quadrant of $\hat{\theta}_2$. The eigenvalues of the discrete system obtained through the numerical interpolation are compared with the real discrete eigenvalues and are shown in Figure 3. It can be seen from the figures that these fitted functions agree well with the real values. Similar approaches can also be applied to obtain the entries of matrices $[\Gamma]$, G_z^u, and G_z^s as the functions of the link coordinate angle $\hat{\theta}_2$.

Using the fitted functions of matrices $[\Gamma]$, F_z^u, F_z^s, G_z^u, and G_z^s, the stable part of the elastic motion can be estimated directly from equation (21), in which $U_z(k)$ is obtained based on the desired elastic joint motions and the estimated nonlinear forces.

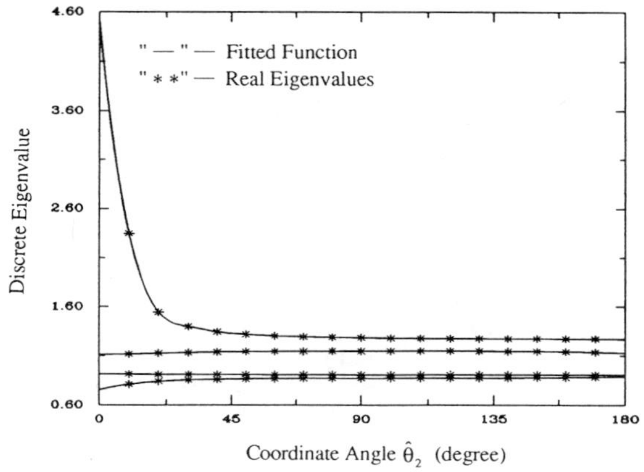

Figure 3 The Discrete Eigenvalues

$$\{U_z\} = \{[B][M_{rr}] - [M_{ur}]\}\begin{Bmatrix}(\ddot{\theta}_1{}^e)_d \\ (\ddot{\theta}_2{}^e)_d\end{Bmatrix} + \{[B][C_{rr}] - [C_{ur}]\}\begin{Bmatrix}(\dot{\theta}_1{}^e)_d \\ (\dot{\theta}_2{}^e)_d\end{Bmatrix} + [B]\begin{Bmatrix}\hat{N}_1 \\ \hat{N}_2\end{Bmatrix} - \begin{Bmatrix}\hat{N}_3 \\ \hat{N}_4\end{Bmatrix} \quad (22)$$

where $\hat{N}_1$, $\hat{N}_2$, $\hat{N}_3$, and $\hat{N}_4$ are the estimated nonlinear forces based on the desired elastic joint motion and the estimated elastic deformation of the previous step.

For the unstable part, direct solution is not possible since the system is unstable. In this paper, by applying the precision tracking control method for discrete time nonminimum phase systems (Menq and Xia, 1990), a digital filter will be formulated to replace the unstable subsystem in estimating the noncausal part of the elastic motion.

4.3 Preview Filter for Unstable Subsystem

In order to apply the precision tracking control method in estimating the noncausal part of the elastic motion, equation (20) is rewritten in the transfer function form.

$$\begin{aligned} Z_u(z) &= [zI_{22} - F_z^u]^{-1} B_z^u\{U_z(z)\} \\ &= \frac{1}{(z - z_1^u)(z - z_2^u)} Adj(zI_{22} - F_z^u) B_z^u\{U_z(z)\} \end{aligned} \quad (22)$$

The instability problem can be clearly identified from equation (22) since the poles z_1^u and z_2^u are outside of the unit circle. This instability problem is due to the noncausal characteristics of the inverse of a non-colocated system. The non-colocated nature of the end-point control of a flexible manipulator results in nonminimum phase zeros of the system transfer function. The poles z_1^u and z_2^u of equation (22) are in fact the two nonminimum phase zeros of the flexible two link manipulators with end-point control. Consequently, equation (22) cannot be directly used to estimate the noncausal part of the elastic deformation.

In this paper, a digital filter is proposed to replace the unstable subsystem in estimating the noncausal part of the elastic motion. The proposed filter is stable, and its frequency response is very close to the unstable subsystem within the desired frequency range. Due to the noncausal characteristics of the elastic motion, the proposed filter requires preview information of the desired end-point motion so as to accurately estimate the elastic motion of the system. For the unstable subsystem as shown in equation (22), such a digital filter is formulated as,

$$\hat{Z}_u = PRV(z^{-1}) \frac{(z^{-1} - z_1^u)(z^{-1} - z_2^u)}{(1 - z_1^u)^2 (1 - z_2^u)^2} Adj(zI_{22} - F_z^u) B_z^u\{U_z\} \quad (23)$$

in equation (23), the second term has the same phase shift and the same dc gain as that of the two unstable poles of equation (22), and the first term $PRV(z^{-1})$ is a zero phase preview filter, which is used to compensate the gain error between the second term of equation (23) and the two unstable poles. Such a digital filter has been successfully applied for the end-point tracking control of a flexible one-link manipulator (Menq and Xia, 1992), and good tracking performance has been obtained for both the computer simulation and the experimental results.

In this paper, the same idea is extended to the case of flexible two-link manipulators. If the proposed digital filter has the same frequency response as that of the unstable subsystem, accurate estimation of the elastic motion is expected. To achieve this objective, the preview filter $PRV(z^{-1})$ is designed as,

$$PRV(z^{-1}) = \sum_{n=0}^{N} \alpha_n (z + z^{-1})^n \quad (24)$$

where N is the order of the preview filter, and it also represents the number of preview steps required. From equations (22) and (23), the frequency response ratio between the proposed digital filter and the unstable subsystem can be written as,

$$R(e^{j\omega T_s}) = \left[\sum_{n=0}^{N} 2^n \alpha_n \cos^n(\omega T_s)\right] \frac{(e^{-j\omega T_s} - z_1^u)(e^{j\omega T_s} - z_1^u)}{(1 - z_1^u)^2} \times \frac{(e^{-j\omega T_s} - z_2^u)(e^{j\omega T_s} - z_2^u)}{(1 - z_2^u)^2} \quad (25)$$

where ω is the frequency, T_s is the sampling time. It can be seen from equation (25) that the phase shift of $R(e^{j\omega T_s})$ is zero. In other words, both the digital filter and the unstable subsystem have the same phase shift. In order to make the gain of the digital filter be close to the unstable subsystem within a given bandwidth ω_b, the coefficients α_i's can be determined by minimizing the penalty function J (Menq and Chen, 1992),

$$J = \int_0^{\omega_b} \varepsilon^2(\omega) d\omega \quad (26)$$

where $\varepsilon(\omega)$ is the gain error, and

$$\varepsilon(\omega) = \left[\sum_{n=0}^{N} 2^n \alpha_n \cos^n(\omega T_s)\right] \frac{(e^{-j\omega T_s} - z_1^u)(e^{j\omega T_s} - z_1^u)}{(1 - z_1^u)^2} \times \frac{(e^{-j\omega T_s} - z_2^u)(e^{j\omega T_s} - z_2^u)}{(1 - z_2^u)^2} - 1 \quad (27)$$

It can be seen from equations (26) and (27) that if the order N of the preview filter $PRV(z^{-1})$ is sufficiently high, the gain error $\varepsilon(\omega)$ can be made arbitrarily small within the given bandwidth ω_b. In other words, the gain of both the digital filter and the unstable subsystem can be made arbitrarily close within the given bandwidth ω_b. For $\hat{\theta}_2 = 60^\circ$, the two unstable poles are $z_1^u = 1.2951$ and $z_2^u = 1.1474$. For these two unstable poles, when the order of the preview filter is selected to be 2, and the desired frequency range is set to be 5 Hz, the preview filter $PRV(z^{-1})$ is designed as,

$$PRV(z^{-1}) = -61.3300 + 3.8486(z + z^{-1}) + 13.6582(z + z^{-1})^2 \quad (28)$$

With this design, the phase shift of the digital filter as shown in equation (23) will be the same as that of the unstable system, and the gain ratio of the proposed filter over the unstable subsystem will be close to one within the desired frequency range as shown in Figure 4. Therefore, the frequency response of the digital filter as shown in equations (23) will be very close to that of the unstable subsystem in the desired frequency range. Due to the fact that the system unstable poles z_1^u and z_2^u are the functions of the link coordinate

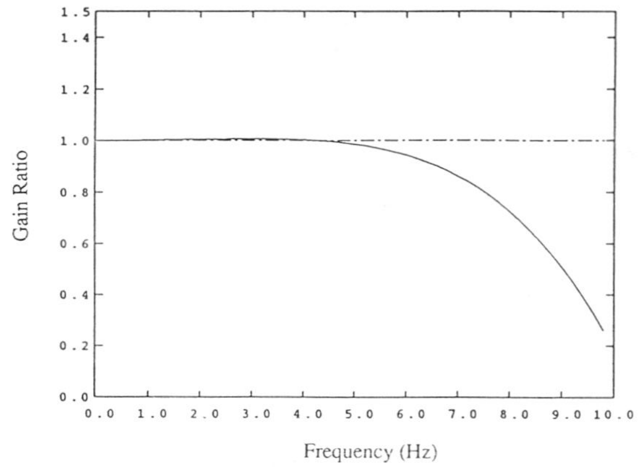

Figure 4 The Gain Ratio between the Elastic Motion Estimator and the Unstable Subsystem

angle $\hat{\theta}_2$, their values will be different at each sampling instant. Since an explicit formula for the design of the optimal preview filter is available (Menq and Chen, 1992), the coefficients of the preview filter $PRV(z^{-1})$ can be updated in each sampling interval according to the configuration-dependent unstable poles.

Using the designed preview filter $PRV(z^{-1})$, the noncausal part of the elastic motion can be estimated using the proposed digital filter. Once both the stable and unstable parts of the elastic motion are obtained, the system elastic motion can be estimated as,

$$\hat{Z} = [\Gamma]\begin{Bmatrix} \hat{Z}_u \\ Z_s \end{Bmatrix} \tag{29}$$

where $\hat{Z} = [(\hat{z}_1, \hat{z}_2), (\dot{\hat{z}}_1, \dot{\hat{z}}_2)]^T$.

5. Tracking Control of End-Point Motion

The control scheme proposed in this paper consists of a feedback controller, a command feedforward controller, and an elastic motion estimator. The control scheme is shown in Figure 5. The command feedforward controller serves to deal with specific commands and the feedback controller takes care of other sources of errors, including those not anticipated by the designer.

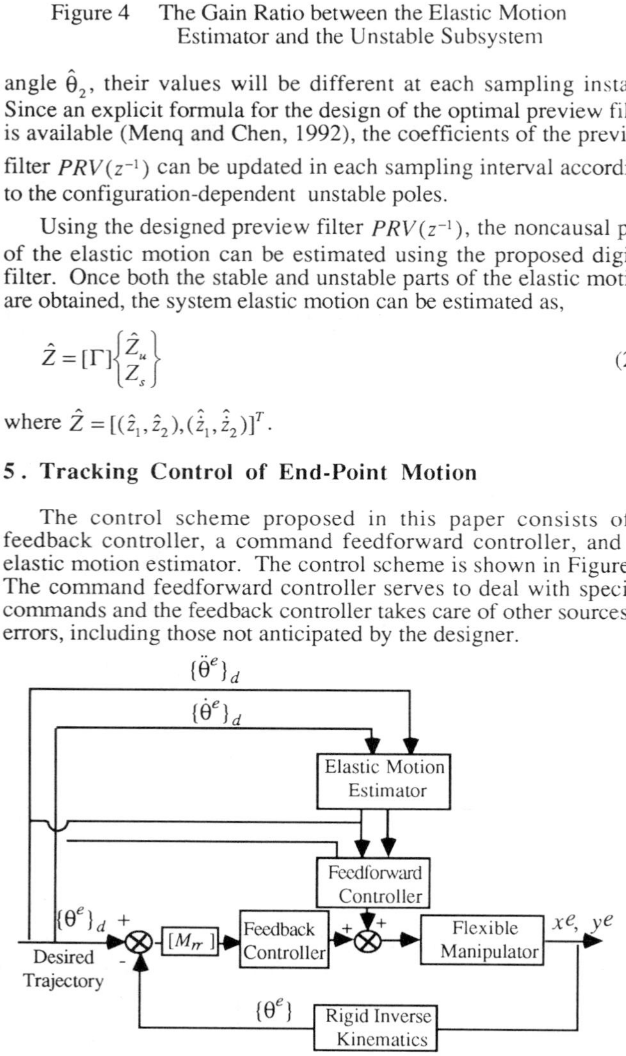

Figure 5 The End-Point Tracking Control Scheme

Based on the given end-point trajectory and the estimated elastic deformation, the feedforward controller is designed such that the desired control torques can be computed as follows,

$$\begin{Bmatrix} T_1 \\ T_2 \end{Bmatrix} = [M_{rr}]\begin{Bmatrix} (\ddot{\theta}_1^e)_d \\ (\ddot{\theta}_2^e)_d \end{Bmatrix} + [M_{ru}]\begin{Bmatrix} \ddot{\hat{z}}_1 \\ \ddot{\hat{z}}_2 \end{Bmatrix} + [C_{rr}]\begin{Bmatrix} (\dot{\theta}_1^e)_d \\ (\dot{\theta}_2^e)_d \end{Bmatrix}$$

$$+[C_{ru}]\begin{Bmatrix} \dot{\hat{z}}_1 \\ \dot{\hat{z}}_2 \end{Bmatrix} + \begin{Bmatrix} \hat{N}_1 \\ \hat{N}_2 \end{Bmatrix} \tag{30}$$

If the system model of the flexible two-link manipulator is exact, the tracking control with the proposed feedforward controller is expected to achieve the performance of near perfection. The possible tracking error is induced by two approximations used in the elastic motion estimator. The first one is the use of one step delay in estimating the nonlinear forces $\hat{N}_1$, $\hat{N}_2$, $\hat{N}_3$, and $\hat{N}_4$. The second approximation is the use of the preview filter, in which small errors of the gain ratio exist between the proposed digital filter and the unstable subsystem as shown in Figure 4. Since only few dominant modes are required for the design of the feedforward controller and the elastic motion estimator, the required computation will not be a major problem when using the state-of-art micro-computer.

A PD type feedback controller is used in this paper. By incorporating the feedback control, the tracking control law becomes,

$$\begin{Bmatrix} T_1 \\ T_2 \end{Bmatrix} = [M_{rr}]\left\{\begin{Bmatrix} (\ddot{\theta}_1^e)_d \\ (\ddot{\theta}_2^e)_d \end{Bmatrix} + [K_P]\left\{[K_D]\{\dot{e}\} + \{e\}\right\}\right\} + [M_{ru}]\begin{Bmatrix} \ddot{\hat{z}}_1 \\ \ddot{\hat{z}}_2 \end{Bmatrix}$$

$$+[C_{rr}]\begin{Bmatrix} (\dot{\theta}_1^e)_d \\ (\dot{\theta}_2^e)_d \end{Bmatrix} + [C_{ru}]\begin{Bmatrix} \dot{\hat{z}}_1 \\ \dot{\hat{z}}_2 \end{Bmatrix} + \begin{Bmatrix} \hat{N}_1 \\ \hat{N}_2 \end{Bmatrix} \tag{31}$$

where the tracking error $\{e\}$ is defined as,

$$\{e\} = \begin{Bmatrix} (\theta_1^e)_d - \theta_1^e \\ (\theta_2^e)_d - \theta_2^e \end{Bmatrix} \tag{32}$$

in which the real elastic joint angles θ_1^e and θ_2^e are obtained from the inverse kinematics of the imaginary manipulator based on the measured end-point positions and equations (3) and (4).

By using the proposed tracking control law, the system error dynamics can be obtained from equation (11) as,

$$\{\ddot{e}\} + \left([M_{rr}]^{-1}[C_{rr}] + [K_P][K_D]\right)\{\dot{e}\} + [K_P]\{e\} = \{D(t)\} \tag{33}$$

where

$$\{D(t)\} = [M_{rr}]^{-1}\left\{ [M_{ru}]\begin{Bmatrix} \ddot{z}_1 - \ddot{\hat{z}}_1 \\ \ddot{z}_2 - \ddot{\hat{z}}_2 \end{Bmatrix} + [C_{ru}]\begin{Bmatrix} \dot{z}_1 - \dot{\hat{z}}_1 \\ \dot{z}_2 - \dot{\hat{z}}_2 \end{Bmatrix} \right.$$

$$\left. + \begin{Bmatrix} N_1 - \hat{N}_1 \\ N_2 - \hat{N}_2 \end{Bmatrix} \right\} \tag{34}$$

When the estimated elastic deformation is accurate, the proposed feedforward controller will be close to the system inverse model. If this is the case, the excitation function $\{D(t)\}$ of the error dynamics should be small and can be viewed as a disturbance function. The feedback controller can then be designed based on equation (33) by placing the poles of the error dynamics to maximize the bandwidth while having sufficient damping. For our case, the PD controller is designed as,

$$[K_P] = \begin{bmatrix} 10 & 0 \\ 0 & 10 \end{bmatrix} \qquad [K_D] = \begin{bmatrix} 2 & 0 \\ 0 & 2 \end{bmatrix}$$

6. Simulation Results

The performance of the proposed end-point tracking control scheme is examined via computer simulation. The desired end-point trajectory of the flexible two-link manipulator is shown in Figure 6. Its time histories in X-coordinate and Y-coordinate are shown by solid lines in Figures 7 and 8, respectively. The real end-point motion of the system by applying the proposed feedforward controller is also shown by "*" in the same figures. It can be seen from the figures that the end-point follows the desired trajectory

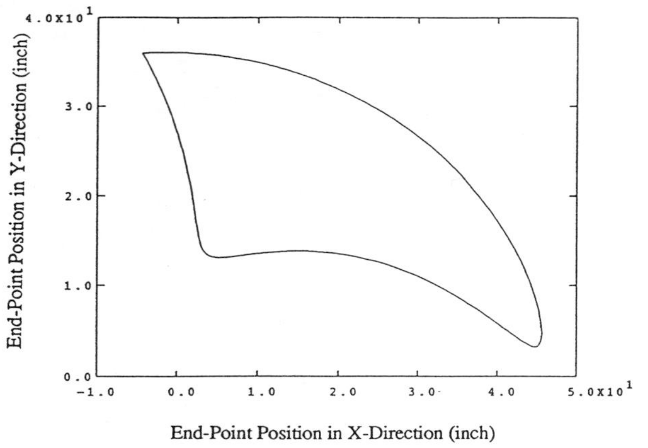

Figure 6 The Desired End-Point Trajectory

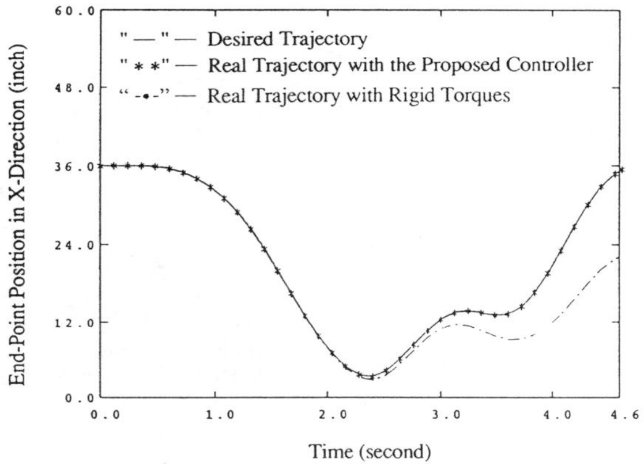

Figure 7 The End-Point Tracking Performance in X-Coordinate

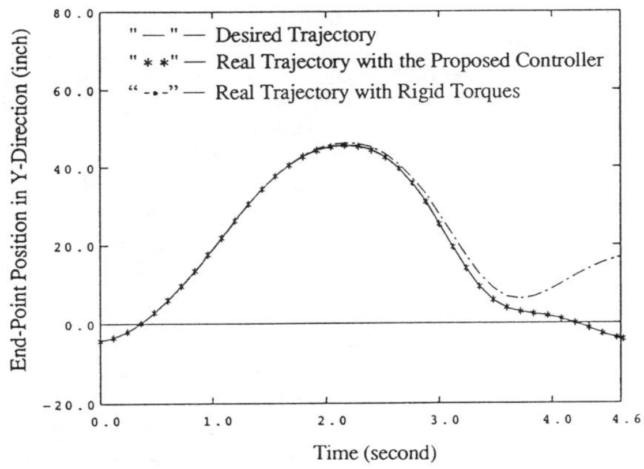

Figure 8 The End-Point Tracking Performance in Y-Coordinate

with good accuracy. It is worth noting that if the elastic motion estimator is turned off, the control torques obtained from equation (30) will be the same as that for rigid manipulators. For the case that the elastic motion estimator is turned off, the end-point motion is shown by dash-dot lines in Figures 7 and 8. It can be seen from the figures that the end-point tracking performance is greatly improved by using the elastic motion estimator. By using equations (3) and (4), the desired trajectories for the elastic joint angles are shown in Figures 9 and 10. The real elastic joint motion by using

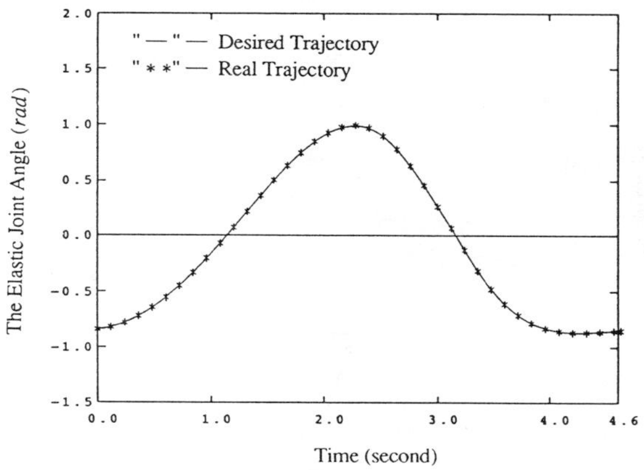

Figure 9 The Tracking Performance of the First Elastic Joint Angle θ_1^e

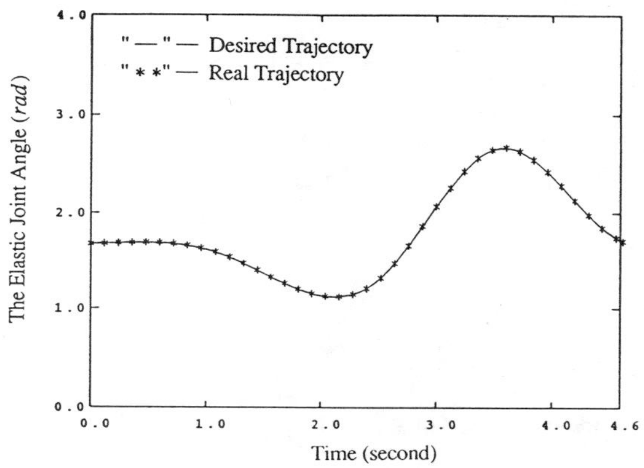

Figure 10 The Tracking Performance of the Second Elastic Joint Angle θ_2^e

the proposed control law, as shown in equation (31), is also shown in the same figures. It can be seen from the figures that very good tracking performance has been achieved

The performance of the elastic motion estimator is also examined in the computer simulation. For the given end-point trajectory, the estimated elastic deformations at the tip of the two links are shown in Figures 11 and 12. The real elastic deformations are also shown in the same figures. It can be seen from the figures that both the estimated and the real elastic deformations agree well. A small delay of the estimated elastic motion exists, which is due to the one-step delay required in estimating the nonlinear forces. It is worth noting that in this paper, in order to test the designed elastic motion estimator and the proposed tracking controller, the acceleration profiles of the elastic joint angles are designed to have discontinuities for the given end-point trajectory, as shown in Figures 13 and 14. From the tracking control law shown in equation (31), the accelerations of the elastic joint angles are directly related to the hub torques, thereby directly related to the elastic deformations of the links. Therefore, the elastic deformations are expected to have abrupt variations at the discontinuities of the accelerations of the elastic joint angles. Such variations are observed in Figures 11 and 12. It can be seen that even when the discontinuities of the acceleration present, the elastic motion estimator still gives good estimation of the elastic deformation.

It can be seen from the simulation results that both the elastic motion estimator and the tracking control scheme proposed in this paper work very well, and a very good end-point tracking

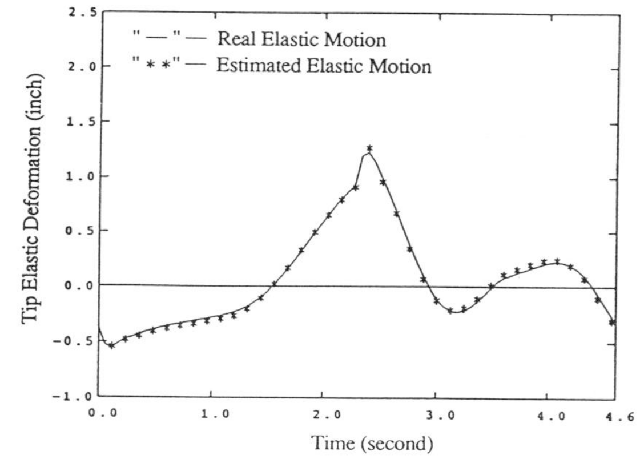

Figure 11 The Tip Elastic Deformation of Link 1

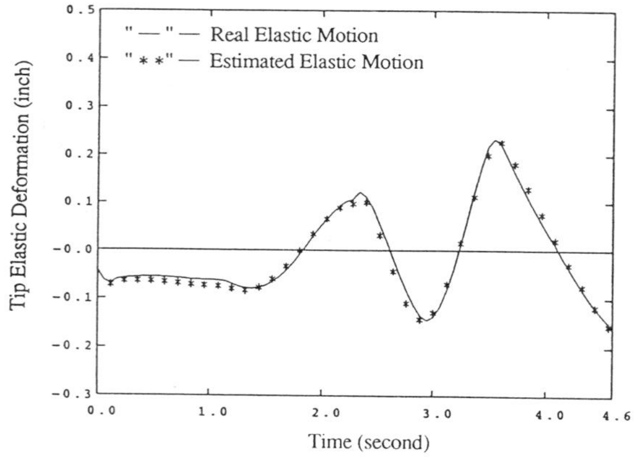

Figure 12 The Tip Elastic Deformation of Link 2

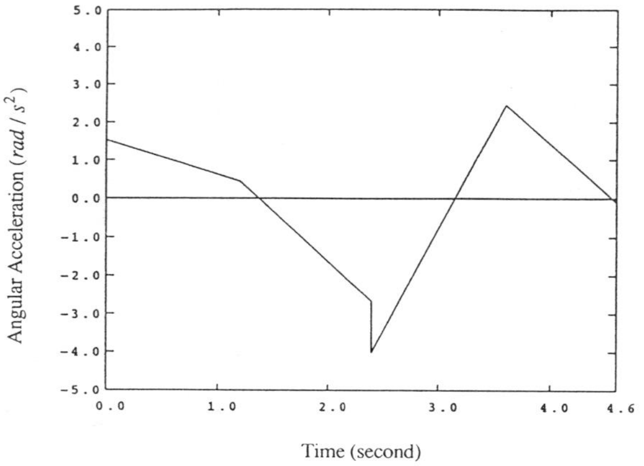

Figure 13 The Acceleration Profile of the First Elastic Joint Angle θ_1^e

performance has been achieved. Moreover, the control algorithm proposed in this paper is computationally efficient. Neither the off-line time integration nor the iterative process in the frequency domain is involved. This tracking control scheme can be easily implemented for real time control.

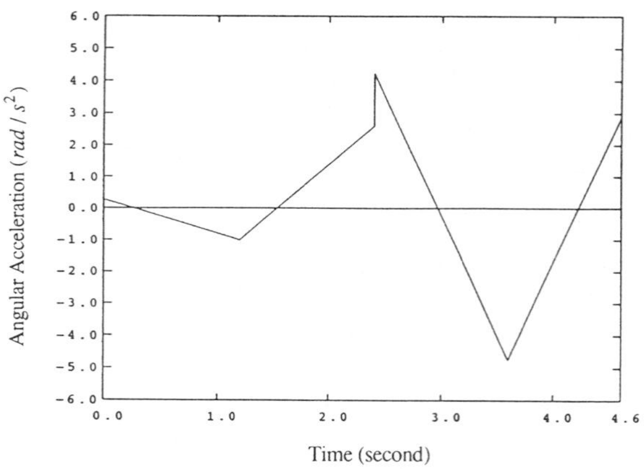

Figure 14 The Acceleration Profile of the Second Elastic Joint Angle θ_2^e

7. Conclusion

The end-point tracking control of a flexible two-link manipulator is studied in this paper. Based on the reduced order model of the flexible two-link manipulator, the characteristics of the system's inverse model is investigated and a tracking control scheme is developed for the end-point tracking control of flexible two link manipulators. The control scheme proposed in this paper consists of a feedback controller, a command feedforward controller, and an elastic motion estimator. The command feedforward controller serves to deal with specific commands based on the inverse model of the system, and the feedback controller takes care of other sources of errors, including those not anticipated by the designer.

Due to the non-colocated nature of the system, the inverse model is divided into two subsystems, namely, the stable subsystem and the unstable one, corresponding to the noncausal part and the causal part of the system's elastic motion. A digital filter is formulated to replace the unstable subsystem so as to estimate the noncausal part of the elastic motion associated with a specified end-point motion. For the design of the filter, the frequency response ratio between the filter and the unstable subsystem is used as the criterion, the objective of which is to have the frequency response ratio have zero phase shift as well as unity gain within a specified frequency range. It is shown that due to the noncausal characteristics of the unstable subsystem, preview information of the input trajectory is required for implementing the proposed digital filter, and the accuracy of the estimation increases as the preview steps increases. Based on the stable subsystem and the proposed digital filter, a time-varying estimator is designed to estimate the elastic motion of the system when the end-point motion is specified. The command feedforward controller is then used to calculate the required control torques based on the estimated elastic deformation and the desired end-point motion.

Simulation results show that very good tracking performance can be achieved by using the proposed tracking controller. The performance of the elastic motion estimator is also examined via computer simulation. It is shown that both the estimated and real elastic deformations agree well. A small delay of the estimated elastic motion exists, which is due to the one-step delay required in estimating the nonlinear forces. The control algorithm proposed in this paper is very computationally efficient and can be easily implemented for real time control.

References

Asada, H., Ma, Z.-D. and Tokumaru, H., 1990, "Inverse Dynamics of Flexible Robot Arms: Modeling and Computation for Trajectory Control," ASME Journal of Dynamic Systems, Measurement, and Control, Vol. 112, pp. 177-185.

Bayo, E., Papadopoulos, P., Stubbe, J. and Serna, M. A., 1989, "Inverse Dynamics and Kinematics of Multi-Link Elastic Robots: An Iterative Frequency Domain Approach," The International Journal of Robotics Research, Vol. 8, pp 49-62.

Cannon, R.H. and Schmitz, E., 1984, "Initial Experiments on the End-Point Control of a Flexible One-Link Robot," The International Journal of Robotics Research, Vol.3, pp. 62-75.

Domroese, M.K. and Hardt, D.E., 1990, "Investigation of Control Schemes and End-point Feedback for Flexible manipulators," Proceeding of 1990 Japan-U.S.A. Symposium on Flexible Automation, pp. 933-943.

Kwon, D. and Book, W.J., 1990, "An Inverse Dynamic Method Yielding flexible Manipulator State Trajectories," Proceedings of American Control Conference, pp. 186-193.

Jonker, B., 1990, "A Finite Element Dynamic Analysis of Flexible Manipulators," The International Journal of Robotics Research, vol. 9, pp. 59-74.

Menq, C.-H. and Chen, J.J., 1992, "Precision Tracking Control of Discrete Time Nonminimum-Phase Systems," to be presented in the 1992 American Control Conference.

Menq, C.-H. and Xia, Z., 1990, "Characterization and Compensation of Nonminimum Phase Zeroes in Digital Systems," Proceeding of 1990 ASME WAM, Robotics Research, pp. 15-23.

Menq, C.-H. and Xia, Z., 1992, "End-Point Tracking Control of A Flexible One-Link Manipulator," to be presented in the 1992 US-Japan Symposium on Flexible Automation and submitted to the ASME Journal of Dynamic Systems, Measurement and Control.

Nagarajan, S. and Turcic, D.A., 1990, "Lagrangian Formulation of the Equations of Motion for Elastic Mechanisms with Mutual Dependence Between Rigid Body and Elastic Motions. Part I: Element Level Equations," ASME, Journal of Dynamic Systems, Measurement, and Control, vol. 112, pp. 203-214.

Nicosia, S., Tomei, P. and Tornambe, A., 1990, "Discrete-Time Modeling of Flexible Robots," Proceedings of the 29th Conference on Decision and Control, pp. 539-544.

Oakley, C.M. and Cannon, R.H., Jr., 1988, "Initial Experiments on the Control of a Two-link Manipulator with a very Flexible forearm," Proceeding of 1988 American Control Conference, pp. 996-1002.

Oakley, C.M. and Cannon, R.H., Jr., 1989, "End-point Control of a Two-link Manipulator with a very Flexible Forarm: Issues and Experiments," Proceeding of 1989 American Control Conference, pp. 1381-1388.

Schoukens, J., 1990, "Modeling of Continuous Time Systems Using a Discrete Time Representation," Automatica, vol. 26, pp. 579-583.

Singh, S.N. and Schy, A.A., 1986, "Control of Elastic Robotic Systems by Nonlinear Inversion and Modal Damping," ASME, Journal of Dynamic Systems, Measurement, and Control, pp. 180-189.

Sunada, W. Dubowsky, S., 1981, "The Applications of Finite Element Methods to the Dynamic Analysis of Flexible Spatial and Co-Planar Linkage systems," ASME Journal of Mechanical Design, Vol. 103, pp. 643-651.

Turcic, D.A., and Midha, A., 1984, "Generalized Equations of Motion for the Dynamic Analysis of Elastic Mechanism Systems," ASME Journal of Dynamic Systems, Measurements, and Control, Vol. 106, pp. 243-248.

Xia, Z., 1988, "A New Approach to the Kineto-Elasto-Dynamic Analysis of Robotic Manipulators," Proceedings of IEEE International Conference on Systems, Man and Cybernetics, pp. 98-102.

Xia, Z., 1992, "Modeling and Control of Flexible Manipulators," Ph.D. dissertation, The Ohio State University.

Young, P., 1979, "Parameter Estimation for Continuous-Time Models-A Survey", Proc. 5th IFAC Symposium Identification on System Parameter Estimation, Darmstadt, F.R.G., pp. 17-41.

Yurkovich, S., Tzes, A.P., Lee, I. and Hillsley, K., 1990, "Control and System Identification of a Two-link Flexible Manipulator," Proceeding of 1990 IEEE International Conference on Robotics and Automation, pp. 1626-1631.

Xia, Z., 1992, "Modeling and Control of Flexible Manipulators," Ph.D. dissertation, The Ohio State University.

AMD-Vol. 141/DSC-Vol. 37, Dynamics of Flexible Multibody Systems:
Theory and Experiment
ASME 1992

AUTOMATIC CONSTRUCTION OF EQUATIONS OF MOTION FOR RIGID-FLEXIBLE MULTIBODY SYSTEMS

Parviz E. Nikravesh
Department of Aerospace and Mechanical Engineering
University of Arizona
Tucson, Arizona

Jorge A. C. Ambrosio
CEMUL, Technical University of Lisbon
Lisbon, Portugal

ABSTRACT

A method for the systematic formulation of the equations of motion for multibody systems containing rigid and flexible bodies is presented. The method of joint coordinates for deriving the minimum number of equations of motion is utilized for rigid bodies, and the finite element method is employed for flexible bodies. The equations of motion for flexible bodies are simplified in several steps: lumped mass assumption, static condensation, and modal superposition. The combined rigid and flexible body formulation can be used to simulate dynamic response in a variety of applications, such as ride handling, rollover, and crash analyses of vehicles; space structure analyses; and biomechanical problems.

1. INTRODUCTION

The equations of motion for multibody systems can be presented in a variety of forms. For rigid multibody systems, some formulations express these equations in terms of a large number of absolute accelerations, usually referred to as the Cartesian coordinate formulation. These formulations are attractive due to their simplicity and ease of manipulation; however, the drawback is that they form a mixed set of differential-algebraic equations. The numerical solution of these equations is computationally inefficient and, furthermore, special procedures must be followed to avoid, or to control, a phenomenon known as the constraint violation. Another approach to deriving the equations of motion is to use a smaller set of accelerations, for example, a minimal set of accelerations equal to the number of degrees of freedom of the system. The numerical integration of these equations is by far more efficient than that of the absolute coordinate formulation. One such method, which has the simplicity of the absolute coordinate formulations and the computational efficiency of a small or even a minimal number equations, is the so-called joint coordinate method.

For flexible bodies or structures, the finite element method is used to discretize the equations of motion. However, most standard finite element formulations do not allow for the gross rigid body motion of the structure, particularly for large rotations. In multibody systems, the finite element formulation of a flexible body requires the inclusion of terms representing the gross motion of the body, the distributed flexibility, and the coupling between them. In a general finite element formulation, each node may have six degrees of freedom, which results in a large set of highly nonlinear equations of motion. Due to the form of these equations and the nature of the nonlinear terms, the popular numerical integration algorithms used in structural dynamics cannot be applied here. The numerical procedures used in multibody dynamics require the explicit solution of the equations of motion for the accelerations. Due to the large number of equations and degrees of freedom, the computation time associated with the integration of the equations of motion can become quite extensive. Therefore, various techniques have been devised to reduce the number of equations in order to gain computational efficiency.

The equations of motion for rigid multibody systems are initially presented in terms of absolute accelerations. Then, a velocity transformation process is applied to transform these equations into a smaller set, which is made up of differential equations without any constraints for open-loop systems. These differential equations are expressed in terms of joint accelerations describing the relative motion between bodies. For closed-loop systems, the equations of motion in terms of the joint acceleration are first expressed as a set of mixed differential-algebraic equations. In a second step, these equations are transformed into a minimal set of differential equations without any kinematic constraints.

The equations of motion for a flexible body are directly obtained from other references without showing the derivation due to their complexity. In these equations, the flexible body is modeled by the finite element method, allowing six degrees of freedom per finite element node. These equations are then simplified using a lumped mass assumption and a transformation of the coordinate system for nodal accelerations. Since in many applications of multibody dynamics a flexible part is attached to a rigid member, the equations of motion for a flexible body and a rigid body are combined to form the equations of motion for a rigid-flexible body. The assumption that a flexible part is attached to a rigid element does not limit the generality of the formulation. Furthermore, two techniques are discussed for the elimination of the nodal rotational degrees of freedom from the equations of motion. A static condensation technique removes the rotational coordinates, but it allows for inclusion of material or geometric nonlinearity. A modal superposition technique, on the other hand, allows for the removal of the rotational degrees of freedom and also the removal of any additional degrees of freedom, but it restricts the use of linear material properties in the flexible body.

The equations presented in this paper have been incorporated in several computer programs and have been used to simulate the dynamics of a variety of mechanical systems. Due to space limitations, no survey of the developments in multibody dynamics is provided in this paper. Only one methodology for formulating the equations of motion for rigid and flexible multibody systems is presented. References are provided for detailed derivations and discussions.

2. RIGID MULTIBODY SYSTEMS

The equations of motion for rigid multibody systems can be presented in several forms. In this paper, the equations of motion are presented in terms of a large set of mixed differential-algebraic equations for open- and closed-loop systems. Then, the equations are reduced to a smaller set, in certain cases to the minimal set, using the concept of joint coordinates and velocity transformation.

2.1 Absolute Coordinate Formulation

In order to describe the position of a rigid body in a global nonmoving xyz coordinate system, it is sufficient to specify the spatial location of the origin and the angular orientation of a body-fixed $\xi\eta\zeta$ coordinate system, as shown in Fig. 1. For the ith body in a multibody system, vector $\mathbf{q}_i$ denotes a vector of coordinates that contains a vector of Cartesian translational coordinates, $\mathbf{r}_i$, and a set of rotational coordinates, such as Euler parameters. Matrix $\mathbf{A}_i$ represents the 3×3 rotational transformation of the $\xi\eta\zeta_i$ axes relative to the xyz axes. A 6-vector of velocities for body i is defined as $\mathbf{v}_i$, which contains a 3-vector of translational velocities, $\dot{\mathbf{r}}_i$, and a 3-vector of angular velocities, $\boldsymbol{\omega}_i$. The components of the angular velocity vector $\boldsymbol{\omega}_i$ are defined in the xyz coordinate system rather than the body-fixed coordinate system. A 6-vector of accelerations for this body is denoted by $\dot{\mathbf{v}}_i$, which contains $\ddot{\mathbf{r}}_i$ and $\dot{\boldsymbol{\omega}}_i$. For a detailed discussion on the absolute coordinate formulation refer to reference 1.

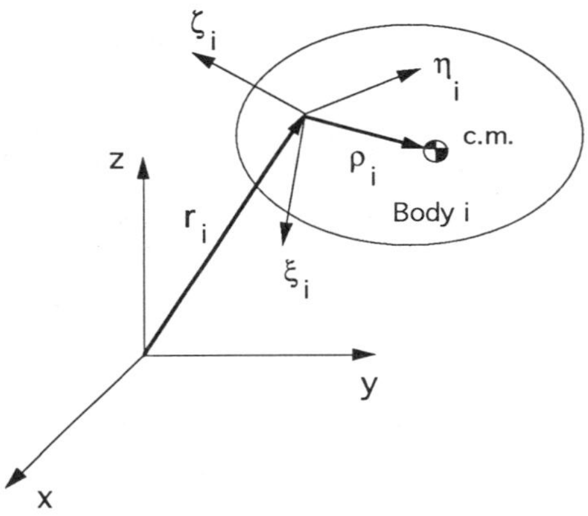

Fig. 1. Spatial configuration of a rigid body in a non-moving reference frame.

A multibody system may contain b interconnected bodies. Therefore, vectors of coordinates, velocities, and accelerations are denoted by $\mathbf{q}$, $\mathbf{v}$, and $\dot{\mathbf{v}}$, which contain the elements of $\mathbf{q}_i$, $\mathbf{v}_i$, and $\dot{\mathbf{v}}_i$, respectively, for $i = 1, ..., b$. The dimension of $\mathbf{v}$ (and $\dot{\mathbf{v}}$) is $n = 6 \times b$.

The translational and rotational equations of motion for a single unconstrained rigid body can be written as

$$\mathbf{M}_i \dot{\mathbf{v}}_i = \mathbf{g}_i \ , \tag{1}$$

where [2]

$$\mathbf{M}_i = \begin{bmatrix} m\mathbf{I} & -m\tilde{\boldsymbol{\rho}} \\ m\tilde{\boldsymbol{\rho}} & \boldsymbol{J} \end{bmatrix}_i \tag{2}$$

$$\mathbf{g}_i = \begin{bmatrix} \mathbf{f}-m\tilde{\boldsymbol{\omega}}\tilde{\boldsymbol{\omega}}\boldsymbol{\rho} \\ \mathbf{n}-\tilde{\boldsymbol{\omega}}\boldsymbol{J}\boldsymbol{\omega} \end{bmatrix}_i \ . \tag{3}$$

In these expressions, m_i is the mass, $\mathbf{f}_i$ is the resultant force acting at the origin, $\mathbf{n}_i$ is the resultant moment, and $\boldsymbol{J}_i$ is the inertia tensor about the body-fixed coordinate axes expressed in the global coordinate system. Vector $\boldsymbol{\rho}_i$ locates the center of the mass of the body relative to its origin. Note that all of the vectors in these equations are expressed in terms of their global xyz components. Furthermore, note that if the origin of the body-fixed coordinate system is at its mass center, i.e., $\boldsymbol{\rho}_i = \mathbf{0}$, then Eq. (1) will become the standard Newton-Euler equations, with the rotational equations being expressed in terms of the global coordinate system instead of the more customary body-fixed coordinate system. Note that an overhead ~ denotes a 3×3 skew-symmetric matrix made out of the components of a 3-vector. For example, $\tilde{\boldsymbol{\omega}}$ is made out of the components of $\boldsymbol{\omega}$.

For a system of b unconstrained bodies (no kinematic joints), the equations of motion of Eq. (1) are repeated for all b bodies and are written in matrix form as

$$\mathbf{M}\dot{\mathbf{v}} = \mathbf{g} \ . \tag{4}$$

Equation (4) represents n second-order differential equations.

For a system of b constrained bodies, the kinematic joints between the bodies can be described by m independent holonomic constraints as

$$\boldsymbol{\Phi}(\mathbf{q}) = \mathbf{0} \ . \tag{5}$$

The first and second time derivatives of the constraints yield the kinematic velocity and acceleration equations,

$$\dot{\boldsymbol{\Phi}} \equiv \mathbf{D}\mathbf{v} = \mathbf{0} \tag{6}$$

$$\ddot{\boldsymbol{\Phi}} \equiv \mathbf{D}\dot{\mathbf{v}} + \dot{\mathbf{D}}\mathbf{v} = \mathbf{0} \ , \tag{7}$$

where $\mathbf{D}$ is the coefficient matrix of the velocity constraints and is a modified form of the constraints' Jacobian. For this system of b bodies, Eq. (4) is modified to [1]

$$\mathbf{M}\dot{\mathbf{v}} - \mathbf{D}^T\boldsymbol{\lambda} = \mathbf{g} \ , \tag{8}$$

where $\boldsymbol{\lambda}$ represents a vector of m Lagrange multipliers and the term $\mathbf{D}^T\boldsymbol{\lambda}$ represents the joint reaction forces and moments. Equations (5)-(8) represent a set of differential-algebraic equations of motion for a constrained multibody system when absolute coordinates are used. The number of degrees of freedom of such a system is $k = n - m$.

The bodies and the kinematic joints in a multibody system may form an open-loop system (a tree structure) or the system may contain one or more closed loops. The equations of motion in terms of the absolute accelerations will have the same form whether the system is open or closed loop.

2.2 Joint Coordinate Formulation: Open-Loop Systems

An open-loop multibody system may be attached to the ground by a kinematic joint, in which case it is a nonfloating system; otherwise, it is a *floating* system. In a floating system, one of the bodies is selected as the reference, or *base*, body; in a nonfloating system, the ground is the reference body. An open-loop system forms a tree structure with the base (or reference) body as the *root* and the bodies farthest away from the root in each branch as the *leaves*. Two adjacent bodies connected to each other by a kinematic joint are referenced to as bodies "j" and "j-1," where body j-1 is closer to the base; the joint between these bodies will carry the index "j."

In an open-loop system, the relative configuration of two adjacent bodies is defined by one or more so-called joint coordinates equal in number to the number of relative degrees of freedom between the two bodies. The vector of joint coordinates for an open-loop system is denoted by $\boldsymbol{\theta}$ containing all of the joint coordinates and, if it is floating, all the absolute coordinates of the base body. The vectors of joint velocities and accelerations are defined as $\dot{\boldsymbol{\theta}}$ and $\ddot{\boldsymbol{\theta}}$. Vector $\dot{\boldsymbol{\theta}}$ has a dimension equal to the number of degrees of freedom of the system.

In an open-loop system, if vectors $\boldsymbol{\theta}$ and $\dot{\boldsymbol{\theta}}$ are known, then vectors $\mathbf{q}$ and $\mathbf{v}$ can be evaluated. This process Ais done by moving from the base body toward the leaves and calculating the absolute coordinates and velocities of body j from the absolute coordinates and velocities of body j-1 and the joint coordinates and velocities of joint j. It can be shown that there is a linear transformation between $\dot{\boldsymbol{\theta}}$ and $\mathbf{v}$ [2, 3]:

$$\mathbf{v} = \mathbf{B}\dot{\boldsymbol{\theta}} . \tag{9}$$

The time derivative of this equation yields a transformation formula for the accelerations:

$$\dot{\mathbf{v}} = \mathbf{B}\ddot{\boldsymbol{\theta}} + \dot{\mathbf{B}}\dot{\boldsymbol{\theta}} . \tag{10}$$

The elements of matrix $\mathbf{B}$ can be found from expressions for the absolute velocities of each body in terms of the joint velocities between that body and the base. An observation of the nonzero entries of $\mathbf{B}$ reveals that its structure follows the connectivity and the topology of the open-loop system and, furthermore, that it is made of small submatrices representing different joints [3-5]. One characteristic of matrix $\mathbf{B}$ is that it is orthogonal to $\mathbf{D}$; i.e.,

$$\mathbf{DB} = \mathbf{0} . \tag{11}$$

Substitution of Eq. (10) into Eq. (8), premultiplying the result by $\mathbf{B}^T$, and using Eq. (11) yields

$$\mathcal{M}\ddot{\boldsymbol{\theta}} = \mathcal{f} , \tag{12}$$

where

$$\mathcal{M} = \mathbf{B}^T\mathbf{MB} \tag{13}$$

$$\mathcal{f} = \mathbf{B}^T(\mathbf{g} - \mathbf{M}\dot{\mathbf{B}}\dot{\boldsymbol{\theta}}) . \tag{14}$$

Equation (12) represents the generalized equations of motion for an open-loop multibody system when the number of the selected coordinates is equal to the number of degrees of freedom.

2.3 Joint Coordinate Formulation: Closed-Loop Systems

To derive the equations of motion for a multibody system containing one or more closed loops, each closed loop is cut at one of the kinematic joints in order to obtain an open-loop system. For this reduced open-loop system, the joint coordinates are defined, as for any open-loop system, without defining any joint coordinates for the cut joints. If the cut joints are reassembled, the joint coordinates within the loops are no longer independent [5].

Assume that a cut joint connects two bodies, i and j. This joint provides algebraic constraints between $\mathbf{q}_i$ and $\mathbf{q}_j$, which can be expressed as [1]

$$\boldsymbol{\Phi}^*(\mathbf{q}_i, \mathbf{q}_j) = \mathbf{0} , \tag{15}$$

where a superscript asterisk (*) denotes a quantity associated with a cut joint. The time derivative of Eq. (15) is

$$\begin{aligned} \dot{\boldsymbol{\Phi}}^* &\equiv \mathbf{D}^*\mathbf{v} = \mathbf{0} \\ &= \mathbf{C}\dot{\boldsymbol{\theta}} = \mathbf{0} , \end{aligned} \tag{16}$$

where Eq. (9) has been used and

$$\mathbf{C} = \mathbf{D}^*\mathbf{B} . \tag{17}$$

Since $\mathbf{q}_i$ and $\mathbf{q}_j$ can be evaluated from the joint coordinates, Eq. (15) can be expressed implicitly as

$$\boldsymbol{\Psi}(\boldsymbol{\theta}) = \mathbf{0} . \tag{18}$$

Hence, Eq. (16) is the time derivative of Eq. (18), i.e.,

$$\dot{\boldsymbol{\Psi}} \equiv \mathbf{C}\dot{\boldsymbol{\theta}} = \mathbf{0} , \tag{19}$$

and its time derivative is

$$\ddot{\boldsymbol{\Psi}} \equiv \mathbf{C}\ddot{\boldsymbol{\theta}} + \dot{\mathbf{C}}\dot{\boldsymbol{\theta}} = \mathbf{0} . \tag{20}$$

Equations (18)-(20) provide the constraints on the joint coordinates, velocities, and accelerations for the closed loops. Note that in these equations only the joints that are within a closed loop are involved.

Due to the topology of a closed loop, some of the equations in Eq. (18) may be redundant and, therefore must be eliminated. Accordingly, some of the rows of the product $\mathbf{D}^*\mathbf{B}$ must be eliminated to obtain the Jacobian of the constraints on the joint coordinates. In our discussion, we assume that Eqs. (18)-(20) and matrix $\mathbf{C}$ represent the constraints and the Jacobian after the elimination of redundant equations. The elements of matrix $\mathbf{C}$ can also be generated directly from the topology of the closed loop instead of multiplying $\mathbf{D}^*$ and $\mathbf{B}$ matrices.

Equation (12) can now be modified for a closed-loop system as

$$\mathcal{M}\ddot{\boldsymbol{\theta}} - \mathbf{C}^T\boldsymbol{\nu} = \mathcal{f} , \tag{21}$$

where $\boldsymbol{\nu}$ is a vector of Lagrange multipliers. Equations (18)-(21) represent the equations of motion for a multibody system when the number of selected joint coordinates is greater than the number of degrees of freedom.

2.4 Minimal Joint Coordinate Formulation: Closed-Loop Systems

The Lagrange multipliers of Eq. (21) can be eliminated in order to obtain a minimal set of equations of motion in terms of a set of independent joint accelerations. For this purpose, a subset of the joint coordinates, vector $\boldsymbol{\theta}$, is selected containing a set of independent joint coordinates, $\boldsymbol{\theta}_{(i)}$. We assume that the velocity constraints of Eq. (19) can be converted to the form [5]

$$\dot{\boldsymbol{\theta}} = \mathbf{E}\dot{\boldsymbol{\theta}}_{(i)} , \tag{22}$$

where $\mathbf{E}$ is a velocity transformation matrix. The joint velocities outside the closed loops and the independent joint velocities within the closed loops are not affected by this transformation; i.e., there are some identity submatrices in $\mathbf{E}$ corresponding to the independent joint velocities. One characteristic of matrix $\mathbf{E}$ is that it is orthogonal to $\mathbf{C}$; i.e.,

$$\mathbf{CE} = \mathbf{0} . \tag{23}$$

The time derivative of Eq. (22) gives

$$\ddot{\boldsymbol{\theta}} = \mathbf{E}\ddot{\boldsymbol{\theta}}_{(i)} + \dot{\mathbf{E}}\dot{\boldsymbol{\theta}}_{(i)} . \tag{24}$$

Matrices $\mathbf{E}$ and $\dot{\mathbf{E}}$ can be found in either explicit or numerical forms for most closed kinematic loops [5].

Substituting Eq. (24) into Eq. (21), premultiplying by $\mathbf{E}^T$, and using Eq. (23) yields

$$\underline{\mathcal{M}}\ddot{\boldsymbol{\theta}}_{(i)} = \underline{\mathcal{f}} \tag{25}$$

where

$$\underline{\mathcal{M}} = \mathbf{E}^T\mathcal{M}\mathbf{E} \tag{26}$$

$$\underline{\mathcal{f}} = \mathbf{E}^T(\mathcal{f} - \mathcal{M}\dot{\mathbf{E}}\dot{\boldsymbol{\theta}}_{(i)}) . \tag{27}$$

Equation (25) represents the minimum number of equations of motion describing the dynamics of a multibody system containing closed kinematic loops.

3. FLEXIBLE BODIES

We can assume that a body in a multibody system is partially deformable and partially rigid. The analysis of the partially deformable body is initially approached by treating the rigid and flexible parts as separate bodies [4]. In this sense, the rigid and flexible bodies are designated by Π and Γ. This system of two bodies represents the partially deformable body provided that the points that belong to the boundary ψ of body Γ and body Π have the same global displacements. Such a result is obtained by imposing kinematic constraints between body Γ and body Π, as will be seen later.

3.1 Flexible Body Formulation

To position a flexible body Γ in the nonmoving reference frame xyz, it is sufficient to specify the spatial location of the origin and the angular orientation of its body-fixed frame, $(\xi\eta\zeta)_\Gamma$, fixed to a point on the body, as shown in Fig. 2. Using the finite element method, $\mathbf{u}'$ is defined as the vector of nodal displacements relative to the body-fixed coordinate frame. Note that, in this paper, a superscript prime (') denotes the components of a vector in a body-fixed coordinate system. For a node k, $\mathbf{u}'_k$ is a 6-vector containing the translational and rotational displacements $\boldsymbol{\delta}'_k$ and $\boldsymbol{\theta}'_k$. The position of node k in the inertial frame is given by

$$\mathbf{d}_k = \mathbf{r} + \mathbf{s}_k + \boldsymbol{\delta}_k , \tag{28}$$

where $\mathbf{s}_k$ is the position vector of the node in the body-fixed system in its undeformed configuration. If the translational acceleration of node k relative to the inertial frame is denoted as $\mathbf{a}_k \equiv \ddot{\mathbf{d}}_k$, then the second time derivative of Eq. (28) yields

$$\mathbf{a}_k = \ddot{\mathbf{r}} + \mathbf{A}\tilde{\boldsymbol{\omega}}'\tilde{\boldsymbol{\omega}}'(\mathbf{s}'_k + \boldsymbol{\delta}'_k) - \mathbf{A}(\tilde{\mathbf{s}}'_k + \tilde{\boldsymbol{\delta}}'_k)\dot{\boldsymbol{\omega}}' + 2\,\mathbf{A}\tilde{\boldsymbol{\omega}}'\dot{\boldsymbol{\delta}}'_k + \mathbf{A}\ddot{\boldsymbol{\delta}}'_k , \tag{29}$$

where $\ddot{\mathbf{r}}$ and $\dot{\boldsymbol{\omega}}'$ are the translational and angular accelerations of the body-fixed reference frame $(\xi\eta\zeta)_\Gamma$. The angular acceleration of node k relative to the inertial reference frame is

$$\boldsymbol{\alpha}_k = \dot{\boldsymbol{\omega}} + \tilde{\boldsymbol{\omega}}\mathbf{A}\dot{\boldsymbol{\theta}}'_k + \mathbf{A}\ddot{\boldsymbol{\theta}}'_k . \tag{30}$$

A vector of nodal accelerations relative to the inertial frame is defined as

$$\ddot{\mathbf{q}}_f = \begin{bmatrix} \mathbf{a} \\ \boldsymbol{\alpha} \end{bmatrix} ,$$

which contains $\mathbf{a}_k$ and $\boldsymbol{\alpha}_k$ for all the nodes in the flexible body.

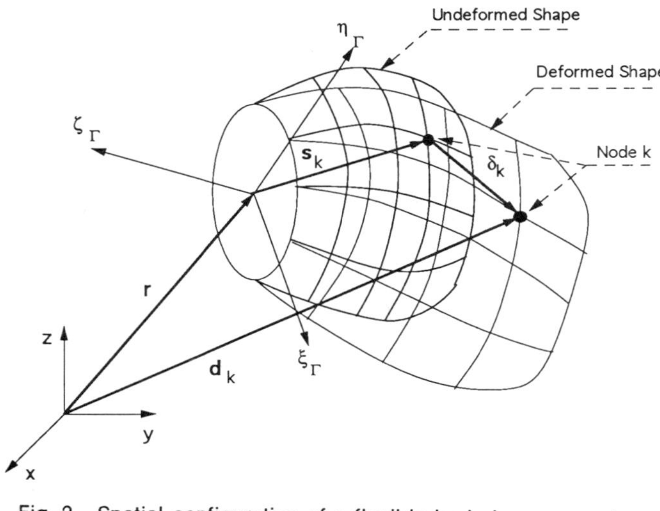

Fig. 2. Spatial configuration of a flexible body in a nonmoving reference frame.

The equations of motion for a flexible body can be derived using the principle of virtual work. If the finite element method is employed to describe the continuum mechanics equations, the result is written as [6-9]

$$\begin{bmatrix} \mathbf{M}_{rr} & \mathbf{M}_{r\phi} & \mathbf{M}_{rf} \\ \mathbf{M}_{\phi r} & \mathbf{M}_{\phi\phi} & \mathbf{M}_{\phi f} \\ \mathbf{M}_{fr} & \mathbf{M}_{f\phi} & \mathbf{M}_{ff} \end{bmatrix} \begin{bmatrix} \ddot{\mathbf{r}} \\ \dot{\boldsymbol{\omega}}' \\ \ddot{\mathbf{u}}' \end{bmatrix} = \begin{bmatrix} \mathbf{g}_r + \mathbf{b}_r \\ \mathbf{g}'_\phi + \mathbf{b}'_\phi \\ \mathbf{g}'_f + \mathbf{b}'_f \end{bmatrix} - \begin{bmatrix} 0 & 0 & 0 \\ 0 & 0 & 0 \\ 0 & 0 & \mathbf{C} \end{bmatrix} \begin{bmatrix} 0 \\ 0 \\ \dot{\mathbf{u}}' \end{bmatrix} - \begin{bmatrix} 0 & 0 & 0 \\ 0 & 0 & 0 \\ 0 & 0 & \mathbf{K} \end{bmatrix} \begin{bmatrix} 0 \\ 0 \\ \mathbf{u}' \end{bmatrix} , \tag{31}$$

where subscripts r, ϕ, and f denote translational, rotational, and nodal degrees of freedom. The damping and stiffness matrices are represented by $\mathbf{C}$ and $\mathbf{K}$. The mass submatrices $\mathbf{M}_{rr}$, $\mathbf{M}_{r\phi}$, $\mathbf{M}_{\phi r}$, and $\mathbf{M}_{\phi\phi}$ are associated with the gross motion of the body-fixed frame. The submatrix $\mathbf{M}_{ff}$ is the mass matrix associated with the deformation of body Γ, and it is the same as the mass matrix of the finite element model for small displacements. $\mathbf{M}_{rf}$, $\mathbf{M}_{\phi f}$, $\mathbf{M}_{fr}$, and $\mathbf{M}_{f\phi}$ are the mass submatrices responsible for the inertial coupling between the reference coordinates and the flexible coordinates. Vectors $\mathbf{g}_r$, $\mathbf{g}'_\phi$, and $\mathbf{g}'_f$ represent the externally applied forces/moments, and vectors $\mathbf{b}_r$, $\mathbf{b}'_\phi$, and $\mathbf{b}'_f$ represent the quadratic velocity terms.

The stiffness matrix may not be a constant due to geometric and material nonlinearities. In such cases, $\mathbf{u}'$ is a vector of incremental displacements and the vector of external applied forces/moments $\mathbf{g}'_f$ contains the nodal forces equivalent to the actual stresses of the finite elements.

3.2 Lumped Mass and Global Accelerations

The equations of motion for a flexible body as depicted by Eq. (31) present a full nonconstant mass matrix. The numerical procedures being used in multibody dynamics require the inversion or factorization of this matrix in every time step, hence the process can be computationally expensive. The form of Eq. (31) can be substantially simplified if (a) a lumped mass assumption is used and (b) the nodal accelerations $\ddot{\mathbf{u}}$ are substituted by a vector of nodal accelerations relative to the nonmoving reference frame. For this purpose, Eqs. (29) and (30) are evaluated for all the nodes and the result is substituted into Eq. (31) to obtain [4, 9, 10]

$$\sum_{i=1}^{n} (m\,\mathbf{a}')_i = \mathbf{g}'_r \tag{32}$$

$$\sum_{i=1}^{n} \left[m\,(\tilde{\mathbf{s}}' + \tilde{\boldsymbol{\delta}}')\,\mathbf{a}' \right]_i = \mathbf{g}'_\phi \tag{33}$$

$$\mathbf{M}_{ff}\,\ddot{\mathbf{q}}'_f = \mathbf{g}'_f - \mathbf{K}\,\mathbf{u}' - \mathbf{C}\,\dot{\mathbf{u}}' , \tag{34}$$

where n is the number of nodes in the finite element mesh and $\mathbf{a}'_i$ represents the absolute translational acceleration of node i expressed in the body-fixed reference frame. Equations (32) and (33) are associated with the gross translational and rotational motion of the flexible bodies while Eq. (34) is the equation of motion for the nodes. Note that due to the use of a lumped mass formulation, the mass matrix is written as $\mathbf{M}_{ff} = \text{diag}\,[m_1\mathbf{I}, \mathbf{0}, ..., m_n\mathbf{I}, \mathbf{0}]$, where m_j is the mass of node j and $\mathbf{I}$ and $\mathbf{0}$ are 3×3 identity and null matrices associated with the translational and rotational degrees of freedom, respectively.

4. RIGID-FLEXIBLE BODY

In the derivation of Eq. (34), no assumption was made on the location of the body-fixed coordinate frame $(\xi\eta\zeta)_\Gamma$. At this point, a new body-fixed coordinate system, $(\xi\eta\zeta)_\Pi$, is defined attached to the center of the mass of body Π. Moreover, it is assumed that $(\xi\eta\zeta)_\Gamma$ and $(\xi\eta\zeta)_\Pi$ coincide as shown in Fig. 3. Furthermore, assume that the flexible body Γ is attached to the rigid body Π by the nodes that belong to the boundary ψ. Then the Newton-Euler equations of motion for body Π can replace the equations of motion of the body-fixed reference frame $(\xi\eta\zeta)_\Gamma$, provided that proper kinematic constraints between the flexible and rigid bodies are enforced. For this purpose, the vectors of nodal accelerations, velocities, and displacements are partitioned as

$$\ddot{q}'_f = \begin{bmatrix} a' \\ \underline{a}' \\ \alpha' \\ \underline{\alpha}' \end{bmatrix}, \quad \dot{u}' = \begin{bmatrix} \dot{\delta}' \\ \dot{\underline{\delta}}' \\ \dot{\theta}' \\ \dot{\underline{\theta}}' \end{bmatrix}, \quad u' = \begin{bmatrix} \delta' \\ \underline{\delta}' \\ \theta' \\ \underline{\theta}' \end{bmatrix},$$

where an underscore denotes displacements of nodes of body Γ that belong to the boundary ψ. Herein, Eq. (34) is written as

$$\begin{bmatrix} M^* & 0 & 0 & 0 \\ 0 & \underline{M}^* & 0 & 0 \\ 0 & 0 & 0 & 0 \\ 0 & 0 & 0 & 0 \end{bmatrix} \begin{bmatrix} a' \\ \underline{a}' \\ \alpha' \\ \underline{\alpha}' \end{bmatrix} + \begin{bmatrix} C_{\delta\delta} & C_{\delta\underline{\delta}} & C_{\delta\theta} & C_{\delta\underline{\theta}} \\ C_{\underline{\delta}\delta} & C_{\underline{\delta}\underline{\delta}} & C_{\underline{\delta}\theta} & C_{\underline{\delta}\underline{\theta}} \\ C_{\theta\delta} & C_{\theta\underline{\delta}} & C_{\theta\theta} & C_{\theta\underline{\theta}} \\ C_{\underline{\theta}\delta} & C_{\underline{\theta}\underline{\delta}} & C_{\underline{\theta}\theta} & C_{\underline{\theta}\underline{\theta}} \end{bmatrix} \begin{bmatrix} \dot{\delta}' \\ \dot{\underline{\delta}}' \\ \dot{\theta}' \\ \dot{\underline{\theta}}' \end{bmatrix} + \begin{bmatrix} K_{\delta\delta} & K_{\delta\underline{\delta}} & K_{\delta\theta} & K_{\delta\underline{\theta}} \\ K_{\underline{\delta}\delta} & K_{\underline{\delta}\underline{\delta}} & K_{\underline{\delta}\theta} & K_{\underline{\delta}\underline{\theta}} \\ K_{\theta\delta} & K_{\theta\underline{\delta}} & K_{\theta\theta} & K_{\theta\underline{\theta}} \\ K_{\underline{\theta}\delta} & K_{\underline{\theta}\underline{\delta}} & K_{\underline{\theta}\theta} & K_{\underline{\theta}\underline{\theta}} \end{bmatrix} \begin{bmatrix} \delta' \\ \underline{\delta}' \\ \theta' \\ \underline{\theta}' \end{bmatrix} = \begin{bmatrix} g'_\delta \\ g'_{\underline{\delta}} \\ g'_\theta \\ g'_{\underline{\theta}} \end{bmatrix}. \qquad (35)$$

Without any loss of generality, it can be assumed that all of the boundary nodes are fixed to body Π. Then, $\underline{\delta}' = \underline{\theta}' = \dot{\underline{\delta}}' = \dot{\underline{\theta}}' = 0$ and $\ddot{\underline{\delta}}' = \ddot{\underline{\theta}}' = 0$ provide the necessary kinematic constraints between the motion of the rigid body and the motion of the boundary nodes. With the aid of the Lagrange multiplier technique, the constraint equations can be appended to the equations of motion of the rigid body Π and to the equations of motion of the boundary nodes in Eq. (35). After some manipulation, the Lagrange multipliers and the accelerations of the boundary nodes can be eliminated to obtain six equations of motion for the rigid portion of the rigid-flexible body [4, 9, 10]

$$(m + \bar{A}\underline{M}^*\bar{A}^T)\ddot{r} - \bar{A}\underline{M}^*S\dot{\omega}' = f + \bar{A}c'_\delta \qquad (36)$$

$$-(\bar{A}\underline{M}^*S)^T\ddot{r} + (J' + S^T\underline{M}^*S)\dot{\omega}' = n' - \tilde{\omega}'J'\omega' - S^Tc'_\delta - \bar{I}^Tc'_\theta\,, \qquad (37)$$

where m and J' are the mass and the inertia tensor of the rigid part expressed in the body-fixed reference frame. Vectors **f** and **n'** are the resultant external force and moment acting on body Π. The equations of motion for the unconstrained nodes are obtained directly from Eq. (35) as

$$\begin{bmatrix} M^* & 0 \\ 0 & 0 \end{bmatrix} \begin{bmatrix} a' \\ \alpha' \end{bmatrix} + \begin{bmatrix} C_{\delta\delta} & C_{\delta\theta} \\ C_{\theta\delta} & C_{\theta\theta} \end{bmatrix} \begin{bmatrix} \dot{\delta}' \\ \dot{\theta}' \end{bmatrix} + \begin{bmatrix} K_{\delta\delta} & K_{\delta\theta} \\ K_{\theta\delta} & K_{\theta\theta} \end{bmatrix} \begin{bmatrix} \delta' \\ \theta' \end{bmatrix} = \begin{bmatrix} g'_\delta \\ g'_\theta \end{bmatrix}. \qquad (38)$$

In these equations, vectors c'_δ and c'_θ represent, respectively, the reaction forces and the reaction moments of the flexible part over the rigid part. They are expressed by

$$c'_\delta = g'_{\underline{\delta}} - \underline{M}^*w - C_{\underline{\delta}\delta}\dot{\delta}' - C_{\underline{\delta}\theta}\dot{\theta}' - K_{\underline{\delta}\delta}\delta' - K_{\underline{\delta}\theta}\theta' \qquad (39)$$

$$c'_\theta = C_{\underline{\theta}\delta}\dot{\delta}' + C_{\underline{\theta}\theta}\dot{\theta}' + K_{\underline{\theta}\delta}\delta' + K_{\underline{\theta}\theta}\theta' - g'_{\underline{\theta}} \qquad (40)$$

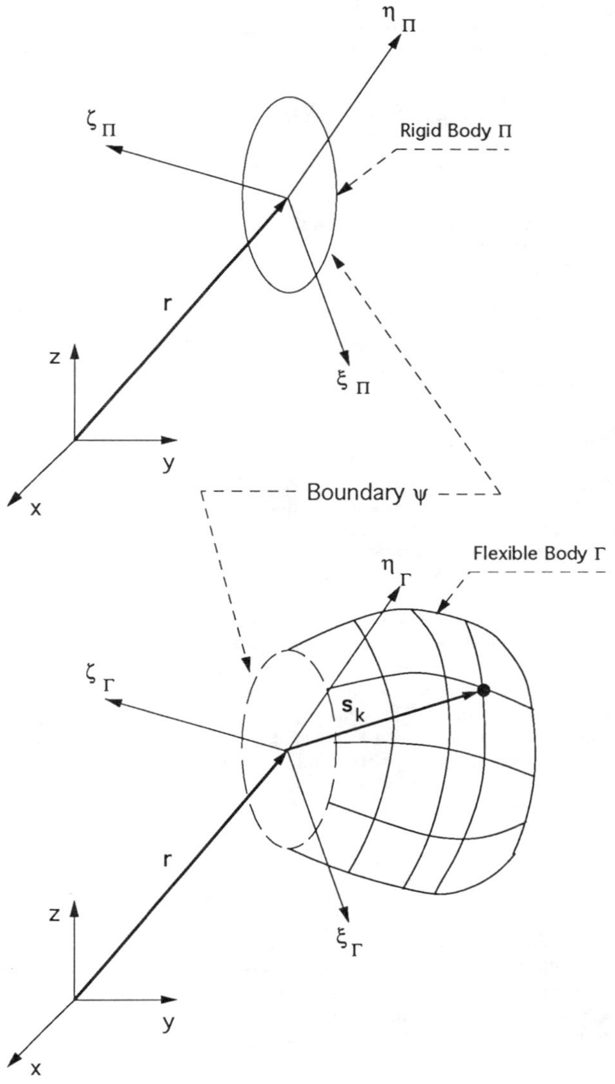

Fig. 3. Body-fixed coordinate system for a rigid-flexible body.

Furthermore,

$$\bar{A}^T = \begin{bmatrix} A^T \\ A^T \\ \dots \\ A^T \end{bmatrix}, \quad S = \begin{bmatrix} \tilde{s}'_1 \\ \tilde{s}'_2 \\ \dots \\ \tilde{s}'_n \end{bmatrix},$$

$$\bar{I} = \begin{bmatrix} I \\ I \\ \dots \\ I \end{bmatrix}, \quad w = \begin{bmatrix} \tilde{\omega}'\tilde{\omega}'s'_1 \\ \tilde{\omega}'\tilde{\omega}'s'_2 \\ \dots \\ \tilde{\omega}'\tilde{\omega}'s'_n \end{bmatrix}.$$

The dimension of $\bar{A}^T$, S, and $\bar{I}$ is ($3n \times 3$), and the dimension of **w** is $3n$, where n is the number of boundary nodes.

Equations (36)-(38) represent the minimum number of equations of motion for a rigid-flexible body, where it is assumed that the boundary nodes do not have any degrees of freedom. These equations can be rearranged and written as

$$\begin{bmatrix} m+\bar{A}\mathcal{M}^*\bar{A}^T & -\bar{A}\mathcal{M}^*S & 0 \\ -(\bar{A}\mathcal{M}^*S)^T & J'+S^T\mathcal{M}^*S & 0 \\ 0 & 0 & M_{ff} \end{bmatrix} \begin{bmatrix} \ddot{r} \\ \dot{\omega}' \\ \ddot{q}'_{ff} \end{bmatrix} = \begin{bmatrix} f + \bar{A}c'_\delta \\ n' - \tilde{\omega}'J'\omega' - S^Tc'_\delta - \bar{I}^Tc'_\theta \\ g'_f - C\dot{u}' - Ku' \end{bmatrix}. \tag{41}$$

By a proper choice for the location and orientation of the body-fixed coordinate frame in the rigid region, the mass matrix in this equation can be turned into an invariant diagonal matrix.

When finite elements with rotational degrees of freedom are used to describe the flexible body, some null coefficients (rotational inertias) appear on the diagonal of the mass matrix M_{ff}. This is the result of the lumped mass assumption, and it prevents the explicit solution of Eq. (41) for the nodal accelerations. To circumvent this problem, either rotational inertias are artificially introduced on the diagonal of the mass matrix or the explicit use of the rotational degrees of freedom is avoided. The introduction of rotational inertias can be done by lumping the off-diagonal terms of a consistent mass matrix M_{ff} [11]. The elimination of the rotational degrees of freedom can be achieved either by a Guyan condensation method or by the modal superposition technique. The advantage of the elimination process is that the number of equations of motion for the flexible body may be reduced considerably.

4.1 Static Condensation

The structure of the mass matrix in Eq. (38) suggests the application of a static condensation procedure in order to eliminate the explicit use of the rotational degrees of freedom [12]. For such purposes, assume that there are no moments g'_θ applied over the nodes of the flexible body Γ. Moreover, for proportional damping ($C = \alpha M + \beta K$), the relation between the translational and rotational coordinates is given by

$$\theta' = -K_{\theta\theta}^{-1} K_{\delta\theta} \delta', \tag{42}$$

which, upon substitution into Eq. (38), yields the reduced equations of motion for the flexible part of the partially deformable body,

$$M^* a' + C^* \dot{\delta}' + K^* \delta' = g'_\delta, \tag{43}$$

where the condensed stiffness and damping matrices are

$$K^* = K_{\delta\delta} - K_{\delta\theta} K_{\theta\theta}^{-1} K_{\delta\theta}^T$$

$$C^* = C_{\delta\delta} - C_{\delta\theta} K_{\theta\theta}^{-1} K_{\delta\theta}^T.$$

The complete set of equations of motion of the partially deformable body is given by Eqs. (36), (37), and (43). This is written in a compact form as

$$\begin{bmatrix} m+\bar{A}\mathcal{M}^*\bar{A}^T & -\bar{A}\mathcal{M}^*S & 0 \\ -(\bar{A}\mathcal{M}^*S)^T & J'+S^T\mathcal{M}^*S & 0 \\ 0 & 0 & M^* \end{bmatrix} \begin{bmatrix} \ddot{r} \\ \dot{\omega}' \\ a' \end{bmatrix} = \begin{bmatrix} f + \bar{A}c'_\delta \\ n' - \tilde{\omega}'J'\omega' - S^Tc'_\delta - \bar{I}^Tc'_\theta \\ g'_\delta - C^*\dot{\delta}' - K^*\delta' \end{bmatrix}. \tag{44}$$

The relation expressed in Eq. (42) can be used in the equations of the reaction forces and reaction moments, Eqs. (39) and (40), in order to replace the explicit use of the rotational degrees of freedom, i.e.,

$$c'_\delta = g'_\delta - M^*w + \left[C_{\delta\theta}K_{\theta\theta}^{-1}K_{\delta\theta} - C_{\delta\delta}\right]\dot{\delta}' + \left[K_{\delta\theta}K_{\theta\theta}^{-1}K_{\delta\theta} - K_{\delta\delta}\right]\delta' \tag{45}$$

$$c'_\theta = \left[C_{\theta\delta} - C_{\theta\theta}K_{\theta\theta}^{-1}K_{\delta\theta}\right]\dot{\delta}' + \left[K_{\theta\delta} - K_{\theta\theta}K_{\theta\theta}^{-1}K_{\delta\theta}\right]\delta'. \tag{46}$$

It should be noted that the application of the static condensation involves the inversion of the submatrix $K_{\theta\theta}$. If the flexible body contains any geometric or material nonlinearity, such a process must be repeated in every time step.

4.2 Modal Superposition

In order to apply the modal superposition to the nodal equations of motion of the flexible body, it is assumed that Eqs. (36) and (37) have been solved for $\ddot{r}$ and $\dot{\omega}'$. Then, $\ddot{q}'_f$ is substituted for $\ddot{u}'$ in Eq. (38), using the relation between the global accelerations and the local accelerations given by Eqs. (29) and (30). The result of such substitution is

$$M_{ff}\ddot{u}' + Ku' = g'_f - C\dot{u}' - f_{in}, \tag{47}$$

where the vector of inertia forces is given by

$$f_{in} = \begin{bmatrix} M^*(\bar{A}^T\ddot{r} - S\dot{\omega}' - w - 2W\dot{\delta}') \\ 0 \end{bmatrix}, \tag{48}$$

where $W = \text{diag}[\tilde{\omega}', \tilde{\omega}', ..., \tilde{\omega}']$. If the material law for the flexible body is linear elastic, the strains are small, and there are no geometric nonlinearities, then the stiffness matrix K is a constant. In this case, the natural frequencies and their respective modes of vibration, associated with the solution of the eigenproblem represented by the left-hand side of Eq. (47), are constants. Then the nodal displacements can be expressed as a linear combination of the modes of vibration [13],

$$\mathbf{u} = \chi \mathbf{z} \,, \tag{49}$$

where χ is the modal matrix. The number of modes of vibration involved in Eq. (49) is n_m, which is normally much smaller than the number of nodal degrees of freedom of the flexible body. Once the modes of vibration are not time dependent, the modal velocities and accelerations are written as

$$\dot{\mathbf{u}} = \chi \dot{\mathbf{z}} \tag{50}$$

$$\ddot{\mathbf{u}} = \chi \ddot{\mathbf{z}} \,. \tag{51}$$

Equations (49)-(51) are substituted into Eq. (47) and the result is premultiplied by χ^T. Using the property of orthonormality of χ with respect to the mass matrix, it is found that

$$\ddot{\mathbf{z}} = \chi^T (\mathbf{g}'_f - \mathbf{f}_{in}) - \Lambda \mathbf{z} - \mathbf{D} \dot{\mathbf{z}} \,, \tag{52}$$

which is an uncoupled system of n_m second-order differential equations. Matrices Λ and $\mathbf{D}$ are constant diagonal matrices containing the squares of the natural frequencies and damping coefficients, respectively.

In this procedure, the most costly single numerical operation is the solution of the eigenproblem. Once the matrices $\mathbf{M}_{ff}$ and $\mathbf{K}$ are constant, this problem needs to be solved once in a preprocessing phase. The drawback of using the modal superposition technique is that a full nonlinear formulation of the flexible body is no longer possible.

5. RIGID AND FLEXIBLE MULTIBODY SYSTEMS

The equations of motion for a system of mixed rigid and flexible bodies are obtained by combining some of the equations derived in the preceding sections. For the rigid bodies, the equations of motion in terms of the absolute accelerations or joint accelerations can be used. The equations of motion for a flexible body or a rigid-flexible body can be in the form of Eq. (31), Eq. (44), or Eq. (52). In this section, the process of combining the joint coordinate formulation for the rigid bodies and the statically condensed equations for a rigid-flexible body is discussed.

The equations of motion for a rigid-flexible body given by Eq. (44) can be expressed as

$$\begin{bmatrix} \mathbf{M}_\Pi & \mathbf{0} \\ \mathbf{0} & \mathbf{M}^* \end{bmatrix} \begin{bmatrix} \dot{\mathbf{v}}_\Pi \\ \mathbf{a}' \end{bmatrix} = \begin{bmatrix} \mathbf{g}_\Pi \\ \mathbf{g}'_\delta - \mathbf{C}^* \dot{\delta}' - \mathbf{K}^* \delta' \end{bmatrix} , \tag{53}$$

where the rotational equations of motion of the rigid body have been transformed from the body-fixed components to the global components, i.e., $\dot{\omega}'$ has been transformed to $\dot{\omega}$. Hence, $\dot{\mathbf{v}}_\Pi$ contains $\ddot{\mathbf{r}}$ and $\dot{\omega}$ of the rigid body Π. This transformation makes the rigid body acceleration of Eq. (53) consistent with the absolute rigid-body accelerations used in the transformations of the joint coordinate method.

For the purpose of simplicity and without any loss of generality, assume that the multibody system is open loop. Depending on the connectivity between the flexible bodies and the rigid bodies, several cases can be considered. In the first case, as shown in Fig. 4(a), the rigid bodies, including the rigid part of the rigid-flexible body, form one branch and the flexible part forms another branch. For this system, Eqs. (14) and (53) are combined as

$$\begin{bmatrix} \mathcal{M} & \mathbf{0} \\ \mathbf{0} & \mathbf{M}^* \end{bmatrix} \begin{bmatrix} \ddot{\theta} \\ \mathbf{a}' \end{bmatrix} = \begin{bmatrix} f \\ \mathbf{g}'_\delta - \mathbf{C}^* \dot{\delta}' - \mathbf{K}^* \delta' \end{bmatrix} , \tag{54}$$

where $\mathcal{M}$ and f contain $\mathbf{M}_\Pi$ and $\mathbf{g}_\Pi$ for body Π, and $\mathbf{M}_i$ and $\mathbf{g}_i$ for other rigid bodies, as given by Eqs. (2) and (3).

In a second case, as shown in Fig. 4(b), a kinematic joint connects the flexible body to a rigid body different from body Π. If this joint is removed (cut-joint technique), the branch is divided

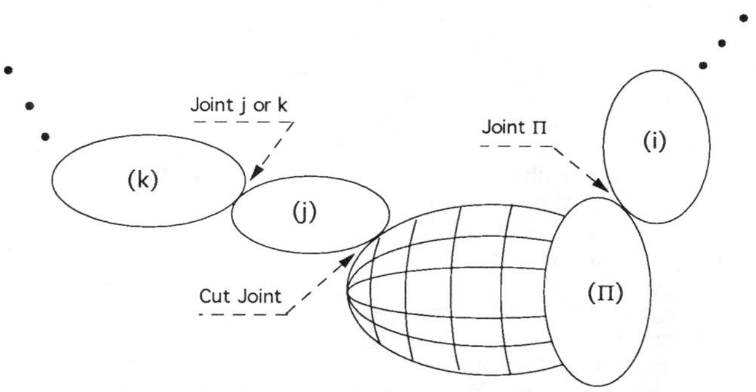

(a) All Kinematic Joints are Between Rigid Bodies

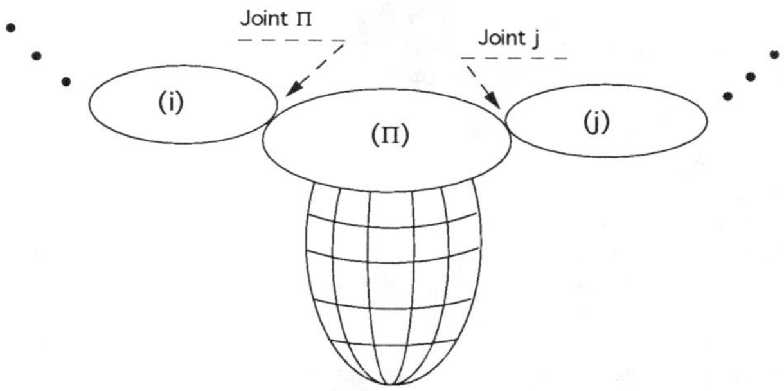

(b) One Kinematic Joint is Between a Rigid and a Flexible Body

Fig. 4. Two examples of rigid and flexible multibody systems.

into two independent branches. Equations of motion for the branch containing the flexible body (branch 1) are written as in Eq. (54). For the other branch (branch 2), assuming there are no other flexible bodies, Eq. (12) describes the equations of motion. If the cut joint is reassembled at this point, the Lagrange multiplier technique can be used to incorporate the reaction forces of this joint into the equations of motion of the two branches. Assume that constraint equations in the form

$$\Phi(\mathbf{q}_f, \mathbf{q}_j) = \mathbf{0} \tag{55}$$

can be written between the absolute nodal coordinates of the flexible body and the absolute coordinates of body j. The second time derivative of these constraints and transformations from absolute accelerations to joint accelerations yield

$$\mathbf{G} \begin{bmatrix} \ddot{\theta}_1 \\ \mathbf{a}' \\ \ddot{\theta}_2 \end{bmatrix} = \gamma \,, \tag{56}$$

where $\mathbf{G}$ is the Jacobian of the constraints, γ contains the quadratic velocity terms, and subscripts 1 and 2 denote the branch numbers. The equations of motion for the two branches are written as

$$\begin{bmatrix} \mathcal{M}_1 & \mathbf{0} & \mathbf{0} \\ \mathbf{0} & \mathbf{M}^* & \mathbf{0} \\ \mathbf{0} & \mathbf{0} & \mathcal{M}_2 \end{bmatrix} - \mathbf{G}^T \lambda = \begin{bmatrix} f_1 \\ \mathbf{g}'_\delta - \mathbf{C}^* \dot{\delta}' - \mathbf{K}^* \delta' \\ f_2 \end{bmatrix} , \tag{57}$$

where λ is the vector of Lagrange multipliers.

The use of Lagrange multipliers can be extended to other cases, for example, a kinematic joint connecting two flexible bodies in a multibody system. If this joint is treated as a cut joint, then constraint equations similar to Eqs. (55) and (56) can be expressed between the nodal coordinates of the two flexible bodies. Two sets of equations, such as Eq. (54), are written for the two branches, with Lagrange multipliers incorporating the effect of the constraints on the nodal coordinates. Furthermore, if the initial assumption that the system is an open loop is relaxed, Eq. (23) or (29) can be used instead of Eq. (12). In all these cases, the Lagrange multipliers are applied at the cut joint(s) to link different sets of equations of motion together.

The constraint equations and their corresponding Lagrange multipliers can be eliminated from the equations of motion by following a velocity transformation process similar in principle to that of the earlier sections. This normally reduces, or even minimizes, the number of equations of motion. However, this elimination process is not recommended in all cases since not much may be gained from it or, in some cases, it may actually increase the amount of computation. This is due to the fact that the number of Lagrange multipliers is normally very low and, in most cases, much smaller than the total number of the joint and flexible coordinates.

6. CONCLUSIONS

The equations of motion for rigid and flexible bodies presented in this paper have been implemented in several computer programs. In these programs, the equations of motion for rigid multibody systems can be generated symbolically and can be manipulated using recursive or nonrecursive algorithms. The rigid and flexible body equations can have either linear or nonlinear materials and can also contain geometric nonlinearity. These programs have been used to model and simulate the transient dynamic response of a variety of applications, such as ride stability and handling of automobiles, roll-over stability and crashworthiness of trucks, the control and maneuverability of spacecraft, helicopter rotor dynamics, and problems in biomechanics.

REFERENCES

1. Nikravesh, P. E., 1988, *Computer-Aided Analysis of Mechanical Systems*, Prentice Hall, Englewood Cliffs, NJ.
2. Jerkovsky, W., 1978, "The Structure of Multibody Dynamics Equations," *J. Guidance and Control*, Vol. 1, pp. 173-182.
3. Kim, S. S. and Vanderploeg, M. J., 1986, "A General and Efficient Method for Dynamic Analysis of Mechanical Systems Using Velocity Transformations," ASME *J. Mech., Trans., and Auto. in Design*, Vol. 108, pp. 176-182.
4. Nikravesh, P. E. and Ambrosio, J. A. C., 1991, "Systematic Construction of Equations of Motion for Rigid-Flexible Multibody Systems Containing Open and Closed Kinematic Loops," *Int. J. Num. Meth. in Eng.*, Vol. 32, pp. 1749-1766.
5. Nikravesh, P. E. and Gim, G., 1989, "Systematic Construction of the Equations of Motion for Multibody Systems Containing Closed Kinematic Loops," *Proceedings, 15th ASME Design Automation Conference, Montreal, Canada*, ASME, New York, pp. 27-28.
6. Shabana, A. and Wehage, R. A., 1983, "Variable Degree of Freedom Component Mode Analysis of Inertia Variant Flexible Mechanical Systems," ASME *J. Mech., Trans., and Auto. in Design*, Vol. 105, pp. 321-328.
7. Thompson, B. S. and Sung, C. K., 1984, "A Variational Formulation for the Nonlinear Finite Element Analysis of Flexible Linkages," ASME *J. Mech., Trans., and Auto. in Design*, Vol. 106, pp. 432-438.
8. Wu, S. C., 1987, "A Substructure Method for Dynamic Simulation of Flexible Mechanical Systems with Geometric Nonlinearities," Ph.D. Thesis, University of Iowa, Iowa City.
9. Ambrosio, J. A. C., 1991, "Elastic-Plastic Large Deformation of Flexible Multibody Systems in Crash Analysis," Ph.D. Thesis, University of Arizona, Tucson.
10. Ambrosio, J. A. C. and Nikravesh, P. E., 1992, "Elasto-Plastic Deformation in Multibody Dynamics," *Nonlinear Dynamics*, Vol. 3, pp. 85-104.
11. Surana, K. S., 1978, "Lumped Mass Matrices with Nonzero Inertia for General Shell Axisymmetric Shell Elements," *Int. J. Num. Meth. in Eng.*, Vol. 12, pp. 1630-1645.
12. Weaver, W. and Johnson, P. R., 1987, *Structural Dynamics by Finite Elements*, Prentice Hall, Englewood Cliffs, NJ.
13. Cook, R. D., Malkus, D. S., and Plesha, M. E., 1989, *Concepts and Applications of Finite Element Analysis*, 3rd ed., John Wiley & Sons, New York.

AMD-Vol. 141/DSC-Vol. 37, Dynamics of Flexible Multibody Systems:
Theory and Experiment
ASME 1992

DYNAMICS AND CONTROL OF A LARGE CLASS OF ORBITING FLEXIBLE STRUCTURES

V. J. Modi, A. C. Ng, and F. Karray
Department of Mechanical Engineering
University of British Columbia
Vancouver, British Columbia, Canada

Key words: Lagrangian formulation, nonlinear dynamics, nonlinear control, Feedback Linearization Technique

ABSTRACT

A rather general Lagrangian formulation for studying dynamics and control of a large class of flexible, deployable and articulating structures, forming a tree-type topology has been presented. The governing equations of motion are highly nonlinear, nonautonomous, and coupled. Nonlinear control, based on the Feedback Linearization Technique, was adopted. Versatility of the general formulation is illustrated through its application to five systems of contemporary interest. They include: evolving Space Station configurations First Element Launch and Permanently Manned Configuration; Japan Space Flyer Unit; Canadian Mobile Servicing System; and India's proposed communications satellite Indian Satellite II. Simulation results suggest trends, and lay foundation for the design and development of control strategies.

NOMENCLATURE

$\bar{d}_i, \bar{d}_{i,j}$	position vectors from O_c to O_i and O_i to $O_{i,j}$, respectively
$dm_c,\ dm_i,\ dm_{i,j}$	elemental mass in body B_c, B_i, and $B_{i,j}$, respectively
$\bar{q}_f,\ \bar{q}_r$	vector representing flexible and rigid generalized coordinates
$(\bar{q}_r)_d$	vector representing the desired rigid generalized coordinates
$\bar{C}^f_{cm}$	position vector from C^i to the instantaneous centre of mass of spacecraft
$\bar{C}^i_{cm}$	position vector from O_c to the centre of mass of undeformed spacecraft
C^i, C^f	centres of mass of the undeformed and deformed configurations of spacecraft, respectively
$F_c, F_i, F_{i,j}$	reference frame for bodies B_c, B_i, and $B_{i,j}$, respectively
$O_c, O_i, O_{i,j}$	origins of the coordinate axes for bodies B_c, B_i, and $B_{i,j}$, respectively
$\bar{Q}_f,\ \bar{Q}_r$	control effort vectors for flexible and rigid coordinates, respectively
$Q_\psi,\ Q_\phi,\ Q_l$	control effort for pitch, roll and yaw degrees of freedom, respectively
$\bar{R}_{cm}$	position vector from the centre of force to the instantaneous centre of mass of spacecraft
$\bar{R}_c, \bar{R}_i, \bar{R}_{i,j}$	position vectors of the mass elements dm_c, dm_i, and $dm_{i,j}$, respectively as measured from the centre of force
$S_x,\ S_y,\ S_z$	pointing errors of the manipulator in the orbit normal, local vertical, and local horizontal directions, respectively
S_{tot}	$\sqrt{S_x^2 + S_y^2 + S_z^2}$
$\alpha_x,\ \alpha_y,\ \alpha_z$	angular accelerations of the system about X_c, Y_c and Z_c axes, respectively
$\bar{\delta}_c, \bar{\delta}_i, \bar{\delta}_{i,j}$	vectors representing transverse vibration of dm_c, dm_i, and $dm_{i,j}$, respectively
$\delta^y_k,\ \delta^z_k$	tip deflection of a beam element in the Y_k and Z_k directions, respectively; $k = c$, i, or i,j;
$\bar{\rho}_c, \bar{\rho}_i, \bar{\rho}_{i,j}$	vectors denoting positions of dm_c, dm_i, and $dm_{i,j}$, respectively, in the undeformed configuration of the spacecraft
$\bar{\tau}_c, \bar{\tau}_i, \bar{\tau}_{i,j}$	vectors denoting thermal deformations of dm_c, dm_i, and $dm_{i,j}$, respectively

Dot ($\dot{}$) represents differentiations with respect to time t. Overbar ($\bar{}$) represents a vector and **boldfaced** symbol, a matrix.

1. INTRODUCTION

With the beginning of the space age in 1957, advent of the Space Shuttle in conducting experiments aimed at structural properties and behaviour in eighties, and the U.S. commitment to the Space Station *Freedom* by late 1990's, dynamics and control of flexible orbiting structures has become a subject of considerable investigation. A vast body of literature already exists which has been reviewed by Likins (1971), Modi and Shrivastava (1974, 1983), Roberson (1979), Lips (1980), Ibrahim (1988), Misra and Modi (1983, 1986), and others (AIAA, 1984; Miura and Matsunaga, 1989; Wada 1989; Matsunaga et al., 1990). Most of this literature, though developed for flexible deployable, articulating (moving, slewing) structures, is quite relevant to the study of so called 'adaptive' structures, the term which remains subjectively defined at best. There is a growing tendency to identify one's specialized activity through a distinctive terminology even at a cost of vaguely identified overlapping boundaries of interacting subject matters.

In a sense all structures must be adaptive to fulfill their intended design missions. Some structural designs tends to be classical in character primarily because of their relatively long history of usage. Even rigid deployable antennae and solar panels are, in this sense, adaptive. Complex character of the structures may accentuate with flexibility, mobility and slewing as would be the case with the Space Station based solar panels, Mobile Servicing System (MSS), and tethered microgravity facility, to cite a few examples. In addition, ability to spatially reshape or configure a structure brings with it the challenging task of storage. With kinematics, dynamics and control of this large class of structures being inherent to their design, the definition of "adaptive" structures becomes progressively fuzzy.

The paper introduces a rather general formulation and its specialized variations for studying dynamics and control of a large class of flexible deployable and articulating structures, forming a tree-type topology. The main features of this highly versatile Lagrangian formulation (Ng, 1992) may be summarized as follows:

(i) It is applicable to an arbitrary number of beam and plate type structural members, in any desired orbit, interconnected to form a tree-type topology.

(ii) The system is free to undergo general three dimensional librational motion. In addition, members can undergo predefined slewing maneuvers to facilitate modeling of the panels tracking the sun, or large angle maneuvers of a robotic arm.

(iii) The solar radiation induced thermal deformations of flexible members are included in the derivation. Thus, it is well suited to study the complex dynamical interactions between librational motion, transverse vibrations, and thermal deformations.

(iv) The governing equations of motion are highly nonlinear, nonautonomous and coupled. They are programmed in such a way that the effects of flexibility, librational motion, thermal deformations, slewing maneuvers, shifting center of mass, higher modes, etc. can be isolated readily.

(v) The nonlinear control, based on the Feedback Linear- ization Technique (FLT) is adopted. The simplicity of the control algorithm and its applicability to both rigid and flexible spacecraft makes the technique quite attractive.

The computer code is written in a modular fashion to help isolate the effects of flexibility, deployment, character and orientation of the appendages, inertia and orbital parameters, number and type of admissible functions, etc. Environmental effects due to aerodynamic forces, earth's magnetic field, etc., can be incorporated easily through generalized forces. The same is true for internal and external dissipation mechanisms.

2. KINEMATICS

Consider the spacecraft model in Fig. 1. The centres of mass of the undeformed and deformed configurations of the system are located at C^i and C^f, respectively. Let X_o,Y_o,Z_o be the inertial coordinate system located at the earth's centre. Attached to each member of the model is a body coordinate system helpful in defining relative motion between the members. Thus reference frame F_c is attached to body B_c at an arbitrary point O_c. Frame F_i, with origin at O_i, is attached to body B_i at the joint between body B_i and B_c. An arbitrary mass element dm_i on body B_i can be reached through a direct path from O_c via O_i. O_c, in turn, is located with respect to the instantaneous center of mass C^f and

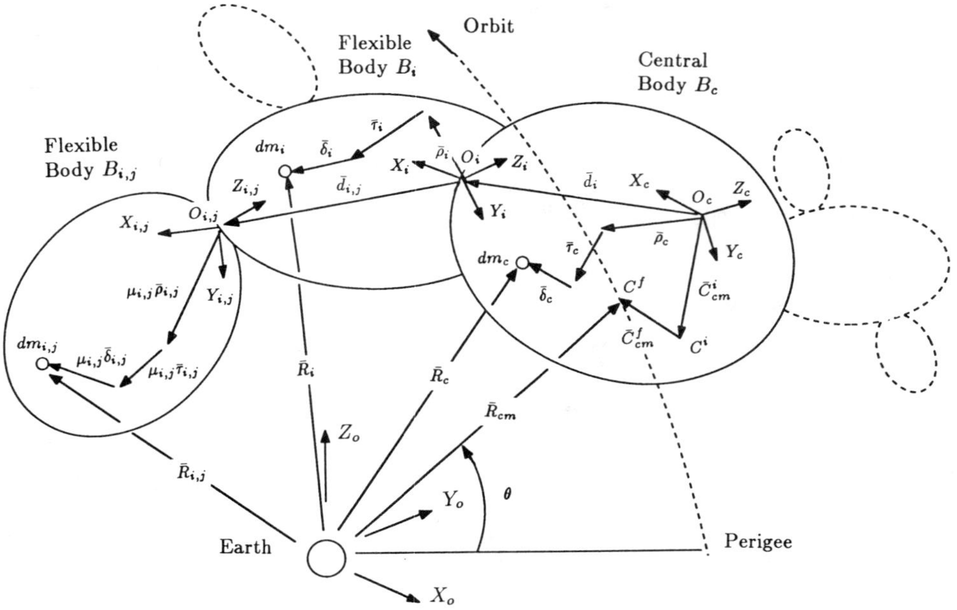

Fig. 1 The system of coordinates used in the formulation.

the inertial reference frame, F_o. Similarly, frame $F_{i,j}$ is attached to body $B_{i,j}$ and has its origin $(O_{i,j})$ at the joint between B_i and $B_{i,j}$. The relative position of O_i with respect to O_c is denoted by the vector $\bar{d}_i$ whereby $\bar{d}_{i,j}$ defines the position of $O_{i,j}$ relative to O_i.

The location of the elemental mass of the central body, dm_c, relative to O_c is defined by a series of vectors. $\bar{\rho}_c$ indicates the undeformed position of the element. Thermal deformation of the element is represented by $\bar{\tau}_c$. Finally, the transverse vibration of the element, $\bar{\delta}_c$, shifts the element to the end position. Similarly, $\bar{\rho}_i$, $\bar{\tau}_i$, and $\bar{\delta}_i$ define the location of the elemental mass dm_i, in body B_i, relative to O_i. For the elemental mass $dm_{i,j}$ of body $B_{i,j}$, its position relative to $O_{i,j}$ is defined by $\bar{\rho}_{i,j}$, $\bar{\tau}_{i,j}$, and $\bar{\delta}_{i,j}$ with $\mu_{i,j}$ denotes the motion of body $B_{i,j}$ relative to body B_i.

As the spacecraft has finite dimensions, i.e. it has mass as well as inertia, in addition to negotiating the trajectory, it is free to undergo librational motion about its center of mass described by three modified Eulerian rotations: pitch (ψ), roll (ϕ), and yaw (λ).

3. KINETICS

3.1 Kinetic Energy

Using Fig. 1, the vectors from the center of the earth to elemental mass dm_c, dm_i, and $dm_{i,j}$ represented by $\bar{R}_c$, $\bar{R}_i$, and $\bar{R}_{i,j}$, respectively, can be determined. The kinetic energy, T, of the spacecraft is given by

$$T = \frac{1}{2}\left\{\int_{m_c} \dot{\bar{R}}_c \cdot \dot{\bar{R}}_c \, dm_c + \sum_{i=1}^{N}\left[\int_{m_i} \dot{\bar{R}}_i \cdot \dot{\bar{R}}_i \, dm_i + \sum_{j=1}^{n_i}\int_{m_{i,j}} \dot{\bar{R}}_{i,j} \cdot \dot{\bar{R}}_{i,j} \, dm_{i,j}\right]\right\},$$

where $\dot{\bar{R}}_c$, $\dot{\bar{R}}_i$, and $\dot{\bar{R}}_{i,j}$ are the time derivatives of $\bar{R}_c$, $\bar{R}_i$, and $\bar{R}_{i,j}$, respectively.

3.2 Potential Energy

The potential energy, U, of the spacecraft has contribution from two sources: gravitational potential energy, U_g, and strain energy due to transverse vibration and thermal deformation, U_e, i.e. $U = U_e + U_g$. The potential energy due to gravity gradient is given by

$$U_g = \mu_e\left\{\int_{m_c} \frac{dm_c}{R_c} + \sum_{i=1}^{N}\left[\int_{m_i} \frac{dm_i}{R_i} + \sum_{j=1}^{n_i}\int_{m_{i,j}} \frac{dm_{i,j}}{R_{i,j}}\right]\right\}.$$

where μ_e is the gravitational constant.

3.3 Lagrangian Formulation

Using the Lagrangian procedure, the governing equations of motion are obtained:

$$\mathbf{M}(q)\ddot{\bar{q}} + \bar{C}(q,\dot{q},\theta) + \bar{K}(q,\theta) = \bar{Q}(\theta). \qquad (1)$$

The total degrees of freedom (N_q) consist of 3 from the libration (ψ, ϕ, and λ) and n_v from the vibrational degrees of freedom. such that $N_q = 3 + n_v$. Here $\mathbf{M}$ is a non-singular symmetric matrix of dimension $N_q \times N_q$. $\bar{C}$ is a $N_q \times 1$ vector representing the gyroscopic terms of the system. $\bar{K}$, also a $N_q \times 1$ vector, denotes the stiffness of the system. $\bar{Q}$, the generalized force vector of dimension $N_q \times 1$, is evaluated using the virtual work principle. Note, nonlinear entries in $\mathbf{M}$ together with nonlinear and time varying components of $\bar{C}$, $\bar{K}$, and $\bar{Q}$ result in a set of coupled, nonlinear, and nonautonomous equations of motion.

4. NONLINEAR CONTROL

Nonlinear control has received considerable attention in the robotics research, particularly during the past decade. Inverse control, based on the Feedback Linearization Technique (FLT), was first investigated by Bejczy (1974) and used by Singh and Schy (1984) for rigid arm control. Spong and Vidyasagar (1985a, 1985b) also used the FLT to formulate a robust control procedure for rigid manipulators. Spong (1985) later extended the method to the control of robots with elastic joints. Advantages of this approach are twofold: (i) the control algorithm based on the FLT is simple; and (ii) the compensator design, using a feedback linearized model, is straightforward. Recently, Karray et al. (1991) and Modi et al. (1991) extended the technique to include structural flexibility for a model of an orbiting manipulator system studied by Chan (1990). The technique is found to provide adequate control for both rigid as well as flexible manipulators.

4.1 Feedback Linearization Technique

Equation (1) can be rewritten with $\bar{q}_r$ and $\bar{q}_f$ corresponding to librational and vibrational generalized coordinates as

$$\begin{bmatrix} \mathbf{M_{rr}} & \vdots & \mathbf{M_{rf}} \\ \cdots & \cdots & \cdots \\ \mathbf{M_{fr}} & \vdots & \mathbf{M_{ff}} \end{bmatrix} \begin{Bmatrix} \ddot{\bar{q}}_r \\ \cdots \\ \ddot{\bar{q}}_f \end{Bmatrix} + \begin{Bmatrix} \bar{F}_r \\ \cdots \\ \bar{F}_f \end{Bmatrix} = \begin{Bmatrix} \bar{Q}_r \\ \cdots \\ \bar{Q}_f \end{Bmatrix}. \qquad (2)$$

Here $\bar{F}_r$ and $\bar{F}_f$ are 3×1 and $(N_q - 3) \times 1$ vectors, respectively, representing first and second order coupling terms. Assuming there is no actuator force for the vibrational degrees of freedom ($\bar{Q}_f = 0$), the objective is to determine $\bar{Q}_r$ such that the closed-loop system is linearized. Select $\bar{Q}_r$ such that

$$\bar{Q}_r(q_r, q_f, \dot{q}_r, \dot{q}_f, t) = \tilde{\mathbf{M}}(q_r, q_f, t)\{(\ddot{\bar{q}}_r)_d + \mathbf{K_v}[(\dot{\bar{q}}_r)_d - \dot{\bar{q}}_r] + \mathbf{K_p}[(\bar{q}_r)_d - \bar{q}_r]\} + \tilde{\mathbf{F}}(q_r, q_f, \dot{q}_r, \dot{q}_f, t);$$

$$\text{where:} \quad \tilde{\mathbf{M}} = \mathbf{M_{rr}} - \mathbf{M_{rf}}\mathbf{M_{ff}}^{-1}\mathbf{M_{fr}}; \quad \tilde{\mathbf{F}} = \bar{F}_r - \mathbf{M_{rf}}\mathbf{M_{ff}}^{-1}\bar{F}_f.$$

Here $(\bar{q}_r)_d$, $(\dot{\bar{q}}_r)_d$ and $(\ddot{\bar{q}}_r)_d$ represent the desired displacement, velocity and acceleration of the librational degrees of freedom. Equation (2) then becomes:

$$\ddot{\bar{q}}_r = (\ddot{\bar{q}}_r)_d + \mathbf{K_v}[(\dot{\bar{q}}_r)_d - \dot{\bar{q}}_r] + \mathbf{K_p}[(\bar{q}_r)_d - \bar{q}_r]; \qquad (3a)$$

$$\ddot{\bar{q}}_f = -\mathbf{M_{ff}}^{-1}\mathbf{M_{fr}}\bar{v} - \mathbf{M_{ff}}^{-1}\bar{F}_f; \qquad (3b)$$

which is linear in the rigid degrees of freedom. Since $\bar{e} = (\bar{q}_r)_d - \bar{q}_r$ denotes the displacement error, $\bar{Q}_r$ can be visualized as a combination of two controllers: the primary ($\bar{Q}_{r,p}$); and the secondary ($\bar{Q}_{r,s}$), where:

$$\bar{Q}_{r,p} = \tilde{\mathbf{M}}(\ddot{\bar{q}}_r)_d + \tilde{\mathbf{F}}; \qquad \bar{Q}_{r,s} = \tilde{\mathbf{M}}(\mathbf{K_v}\dot{\bar{e}} + \mathbf{K_p}\bar{e}).$$

The function of the primary controller is to offset nonlinear effects inherent in the rigid degrees of freedom; whereas the secondary controller ensures a robust behaviour.

Eq. (3a) can be rewritten as

$$\ddot{\bar{e}} + \mathbf{K_v}\dot{\bar{e}} + \mathbf{K_p}\bar{e} = 0.$$

The role of $\mathbf{K_p}$ and $\mathbf{K_v}$ is now apparent; they are position and velocity gains to insure asymptotic behaviour of the closed-loop system. A suitable candidate for $\mathbf{K_p}$ and $\mathbf{K_v}$ would be diagonal matrices such that

$$\mathbf{K_p} = \begin{bmatrix} \tilde{\omega}_1^2 & & \\ & \ddots & \\ & & \tilde{\omega}_n^2 \end{bmatrix}, \qquad \mathbf{K_v} = \begin{bmatrix} 2\tilde{\omega}_1 & & \\ & \ddots & \\ & & 2\tilde{\omega}_n \end{bmatrix},$$

which results in a globally decoupled system with each generalized coordinate behaving as a critically damped system. For a rigid spacecraft libration control, $\mathbf{K_p}$ and $\mathbf{K_v}$ are 3×3 matrices for pitch, roll, and yaw degrees of freedom. In general, a larger value of $\tilde{\omega}_n$ gives rise to a faster response of the n-th generalized coordinate. Block diagram of the control based on the FLT is shown in Fig. 2.

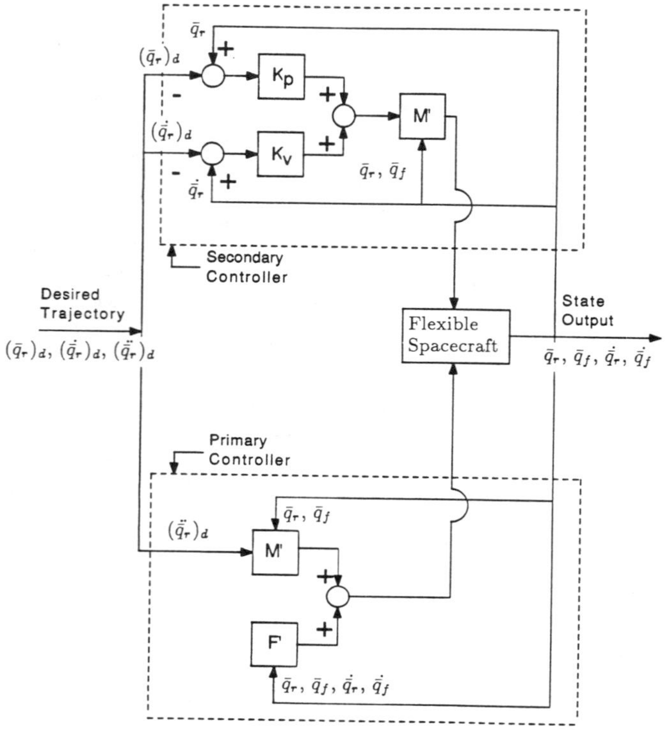

Fig. 2 Block diagram of the FLT as applied to a flexible spacecraft.

5. PARAMETRIC STUDIES

To illustrate versatility of the relatively general formulation, dynamic simulations of five spacecraft models, of contemporary interest, were carried out. The amount of information obtained is enormous and only a sample of it is presented here in a condensed form.

5.1 First Element Launch (FEL)

The Space Station will be constructed utilizing around thirty Space Transportation System (STS) flights. The first flight would result in construction of the FEL configuration. It will have an overall length of 60 m and a mass of 17,680 kg. Major equipment installed in the FEL configuration includes two PV arrays, radiator, and stinger. Figure 3(a) shows the design configuration of the FEL such that the axial directions of the power boom and PV arrays are parallel to the orbit normal and local vertical, respectively. The FEL is simulated here by the power boom as a free-free beam and the stinger as a cantilever beam. The PV arrays and PV array radiator are represented as cantilevered plates.

Simulation results show that flexibility effects on the FEL response cannot be overlooked. A small disturbance applied to any flexible member can affect the rigid body motion significantly. The power boom disturbance in the local vertical direction is the most critical one as the resulting high frequency modulated pitch and roll responses would require high bandwidth controllers. Figure 3(b) shows the response of the system with the power boom initially deformed in the first mode with a tip deflection of 1 cm in the local vertical direction. Even with this small disturbance, the pitch response is excited significantly with high frequency harmonics of 0.02° in amplitude. Similar trend is observed in the roll response. Although the amplitude is lower than that in the pitch response, the roll motion is excited at a higher frequency. The power boom response, δ_c^y, and roll have similar response trend with the distinct beat phenomenon present. The beat response is attributed to asymmetrical loading on the power boom coupled with the gravitational disparity. The flexible element experiencing the largest deflection is the stinger. Its maximum amplitude is about 3 cm in the local vertical direction.

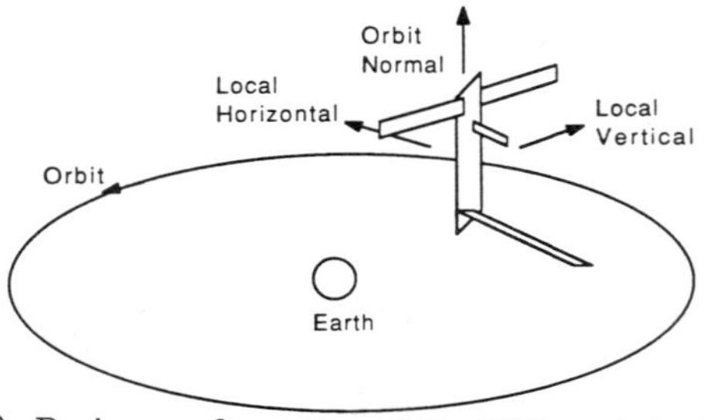

Fig. 3(a) Design configuration of the FEL used in the numerical simulation.

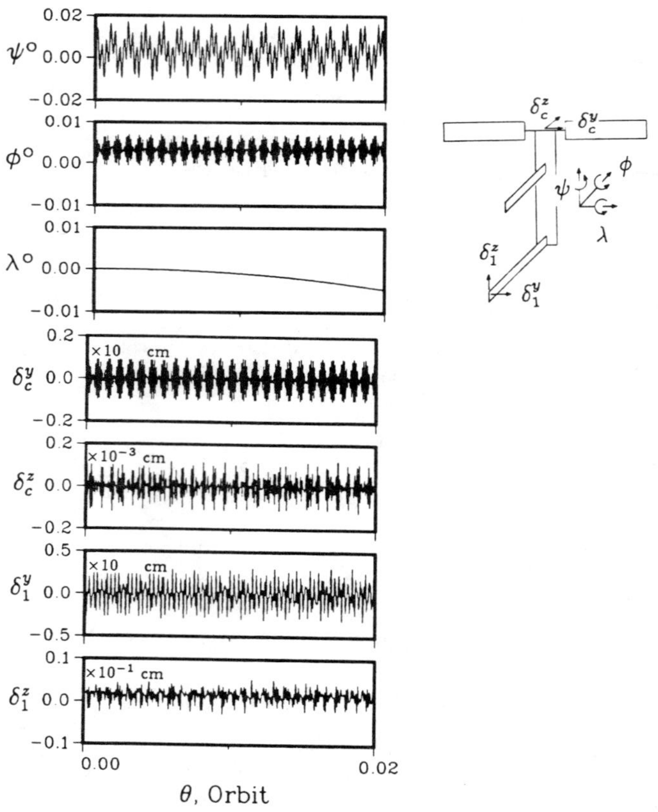

Fig. 3(b) Effect of an initial 1 cm tip deflection of the power boom in the local vertical direction on the FEL dynamics.

5.2 Permanently Manned Configuration (PMC)

The PMC will be established after fifteen STS flights. It will be 115 m in length and 160,972 kg in mass. The major difference from the FEL configuration is an additional pair of PV arrays and their radiator, two station radiators, and the modules. The orientation of the PMC will be similar to that of the FEL as shown in Fig. 4(a).

One of the stated missions of the Space Station is to provide microgravity environment for the purpose of scientific research. As stated in the NASA (1987) report , the objective of the station is to provide a $1 - 10$ μg environment in some portion of the Laboratory Module. Furthermore, a drift rate, apart from the orbital rate, below 0.005 °/s is desired. Simulation results indicate that the requirements on velocity and acceleration are stringent. Even with a small disturbance applied to any flexible member in the local vertical or local horizontal direction, the system velocities and accelerations easily reach or exceed the acceptable value. For instance, Fig. 4(b) shows that with an initial tip deflection of 1 cm applied to the power boom in the local vertical direction, the resulting angular velocities and accelerations approach or exceed the specified limit. Maximum ω_x attains a value of 0.004 °/s. The microgravity level near the power boom centre is over 1,000 μg.

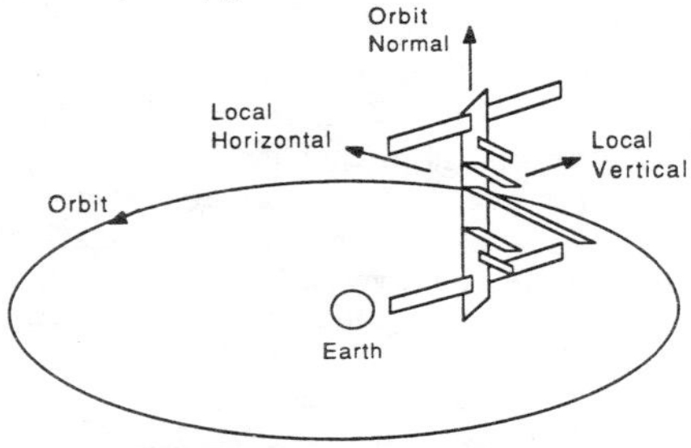

Fig. 4(a) Schematic diagram of the PMC design configuration used in the numerical simulations.

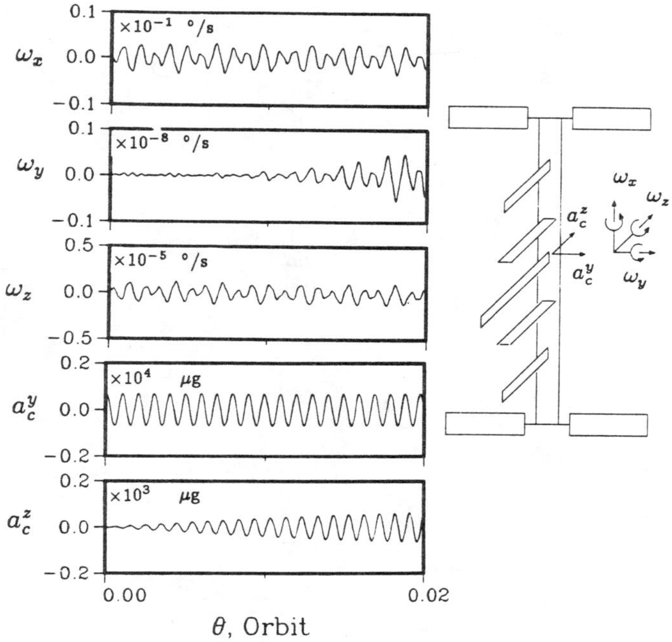

Fig. 4(b) The effect of an initial power boom tip displacement of 1 cm in the local vertical direction on the PMC velocities and accelerations.

5.3 Space Flyer Unit (SFU)

The Japan SFU is an unmanned, reusable and free-flying platform for multipurpose use. The SFU consists of an octagonal shaped central body which includes eight modules of scientific experiments. Two solar array pedals (SAPs), each 9.7 m × 2.4 m, are deployed at either end of the central body. Here, the present formulation is used to simulate the dynamics of the SFU during symmetric and asymmetric deployment/retrieval of the pedals inplane or out-of-plane (Fig. 5). The study shows that the spacecraft remains stable under both symmetric and asymmetric deployment of the SAPs. The retrieval of the SAPs may pose some problems. The symmetric out-of-plane retrieval poses no difficulty as shown in Fig. 6. Here, the SAPs are retrieved in 5 minutes. Even with this fast retrieval, the SFU remains stable with a maximum pitch angle of 6°, which is about the same in magnitude as the deployment case. The roll and yaw angles remain small. As the SAPs are becoming more and more rigid with the progress of retrieval, it is reasonable that they attain larger deflection during retrieval and oscillate with a peak-to-peak amplitude of less than 10^{-7} cm afterwards. However, if only one SAP is retrieved, the spacecraft becomes unstable even when the retrieval time is increased. The SAP retrieval in the inplane direction is found to be undesirable. Regardless of symmetry and period of retrieval, the spacecraft starts to tumble in a short time.

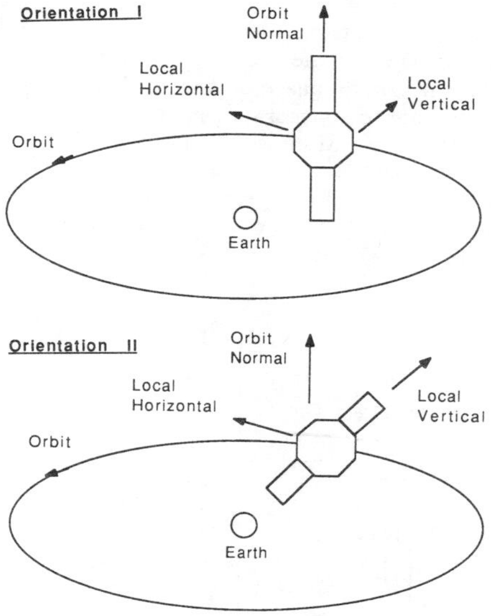

Fig. 5 The two possible design configurations for the SFU central body and arrays.

5.4 Mobile Servicing System (MSS)

On board the space shuttle is a remote manipulator system designed mainly for satellite retrieval and release. A similar system, known as the MSS, is planned for the Space Station. The MSS is essentially a two link manipulator attached to a mobile base which traverses along the station power boom. The simulation is based on the coordinate systems shown in Fig. 7. Both InPlane (IP) and Out-of-Plane (OP) maneuvers of the lower link are compared. In addition, the effects of maneuver period, link stiffness and offset are investigated. Simulation results indicate that the OP maneuver results in relatively smaller tip deflections of the links. However, it excites a large amplitude yaw motion. Since the primary function of a manipulator is to position a payload at a desired location, the yaw motion results in an unacceptable magnitude of pointing error. A typical system response subjected to a 5-minute OP maneuver is shown in Fig. 8.

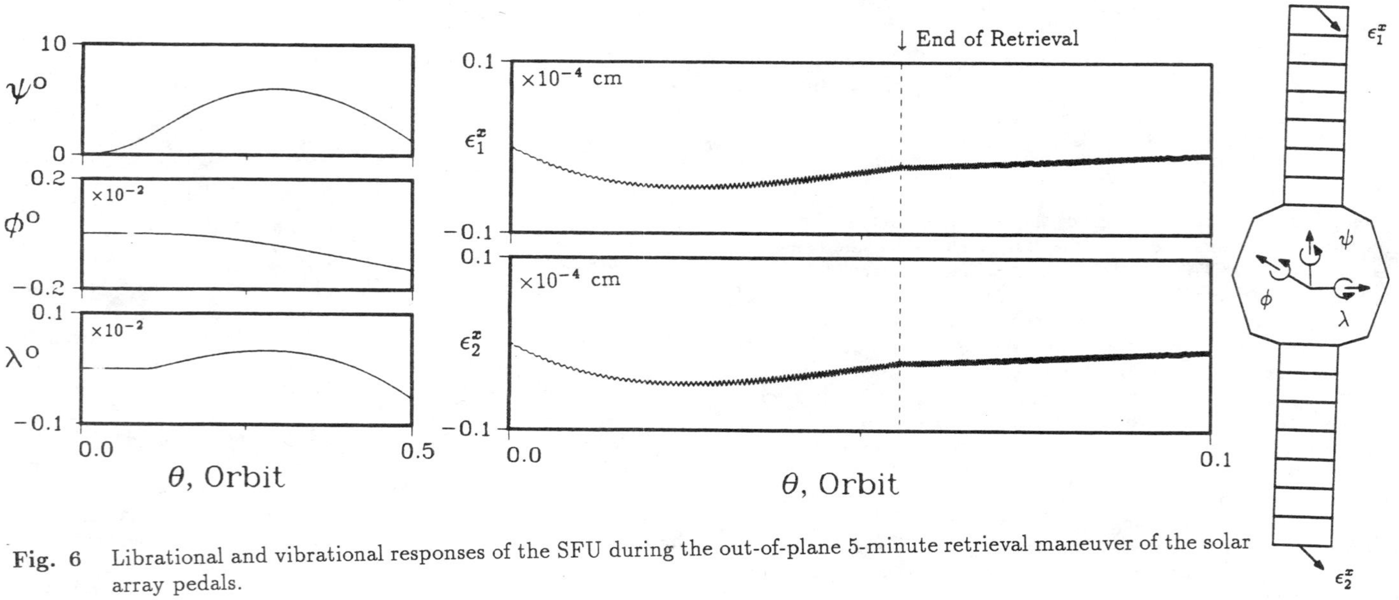

Fig. 6 Librational and vibrational responses of the SFU during the out-of-plane 5-minute retrieval maneuver of the solar array pedals.

It is apparent that to improve the performance of the OP maneuver, the attitude of the MSS has to be first controlled. To this end, nonlinear control based on the FLT is applied. Application of the technique to spacecraft with flexible appendages was recently put forward by Modi et al. (1991). The strategy can be regarded as a combination of two controllers: primary and secondary. The function of the primary controller is to offset the nonlinear effects inherent in the rigid degrees of freedom; whereas the secondary controller ensures robust behaviour against the error.

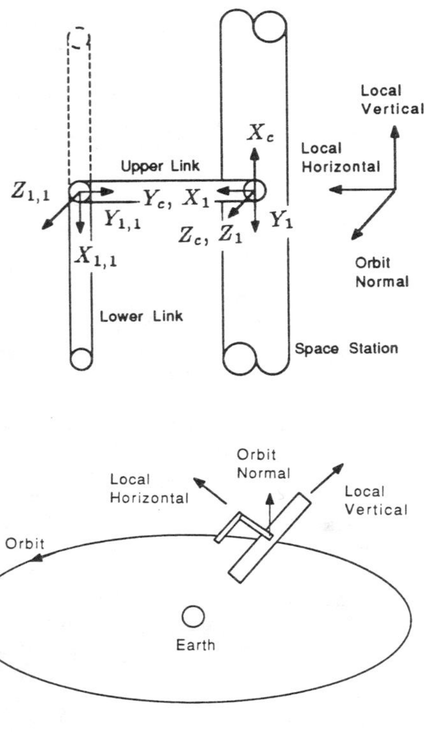

Fig. 7 Coordinate systems and design configuration used in the simulations of MSS slewing maneuvers.

Figure 9 shows the controlled system performance for a 5-minute OP maneuvers. The controller is turned on 2.5 minutes after the robotic arm begins the maneuver. The controlled attitude response is shown in Fig. 9(a). A comparison with Fig. 8(a) shows that with application of the control, the relatively large yaw angle is damped quite effectively. At the end of 0.25 orbit, the yaw angle is only $-1.1°$ as compared to $8.2°$ for the uncon-

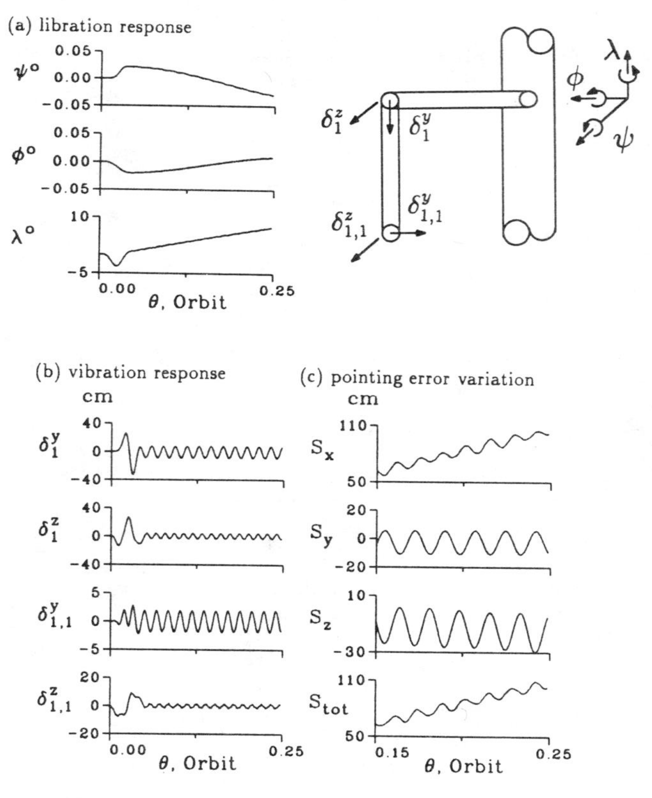

Fig. 8 Effect of a 5-minute 180° out-of-plane slew of the MSS lower link on the response of the system.

trolled case. Note, influence of the controller on the appendage vibration is rather small. Considering the mass and inertia of the Space Station, the peak torques Q_ψ (−34.5 Nm), Q_ϕ (12.1 Nm), and Q_λ (25.5 Nm) are reasonable (Fig. 9c). The pointing accuracy has improved remarkably with the application of the control (Fig. 9d). With the yaw attitude damped, the error in the orbit normal direction (S_x) no longer increases progressively as in Fig. 8(c). At 0.25 orbit, the mean error is 25.7 cm with 9.8 cm p-p superposed. This is a significant improvement compared to the error for the uncontrolled system which is 103.4 cm.

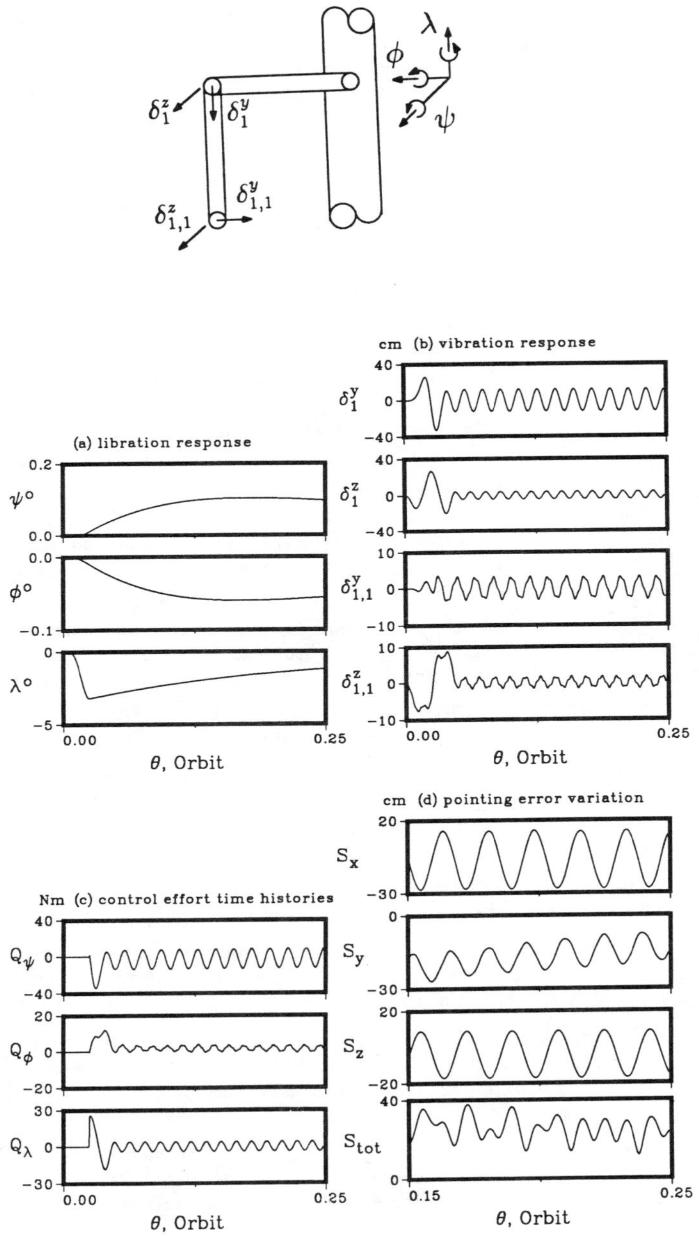

Fig. 9 Controlled system performance of the MSS undergoing a 5-minute OP maneuver.

5.5 Indian Satellite II (INSAT–II)

The satellite, designed by the Indian Space Research Organization, is scheduled to be launched in 1992. INSAT–II is a telecommunications satellite orbiting at the geosynchronous orbit. It has two flexible components attached to the main body. An array extending to 9 m collects solar energy to power the electronic components onboard. A 15 m solar boom attached on the opposite end is used to counterbalance the torque produced by the solar radiation pressure exerted on the array and the main body. The simulation carried out here is based on the design configuration of Fig. 10.

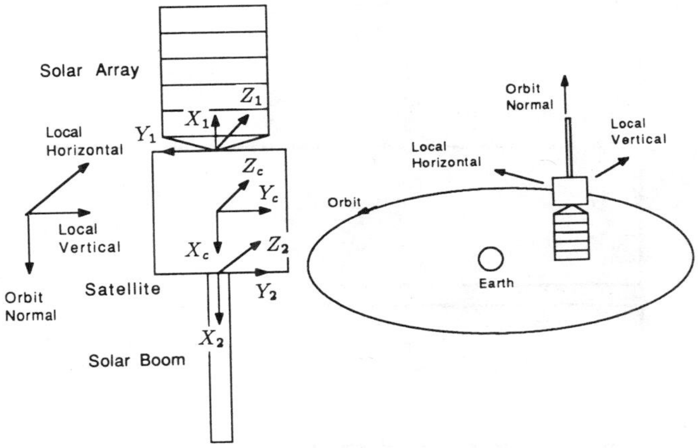

Fig. 10 Coordinate systems and design configuration of the INSAT–II used in simulations.

The performance of the satellite under thermal disturbances on the array and boom is investigated here. Figure 11 shows uncontrolled response of the satellite with the array and boom under the influence of thermal deformation. In addition, the array and boom are subjected to time varying thermal deformation. An initial deflection of 1 cm in the local horizontal direction of the array and boom are assumed here. As can be seen, in less than 0.05 orbit, the satellite starts to tumble (the pitch angle reaches 90°). The fast deterioration of the attitude is due to large amplitudes of array and boom vibration of magnitudes 1.7 and 3.6 cm, respectively. The inclusion of thermal deformations do not adversely affect the FLT controller performance. The attitude of the satellite can be restored in less than 2 minutes with a peak control torque of about 5.7 Nm (not shown).

6. CONCLUSIONS

Application of the general formulation has been demonstrated through parametric studies of five spacecraft models. Based on the simulation, the following conclusions can be drawn:

(i) Flexibility effect of the Space Station cannot be overlooked especially in maintaining the microgravity environment of the station.

(ii) Deployment of the SAPs of the SFU poses no problem; however, their retrieval requires careful planning and the application of control.

(iii) The OP maneuver of the MSS results in large pointing errors. By controlling the attitude of the Space Station using the FLT-controller, the pointing accuracy improves dramatically.

(iv) Thermal deformations of the solar array and boom on the INSAT–II tend to destabilize the satellite. Once again, controller based on the FLT is proved effective.

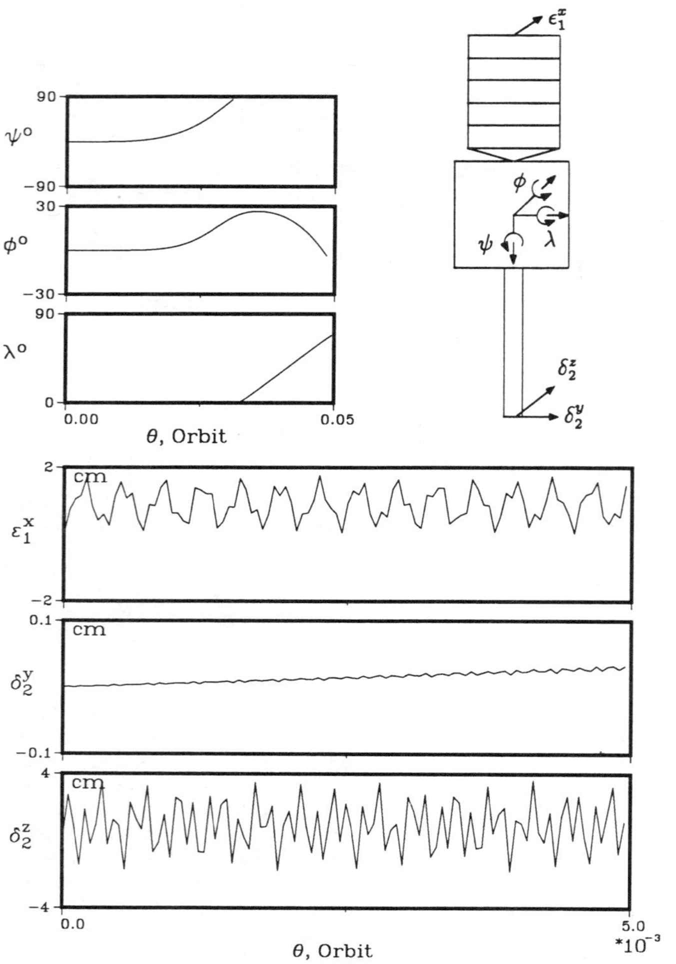

Fig. 11 Dynamics of the INSAT–II with thermally deformed appendages.

7. ACKNOWLEDGMENTS

Investigation reported here was supported by the Natural Sciences and Engineering Research Council of Canada, Grants A–2181, and the Network of Centers of Excellence Program, IRIS/C–8, Grant No. 5–55380.

8. REFERENCE

Bejczy, A.K., 1974, *Robot Arm Dynamics and Control*, JPL TM 33-669, California Institute of Technology, Pasadena, California.

Chan, J.K., 1990, *Dynamics and Control of an Orbiting Space Platform Based Mobile Flexible Manipulator*, M.A.Sc. Thesis, The University of British Columbia.

Ibrahim, A.M., 1988, *Mathematical Modelling of Flexible Multibody Dynamics with Application to Orbiting Systems*, Ph.D. Dissertation, The University of British Columbia.

Karray, F., Modi, V.J., and Chan, J.K., 1991, "Inverse Control of Flexible Orbiting Manipulators," *Proceedings of the American Control Conference*, Boston, Mass., Editor: A.G. Ulsoy, Vol. 2, pp. 1909–1912.

Likins, P.W., and Bouvier, M.K., 1971, " Attitude Control of Nonrigid Spacecraft," *Astronautics and Aeronautics*, Vol. 9, pp. 64–71.

Lips, K.W., 1980, *Dynamics of a Large Class of Satellites with Deploying Flexible Appendages*, Ph.D. Dissertation, The University of British Columbia.

Matsunaga, S., Miura, K., and Natori, M., 1990, "A Constructional Concept for Large Space–Based Intelligent/Adaptive Structures," *30th AIAA/ASME/ASCE/AHS/ASC Structures, Structural Dynamics and Materials Conference*, Long Beach, California, U.S.A., Paper No. 90–1128.

Misra, A.K., and Modi, V.J., 1983, "Dynamics and Control of Tether Connected Two Body Systems – A Brief Review," *Space 2000*, AIAA, New York, pp. 473–514.

Misra, A.K., and Modi, V.J., 1986, "A Survey on the Dynamics and Control of Tethered Satellite Systems," *NASA/AIAA/PSN International Conference on Tethers in Space*, Arlington, Virginia, U.S.A., Paper No. AAS–86–246; also *Advances in the Astronautical Sciences*, Editors: P.M. Bainum et al., Vol. 62, pp. 667–719.

Miura, K., and Matsunaga, S., 1989, "An Attempt to Introduce Intelligence in Space Structure," *30th AIAA/ASME/ASCE/AHS/ASC Structures, Structural Dynamics and Materials Conference*, Mobile Alabama, U.S.A., Paper No. 89–1289.

Modi, V.J., 1974, "Attitude Dynamics of Satellites with Flexible Appendages – A Brief Review," *Journal of Spacecraft and Rockets,*, Vol. 11, pp. 743–751.

Modi, V.J., and Shrivastava, S.K., 1983, "Satellite Attitude Dynamics and Control in the Presence of Environmental Torques – A Brief Review," *Journal of Guidance, Control, and Dynamics*, Vol. 6, No. 6, pp. 461–471.

Modi, V.J., Karray, F., and Chan, J.K., 1991, "On the Control of a Class of Flexible Manipulators Using Feedback Linearization Approach," *42nd Congress of the International Astronautical Federation*, Montreal, Canada, Paper No. IAF–91–324; also *Acta Astronautica*, in press.

Ng, A.C., 1992, *Dynamics and Control of Orbiting Flexible Systems: A Formulation with Applications*, Ph.D. Dissertation, University of British Columbia.

Roberson, R.E., 1979, "Two Decades of Spacecraft Attitude Control," *Journal of Guidance and Control*, Vol. 2, pp. 3–8.

Singh, S.N. and Schy, A.A., 1984, "Invertibility and Robust Nonlinear Control of Robotic Systems," *Proceedings of 23rd Conference on Decision and Control*, Las Vegas, Nevada, pp. 1058–1063.

Space Station Engineering Data Book, 1987, NASA SSE-E-87-R1, NASA Space Station Program Office.

Special Section, 1984, "Large Space Structure Control: Early Experiments," *Journal of Guidance, Control, and Dynamics,*, AIAA, Vol. 7, No. 5, pp. 513–562.

Spong, M.W. and Vidyasagar, M., 1985a, "Robust Linear Compensator Design for Nonlinear Robotic Control," *Proceedings of IEEE Conference on Robotics and Automation*, St. Louis, Missouri, pp. 954–959.

Spong, M.W. and Vidyasagar, M., 1985b, "Robust Nonlinear Control of Robot Manipulator," *Proceedings of the 24th IEEE Conference on Decision and Control*, Fort Lauderdale, Florida, pp. 1767–1772.

Spong, M.W., 1987, "Modelling and Control of Elastic Joint Robots," *Journal of Dynamic Systems, Measurement and Control*, Vol. 109, pp. 310–319.

Wada, B.K., 1989, "Adaptive Structures," *30th AIAA/ASME/ASCE/AHS/ASC Structures, Structural Dynamics and Materials Conference*, Mobile, Alabama, Paper No. 89–1160.

AMD-Vol. 141/DSC-Vol. 37, Dynamics of Flexible Multibody Systems:
Theory and Experiment
ASME 1992

DYNAMIC AND STRESS ANALYSIS OF SPATIAL SYSTEMS CONTAINING COMPOSITE PLATES

J. M. Kremer
Department of Mechanical Engineering
University of Illinois, Chicago
Chicago, Illinois

ABSTRACT

In this investigation, a finite element formulation is developed for the stress and large displacement analysis of composite plates that undergo finite rotation displacements. The composite structure is assumed to consist of layers of orthotropic laminae. The bonds between the laminae are assumed to be infinitesimally thin and shear nondeformable. The mass and stiffness matrices of the laminae are identified using the expressions of the kinetic and strain energies, respectively. By summing the kinetic and strain energies of an element, the element mass and stiffness matrices are identified in terms of a set of element invariants that depend on the assumed displacement field. It is shown in this investigation that the invariants of the element can be obtained by assembling the invariants of its laminae. In the finite element formulation presented in this investigation, it is sufficient that the element shape function describe only large rigid body translations.

1 INTRODUCTION

The use of laminated composite structures in weight-sensitive applications is ever increasing in industry (Schwartz, 1984). This is due in part to the high strength-to-weight and stiffness-to-weight ratios of composite materials. Analysis of laminated composite structures, such as plates and shells, has never been an easy task, due to the bending-extension, twisting-extension and bending-twisting coupling effects. These coupling effects are a result of the fiber orientation of the various laminae that comprise the laminate. Developing means to accurately predict the response of a laminated structure is of fundamental importance to the reliability of a design. This need for new and more accurate analysis techniques to accommodate the wide and varied application of composite materials is reflected in the published literature.

It is the objective this investigation to develop a formulation that will include laminated plates consisting of layers of orthotropic laminae in the analysis of spatial flexible mechanical systems. The plates in this analysis are subject to large arbitrary translation and rotation displacements. In this investigation the laminated composite plate is examined without regards to the lamina constituents, that is, without regards to the individual properties of the fibers or the polymer matrix. Rather, the resulting properties of the fiber/matrix combination are used in modeling each lamina. The layers are assumed to be bonded together, and the bonds are assumed to be infinitesimally thin and perfect so that there is no relative motion between the laminae. A variety of elements and shape functions could have been used in developing the plate elements, as shown by Hrabok and Hrudey (1984); however, an eight-noded quadrilateral element is developed for the numerical investigation presented in this paper. The laminate is permitted to have laminae with different orthotropic material properties, different fiber orientations, and different locations within the element. In the analysis presented in this investigation it is assumed that the element shape function can be used to describe large rigid body translations. By using this assumption and by introducing an *intermediate element coordinate system*, the kinematic equations of each lamina is defined in the global coordinate system in terms of a fixed set of absolute reference and relative elastic coordinates. The mass matrix of a lamina that undergoes large reference displacement is identified in terms of a set of invariants that depend on the assumed displacement field. The invariants of the elements are obtained by assembling the invariants of its laminae. The invariants of the body can then be obtained by assembling the invariants of its elements, using standard finite element procedures. A preprocessor finite element computer program is used to evaluate the invariants of motion as well as the stiffness matrix of each lamina. A post-processing stress analysis capability is also developed in this investigation.

2 COORDINATE SYSTEMS

In the analysis of thin plates and shells, it is assumed that the thickness dimension of the plate is much less than the length and width dimensions. This assumption leads to the conclusion that the normal strain in the thickness direction is negligible. Imposing this assumption also allows the approximation of the transverse coordinate of an arbitrary point in lamina l with the midsurface coordinate of that lamina w_o^{ijl}.

In the finite element formulation, the nodal coordinates of an element can be expressed as a sum of the undeformed coordinates and the deformed coordinates of the point. For node k of lamina l in element j of body i, this sum can be written as

$$\hat{\mathbf{q}}_{nk}^{ijl} = \hat{\mathbf{q}}_{ok}^{ijl} + \hat{\mathbf{q}}_{fk}^{ij} = \begin{bmatrix} u_{ok}^{ij} \\ v_{ok}^{ij} \\ w_{ok}^{ijl} \\ 0 \\ 0 \end{bmatrix} + \begin{bmatrix} u_{fk}^{ij} \\ v_{fk}^{ij} \\ w_{fk}^{ij} \\ \alpha_{fk}^{ij} \\ \beta_{fk}^{ij} \end{bmatrix} \tag{1}$$

where $\hat{\mathbf{q}}_{nk}^{ijl}$ is the total coordinate vector for node k, $\hat{\mathbf{q}}_{ok}^{ijl}$ is the vector of undeformed coordinates, and $\hat{\mathbf{q}}_{fk}^{ij}$ is the vector of nodal deformation for node k. The cartesian coordinates u_{ok}^{ij}, v_{ok}^{ij}, and w_{ok}^{ijl} are the non-zero entries of the vector of undeformed coordinates for node k. The cartesian coordinates u_{fk}^{ij}, v_{fk}^{ij}, w_{fk}^{ij} and the rotational coordinates α_{fk}^{ij}, β_{fk}^{ij} comprise the vector of deformation coordinates for node k. Note that in the above equation, w_{ok}^{ijl} is the only nodal coordinate that uniquely defines lamina l. All other coordinates are shared by all the laminae of the element.

In the finite element method of analysis, interpolating functions are selected to closely approximate the actual deformations in the element. To assure that the interpolating functions pass through the nodal points, the functions must be defined in a coordinate system rigidly attached to the element. This coordinate system is referred to as the *element coordinate system*, $\hat{\mathbf{X}}^{ij}\ \hat{\mathbf{Y}}^{ij}\ \hat{\mathbf{Z}}^{ij}$. These interpolating functions are commonly referred to as the element shape function.

Whereas the element shape function must be defined in the element coordinate system, the nodal coordinates may be defined in a more convenient coordinate system, called the *intermediate element coordinate system*, $\bar{\mathbf{X}}^{ij}\ \bar{\mathbf{Y}}^{ij}\ \bar{\mathbf{Z}}^{ij}$. If the element shape function can describe large rigid body translations, the origin of the intermediate element coordinate system need not coincide with the origin of the element coordinate system. It is only required that the two coordinate systems be parallel to one another before deformation and that the developed deformations in the element are small. Figure 1 depicts the relationship between these two coordinate systems.

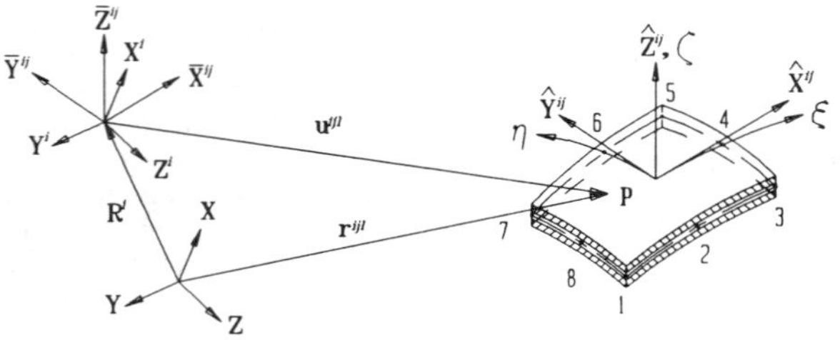

Figure 1: Position vector of an arbitrary point in lamina l of element j.

To facilitate the development of the governing equations of motion, one must be able to define the position vector of any arbitrary point in the element. To this end, the position vector is defined as

$$\hat{\mathbf{w}}^{ijl} = \mathbf{S}^{ijl}\hat{\mathbf{q}}_n^{ijl} \tag{2}$$

where $\mathbf{S}^{ijl}$ is the element shape function matrix, $\hat{\mathbf{w}}^{ijl}$ is the coordinate vector of an arbitrary point in element j, and $\hat{\mathbf{q}}_n^{ijl}$ is the total vector of nodal coordinates for element j. If the nodal coordinates in eq. 2 are defined in the element coordinate system then the position vector will also be defined in the element coordinate system. If, however, the nodal coordinates are defined in the intermediate element coordinate system, then $\hat{\mathbf{w}}^{ijl}$ will also be defined in the intermediate element coordinate system. We note the difference by writing the position vector, as defined in the intermediate element coordinate system, as

$$\bar{\mathbf{w}}^{ijl} = \mathbf{S}^{ijl}\bar{\mathbf{q}}_n^{ijl}. \tag{3}$$

3 FINITE ELEMENT

In this investigation an eight-noded composite plate element based on the *Mindlin plate theory* is developed and used in the analysis of spatial flexible mechanical systems. This element is depicted in Fig. 1. The strain displacement relationship defined by the *Mindlin* hypothesis is

$$\bar{\epsilon}^{ijl} = \mathbf{D}\bar{\mathbf{w}}_f^{ijl} = \begin{bmatrix} \partial/\partial x & 0 & 0 \\ 0 & \partial/\partial y & 0 \\ \partial/\partial y & \partial/\partial x & 0 \\ \partial/\partial z & 0 & \partial/\partial x \\ 0 & \partial/\partial z & \partial/\partial y \end{bmatrix} \begin{bmatrix} u^{ijl} \\ v^{ijl} \\ w^{ijl} \end{bmatrix} \tag{4}$$

where $\bar{\epsilon}^{ijl}$ is the strain vector for lamina l, $\mathbf{D}$ is the spatial differential operator as defined by *Mindlin's* hypothesis, and $\bar{\mathbf{w}}_f^{ijl} = \mathbf{S}^{ijl}\bar{\mathbf{q}}_f^{ij}$ is the displacement field for the lamina. The shape function matrix for this element is defined as (Panda and Natarajan, 1979)

$$\mathbf{S}^{ijl} = \begin{bmatrix} N_1 & 0 & 0 & N_1\zeta\frac{t_e}{2} & 0 & N_2 & \cdots \\ 0 & N_1 & 0 & 0 & N_1\zeta\frac{t_e}{2} & 0 & \cdots \\ 0 & 0 & N_1 & 0 & 0 & 0 & \cdots \end{bmatrix.$$
$$\left.\begin{matrix} \cdots & N_8 & 0 & 0 & N_8\zeta\frac{t_e}{2} & 0 \\ \cdots & 0 & N_8 & 0 & 0 & N_8\zeta\frac{t_e}{2} \\ \cdots & 0 & 0 & N_8 & 0 & 0 \end{matrix}\right]^{ijl} \tag{5}$$

where N_k $(k = 1, 2, ..., 8)$ are the nodal shape functions and are listed in the reference. Note that the shape function matrix is defined in terms of the curvilinear coordinates ξ and η which range from -1 to 1, and the linear coordinate ζ which spans the thickness of the element. By introducing the parameter ζ_l, a linear coordinate that ranges from -1 to 1 and spans the thickness of the *lamina*, the parameter ζ can be written as (Panda and Natarajan, 1979)

$$\zeta = -1 + \frac{1}{t_e}\left[2\sum_{m=1}^{l} h_m - h_l(1 - \zeta_l)\right] \tag{6}$$

where h_m is the thickness of lamina m, h_l is the thickness of layer l, and t_e is the thickness of the element. With this expression, ζ will range from the bottom of lamina l to the top of lamina l, as measured from the midsurface of the element. Equation 6 yields the relationship

$$d\zeta = \frac{h_l}{t_e}d\zeta_l \tag{7}$$

The matrix of orthotropic material properties for lamina l of the *Mindlin* plate element is given as (Jones, 1975)

$$\bar{\mathbf{E}}_1^{ijl} = \begin{bmatrix} E_{11} & E_{12} & 0 & 0 & 0 \\ E_{12} & E_{22} & 0 & 0 & 0 \\ 0 & 0 & E_{44} & 0 & 0 \\ 0 & 0 & 0 & E_{55} & 0 \\ 0 & 0 & 0 & 0 & E_{66} \end{bmatrix}^{ijl} \tag{8}$$

where the E_{mn} $(n, m = 1, ..., 6)$ are defined in the reference.

4 MASS MATRIX AND INVARIANTS OF MOTION

Using the fact that the shape function matrix can be used to describe large rigid body translations, the global position vector of an arbitrary point in lamina l on the plate element is defined as (Shabana, 1989; Kremer, 1991)

$$\begin{aligned} \mathbf{r}^{ijl} &= \mathbf{R}^i + \mathbf{A}^i\mathbf{u}^{ijl} = \mathbf{R}^i + \mathbf{A}^i\mathbf{C}^{ij}\bar{\mathbf{w}}^{ijl} \\ &= \mathbf{R}^i + \mathbf{A}^i\mathbf{C}^{ij}\mathbf{S}^{ijl}\bar{\mathbf{C}}^{ij}(\mathbf{q}_o^{ijl} + \mathbf{B}^{ij}\mathbf{q}_f^i) \end{aligned} \tag{9}$$

where, as shown in Fig. 1, $\mathbf{r}^{ijl}$ is the global position vector of the arbitrary point on the lth lamina, $\mathbf{u}^{ijl}$ is the position vector of the arbitrary point defined with respect to the body coordinate system, $\mathbf{R}^i$ is the global position vector of the origin of the body coordinate system, $\mathbf{A}^i$ is the orthogonal transformation matrix that transforms the displacement field from the body coordinate system to the global coordinate system, $\mathbf{C}^{ij}$ is a 3×3 constant orthogonal matrix that transforms the displacement field from the intermediate element coordinate system to the body coordinate system, $\mathbf{S}^{ijl}$ is the shape function matrix for the lamina, $\bar{\mathbf{C}}^{ij}$ is an $n_{dfe} \times n_{dfe}$ constant orthogonal matrix (where n_{dfe} is the number of degrees of freedom of the element) that transforms the nodal coordinates from the body coordinate system to the intermediate element coordinate system, and $\mathbf{B}^{ij}$ is the element *Boolean* transformation matrix.

Taking the time derivative of eq. 9, the velocity vector of the arbitrary point is obtained and used in the expression for the kinetic energy of the lamina

$$T^{ijl} = \frac{1}{2}\int_{V^{ijl}} \rho^{ijl}\dot{\mathbf{r}}^{ijl^T}\dot{\mathbf{r}}^{ijl}dV^{ijl} = \frac{1}{2}\dot{\mathbf{q}}^{i^T}\mathbf{M}^{ijl}\dot{\mathbf{q}}^{i} \tag{10}$$

where V^{ijl} is the volume, ρ^{ijl} is the mass density, and $\mathbf{M}^{ijl}$ is the mass matrix of the lth lamina. The lamina mass matrix can be written as

$$\mathbf{M}^{ijl} = \begin{bmatrix} \mathbf{m}_{RR}^{ijl} & \mathbf{m}_{R\theta}^{ijl} & \mathbf{m}_{Rf}^{ijl} \\ \mathbf{m}_{\theta R}^{ijl} & \mathbf{m}_{\theta\theta}^{ijl} & \mathbf{m}_{\theta f}^{ijl} \\ \mathbf{m}_{fR}^{ijl} & \mathbf{m}_{f\theta}^{ijl} & \mathbf{m}_{ff}^{ijl} \end{bmatrix} \tag{11}$$

where the submatrices $\mathbf{m}^{ijl}$ are defined by Kremer (1991). It can be shown that the mass matrix is an implicit function of time. That is, this matrix contains the vector $\mathbf{u}^{ijl}$, which is an implicit function of time.

An examination of the mass submatrices reveals that the following time invariant matrices also exist as components of the lamina mass matrix (Kremer, 1991)

$$\bar{\mathbf{S}}^{ijl} = \int_{V^{ijl}} \rho^{ijl}\mathbf{S}^{ijl}dV^{ijl} \tag{12}$$

$$\mathbf{S}_{mn}^{ijl} = \int_{V^{ijl}} \rho^{ijl}\mathbf{S}_m^{ijl^T}\mathbf{S}_n^{ijl}dV^{ijl} \qquad m,n = 1,2,3 \tag{13}$$

where $\mathbf{S}_1^{ijl}$, $\mathbf{S}_2^{ijl}$, and $\mathbf{S}_3^{ijl}$ are the row vectors of the shape function matrix. Clearly, the time-invariants of motion of eq. 12 and eq. 13 depend on the mass density, volume, and the assumed displacement field of each layer. Hence, each lamina may have a unique set of invariants. Observe that in the case of the eight-noded plate element used in this investigation, the matrix $\bar{\mathbf{S}}^{ijl}$ is a 3×40 matrix while $\mathbf{S}_{mn}^{ijl}$ are 40×40 matrices. The explicit form of these matrices are presented by Kremer (1991). Equations 12 and 13 can be re-written as (Cook, 1981)

$$\bar{\mathbf{S}}^{ijl} = \sum_{r=1}^{n_{\zeta_l}}\sum_{q=1}^{n_\eta}\sum_{p=1}^{n_\xi} W_{\xi_r}W_{\eta_q}W_{\zeta_{lr}}\mathbf{S}_{\xi_p\eta_q\zeta_{lr}}^{ijl}J\frac{h_l}{t} \tag{14}$$

$$\mathbf{S}_{mn}^{ijl} = \sum_{r=1}^{n_{\zeta_l}}\sum_{q=1}^{n_\eta}\sum_{p=1}^{n_\xi} W_{\xi_r}W_{\eta_q}W_{\zeta_{lr}}\mathbf{S}_{m\xi_p\eta_q\zeta_{lr}}^{ijl^T}\mathbf{S}_{n\xi_p\eta_q\zeta_{lr}}^{ijl}J\frac{h_l}{t} \qquad m,n = 1,2,3 \tag{15}$$

and evaluated using numerical integration techniques. In the above equations, n_ξ, n_η and n_{ζ_l} are the number of Gauss points in the ξ, η and ζ_l directions, respectively, W_{ξ_p}, W_{η_q} and $W_{\zeta_{lr}}$ are the Gauss weights associated with the Gauss points, and, for example, $\mathbf{S}_{\xi_p\eta_q\zeta_{lr}}^{ijl}$ is the matrix $\mathbf{S}^{ijl}$ evaluated at Gauss points ξ_p, η_q and ζ_{lr}. The evaluation of the invariants of eq. 14 and eq. 15 is carried out using a $3 \times 3 \times 2$ sampling point scheme in the ξ, η and ζ_l directions, respectively. Using this sampling point scheme, the exact closed form integration results are obtained for rectangular elements (Kremer, 1991).

It can be shown that by summing the kinetic energies of the laminae of an element, the element mass matrix can be obtained, and by summing the kinetic energies of the elements of a body, the mass matrix of the body can be obtained (Kremer, 1991). This is expressed compactly as

$$T^i = \sum_{j=1}^{n_e} T^{ij} = \sum_{j=1}^{n_e}\sum_{l=1}^{n_l} T^{ijl} = \frac{1}{2}\dot{\mathbf{q}}^{i^T}\left[\sum_{j=1}^{n_e}\sum_{l=1}^{n_l}\mathbf{M}^{ijl}\right]\dot{\mathbf{q}}^i \tag{16}$$

where n_e is the number of elements in the body, and n_l is the number of laminae in the element.

5 STIFFNESS MATRIX

The stiffness matrix for the composite plate element is derived using standard finite element procedures for structures composed of layers of orthotropic laminae. The strain energy for lamina l of element j in body i is given as

$$U^{ijl} = -\frac{1}{2}\int_{V^{ijl}} \bar{\sigma}^{ijl^T}\bar{\epsilon}^{ijl}dV^{ijl} \tag{17}$$

where V^{ijl} is the volume of lamina l, $\bar{\sigma}^{ijl}$ is the stress vector, and $\bar{\epsilon}^{ijl}$ is the strain vector for lamina l. The strain vector is defined by eq. 4. Assuming linear elastic material behavior, the constitutive relations are defines as

$$\bar{\sigma}^{ijl} = \bar{\mathbf{E}}^{ijl}\bar{\epsilon}^{ijl} \tag{18}$$

where $\bar{\mathbf{E}}^{ijl}$ is the matrix of elastic coefficients for lamina l, and is defined by Jones (1975). Substituting for $\bar{\sigma}^{ijl}$ and $\bar{\epsilon}^{ijl}$ into eq. 17 one obtains the strain energy of the lamina in terms of the nodal degrees of freedom as

$$U^{ijl} = -\frac{1}{2}\bar{\mathbf{q}}_f^{ij^T}\bar{\mathbf{K}}_{ff}^{ijl}\bar{\mathbf{q}}_f^{ij} \tag{19}$$

where $\bar{\mathbf{q}}_f^{ij}$ is the vector of deformation degrees of freedom defined in the intermediate element coordinate system, and $\bar{\mathbf{K}}_{ff}^{ijl}$ is the lamina stiffness matrix and is defined as

$$\bar{\mathbf{K}}_{ff}^{ijl} = \int_{V^{ijl}} (\mathbf{DS}^{ijl})^T(\mathbf{T}^{ijl^T}\bar{\mathbf{E}}_1^{ijl}\mathbf{T}^{ijl})(\mathbf{DS}^{ijl})dV^{ijl} \tag{20}$$

where $\mathbf{T}^{ijl}$ is the strain transformation matrix (Cook, 1981), and $\bar{\mathbf{E}}_1^{ijl}$ is the matrix of orthotropic material properties for lamina l. Employing numerical integration techniques to evaluate the volume integral of eq. 20, this equation can be written as (Cook, 1981)

$$\bar{\mathbf{K}}_{ff}^{ijl} = \sum_{r=1}^{n_{\zeta_l}}\sum_{q=1}^{n_\eta}\sum_{p=1}^{n_\xi} W_{\xi_r}W_{\eta_q}W_{\zeta_{lr}}(\mathbf{DS}^{ijl^T})_{\xi_p\eta_q\zeta_{lr}} (\mathbf{T}^{ijl^T}\bar{\mathbf{E}}_1^{ijl}\mathbf{T}^{ijl})(\mathbf{DS}^{ijl})_{\xi_p\eta_q\zeta_{lr}}J\frac{h_l}{t} \tag{21}$$

where $(\mathbf{DS}^{ijl})_{\xi_p\eta_q\zeta_{lr}}$ is the matrix $(\mathbf{DS}^{ijl})$ evaluated at Gauss points ξ_p, η_q and ζ_{lr}, and W_{ξ_p}, W_{η_q}, $W_{\zeta_{lr}}$, n_ξ, n_η, and n_{ζ_l} are as defined in the previous section. The evaluation of the stiffness matrix of eq. 21 in the finite element preprocessor code is accomplished using a $2 \times 2 \times 2$ sampling point scheme in the ξ, η and ζ_l directions, respectively. This is done to avoid the effects of shear locking (Cook, 1981). Note that $\bar{\mathbf{K}}_{ff}^{ijl}$ is defined in the intermediate element coordinate system. Using the transformation matrices discussed in the previous section, the stiffness matrix, as defined in the body coordinate system, is given by

$$\mathbf{K}_{ff}^{ijl} = \mathbf{B}^{ij^T}\bar{\mathbf{C}}^{ij^T}\bar{\mathbf{K}}_{ff}^{ijl}\bar{\mathbf{C}}^{ij}\mathbf{B}^{ij} \tag{22}$$

It can be shown that by summing the strain energies of the laminae of an element, the stiffness matrix for the element is obtained, and by summing the strain energies of the elements of a body, the body stiffness matrix is obtained (Kremer, 1991). In matrix form this summation is defined as

$$\mathbf{K}_{ff}^{i} = \sum_{j=1}^{n_e}\mathbf{K}_{ff}^{ij} = \sum_{j=1}^{n_e}\sum_{l=1}^{n_l}\mathbf{K}_{ff}^{ijl} = \sum_{j=1}^{n_e}\left[\mathbf{B}^{ij^T}\bar{\mathbf{C}}^{ij^T}\left(\sum_{l=1}^{n_l}\bar{\mathbf{K}}_{ff}^{ijl}\right)\bar{\mathbf{C}}^{ij}\mathbf{B}^{ij}\right] \tag{23}$$

6 DYNAMIC EQUATIONS OF MOTION

In this investigation the dynamic equations of the deformable body are formulated using the absolute reference and the relative elastic coordinates. In constrained multibody dynamics, the absolute coordinates are not independent, because of the kinematic joint constraints and the specified motion trajectories. These kinematic constraints can be formulated using a set of nonlinear algebraic constraint equations, which can be written compactly as (Shabana, 1989)

$$\mathbf{C}(\mathbf{q}, t) = \mathbf{0} \tag{24}$$

where $\mathbf{C}$ is the vector of constraint functions, $\mathbf{q}$ is the vector of system generalized coordinates, and t is time. Using the technique of Lagrange multipliers, the dynamic equations of the system can be written as (Shabana, 1989)

$$\mathbf{M}\ddot{\mathbf{q}} + \mathbf{K}\mathbf{q} + \mathbf{C}_{\mathbf{q}}^T\boldsymbol{\lambda} = \mathbf{Q}_e + \mathbf{Q}_v \tag{25}$$

where $\mathbf{M}$ is the mass matrix, $\mathbf{K}$ is the stiffness matrix, $\mathbf{q}$ is the vector of generalized coordinates, both independent and dependent, $\mathbf{C}_{\mathbf{q}}$ is the constraint Jacobian matrix which is obtained by taking the partial derivative of eq. 24 with respect to the generalized coordinates, and $\boldsymbol{\lambda}$ is the vector of Lagrange multipliers. The total vector of generalized forces is $\mathbf{Q}_e$, and $\mathbf{Q}_v$ is a vector that contains the Coriolis and the centrifugal forces. Taking the second derivative of eq. 24 with respect to time, we get

$$\mathbf{C}_{\mathbf{q}}\ddot{\mathbf{q}} = -\mathbf{C}_{tt} - 2\mathbf{C}_{\mathbf{q}t}\dot{\mathbf{q}} - (\mathbf{C}_{\mathbf{q}}\dot{\mathbf{q}})_{\mathbf{q}}\,\dot{\mathbf{q}} = \mathbf{Q}_c \tag{26}$$

where $\mathbf{Q}_c$ is a vector that depends on the generalized coordinates and velocities, and possibly on time. Combining eq. 25 and eq. 26, one obtains

$$\begin{bmatrix} \mathbf{M} & \mathbf{C}_{\mathbf{q}}^T \\ \mathbf{C}_{\mathbf{q}} & \mathbf{0} \end{bmatrix} \begin{bmatrix} \ddot{\mathbf{q}} \\ \boldsymbol{\lambda} \end{bmatrix} = \begin{bmatrix} \mathbf{Q}_e + \mathbf{Q}_v - \mathbf{K}\mathbf{q} \\ \mathbf{Q}_c \end{bmatrix}. \tag{27}$$

This system of equations can be solved for the vector of Lagrange multipliers and the vector of generalized accelerations, which include the reference as well as flexible coordinate accelerations. The generalized reaction forces can be obtained through the product $\mathbf{C}_{\mathbf{q}}^T\,\boldsymbol{\lambda}$. The vectors $\dot{\mathbf{q}}$ and $\ddot{\mathbf{q}}$ can be integrated forward in time to find successive position $\mathbf{q}$ and velocity $\dot{\mathbf{q}}$ vectors of the system (Shabana, 1989).

7 APPLICATION

In this section the dynamics of a spatial RSSR (Revolute-Spherical-Revolute-Spherical) mechanism is examined. The coupler of the mechanism is modeled as a flexible plate using the formulation developed herein. The initial configuration of the mechanism is shown in Fig. 2. The original configuration data for the RSSR mechanism is given by Kremer, et al. (1991) The coupler, body 3, is modeled as a flexible body. Three different coupler models are examined utilizing properties of the following materials:

1. Homogeneous, isotropic steel;
2. Orthotropic graphite/epoxy with the fibers aligned to the $\mathbf{X}^3$-axis (the $\mathbf{X}^3$-axis is along the line that is defined by the two spherical joints); and
3. Cross-ply graphite/epoxy having symmetric layering of 0-90-90-0 with respect to the $\mathbf{X}^3$-axis.

The material properties used for the couplers, as well as the inertia properties of the couplers, are given by Kremer (1991).

Eight elements and thirty seven nodes are used in modeling the coupler. Each element is rectangular, having dimensions of 0.09525 × 0.09525 × 0.004 meters. To determine the mode shapes of vibration, the degrees of freedom u, v, w, and β at spherical joint A are suppressed, and the translational degrees of freedom v and w at joint B are fixed.

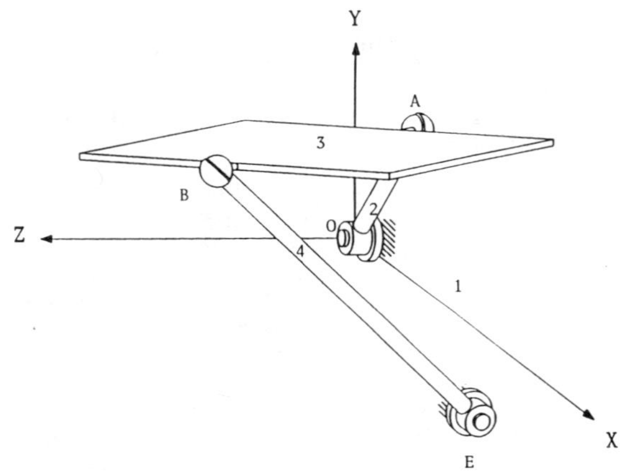

Figure 2: The RSSR mechanism.

The input crank of the RSSR mechanism is specified to oscillate about point O according to the function

$$\phi^2 = \frac{\pi}{9}\sin(15\pi t) + \phi_o^2 \tag{28}$$

where the superscript 2 refers to the input crank, body 2, t is time, ϕ^2 is the crank angle at any given time t, and ϕ_o^2 is the crank angle at time $t = 0$. This function specifies the crank to oscillate between 0.7 and 1.4 radians.

The nonlinear dynamic equations of this mechanism are developed using the augmented formulation (Shabana, 1989; Kremer, 1991) in which the nonlinear algebraic constraint equations that describe mechanical joints and specified motion trajectories are adjoined to the differential equations using Lagrange multipliers. The resulting mixed system of differential and algebraic equations is solved numerically using a direct numerical integration method coupled with a Newton Raphson algorithm to check on constraint violations.

In Fig. 3 a plot of the the transverse displacement of the coupler midpoint versus time is given for each of the coupler models. Note the increase in flexibility of the composite coupler models. Also note that there is a decrease in deflection for the cross-ply model as compared to the orthotropic model. This indicates that the stiffness in the $\mathbf{Y}^3$-direction can notably affect the transverse deflection of the coupler.

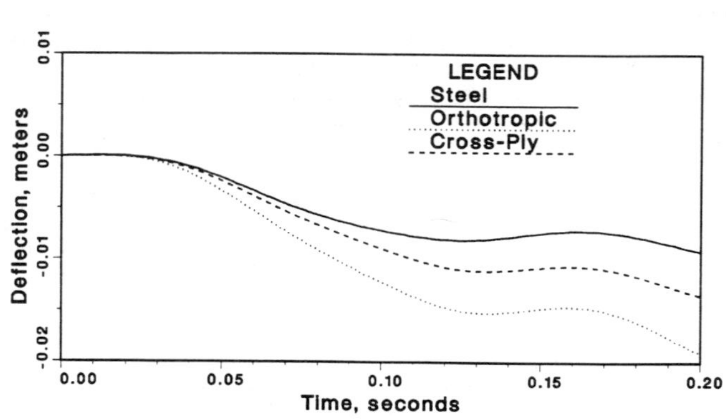

Figure 3: Transverse deflection of the coupler midpoint.

8 DYNAMIC STRESSES

Checking the integrity of a structural component is a basic step in the design process. A common method to check the soundness of a design is to perform a stress analysis of the component. Commercially available finite element computer codes have made this analysis easier. The wide range of element types available make virtually any type of analysis possible. In addition, these codes make the stress analysis of very complex geometries possible. Stress analysis of dynamic components is as important as for static components. The process of such an analysis is not straight forward. If the component is part of a multibody system, joint reaction forces and inertia forces must first be computed in order to be able to evaluate the deformations, strains and stresses. Through the use of the finite element models developed here, the ability to compute stresses directly from a dynamic simulation becomes available.

The computation of dynamic stresses is accomplished using a post processing stress analysis program, and the stresses are computed via

$$\sigma^{ijl} = \mathbf{E}^{ijl}\epsilon^{ijl} = \mathbf{E}^{ijl}\mathbf{D}\mathbf{S}^{ijl}\mathbf{q}_f^{ijl} \quad (29)$$

The dynamic stresses σ_x developed in the flexible steel coupler of the RSSR mechanism are shown in Fig. 4. To avoid extrapolation errors, the stresses reported are for the $2 \times 2 \times 2$ Gauss point locations. The Gauss points at which the stress are computed form a line, near the top surface of the coupler, that spans the distance between the two spherical joints. It is evident from Fig. 4 that complex bending is taking place in the steel coupler, and that the extent of the bending is increasing with time.

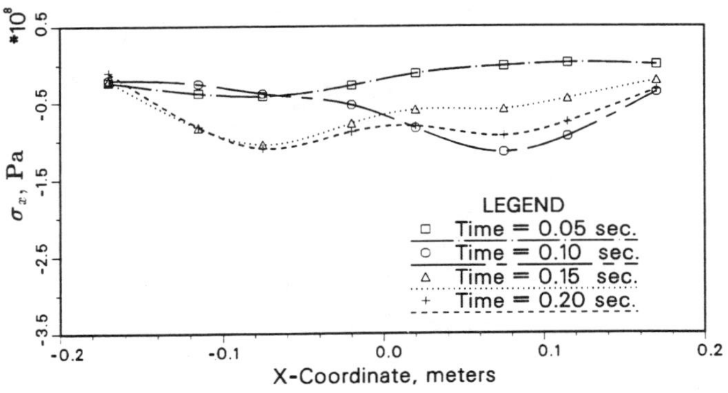

Figure 4: Normal stress σ_x in the steel coupler.

9 SUMMARY

The use of laminated plates consisting of layers of orthotropic laminae in the analysis of deformable bodies that undergo large arbitrary reference displacement and small elastic deformation is demonstrated in this investigation. A finite element formulation is developed for determining the inertia and stiffness characteristics of laminated composite plates. In this general formulation, the plate may consist of an arbitrary number of laminae which may have different orthotropic material properties and may be arbitrarily oriented. A consistent mass formulation is used to define the kinetic energy of each lamina in the composite finite element. The resulting expression for the kinetic energy is used to identify the nonlinear lamina mass matrix. It is noted that the lamina mass matrix can be expressed in terms of a set of time invariant matrices that depend on the assumed displacement field and on the inertia properties of the lamina. The invariants of the element can be obtained by assembling the invariants of its laminae. These element invariants can be used to define the nonlinear element mass matrix. By summing the expressions for the kinetic energies of the element, an expression for the body kinetic energy can be obtained. This leads to the definition of the nonlinear body mass matrix. It is noted that this matrix can be expressed in terms of the invariants of the laminae of its elements. The finite element formulation presented in this investigation can be used in the analysis of spatial multibody mechanical systems that consist of interconnected bodies having composite materials properties. The general formulation presented in this paper is demonstrated by the development of an eight-noded *Mindlin* plate element. A dynamic simulation of a multibody spatial mechanism (the RSSR mechanism) showed the effect of the use of composite materials. The results indicate that the inertia properties of the plate are of significance in determining the transverse deflection of the plate. The examples conducted using the orthotropic laminated coupler and the cross-ply laminated coupler demonstrate the effect of laminate stiffness on the transverse deflection of the plate. Through the use of a stress analysis postprocessor, it has been demonstrated that the computation of stresses in the analysis of flexible multibody dynamic systems is feasible.

REFERENCES

Cook, R.D., 1981, *Concepts and Applications of Finite Element Analysis*, 2nd edn. John Wiley and Sons, NY.

Hrabok, M.M. and Hrudey, T.M., 1984, "A Review and Catalogue of Plate Bending Finite Elements", *Computer and Structures*, vol. 19, pp. 479–495.

Jones, R.M., 1975, *Mechanics of Composite Materials*, 1st edn. McGraw-Hill, NY.

Kremer, J.M., 1991, "Large Reference Displacement Analysis of Composite Plates", Ph.D. dissertation, University of Illinois at Chicago, Department of Mechanical Engineering.

Kremer, J.M., Shabana, A.A. and Widera, G.E.O., 1991, "Application of the Composite Plate Theory and the Finite Element Method to the Dynamics and Stress Analysis of Spatial Flexible Mechanical Systems", Submitted to *ASME Journal of Mechanical Design*.

Panda, S.C. and Natarajan, R., "Finite Element Analysis of Laminated Composite Plates", *International Journal for Numerical Methods in Engineering*, vol. 14, pp. 69–79, (1979).

Schwartz, M.M., 1984, *Composite Materials Handbook*, 1st edn. McGraw-Hill, NY.

Shabana, A.A., 1989, *Dynamics of Multibody Systems*, 1st edn. John Wiley and Sons, NY.

AMD-Vol. 141/DSC-Vol. 37, Dynamics of Flexible Multibody Systems:
Theory and Experiment
ASME 1992

FLEXIBLE MULTIBODY DYNAMICS TECHNIQUES FOR SHAPE AND NONRIGID MOTION ESTIMATION AND SYNTHESIS

Dimitri Metaxas and Demetri Terzopoulos
Department of Computer Science
University of Toronto
Toronto, Ontario, Canada

ABSTRACT

This paper develops a general approach for the efficient modeling and animation of nonrigid multibody objects. We develop a new family of physics-based modeling primitives that can undergo free motions as well as parameterized and free-form deformations. Starting with parameterized geometric primitives and adding global geometric deformations plus local finite-element deformations, mass distributions, and elasticities, we create deformable models that display intuitively correct physical behaviors. Furthermore, the paper develops efficient constraint methods for connecting these physical primitives together to make articulated multibody objects. We demonstrate the computationally efficient performance of our algorithms for the synthesis and visualization of deformable models, as well as for estimating the shapes and motions of deformable objects.

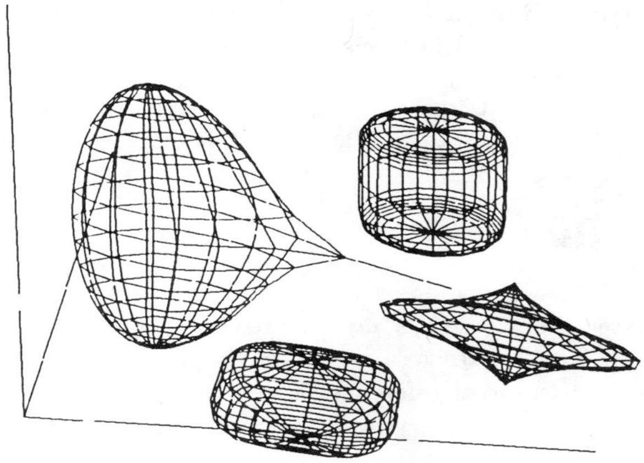

Figure 1: Interaction with deformable superquadrics.

1 INTRODUCTION

This paper develops an an efficient approach to approximating the dynamic behavior of nonrigid multibody objects for simulation, visualization, and estimation. Our approach draws upon the parameterized and free-form shape primitives which are the mainstay of geometric modeling methodology (Hoffmann 1989). Through modification of parameters or control points, geometric primitives can deform into a family of shapes; however, the deformations are purely kinematic. We propose an approach for creating dynamic deformable models with intuitive physical behaviors by combining parameterized geometric primitives (e.g., superquadrics), parameterized geometric deformations (e.g., tapering, bending, shearing, twisting), and local free-form deformations (e.g., finite element shape functions).

Our approach makes use of some of the features of Shabana's formulation of multibody dynamics (Shabana 1989). Applying Lagrangian dynamics and the finite element method, we convert the translational, rotational, and geometric degrees of freedom of our models into generalized coordinates of equations of motion governing the model's response to applied forces. The method is general and does not depend on the particular choice of geometric primitives and deformations, as long as the Jacobian of the geometric functions is computable.

For example, Figure 1 shows several deformable superquadrics. It illustrates the response of a deformable superellipsoid to the traction from a Hookean spring attached to its surface and pulled interactively. In general, the models are abstract viscoelastic solids. The various models shown in the figure can be obtained by applying forces to the surface of a viscoelastic supersphere.

We also develop a constraint algorithm (Metaxas and Terzopoulos 1992 to appear), based on Baumgarte's stabilization method (Baumgarte 1972), that allows point-to-point constraints among the dynamic models. Our algorithm computes the unknown constraint forces by solving a linear system of equations whose size is proportional to the number of constraints present in the articulated object. This allows the efficient construction and dynamic simulation of articulated objects with rigid or flexible parts. The constraint algorithm computes generalized constraint forces and adds them to the generalized external forces in the Lagrange equations of motion of the multibody system. Finally, we integrate these equations using standard numerical techniques.

We demonstrate the efficiency of our techniques for

the synthesis and visualization of shape and nonrigid motion, as well as for the estimation of elastically deformable models from visual data. For efficiency, we can lump masses to obtain diagonal mass matrices and we can employ mass-proportional damping. For the purposes of computer animation, we have simulated flexible multibody objects in gravitational fields, including elastic collisions with obstacles and friction effects. Our constraint algorithm can assemble complex objects satisfying point-to-point constraints from inappropriately shaped, mispositioned parts that do not initially satisfy the constraints. We have applied our models to visual analysis to recover the shape and motion of a deformable object (e.g., parts of a human body) from sensor data (e.g., range data).

The body of the paper develops our approach the general case, while the appendices give mathematical details for a specific physics-based primitive: deformable superquadrics.

2 GEOMETRY

We consider solid models whose intrinsic (material) coordinates are $\mathrm{u} = (u, v, w)$, defined on a bounded domain Ω. The positions of points on the model relative to an inertial frame of reference Φ in space are given by a vector-valued, time varying function of u:

$$\mathbf{x}(\mathrm{u}, t) = (x_1(\mathrm{u}, t), x_2(\mathrm{u}, t), x_3(\mathrm{u}, t))^T, \tag{1}$$

where ^{T} is the transpose operator. We set up a noninertial, model-centered reference frame ϕ, and express these positions as

$$\mathbf{x} = \mathbf{c} + \mathbf{R}\mathbf{p}, \tag{2}$$

where $\mathbf{c}(t)$ is the origin of ϕ at the center of the model and the orientation of ϕ is given by the rotation matrix $\mathbf{R}(t)$. Thus, $\mathbf{p}(\mathrm{u}, t)$ denotes the positions of points on the model relative to the model frame. We further express $\mathbf{p}$ as the sum of a reference shape $\mathbf{s}(\mathrm{u}, t)$ and a displacement function $\mathbf{d}(\mathrm{u}, t)$:

$$\mathbf{p} = \mathbf{s} + \mathbf{d}. \tag{3}$$

Figure 2 illustrates the model geometry.

We define the reference shape as

$$\mathbf{s} = \mathbf{T}(\mathbf{e}; b_1, b_2, \ldots) \tag{4}$$

where $\mathbf{T}$ is a global deformation, a function dependent on the parameters b_i. This is applied to a parametric function of material coordinates

$$\mathbf{e} = \mathbf{e}(\mathrm{u}; a_1, a_2, \ldots) \tag{5}$$

which defines a geometric primitive whose parameters are a_i. [1] We collect both sets of global parameters into the

[1]Both $\mathbf{T}$ and $\mathbf{e}$ are assumed to be differentiable. Note that $\mathbf{T}$ will usually be a composite deformation which can be a sequence of simpler deformations.

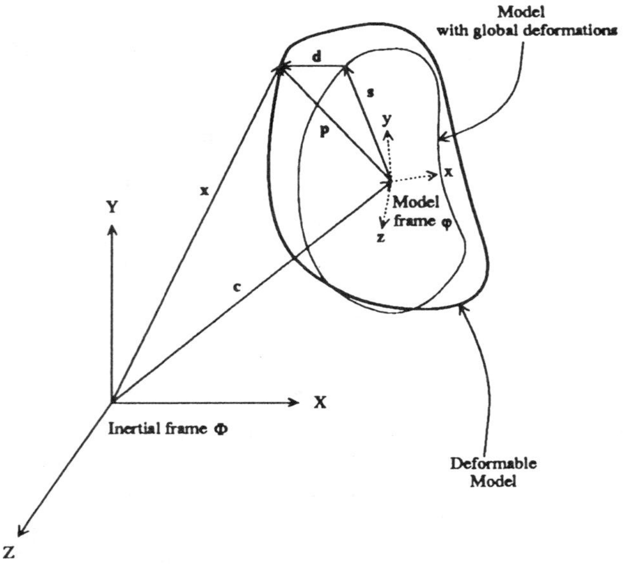

Figure 2: Geometry of deformable model.

vector of global generalized coordinates

$$\mathbf{q}_s = (a_1, a_2, \ldots, b_1, b_2, \ldots)^T. \tag{6}$$

For concreteness, Appendix A gives the parametric equation of a superquadric ellipsoid with tapering, bending, shearing, and twisting deformations.

We express the displacement $\mathbf{d}$ as a linear combination of basis functions $\mathbf{b}_i(\mathrm{u})$

$$\mathbf{d} = \sum_i \mathbf{b}_i^T \mathbf{q}_i, \tag{7}$$

where $\mathbf{b}_i$ is a vector formed from the basis functions while $\mathbf{q}_i$ are time dependent generalized coordinates. The basis functions can be local or global; however, finite element basis functions (Kardestuncer 1987) are the natural choice for representing local deformations. We associate a displacement vector $\mathbf{q}_i$ with each node i of the model. Collecting the generalized coordinates into a vector $\mathbf{q}_d = (\ldots, \mathbf{q}_i, \ldots)^T$, we write

$$\mathbf{d} = \mathbf{S}\mathbf{q}_d, \tag{8}$$

where $\mathbf{S}$ is the shape matrix whose entries are the finite element basis functions. A forthcoming section describes the finite elements that we employ to create deformable superquadrics.

3 KINEMATICS AND DYNAMICS

To convert parameterized geometric models into physical models that respond dynamically to forces, we first consider the kinematics implied by the geometry, then we introduce mass, damping, and elasticity into the model to derive its mechanics.

The velocity of points on the model is given by,

$$\dot{\mathbf{x}} = \dot{\mathbf{c}} + \dot{\mathbf{R}}\mathbf{p} + \mathbf{R}\dot{\mathbf{p}} = \dot{\mathbf{c}} + \mathbf{B}\dot{\boldsymbol{\theta}} + \mathbf{R}\dot{\mathbf{s}} + \mathbf{R}\mathbf{S}\dot{\mathbf{q}}_d, \quad (9)$$

where $\boldsymbol{\theta} = (\ldots, \theta_i, \ldots)^T$ is the vector of rotational coordinates and $\mathbf{B} = [\ldots \partial(\mathbf{Rp})/\partial\theta_i \ldots]$. Furthermore, $\dot{\mathbf{s}} = [\partial \mathbf{s}/\partial \mathbf{q}_s]\dot{\mathbf{q}}_s = \mathbf{J}\dot{\mathbf{q}}_s$, where $\mathbf{J}$ is the Jacobian of s with respect to the global parameter vector $\mathbf{q}_s$.

We can therefore write the model kinematics compactly as

$$\mathbf{x} = \mathbf{c} + \mathbf{R}(\mathbf{s} + \mathbf{d}) = \mathbf{h}(\mathbf{q}), \quad (10)$$

and

$$\dot{\mathbf{x}} = [\mathbf{I}\ \mathbf{B}\ \mathbf{RJ}\ \mathbf{RS}]\dot{\mathbf{q}} = \mathbf{L}\dot{\mathbf{q}}, \quad (11)$$

where $\mathbf{q} = (\mathbf{q}_c^T, \mathbf{q}_\theta^T, \mathbf{q}_s^T, \mathbf{q}_d^T)^T$, with $\mathbf{q}_c = \mathbf{c}$ and $\mathbf{q}_\theta = \boldsymbol{\theta}$ is the total generalized coordinate vector for the model.

Next, we introduce a mass distribution over the model's material domain $\mathrm{u} \in \Omega$ and assume that the material is subject to velocity-dependent damping while in motion. We also assume that the material is subject to elastic or viscoelastic deformations. Applying Lagrangian dynamics, we obtain second-order equations of motion which take the general form

$$\mathbf{M}\ddot{\mathbf{q}} + \mathbf{D}\dot{\mathbf{q}} + \mathbf{K}\mathbf{q} = \mathbf{g}_\mathbf{q} + \mathbf{f}_\mathbf{q}, \quad (12)$$

where $\mathbf{M}$, $\mathbf{D}$, and $\mathbf{K}$ are the mass, damping, and stiffness matrices, respectively, where $\mathbf{g}_\mathbf{q}$ are generalized inertial centrifugal and Coriolis forces arising from the dynamic coupling between the local and global deformation generalized coordinates, and where $\mathbf{f}_\mathbf{q}(\mathrm{u}, t)$ are the generalized external forces associated with the generalized coordinates of the model. Appendix B derives (12) along with expressions for the relevant matrices and force vectors.

4 CONSTRAINED NONRIGID MOTION

We now extend (12) to account for the motions of composite models with interconnected deformable parts which are constrained not to separate. Shabana (Shabana 1989) describes the well-known Lagrange multiplier method for multibody objects. The composite generalized coordinate vector $\mathbf{q}$ and force vectors $\mathbf{g}_\mathbf{q}$ and $\mathbf{f}_\mathbf{q}$ for an n-body object are formed by concatenating the vectors $\mathbf{q}_i$, $\mathbf{g}_{\mathbf{q}_i}$, $\mathbf{f}_{\mathbf{q}_i}$ associated with part i and $i = 1, \ldots, n$. Similarly, the composite matrices $\mathbf{M}$, $\mathbf{D}$, and $\mathbf{K}$ for the n-body object are block diagonal matrices whose components are $\mathbf{M}_i$, $\mathbf{D}_i$, and $\mathbf{K}_i$ associated with each part i. The method augments the composite equations of motion as follows:

$$\mathbf{M}\ddot{\mathbf{q}} + \mathbf{D}\dot{\mathbf{q}} + \mathbf{K}\mathbf{q} = \mathbf{g}_\mathbf{q} + \mathbf{f}_\mathbf{q} - \mathbf{C}_\mathbf{q}^T\boldsymbol{\lambda}, \quad (13)$$

where the holonomic constraints are expressed by

$$\mathbf{C}(\mathbf{q}, t) = \mathbf{0}, \quad (14)$$

where $\mathbf{C} = [\mathbf{C}_1^T, \mathbf{C}_2^T, \ldots, \mathbf{C}_k^T]^T$ is the vector of k linearly independent constraint equations among the n parts of the object. Hence, $\mathbf{C}_\mathbf{q}^T$ is the transpose of the constraint Jacobian matrix and $\boldsymbol{\lambda} = (\boldsymbol{\lambda}_1^T, \ldots, \boldsymbol{\lambda}_n^T)^T$ is the vector of Lagrange multipliers. The additional term $\mathbf{f}_{g_c} = -\mathbf{C}_\mathbf{q}^T\boldsymbol{\lambda}$ represents the unknown generalized forces on the parts due to the constraints.

Equation (13) comprises fewer equations than unknowns. To obtain additional equations, we differentiate (14) twice with respect to time, yielding

$$\boldsymbol{\gamma} = \mathbf{C}_\mathbf{q}\ddot{\mathbf{q}} = -\mathbf{C}_{tt} - (\mathbf{C}_\mathbf{q}\dot{\mathbf{q}})_\mathbf{q}\dot{\mathbf{q}} - 2\mathbf{C}_{\mathbf{q}_t}\dot{\mathbf{q}}. \quad (15)$$

Appending this equation to (13) and rearranging terms, we arrive at the augmented equations of motion

$$\begin{bmatrix} \mathbf{M} & \mathbf{C}_\mathbf{q}^T \\ \mathbf{C}_\mathbf{q} & \mathbf{0} \end{bmatrix} \begin{bmatrix} \ddot{\mathbf{q}} \\ \boldsymbol{\lambda} \end{bmatrix} = \begin{bmatrix} -\mathbf{D}\dot{\mathbf{q}} - \mathbf{K}\mathbf{q} + \mathbf{g}_\mathbf{q} + \mathbf{f}_\mathbf{q} \\ \boldsymbol{\gamma} \end{bmatrix}. \quad (16)$$

Given initial conditions $\mathbf{q}(0)$ and $\dot{\mathbf{q}}(0)$ that satisfy $\mathbf{C}(\mathbf{q}(0), 0) = \mathbf{0}$ and $\dot{\mathbf{C}}(\mathbf{q}(0), \dot{\mathbf{q}}(0), 0) = \mathbf{0}$, these equations may be integrated, in principle. At each time step, we may solve (16) for $\ddot{\mathbf{q}}$ and $\boldsymbol{\lambda}$ with known $\mathbf{q}$ and $\dot{\mathbf{q}}$, and then we integrate $\dot{\mathbf{q}}$ and $\ddot{\mathbf{q}}$ from t to $t + \Delta t$ to obtain $\mathbf{q}$ and $\dot{\mathbf{q}}$, respectively.

There are two practical problems with (16) in animation applications. The first is that the constraints must be satisfied initially; however, the initial placement of parts and/or their parameter values may not be preset such that the parts satisfy the constraints. Secondly the numerical solution of (16) is problematic, because although the constraints may be satisfied exactly at a given time step of the animation (i.e., $\mathbf{C}(\mathbf{q}, t) = \mathbf{0}$), they may not be at the next time step (i.e., $\mathbf{C}(\mathbf{q}, t + \Delta t) \neq \mathbf{0}$).

Baumgarte remedied these two problems by applying linear feedback control to stabilize the constraint equations in the sense of Ljapunov (Baumgarte 1972). The contribution of (14) to (16) is replaced by the damped second-order differential equation

$$\ddot{\mathbf{C}} + 2\alpha\dot{\mathbf{C}} + \beta^2\mathbf{C} = \mathbf{0}, \quad (17)$$

where α and β are stabilization factors. This replaces the lower entry of the vector on the right-hand-side of (16) with $\boldsymbol{\gamma} - 2\alpha\dot{\mathbf{C}} - \beta^2\mathbf{C}$. Fast stabilization suggests choosing $\beta = \alpha$ to obtain the non-oscillatory, critically damped solution $\mathbf{C}(\mathbf{q}, 0)e^{-\alpha t}$, which will have the quickest asymptotic decay towards constraint satisfaction $\mathbf{C} = \mathbf{0}$ for a given α. A potential problem with the constraint stabilization method is that it introduces additional eigenfrequencies into the dynamical system. Setting α in an

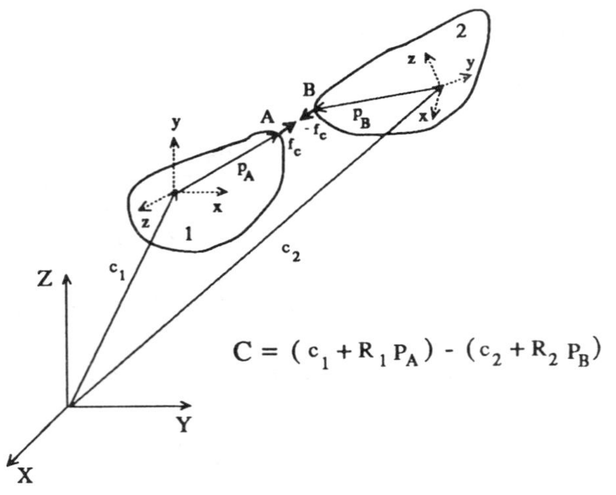

Figure 3: Constraint force computation.

attempt to increase the constraint satisfaction rate can result in numerically stiff equations and instability.

The general Lagrange multiplier method is potentially expensive, since the matrix in (16) is large. We have devised a specialized method to solve for the unknown constraint forces $\mathbf{f}_{g_c}$ arising from point-to-point constraints. The method involves the solution of a linear system whose size is proportional to the number of constraints, which is usually small. In this sense, it is similar to the dynamic constraint technique of (Barzel and Barr 1988); however, it is suitable for nonrigid parts. In the interest of space, we describe the constraint force calculation for a single constraint between the two deformable parts of an object.

4.1 Fast Constraint Force Computation

Figure 3 shows two parts 1 and 2 of an object whose points A and B are required to be in contact. We consequently want to calculate the constraint forces $\mathbf{f}_c(t)$ at point A and $-\mathbf{f}_c(t)$ at point B that are appropriate to keep the two parts connected at the above points. From (12), the motion equations of parts 1 and 2 are

$$\ddot{\mathbf{q}}_1 = \mathbf{M}_1^{-1}(\mathbf{g}_{\mathbf{q}_1} + \mathbf{f}_{\mathbf{q}_1} + \mathbf{f}\mathbf{g}_{c_A} - \mathbf{K}_1\mathbf{q}_1 - \mathbf{D}_1\dot{\mathbf{q}}_1) \quad (18)$$

and

$$\ddot{\mathbf{q}}_2 = \mathbf{M}_2^{-1}(\mathbf{g}_{\mathbf{q}_2} + \mathbf{f}_{\mathbf{q}_2} + \mathbf{f}\mathbf{g}_{c_B} - \mathbf{K}_2\mathbf{q}_2 - \mathbf{D}_2\dot{\mathbf{q}}_2) \quad (19)$$

where the generalized constraint forces at points A and B are, respectively,

$$\mathbf{f}\mathbf{g}_{c_A} = \mathbf{L}_A^T\mathbf{f}_c, \qquad \mathbf{f}\mathbf{g}_{c_B} = -\mathbf{L}_B^T\mathbf{f}_c. \quad (20)$$

Using (11), the value of the point-to-point constraint and its time derivatives are

$$\begin{aligned} \mathbf{C} &= \mathbf{x}_A - \mathbf{x}_B = (\mathbf{c}_1 + \mathbf{R}_1\mathbf{p}_A) - (\mathbf{c}_2 + \mathbf{R}_2\mathbf{p}_B) \\ \dot{\mathbf{C}} &= \mathbf{L}_A\dot{\mathbf{q}}_1 - \mathbf{L}_B\dot{\mathbf{q}}_2 \\ \ddot{\mathbf{C}} &= \mathbf{L}_A\ddot{\mathbf{q}}_1 + \dot{\mathbf{L}}_A\dot{\mathbf{q}}_1 - \mathbf{L}_B\ddot{\mathbf{q}}_2 - \dot{\mathbf{L}}_B\dot{\mathbf{q}}_2. \end{aligned} \quad (21)$$

Replacing the above formulas into Baumagarte's differential equation (17) of the constraint (with $\alpha = \beta = 6$), we arrive at the following linear system for the unknown constraint force $\mathbf{f}_c$

$$(\mathbf{L}_A\mathbf{M}_1^{-1}\mathbf{L}_A^T + \mathbf{L}_B\mathbf{M}_2^{-1}\mathbf{L}_B^T)\mathbf{f}_c + \mathbf{V} = \mathbf{0}, \quad (22)$$

where the 3×1 vector $\mathbf{V}$ is,

$$\begin{aligned} \mathbf{V} = {} & \dot{\mathbf{L}}_A\dot{\mathbf{q}}_1 - \dot{\mathbf{L}}_B\dot{\mathbf{q}}_2 + 2\alpha\dot{\mathbf{C}} + \alpha^2\mathbf{C} + \\ & \mathbf{L}_A\mathbf{M}_1^{-1}(\mathbf{g}_{\mathbf{q}_1} + \mathbf{f}_{\mathbf{q}_1} - \mathbf{K}_1\mathbf{q}_1 - \mathbf{D}_1\dot{\mathbf{q}}_1) - \\ & \mathbf{L}_B\mathbf{M}_2^{-1}(\mathbf{g}_{\mathbf{q}_2} + \mathbf{f}_{\mathbf{q}_2} - \mathbf{K}_2\mathbf{q}_2 - \mathbf{D}_2\dot{\mathbf{q}}_2). \end{aligned} \quad (23)$$

For k constraints we proceed in a similar way and arrive at a linear system with dimension $3k \times 3k$ whose solution yields the unknown constraint forces $\mathbf{f}_c = (\mathbf{f}_{c_1}^T, \mathbf{f}_{c_2}^T, \ldots, \mathbf{f}_{c_k}^T)^T$.

5 IMPLEMENTATION

The equations of motion of the model are numerically well-conditioned. Our approach partitions complicated nonrigid shapes and motions into rigid-body motions and local deformations away from globally deforming reference shapes. This partitioning leads to inherently stable simulation algorithms. We achieve interactive response by employing a simple Euler method to integrate (12). The Euler procedure updates the generalized coordinates $\mathbf{q}$ of the model at time $t + \Delta t$ according to the formulas

$$\begin{aligned} \ddot{\mathbf{q}}^{(t)} &= \mathbf{M}^{-1}(\mathbf{g_q}^{(t)} + \mathbf{f_q}^{(t)} + \mathbf{f}_{g_c}^{(t)} - \mathbf{K}\mathbf{q}^{(t)} - \mathbf{D}\dot{\mathbf{q}}^{(t)}) \\ \dot{\mathbf{q}}^{(t+\Delta t)} &= \dot{\mathbf{q}}^{(t)} + \Delta t\, \ddot{\mathbf{q}}^{(t)} \\ \mathbf{q}^{(t+\Delta t)} &= \mathbf{q}^{(t)} + \Delta t\, \dot{\mathbf{q}}^{(t+\Delta t)}, \end{aligned} \quad (24)$$

where Δt is the time step size.

We represent the rotation component of the models using quaternions, which simplifies the updating of $\mathbf{q}_\theta$. It is important to realize that we need not assemble and factorize a finite element stiffness matrix, as is normally the case when applying the finite element method. Rather, in the explicit Euler method we compute $\mathbf{Kq}$ very efficiently in an element-by-element fashion for the local deformation coordinates $\mathbf{q}_d$. For added efficiency, we may lump masses to obtain a diagonal $\mathbf{M}$, and we may assume mass-proportional damping, i.e. $\mathbf{D} = \nu\mathbf{M}$ where ν is the damping coefficient (Kardestuncer 1987).

5.1 Implementing Deformable Shells

To achieve the goal of interactive-time on available graphics workstations, we have implemented deformable shells, hollow primitives with a reduced number of finite elements. We describe here the formulation of deformable superquadric shells, where the material domain is restricted to u $= (u, v, 1)$. The formulation of their solid counterparts is straightforward (see Appendix A).

To create a shell, set the mass density to zero in the interior $0 \leq w < 1$, and tessellate the material domain into finite element subdomains. Many tessellations are possible. Triangular elements afford us the flexibility of regular or irregular tessellations. Appendix C presents the details of modeling an elastic membrane surface. Appendix D describes the the discretization of the associated deformation energy using linear triangular finite elements, which yields the local deformational submatrix $\mathbf{K}_{dd}$ of the stiffness matrix $\mathbf{K}$ in (34).

6 ANIMATION AND ESTIMATION EXAMPLES

We have created several physics-based animations and shape estimation examples involving models made from deformable superquadric primitives. The primitives interact with one another and with their simulated physical environments through constraints, collisions, gravity, and friction.

Figure 4 shows several frames from an animation of two deformable superquadric "balloons" suspended in gravity by flexible but inextensible strings attached to a ceiling using point-to-point constraints. Point-to-point constraints also connect the balloons to the strings. [2] Figure 4a shows the initial configuration with the left balloon pulled to the side. Gravity is activated in Figure 4b. The balloons deform under their own weight. The left balloon swings to the right, colliding inelastically with its neighbor in Figure 4c, thereby transferring some of its kinetic energy. In Figures 4d–f the balloons collide repeatedly until all the kinetic energy is dissipated. The collisions are implemented using reaction constraints between multiple deformable bodies. Figure 5 shows a similar scenario involving three balloons. By deforming, the middle balloon cushions the left balloon from the blow of the collision. It therefore swings a shorter distance than the left balloon did in the 2-balloon animation.

Figure 6 shows the automatic construction of a minimalist dragonfly from its constituent deformable superquadric parts. Figure 6a shows the disjoint parts in their initial configurations. [3] After activating our constraint algorithm, the model self-assembles (in a similar fashion to the self-assembling rigid part models in (Barzel

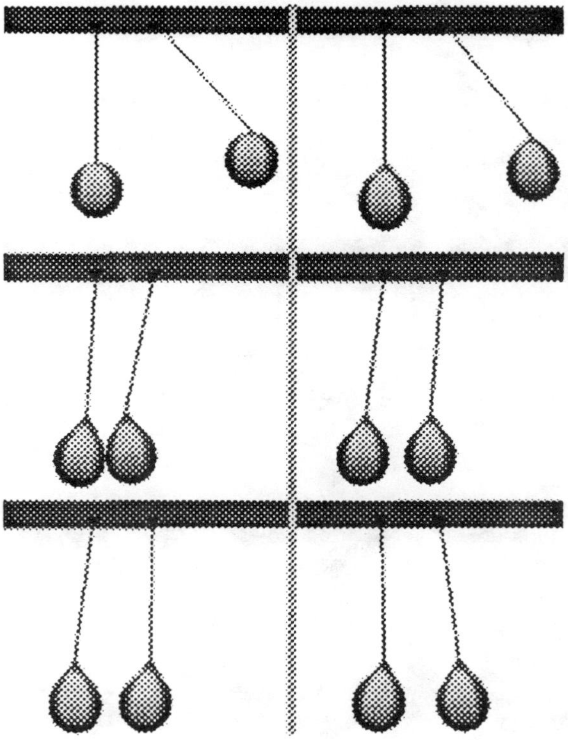

Figure 4: Two balloon pendulums.

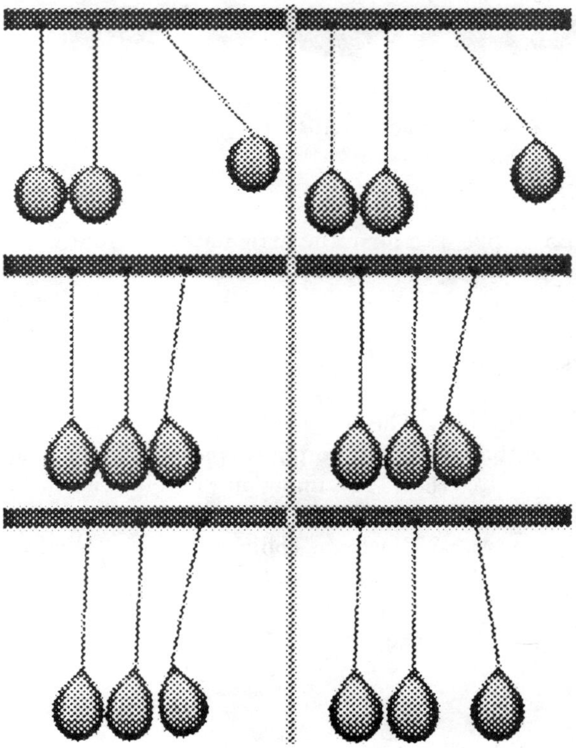

Figure 5: Three balloon pendulums.

[2]The elastic parameters were $w_0 = 60.0$ and $w_1 = 90.0$, the Euler step was 0.003, the nodal mass 10.0 and the damping coefficient $\nu = 1.6$.

[3]The elastic parameters were $w_0 = 0.1$ and $w_1 = 0.5$, the Euler step 0.0031, the nodal mass 2.0 and the damping coefficient $\nu = 1.6$.

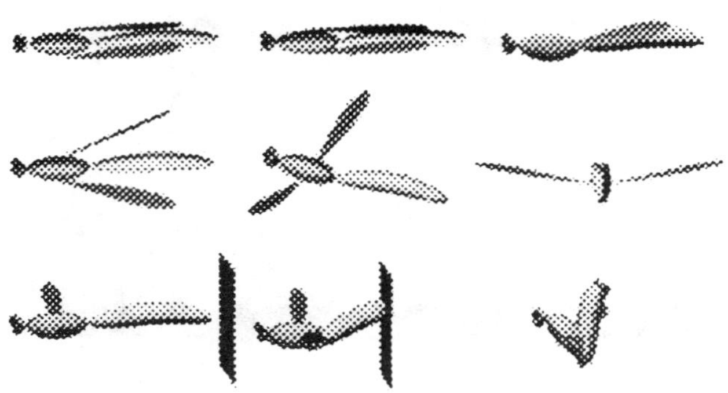

Figure 6: Self-assembly, articulation, and swatting of a dragonfly.

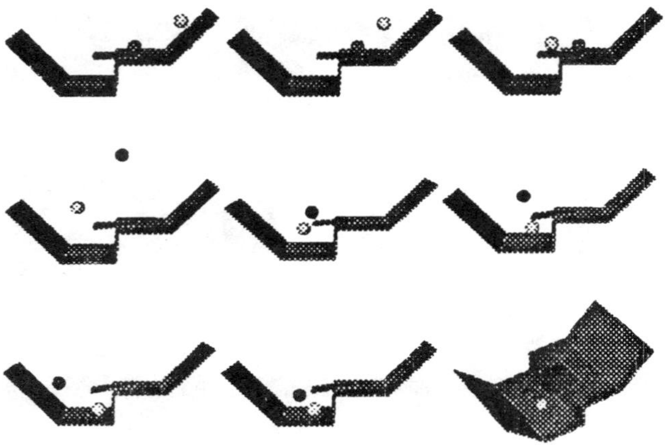

Figure 7: Collisions of deformable balls with planes and a spring loaded see-saw

and Barr 1988)) to form the articulated dragonfly shown in Figures 6b-c. Four point-to-point constraints hold the deformable body parts together. The dragonfly "works" inasmuch as forces can induce opening and flapping of the wings, as is illustrated in Figures 6d–f. An impenetrable plane appears in Figure 6g to swat the dragonfly in the rear (Figure 6h). The body parts deform in response to the blow, but the point-to-point constraints continue to hold them together. The mangled dragonfly is shown in Figure 6i.

Figures 7a-i illustrate collisions of two locally deformable balls with planes and a spring loaded see-saw. [4]

Figures 8a-i show an elastic "banana" dropping on

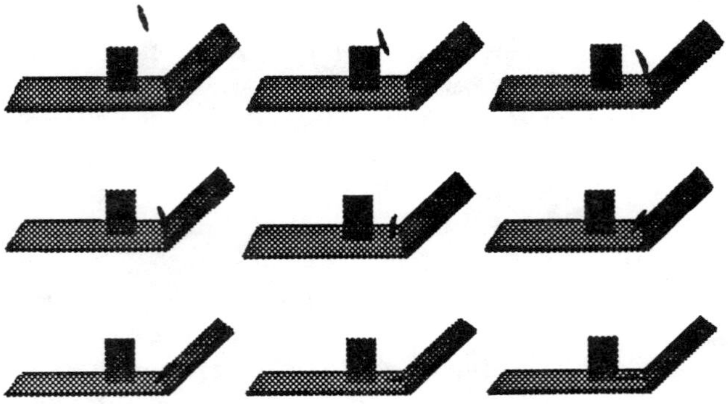

Figure 8: Elastic "banana" dropped on a box

a box and subsequently colliding with two planes. [5]

In Figure 9 we fit 5 deformable superquadrics to data collected from the raising and flexing motion of the two arms of a human subject (approximately 120 frames). The human motion 3D data were collected using WATSMART, a non-contact, three-dimensional motion digitizing and analysis system. [6] Figure 9(a) shows a view of the range data and the initial models. Figure 9(b) shows an intermediate step of the model estimation process driven by data forces from the first frame of the motion sequence, while Figures 9(c) and (d) show the models estimated from the initial data. Figures 9(e) and (f) show intermediate frames of the models tracking the nonrigid motion of the arms, while Figures 9(g) and (h) show two views of the final position of the estimated models.

7 CONCLUSION

This paper has presented a new class of dynamic solid models that have direct ties to standard, parameterized geometric primitives. Unlike their geometric predecessors, however, our models will deform in physically intuitive ways under the control of Lagrange equations of motion. These equations incorporate the geometric parameters of a conventional model such as a superquadric as global deformational degrees of freedom and finite element basis functions as local deformational degrees of freedom, thereby taking advantage of both the parameterized and free-form modeling paradigms.

We have also described a constraint algorithm suitable for our deformable multibody models, which is based on Baumgarte's stabilization method. For point-to-point

[4] The elastic parameters of each ball were $w_0 = 65.0$ and $w_1 = 85.0$, the Euler step 0.0031, the nodal mass 6.8 and the damping coefficient $\nu = 1.2$.

[5] The elastic parameters of the "banana" were $w_0 = 65.0$ and $w_1 = 85.0$, the Euler step 0.0031, the nodal mass 6.8 and the damping coefficient $\nu = 1.2$.

[6] We used an Euler step equal to 4.0×10^{-5}, a zero mass matrix **M** and a unit damping matrix **D**. In this way the deformable models have no inertia and therefore stop moving as soon as they fit the data and no external forces are exerted on them.

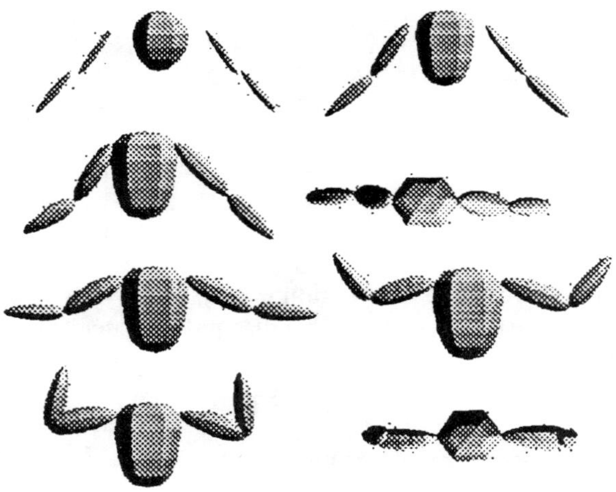

Figure 9: Tracking of raising and flexing human arm motion.

constraints, the algorithm computes the unknown constraint forces by solving a linear system of equations whose size is of the order of the number of constraints. This permits the efficient construction and dynamic simulation of articulated objects with rigid or flexible parts.

Although we have demonstrated our techniques for the specific case of superquadrics with global twisting, bending, tapering, and shearing deformations, a fairly complicated parameterized model, we emphasize that our approach is generally applicable to a wide variety of parameterized models, global deformations, and finite element basis functions.

A Superquadrics with Deformations

The parametric equations of a superquadric ellipsoid solid $\mathbf{e} = (e_1, e_2, e_3)$ are

$$\mathbf{e} = aw \begin{pmatrix} a_1 C_u^{\epsilon_1} C_v^{\epsilon_2} \\ a_2 C_u^{\epsilon_1} S_v^{\epsilon_2} \\ a_3 S_u^{\epsilon_1} \end{pmatrix}, \qquad (25)$$

where $-\pi/2 \leq u \leq \pi/2$, $-\pi \leq v < \pi$, $0 \leq w \leq 1$, and $S_u^{\epsilon} = \mathrm{sgn}(\sin u)|\sin u|^{\epsilon}$, $C_u^{\epsilon} = \mathrm{sgn}(\cos u)|\cos u|^{\epsilon}$, and similarly for S_v^{ϵ} and C_v^{ϵ}. Here, $a \geq 0$ is a scale parameter, $0 \leq a_1, a_2, a_3 \leq 1$ are aspect ratio parameters, and $\epsilon_1, \epsilon_2 \geq 0$ are "squareness" parameters (Barr 1981).

We define a global transformation that includes tapering, bending, shearing, and twisting along each of the three principal axes of the solid (cf. (Barr 1984)). The transformation along axis 3 is given by

$$\mathbf{T}(\mathbf{e}) = \begin{pmatrix} A_3 \cos(\phi) - B_3 \sin(\phi) \\ A_3 \sin(\phi) + B_3 \cos(\phi) \\ e_3 \end{pmatrix} \qquad (26)$$

where $A_3 = [(t^1_3 e_3/aa_3w) + 1]\ e_1 + b^{a}{}^1_3 \cos[(e_3 + b^{l}{}^1_3)/(aa_3w)\pi b^{r}{}^1_3] + e_3 s_1$, and $B_3 = [(t^2_3 e_3/aa_3w) + 1]\ e_1 + b^{a}{}^2_3 \cos[(e_3 + b^{l}{}^2_3)/(aa_3w)\pi b^{r}{}^2_3] + e_3 s_2$, and $\phi = (e_3/aa_3w)\pi\tau_3$. Here τ_3 is a twisting parameter along axis 3, t^j_3 are tapering parameters along axes $j \neq 3$, s^j_3 are shearing parameters, and $b^{a}{}^j_3$, $b^{l}{}^j_3$, $b^{r}{}^j_3$ define the amount, location, and range of bending in the plane spanned by axis 3 and axis j. The transformations along axes 1 and 2 are obtained by appropriately permuting the above formulas. We collect the 39 global deformation parameters associated with $\mathbf{s}$ into the vector

$$\mathbf{q}_s = (a, a_1, a_2, a_3, \epsilon_1, \epsilon_2, \tau_i, t^j_i, s^j_i, b^{a}{}^j_i, b^{l}{}^j_i, b^{r}{}^j_i)^T, \qquad (27)$$

where $i, j = 1, 2, 3,\ j \neq i$.

B Derivation of Motion Equations

B.1 Kinetic Energy: Mass Matrix

The kinetic energy of the model is given by

$$\mathcal{T} = \frac{1}{2}\int \mu\dot{\mathbf{x}}^T\dot{\mathbf{x}}\,du = \frac{1}{2}\dot{\mathbf{q}}^T\left[\int \mu\mathbf{L}^T\mathbf{L}\,du\right]\dot{\mathbf{q}} = \frac{1}{2}\dot{\mathbf{q}}^T\mathbf{M}\dot{\mathbf{q}}, \qquad (28)$$

where $\mathbf{M} = \int \mu\mathbf{L}^T\mathbf{L}\,du$ is the symmetric mass matrix of the object and $\mu(\mathrm{u})$ is the mass density. Using (11),

$$\mathbf{M} = \begin{bmatrix} \mathbf{M}_{cc} & \mathbf{M}_{c\theta} & \mathbf{M}_{cs} & \mathbf{M}_{cd} \\ & \mathbf{M}_{\theta\theta} & \mathbf{M}_{\theta s} & \mathbf{M}_{\theta d} \\ & & \mathbf{M}_{ss} & \mathbf{M}_{sd} \\ \text{symmetric} & & & \mathbf{M}_{dd} \end{bmatrix}, \qquad (29)$$

where

$$\begin{array}{lllll} \mathbf{M}_{cc} & = & \int \mu\mathbf{I}\,du & \mathbf{M}_{\theta s} & = & \int \mu\mathbf{B}^T\mathbf{R}\mathbf{J}\,du \\ \mathbf{M}_{c\theta} & = & \int \mu\mathbf{B}\,du & \mathbf{M}_{\theta d} & = & \int \mu\mathbf{B}^T\mathbf{R}\mathbf{S}\,du \\ \mathbf{M}_{cs} & = & \mathbf{R}\int \mu\mathbf{J}\,du & \mathbf{M}_{ss} & = & \int \mu\mathbf{J}^T\mathbf{J}\,du \\ \mathbf{M}_{cd} & = & \mathbf{R}\int \mu\mathbf{S}\,du & \mathbf{M}_{sd} & = & \int \mu\mathbf{J}^T\mathbf{S}\,du \\ \mathbf{M}_{\theta\theta} & = & \int \mu\mathbf{B}^T\mathbf{B}\,du & \mathbf{M}_{dd} & = & \int \mu\mathbf{S}^T\mathbf{S}\,du \end{array} \qquad (30)$$

B.2 Energy Dissipation: Damping Matrix

We assume velocity dependent kinetic energy dissipation; i.e., the (Raleigh) dissipation functional

$$\mathcal{F} = \frac{1}{2}\int \gamma\dot{\mathbf{x}}^T\dot{\mathbf{x}}\,du, \qquad (31)$$

where $\gamma(\mathrm{u})$ is a damping density. We can rewrite (31) as

$$\mathcal{F} = \frac{1}{2}\dot{\mathbf{q}}^T\mathbf{D}\dot{\mathbf{q}}, \qquad (32)$$

where the damping matrix $\mathbf{D}$ has the same form as $\mathbf{M}$, but with γ replacing μ.

B.3 Strain Energy: Stiffness Matrix

We define the deformation characteristics of the model in terms of a deformation strain energy

$$\mathcal{E}(\mathbf{x}) = \frac{1}{2}\mathbf{q}^T\mathbf{K}\mathbf{q}, \tag{33}$$

with stiffness matrix

$$\mathbf{K} = \begin{pmatrix} \mathbf{0} & \mathbf{0} & \mathbf{0} & \mathbf{0} \\ & \mathbf{0} & \mathbf{0} & \mathbf{0} \\ & & \mathbf{K}_{ss} & \mathbf{K}_{sd} \\ & \text{symmetric} & & \mathbf{K}_{dd} \end{pmatrix}. \tag{34}$$

The zero submatrices indicate that the generalized coordinates representing rigid-body motions have no associated strain energies; only the global $\mathbf{q}_s$ and local $\mathbf{q}_d$ deformation generalized coordinates contribute to the strain energy.

B.4 External Forces and Virtual Work

The external forces $\mathbf{f}(\mathrm{u}, t)$ applied to the model do virtual work which can be written as

$$\delta \mathbf{W}_F = \int \mathbf{f}^T\mathbf{L}\delta\mathbf{q}\,d\mathrm{u} = \int \mathbf{f}_q^T\delta\mathbf{q}\,d\mathrm{u}, \tag{35}$$

where the vector of generalized external forces associated with the generalized coordinates of the model is

$$\mathbf{f}_q^T = \mathbf{f}^T\mathbf{L} = (\mathbf{f}_c^T, \mathbf{f}_\theta^T, \mathbf{f}_s^T, \mathbf{f}_d^T), \tag{36}$$

with

$$\begin{array}{lllll} \mathbf{f}_c^T & = & \int \mathbf{f}^T\,d\mathrm{u}, & \mathbf{f}_s^T & = & \int \mathbf{f}^T\mathbf{R}\mathbf{J}\,d\mathrm{u}, \\ \mathbf{f}_\theta^T & = & \int \mathbf{f}^T\mathbf{B}\,d\mathrm{u}, & \mathbf{f}_d^T & = & \int \mathbf{f}^T\mathbf{R}\mathbf{S}\,d\mathrm{u}. \end{array} \tag{37}$$

B.5 Lagrange Equations of Motion

The Lagrange equations of motion for the model take the form

$$\frac{d}{dt}\left(\frac{\partial \mathcal{T}}{\partial \dot{\mathbf{q}}}\right)^T - \left(\frac{\partial \mathcal{T}}{\partial \mathbf{q}}\right)^T + \left(\frac{\partial \mathcal{F}}{\partial \dot{\mathbf{q}}}\right)^T + \delta_{\mathbf{x}}\mathcal{E} = \mathbf{f}_q. \tag{38}$$

The inertial forces and can be written as

$$\frac{d}{dt}\left(\frac{\partial \mathcal{T}}{\partial \dot{\mathbf{q}}}\right)^T - \left(\frac{\partial \mathcal{T}}{\partial \mathbf{q}}\right)^T = \mathbf{M}\ddot{\mathbf{q}} - \mathbf{g}_q, \tag{39}$$

where

$$\mathbf{g}_q = -\int \mu\mathbf{L}^T\dot{\mathbf{L}}\dot{\mathbf{q}}\,d\mathrm{u}, \tag{40}$$

with

$$\dot{\mathbf{L}}\dot{\mathbf{q}} = \boldsymbol{\omega} \times (\boldsymbol{\omega} \times \mathbf{R}\mathbf{p}) + 2\boldsymbol{\omega} \times \mathbf{R}\dot{\mathbf{p}}, \tag{41}$$

$\omega \times (\omega \times \mathbf{R}\mathbf{p})$ and $2\omega \times \mathbf{R}\dot{\mathbf{p}}$ are the centrifugal and the Coriolis accelerations respectively (Shabana 1989). We calculate

$$\dot{\mathbf{p}} = \dot{\mathbf{s}} + \dot{\mathbf{d}} = \mathbf{J}\dot{\mathbf{q}}_s + \mathbf{S}\dot{\mathbf{q}}_d, \tag{42}$$

and the angular velocity $\boldsymbol{\omega}$ of the model with respect to the world coordinate system as

$$\boldsymbol{\omega} = \mathbf{Q}\dot{\boldsymbol{\theta}}, \tag{43}$$

where the matrix

$$\mathbf{Q} = 2\begin{bmatrix} -v_1 & s & -v_3 & v_2 \\ -v_2 & v_3 & s & -v_1 \\ -v_3 & -v_2 & v_1 & s \end{bmatrix} \tag{44}$$

is derived from the quaternion $\boldsymbol{\theta} = \mathbf{q}_\theta = [s, \mathbf{v}]$ representing the rotation at time t. The dissipation force takes the form

$$\frac{\partial \mathcal{F}}{\partial \dot{\mathbf{q}}} = \mathbf{D}\dot{\mathbf{q}}. \tag{45}$$

The variation of $\mathcal{E}$ with respect to $\mathbf{x}$ in (38) expresses the elastic forces

$$\delta_{\mathbf{x}}\mathcal{E} = \mathbf{K}\mathbf{q}. \tag{46}$$

Substituting the above expressions into (38) yields the equations of motion (12).

C $\mathbf{K}_{dd}$ for Deformable Shells

We discretize the surface of the model in material coordinates $\mathrm{u} = (u, v, 1)$ using finite elements and derive $\mathbf{K}_{dd}$ as an assembly of the local stiffness matrices $\mathbf{K}_{dd}^j$ associated with each element domain $E_j \subset \mathrm{u}$. Since $\mathbf{d}(\mathrm{u}, t) = [d_1(\mathrm{u}, t), d_2(\mathrm{u}, t), d_3(\mathrm{u}, t)]^T$, we can write on E_j the deformation energies

$$\mathcal{E}^j(\mathbf{d}) = \mathcal{E}^j(d_1) + \mathcal{E}^j(d_2) + \mathcal{E}^j(d_3), \tag{47}$$

where for $k = 1, 2, 3$,

$$\mathcal{E}^j(d_k) = \int_{E_j} w_1^j\left(\left(\frac{\partial d_k}{\partial u}\right)^2 + \left(\frac{\partial d_k}{\partial v}\right)^2\right) + w_0^j d_k^2\,d\mathrm{u}. \tag{48}$$

These integrals specify a loaded elastic membrane deformation energy, where the physical parameters $w_0 \geq 0$ and $w_1 \geq 0$ control the local magnitude and the local variation of the deformation respectively.

In accordance to the theory of elasticity, (48) can be written in the form

$$\mathcal{E}^j(d_k) = \int_{E_j} \boldsymbol{\sigma}_k^{j\,T}\boldsymbol{\epsilon}_k^j\,d\mathrm{u}, \tag{49}$$

where the strain vector associated with component k of $\mathbf{d}$ is

$$\boldsymbol{\epsilon}_k^j = \left[\frac{\partial d_k}{\partial u}, \frac{\partial d_k}{\partial v}, d_k\right]^T \tag{50}$$

and the stress vector is

$$\boldsymbol{\sigma}_k^j = \mathbf{D}_k^j\boldsymbol{\epsilon}_k^j = \begin{pmatrix} w_1^j & 0 & 0 \\ 0 & w_1^j & 0 \\ 0 & 0 & w_0^j \end{pmatrix}\boldsymbol{\epsilon}_k^j. \tag{51}$$

Therefore, the element stress vector is

$$\sigma^j = \mathbf{D}^j \epsilon^j = \begin{pmatrix} \mathbf{D}_1^j & 0 & 0 \\ 0 & \mathbf{D}_2^j & 0 \\ 0 & 0 & \mathbf{D}_3^j \end{pmatrix} \begin{pmatrix} \epsilon_1^j \\ \epsilon_2^j \\ \epsilon_3^j \end{pmatrix}, \qquad (52)$$

where $\mathbf{D}_1^j = \mathbf{D}_2^j = \mathbf{D}_3^j$.

We denote the finite element nodal shape functions by N_i^j, i=1, ..., n, where n is the number of nodes associated with element E_j. Hence, we can write (50) as

$$\epsilon_k^j = \sum_{i=1}^{n} \gamma_i^j (q_{d_k}^j)_i = \mathbf{\Gamma}_k^j \mathbf{q}_{d_k}^j, \qquad (53)$$

where $\gamma_i^j = \left[\frac{\partial N_i^j}{\partial u}, \frac{\partial N_i^j}{\partial v}, 1\right]^T$, $\mathbf{\Gamma}_k^j = \left(\gamma_1^j \gamma_2^j \ldots \gamma_n^j\right)$ and $\mathbf{q}_{d_k}^j = [(q_{d_k}^j)_1, (q_{d_k}^j)_2, \ldots, (q_{d_k}^j)_n]^T$. The element strain vector ϵ^j is

$$\epsilon^j = \begin{pmatrix} \epsilon_1^j \\ \epsilon_2^j \\ \epsilon_3^j \end{pmatrix} = \begin{pmatrix} \mathbf{\Gamma}_1^j & 0 & 0 \\ 0 & \mathbf{\Gamma}_2^j & 0 \\ 0 & 0 & \mathbf{\Gamma}_3^j \end{pmatrix} \begin{pmatrix} \mathbf{q}_{d_1}^j \\ \mathbf{q}_{d_2}^j \\ \mathbf{q}_{d_3}^j \end{pmatrix} = \mathbf{\Gamma}^j \mathbf{q}_d^j, \qquad (54)$$

where $\mathbf{\Gamma}_1^j = \mathbf{\Gamma}_2^j = \mathbf{\Gamma}_3^j$. Thus the element stiffness matrix is

$$\mathbf{K}_{dd}^j = \int_{E_j} \mathbf{\Gamma}^{j^T} \mathbf{D}^j \mathbf{\Gamma}^j \, du = \mathrm{diag}\left(\int_{E_j} \mathbf{\Gamma}_k^{j^T} \mathbf{D}_k^j \mathbf{\Gamma}_k^j \, du \right). \qquad (55)$$

D Linear Triangular Finite Elements

The nodal shape functions for a linear triangular element (Kardestuncer 1987) (Fig. 10) are

$$\begin{aligned} N_1(\xi, \eta) &= 1 - \xi - \eta, & (56) \\ N_2(\xi, \eta) &= \xi, & (57) \\ N_3(\xi, \eta) &= \eta. & (58) \end{aligned}$$

The relationship between the uv and $\xi\eta$ coordinates is

$$\begin{aligned} \xi &= (u - u_1)/a, & (59) \\ \eta &= (v - v_1)/b, & (60) \end{aligned}$$

where (u_1, v_1) are the coordinates of node 1 at which $(\xi_1, \eta_1) = (0, 0)$. Computing the derivatives of the shape functions yields

$$\frac{\partial N_1}{\partial u} = -\frac{1}{a}; \quad \frac{\partial N_2}{\partial u} = \frac{1}{a}; \quad \frac{\partial N_3}{\partial u} = 0; \qquad (61)$$

$$\frac{\partial N_1}{\partial v} = -\frac{1}{b}; \quad \frac{\partial N_2}{\partial v} = 0; \quad \frac{\partial N_3}{\partial v} = \frac{1}{b}; \qquad (62)$$

and we may integrate a function $f(u, v)$ over the E_j according to

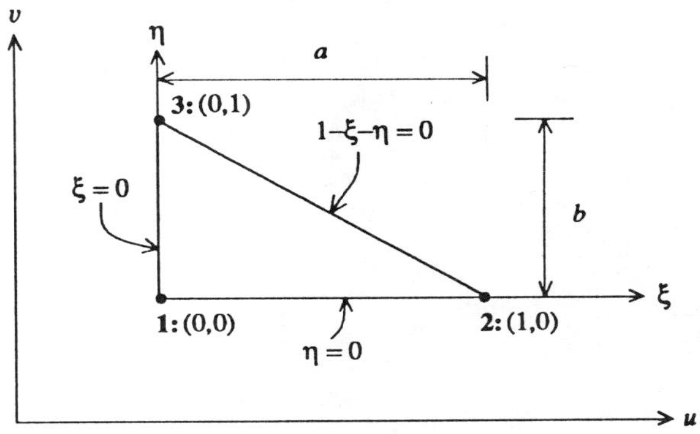

Figure 10: Linear triangular element. The three nodes are numbered.

$$\iint_{E_j} f(u, v) \, du \, dv = \int_0^1 \int_0^{(1-\eta)} f(\xi, \eta) ab \, d\xi \, d\eta. \qquad (63)$$

We approximate such integrals using Radau quadrature rules (Kardestuncer 1987).

Using the above formulas, we can compute the matrix $\int \mathbf{\Gamma}_k^{j^T} \mathbf{D}_k^j \mathbf{\Gamma}_k^j \, du$ and, hence, $\mathbf{K}_{dd}^j$ for the linear triangular element.

REFERENCES

(Barr 1981) A. Barr. Superquadrics and Angle-Preserving Transformations. *IEEE Computer Graphics and Applications*, 1:11–23, 1981.

(Barr 1984) A. Barr. Global and Local Deformations of Solid Primitives. *Computer Graphics*, 18:21–30, 1984.

(Barzel and Barr 1988) R. Barzel and A. Barr. A Modeling System Based on Dynamic Constraints. *Computer Graphics*, 22(4):179–188, 1988. Proc. ACM SIGGRAPH'88.

(Baumgarte 1972) J. Baumgarte. Stabilization of Constraints and Integrals of Motion in Dynamical Systems. *Computer Methods in Applied Mechanics and Engineering*, 1:1–16, 1972.

(Hoffmann 1989) C.M. Hoffmann. *Geometric and Solid Modeling*. Morgan-Kaufmann, Palo Alto, 1989.

(Kardestuncer 1987) H. Kardestuncer. *Finite Element Handbook*. McGraw–Hill, New York, 1987.

(Metaxas and Terzopoulos 1992 to appear) D. Metaxas and D. Terzopoulos. Shape and Nonrigid Motion Estimation through Physics-Based Synthesis. *IEEE Trans. Pattern Analysis and Machine Intelligence*, 1992, to appear.

(Shabana 1989) A.A. Shabana. *Dynamics of Multibody Systems*. Wiley, New York, 1989.

AMD-Vol. 141/DSC-Vol. 37, Dynamics of Flexible Multibody Systems:
Theory and Experiment
ASME 1992

MODE-ACCELERATION METHOD IN FLEXIBLE MULTIBODY DYNAMICS

Jeha Ryu
BMY Combat Systems
Division of Harsco Company
York, Pennsylvania

Sung-Soo Kim and Sang Sup Kim
Department of Mechanical Engineering
Center for Simulation and Design Optimization
of Mechanical Systems
University of Iowa
Iowa City, Iowa

ABSTRACT

This paper presents a mode-acceleration method (MAM) for the dynamic simulation of flexible multibody systems in order to predict more accurately dynamic stresses and elastic displacements. Mode-acceleration equations for each flexible body are derived from the variational equations of motion of the body. These mode-acceleration equations are quasi-static equilibrium equations in which internal elastic load is self-equilibrated with time-varying external loads, which consist of inertial, applied, and constraint reaction loads. Time-histories of these loads are approximated by a flexible multibody dynamic simulation that is based on the mode-displacement method (MDM). Since the load time-histories can be represented as a combination of space-dependent and time-dependent terms, dynamic stresses or elastic displacements are obtained by multiplying time-dependent terms with dynamic stress or elastic displacement influence coefficients. These coefficients are generated from quasi-static structural analyses with the loads corresponding to unit values of the time-dependent terms. The overall computational procedure of the proposed method is discussed with a numerical example. This example shows that dynamic solutions can be improved with a small amount of additional effort.

1. INTRODUCTION

Dynamic analyses of flexible multibody systems have been carried out to obtain dynamic solutions such as dynamic stress and elastic displacement histories at points of interest in a flexible component. Dynamic stress histories can be used for fatigue life prediction and stress-safe design of mechanical systems. Fatigue failure may result from high stress levels together with the large number of stress reversals when millions of cycles of dynamic loads are applied to a machine component in a dynamic system. Even in a few cycles of dynamic loads, a machine component may fail because of the excessive stresses resulting from large magnitude loads. Therefore, accurate prediction of dynamic stress histories is a keystone for fatigue life prediction and stress-safe design.

Dynamic stress histories have been computed by high speed digital computers to accelerate design cycles in the design of flexible multibody systems. These computer-based approaches can be categorized into four groups: *Rigid Body Dynamic Simulation combined with Quasi-static Finite Element Method* (Winfrey, 1971; and Sadler and Sandor, 1973), *Finite Element Nodal Approach* (Turcic and Midha, 1984; Nagarajan and Turcic, 1990; and Yang and Sadler, 1990), *Modal Stress Superposition Method* (Hurty and Rubinstein, 1966; Iman et al., 1973; and Liu, 1987), and *Hybrid Method* (Yim, 1990). Merits and demerits of these methods are discussed by Ryu et al. (1992). Among the methods, the finite element nodal approach is the most accurate but the least efficient, while the hybrid method is accurate and efficient.

In addition to dynamic stresses, accurate prediction of elastic displacements is also required to increase the positioning accuracy or to avoid interference resulting from excessive deformation. Several methods have been developed for the prediction of elastic displacements (Yoo and Haug, 1986; and Wu et al., 1989).

In the area of linear structural dynamics (Craig, 1981), the *Mode-Acceleration Method* (MAM) has been used to improve dynamic solutions by static structural analyses after modal velocities and accelerations are obtained by the *Mode-Displacement Method* (MDM). This method uses a truncated set of vibration normal modes to approximate the dynamic responses. In the area of nonlinear flexible multibody dynamics, on the other hand, the hybrid method improves dynamic stresses by utilizing the dynamic loads obtained from the flexible multibody dynamic simulation that uses the assumed mode superposition method (Wu et al., 1989; and Yoo and Haug, 1986). However, the main concern of the hybrid method was development of a computational procedure to obtain more accurate dynamic stresses. No theoretical support was presented to explain why the hybrid method improved the accuracy of dynamic stresses.

The objective of this paper is to present a mode-acceleration method to improve dynamic solutions that are obtained from the conventional flexible multibody dynamic analysis that uses the assumed mode superposition method. By following the basic procedures of the MAM in linear structural dynamics, we explain theoretically why and how the proposed MAM improves elastic displacements as well as dynamic stresses. These explanations also give theoretical support for the hybrid method. Finally, an efficient method is summarized to solve the mode-acceleration equations that are used in the proposed MAM.

Section 2 summarizes basic ideas and mathematical derivations of the MDM and the MAM in linear structural dynamics and explains the reason why the MAM improves the dynamic solutions. Section 3 derives mode-acceleration equations for a flexible body in the nonlinear flexible multibody dynamic systems from the variational Cartesian equations of motion of the body. This section also theoretically explains the improvement made by applying the MAM of linear structural dynamics to nonlinear flexible multibody dynamics. Section 4 summarizes a computationally efficient method of solving the mode-acceleration equations. Section 5 presents a numerical example to show the improvement in the dynamic stresses and the elastic displacements. Finally, conclusions are made in section 6.

2. MDM *versus* MAM IN LINEAR STRUCTURAL DYNAMICS

Two versions of the mode superposition method have been used in the dynamic analysis of linear structures: (1) the *mode-displacement method* (MDM), and (2) the *mode-acceleration method* (MAM). Both methods use a truncated set of eigenvectors as a basis to approximate the response of the structure.

Consider an undamped n degree-of-freedom linear dynamic system. In matrix form, the dynamic equilibrium equations are given by

$$\mathbf{M}\ddot{\mathbf{U}} + \mathbf{K}\mathbf{U} = \mathbf{F}(t) \tag{1}$$

where $\mathbf{M}$ and $\mathbf{K}$ are the (n x n) mass and stiffness matrices of the structure, respectively; and $\ddot{\mathbf{U}}$, $\mathbf{U}$, and $\mathbf{F}(t)$ are the nodal acceleration, displacement, and force vectors, respectively.

Equation (1) can be transformed to a set of n uncoupled equations by the transformation

$$\mathbf{U} = \mathbf{\Phi}\mathbf{a}(t) = \sum_{i=1}^{n} \phi_i a_i(t) \tag{2}$$

where $\mathbf{a}(t)$ is the modal coordinate vector and $\mathbf{\Phi}$ is the modal matrix whose n column vectors ϕ_i are the orthonormalized eigenvectors of the system satisfying the eigenvalue problem:

$$\mathbf{K}\phi_i = \omega_i^2 \mathbf{M}\phi_i \tag{3}$$

where ω_i is the i-th natural frequency.

Then, Eq. (1) can be rewritten as

$$\ddot{\mathbf{a}} + \mathbf{\Lambda}\mathbf{a} = \mathbf{f}(t) \tag{4}$$

where matrix $\mathbf{\Lambda}$ is a diagonal matrix that is composed of the square of natural frequencies. The following equations are used in deriving Eq. (4) from Eq. (1).

$$\mathbf{\Phi}^T\mathbf{M}\mathbf{\Phi} = \mathbf{I} \tag{5}$$

$$\mathbf{K}\mathbf{\Phi} = \mathbf{M}\mathbf{\Phi}\mathbf{\Lambda} \tag{6}$$

$$\mathbf{\Phi}^T\mathbf{F}(t) = \mathbf{f}(t) \tag{7}$$

In the traditional MDM, after modal coordinate history $a_i(t)$ (i = 1, ..., m) is computed from Eq. (4), the mode-displacement solution is approximately obtained as

$$\hat{\mathbf{U}}(t) = \sum_{i=1}^{m} \phi_i a_i(t) \tag{8}$$

where m is the number of vibration normal modes, and is generally much smaller than the total degrees of freedom n. Therefore, the MDM neglects completely the vibration normal modes from (m+1) to n.

The MAM, first proposed by Williams (1945), compensates for the effect of neglected high frequency modes. Bisplinghoff and Ashley (1955) and Thomson (1972) argued that the MAM has better convergence characteristics and requires fewer modes than the MDM. More recent versions of the MAM in the literature include Maddox (1975), Hangsteen and Bell (1979), and Cornwell, Craig, and Johnson (1983). The classical mode-acceleration method (Cornwell et al., 1983) can be derived by solving Eq. (1) for $\mathbf{U}(t)$:

$$\mathbf{U}(t) = \mathbf{K}^{-1}\mathbf{F}(t) - \mathbf{K}^{-1}\mathbf{M}\ddot{\mathbf{U}}(t) \tag{9}$$

where it is assumed without loss of generality that the stiffness matrix is nonsingular. In other words, rigid body modes are not considered. If a truncated set of m vibration normal modes is used to approximate the nodal acceleration, the second term of the right side in Eq. (9) can be written as

$$\mathbf{K}^{-1}\mathbf{M}\ddot{\mathbf{U}}(t) \cong \mathbf{K}^{-1}\mathbf{M}\hat{\mathbf{\Phi}}\ddot{\hat{\mathbf{a}}}(t) \tag{10}$$

where $\hat{\mathbf{\Phi}}$ and $\ddot{\hat{\mathbf{a}}}(t)$ are the truncated (n x m) modal matrix and the (m x 1) modal coordinate acceleration vector, respectively. Using the relationship derived from Eq. (6)

$$\mathbf{K}^{-1}\mathbf{M}\hat{\mathbf{\Phi}} = \hat{\mathbf{\Phi}}\hat{\mathbf{\Lambda}}^{-1} \tag{11}$$

where $\hat{\mathbf{\Lambda}}$ is the (m x m) diagonal matrix that is composed of the square of m natural frequencies, the mode-acceleration solution $\tilde{\mathbf{U}}$ is given by

$$\tilde{\mathbf{U}} = \mathbf{K}^{-1}\mathbf{F}(t) - \hat{\mathbf{\Phi}}\hat{\mathbf{\Lambda}}^{-1}\ddot{\hat{\mathbf{a}}}(t) \tag{12}$$

The first term in the right side of Eq. (12) accounts for the contribution of higher modes by implicitly considering a complete set of n modes. This term is the pseudo-static response that is equivalent to carrying out a static analysis at each time step. Because of this term, Cornwell et al. (1983) concluded that anything that tends to make the system act more statically favors the use of the MAM in the linear structural dynamics. The second term in the right side of Eq. (12) represents a dynamic correction applied to the pseudo-static response and gives the method its name, the *mode-acceleration method*. Once the (m x 1) vector $\ddot{\hat{\mathbf{a}}}(t)$ is obtained by Eq. (4), it is easy to obtain the mode-acceleration solution by Eq. (12), which is a quasi-static equation.

In order to more explicitly show that the MAM improves dynamic solutions obtained by the MDM, consider a computational variant that is obtained by expanding the flexibility matrix in terms of a truncated eigenbasis (Léger and Wilson, 1988). Solution of Eq. (4) for $\ddot{\hat{\mathbf{a}}}(t)$ gives

$$\ddot{\hat{\mathbf{a}}}(t) = \hat{\mathbf{\Phi}}^T\mathbf{F}(t) - \hat{\mathbf{\Lambda}}\hat{\mathbf{a}}(t) \tag{13}$$

Inserting Eq. (13) into Eq. (12) gives

$$\tilde{\mathbf{U}} = \mathbf{K}^{-1}\mathbf{F}(t) + \hat{\mathbf{\Phi}}\hat{\mathbf{a}}(t) - \hat{\mathbf{\Phi}}\hat{\mathbf{\Lambda}}^{-1}\hat{\mathbf{\Phi}}^T\mathbf{F}(t) \tag{14}$$

This equation can be written as

$$\tilde{\mathbf{U}} = \hat{\mathbf{\Phi}}\hat{\mathbf{a}}(t) + (\mathbf{K}^{-1} - \hat{\mathbf{\Phi}}\hat{\mathbf{\Lambda}}^{-1}\hat{\mathbf{\Phi}}^T)\mathbf{F}(t) \tag{15}$$

Equation (15) is then rewritten as

$$\tilde{\mathbf{U}} = \hat{\mathbf{U}}(t) + (\mathbf{K}^{-1} - \mathbf{K}_m^{-1})\mathbf{F}(t) \tag{16}$$

In this final expression the first term corresponds to the usual mode-displacement solution, and the second term represents the additional static correction, i.e., the amount of improvement by the MAM in deflections, where $\mathbf{K}_m^{-1}$ is a symbolic representation for the truncated expansion of the flexibility matrix using a reduced set of m eigenvectors. Note that this static correction concept has been used in the context of dynamic substructuring to improve the convergence of structural responses computed by component mode synthesis (Craig and Chang, 1977). The static correction vector is defined as the residual attachment mode obtained from the application of the residual flexibility matrix $(\mathbf{K}^{-1} - \mathbf{K}_m^{-1})$ to the specified unit magnitude loading.

In addition to determining the displacement history, a dynamic analysis usually includes the determination of stress histories (e.g., moment, shear, axial stress, etc.) or at least the determination of maximum values of stress at specified locations in the structure. The following are symbolic representations of stresses obtained by mode-superposition. For the MDM,

$$\hat{\sigma}(t) = \sum_{i=1}^{m} \mathbf{s}_i a_i \tag{17}$$

where $\mathbf{s}_i$ is the stress influence coefficient that is the contribution to the stress vector $\boldsymbol{\sigma}$ due to a unit displacement of the i-th mode, that is, for $a_i = 1$. Note that the modal stress superposition method (Liu, 1987) used the same principle in the flexible multibody dynamics. For the MAM, the displacement approximation in Eq. (12) leads to the stress approximation

$$\tilde{\sigma}(t) = \sigma_{pseudostatic} - \sum_{i=1}^{m} \mathbf{s}_i \frac{\ddot{a}_i}{\omega_i^2} \tag{18}$$

Higher modes are increasingly more important for moments and shears than for deflections (Hurty and Rubinstein, 1966). Therefore, the MAM is particularly beneficial in speeding convergence of the internal stresses.

3. MDM *vs* MAM IN FLEXIBLE MULTIBODY DYNAMICS

3.1 Variational Equations of Motion of a Flexible Body

Consider a flexible body that is in dynamic equilibrium in the deformed configuration, as shown in Fig. 1. In this figure, the $\underline{X}$-$\underline{Y}$-$\underline{Z}$ coordinate system is the inertial reference frame and the x-y-z coordinate system is the body reference frame in an undeformed configuration. An underlined variable is measured in the inertial reference frame, while other variables are measured in the local body reference frame.

The variational equations of motion of a flexible body are given as (Wu et al., 1989)

$$\int_S \delta\mathbf{r}^{pT}\mathbf{T}^p dS - \int_V \delta\mathbf{r}^{pT}\{\rho\ddot{\mathbf{r}}^p - \mathbf{f}^p\}\,dV = \int_V \delta\boldsymbol{\varepsilon}^{pT}\boldsymbol{\tau}^p dV \tag{19}$$

where $\delta\mathbf{r}^p$ is the virtual displacement vector of point p that is consistent with constraints; $\mathbf{T}^p$ is a surface traction vector at point p; ρ is the mass density; $\ddot{\mathbf{r}}^p$ is the acceleration vector of point p; $\mathbf{f}^p$ is the body force density vector at point p; τ^p is the (6 x 1) stress vector; $\delta\boldsymbol{\varepsilon}^p$ is the variation of the (6 x 1) strain vector consistent with given boundary conditions; and V and S are the volume and surface of the body before it is deformed. Dots over a vector in Eq. (19) and in the following derivations denote the total differentiation with respect to time. Note that every vector is represented with respect to the body reference frame; i.e., for any arbitrary vector, $\mathbf{v} = \mathbf{A}^T\underline{\mathbf{v}}$, where $\mathbf{A}$ is the orientation matrix of the body reference frame.

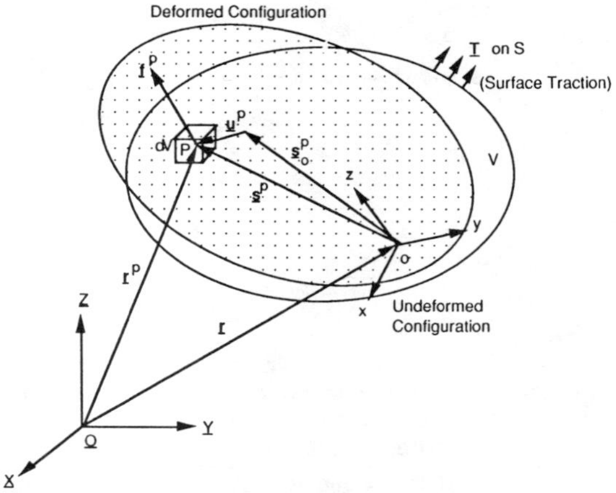

Figure 1 Undeformed and Deformed Configurations of a Flexible Body

By using the relationship $\mathbf{s}^p = \mathbf{s}_0^p + \mathbf{u}^p$, the virtual displacement and the acceleration vectors of point p are represented in the body reference frame as

$$\delta\mathbf{r}^p = \delta\mathbf{r} - \tilde{\mathbf{s}}^p\delta\boldsymbol{\pi} + \delta\mathbf{u} \tag{20}$$

$$\ddot{\mathbf{r}}^p = \ddot{\mathbf{r}} + (\tilde{\omega}\tilde{\omega} + \dot{\tilde{\omega}})\mathbf{s}_o^p + (\tilde{\omega}\tilde{\omega} + \dot{\tilde{\omega}})\mathbf{u} + 2\tilde{\omega}\dot{\mathbf{u}} + \ddot{\mathbf{u}} \tag{21}$$

where $\delta\mathbf{r}$, $\delta\boldsymbol{\pi}$, and $\delta\mathbf{u}$ are the variation of the position, orientation, and deformation vectors, respectively; and $\ddot{\mathbf{r}}$, $\dot{\boldsymbol{\omega}}$, and $\boldsymbol{\omega}$ are the translational and rotational acceleration and angular velocity vectors of the body reference frame, respectively. Note that the superscript p on $\mathbf{u}$ is omitted to simplify notation.

The variational strain energy can be represented as

$$\delta\boldsymbol{\varepsilon}^{pT}\boldsymbol{\tau}^p = \delta\mathbf{u}^T\mathbf{L}^T\mathbf{D}\mathbf{L}\mathbf{u} \tag{22}$$

where $\mathbf{L}$ and $\mathbf{D}$ are the strain-displacement and the stress-strain matrices (Bathe, 1982), respectively.

Substitution of Eqs. (20), (21), and (22) into Eq. (19) gives

$$\begin{aligned} &\delta\mathbf{q}_R^T[\int_S \mathbf{Q}_{RT}\,dS - \int_V \{\rho(\mathbf{M}_{RR}\ddot{\mathbf{q}}_R + \mathbf{M}_{RF}\ddot{\mathbf{u}} + \mathbf{S}_{RR}) - \mathbf{Q}_{Rf}\}dV] \\ &+ \int_S \delta\mathbf{u}^T\mathbf{T}^p dS - \int_V \delta\mathbf{u}^T\{\rho(\ddot{\mathbf{u}} + \mathbf{M}_{RF}^T\ddot{\mathbf{q}}_R + \mathbf{S}_{FF}) - \mathbf{f}^p\}\,dV \\ &= \int_V \delta\mathbf{u}^T\mathbf{L}^T\mathbf{D}\mathbf{L}\mathbf{u}\,dV \end{aligned} \tag{23}$$

Terms in Eq. (23) are defined as

$$\delta\mathbf{q}_R^T = [\delta\mathbf{r}^T, \delta\boldsymbol{\pi}^T] \tag{24}$$

$$\ddot{\mathbf{q}}_R^T = [\ddot{\mathbf{r}}^T, \dot{\boldsymbol{\omega}}^T] \tag{25}$$

$$\mathbf{M}_{RR} = \begin{bmatrix} \mathbf{I} & -\tilde{\mathbf{s}}^p \\ \tilde{\mathbf{s}}^p & -\tilde{\mathbf{s}}^p\tilde{\mathbf{s}}^p \end{bmatrix} \tag{26}$$

$$\mathbf{M}_{RF} = \begin{bmatrix} \mathbf{I} \\ \tilde{\mathbf{s}}^p \end{bmatrix} \tag{27}$$

$$S_{RR} = \begin{bmatrix} \tilde{\omega}\tilde{\omega}\mathbf{s}^P + 2\tilde{\omega}\dot{\mathbf{u}} \\ \tilde{\mathbf{s}}^P\tilde{\omega}\tilde{\omega}\mathbf{s}^P + 2\tilde{\mathbf{s}}^P\tilde{\omega}\dot{\mathbf{u}} \end{bmatrix} \quad (28)$$

$$Q_{Rf} = \begin{bmatrix} \mathbf{f}^P \\ \tilde{\mathbf{s}}^P\mathbf{f}^P \end{bmatrix} \quad (29)$$

$$Q_{RT} = \begin{bmatrix} \mathbf{T}^P \\ \tilde{\mathbf{s}}^P\mathbf{T}^P \end{bmatrix} \quad (30)$$

$$S_{FF} = \tilde{\omega}\tilde{\omega}\mathbf{s}^P + 2\tilde{\omega}\dot{\mathbf{u}} \quad (31)$$

If the surface traction vector $\mathbf{T}^P$ includes constraint reaction loads that are obtained through Lagrangian multipliers associated with constraints, then $\delta\mathbf{q}_R^T$ in Eq. (23) is arbitrary. Therefore, Eq. (23) is divided into (6 x 1) vector equations for gross body motion and a variational equation for deformation as

$$\int_S Q_{RT}dS - \int_V \{\rho(\mathbf{M}_{RR}\ddot{\mathbf{q}}_R + \mathbf{M}_{RF}\ddot{\mathbf{u}} + \mathbf{S}_{RR}) - Q_{Rf}\}dV = \mathbf{0} \quad (32)$$

$$\int_S \delta\mathbf{u}^T\mathbf{T}^P dS - \int_V \delta\mathbf{u}^T\{\rho(\ddot{\mathbf{u}} + \mathbf{M}_{RF}^T\ddot{\mathbf{q}}_R + \mathbf{S}_{FF}) - \mathbf{f}^P\}dV$$
$$= \int_V \delta\mathbf{u}^T\mathbf{L}^T\mathbf{D}\mathbf{L}\mathbf{u}dV \quad (33)$$

Rearrangement of Eq. (33) yields

$$\int_V \delta\mathbf{u}^T\rho\{\ddot{\mathbf{u}} + 2\tilde{\omega}\dot{\mathbf{u}} + (\tilde{\omega}\tilde{\omega} + \dot{\tilde{\omega}})\mathbf{u}\}dV + \int_V \delta\mathbf{u}^T\mathbf{L}^T\mathbf{D}\mathbf{L}\mathbf{u}dV$$
$$= \int_S \delta\mathbf{u}^T\mathbf{T}^P dS + \int_V \delta\mathbf{u}^T\rho\,\mathbf{f}^P dV - \int_V \delta\mathbf{u}^T\rho\{\ddot{\mathbf{r}} + (\tilde{\omega}\tilde{\omega} + \dot{\tilde{\omega}})\mathbf{s}_0^P\}dV \quad (34)$$

Equation (34) means that, for a flexible component in a flexible multibody system, internal loads, D'Alembert inertial loads, applied loads, and constraint reaction loads are self-equilibrated at each time step t during the dynamic simulation of the system. Note that Eq. (32), which is the equilibrium equation for the gross body motion, is also satisfied during the dynamic simulation.

3.2 Mode-displacememt and mode-acceleration solutions in flexible multibody dynamics

Dynamic simulation of a flexible multibody system can be performed by representing the elastic displacement field $\mathbf{u}$ of a flexible component as the sum of modal coordinates multiplied by assumed modes, which may be a truncated set of vibration normal modes and/or static correction modes (Yoo and Haug, 1986). Vibration normal modes of a flexible component are obtained bv solving the eigenvalue problem in Eq. (3) for a few of the lowest frequency vibration modes. These free vibration modes represent deformation shapes of the component due to its mass and stiffness distribution. To represent the local deformation at the action points of constraint and applied loads, static correction modes, such as attachment modes (Craig and Chang, 1977) or constraint modes (Craig and Bampton, 1968), can be used together with the vibration normal modes. For example, an attachment mode is defined by imposing a unit force in the direction of one physical coordinate of the flexible component, and zero forces elsewhere.

With the assumed-mode representation of the elastic displacement field of each flexible component, system equations of motion of a flexible multibody system are constructed by assembling variational equations of motion (Eqs. (32) and (33)) of each flexible body in the system. The mode-displacement solutions are then obtained from the dynamic simulation of the system by solving and by integrating the system equations of motion (Wu et al., 1989). In this case, elastic displacements ($\hat{\mathbf{U}}$) and dynamic stresses ($\hat{\sigma}$) take the form of Eqs. (8) and (17), respectively. If only a truncated set of vibration normal modes is used in the dynamic simulation of the system, then the effects from the truncated higher frequency modes are totally neglected, as in linear structural dynamics.

On the other hand, mode-acceleration solutions can be obtained as follows. If Eq. (34) is projected into a finite element equilibrium equation, then it can be expressed symbolically as

$$\mathbf{M}\ddot{\mathbf{U}} + \mathbf{C}_c(t)\dot{\mathbf{U}} + \mathbf{M}_c(t)\mathbf{U} + \mathbf{K}\mathbf{U} = \mathbf{F}$$
$$= \mathbf{F}_c(t) + \mathbf{F}_a(t) + \mathbf{F}_i(\mathbf{s}_0^P, \ddot{\mathbf{r}}, \omega, \dot{\omega}, t) \quad (35)$$

where $\mathbf{M}$ is the mass matrix of a flexible body; $\mathbf{C}_c(t)$ and $\mathbf{M}_c(t)$ are the time-varying matrices resulting from the coupling terms between the gross body motion and the elastic deformation; $\mathbf{K}$ is the stiffness matrix of the body; $\ddot{\mathbf{U}}$, $\dot{\mathbf{U}}$, and $\mathbf{U}$ are the time-varying nodal acceleration, velocity, and displacement vectors, respectively; and $\mathbf{F}_c$, $\mathbf{F}_a$, and $\mathbf{F}_i$ are the nodal force vectors generated from the constraint reaction loads, the applied loads, and the distributed D'Alembert inertial loads resulting from the gross body motion, respectively. Note that the mass matrix $\mathbf{M}$ and the stiffness matrix $\mathbf{K}$ are time-invariant because they are constructed for only one flexible component of the multibody system and because they are defined with respect to the body reference frame, which is fixed to the undeformed configuration of the flexible body.

Equation (35) can be rewritten as

$$\mathbf{U} = \mathbf{K}^{-1}(\mathbf{F} - \mathbf{M}\ddot{\mathbf{U}} - \mathbf{C}_c(t)\dot{\mathbf{U}} - \mathbf{M}_c(t)\mathbf{U}) \quad (36)$$

If the nodal forces, accelerations, velocities, and displacements in the right side of Eq. (36) are approximated by the mode-displacement solutions that are obtained by the dynamic simulation of a flexible multibody system, the mode-acceleration solutions in flexible multibody dynamics can be written as

$$\tilde{\mathbf{U}} = \mathbf{K}^{-1}(\mathbf{F}' - \mathbf{M}\ddot{\hat{\mathbf{U}}} - \mathbf{C}_c'\dot{\hat{\mathbf{U}}} - \mathbf{M}_c'\hat{\mathbf{U}}) \quad (37)$$

Note that the coupling matrices $\mathbf{C}_c'$ and $\mathbf{M}_c'$ and the nodal force vector $\mathbf{F}'$ in Eq. (37) are approximate because they are functions of gross body motion as well as constraint reaction loads, both of which in turn are functions of the deformation field that is represented with the truncated set of basis vectors. Note that Eq. (37) is a quasi-static equation where every load is self-equilibrated. Dynamic stresses ($\tilde{\sigma}$) are also computed from this quasi-static equation.

When only a truncated set of vibration normal modes is used to describe an elastic deformation field of a flexible component in the dynamic simulation of a flexible multibody system, the fact that the MAM improves the dynamic solutions obtained by the MDM can be easily explained in the same way as in linear structural dynamics. Following the procedure in Eqs. (12) to (16), Eq. (37) can be rewritten as

$$\tilde{\mathbf{U}} = \hat{\mathbf{U}} + (\mathbf{K}^{-1} - \mathbf{K}_m^{-1})(\mathbf{F}' - \mathbf{C}_c'\dot{\hat{\mathbf{U}}} - \mathbf{M}_c'\hat{\mathbf{U}}) \quad (38)$$

Therefore, the mode-acceleration solutions improves the mode-displacement solutions by the addition of the static correction terms $(\mathbf{K}^{-1} - \mathbf{K}_m^{-1})(\mathbf{F}' - \mathbf{C}_c'\dot{\hat{\mathbf{U}}} - \mathbf{M}_c'\hat{\mathbf{U}})$. Notice that if all vibration normal modes are used, then the residual flexibility matrix ($\mathbf{K}^{-1}$-$\mathbf{K}_m^{-1}$) becomes zero, meaning that no improvement is achieved because the mode-acceleration solutions are the same as the mode-displacement solutions.

If some of the static correction modes, for example residual attachment modes, are included together with m kept vibration normal modes in the dynamic simulation of a flexible multibody system, Eq. (37) still gives mode-acceleration solutions, this time with more accurate nodal forces ($\mathbf{F}' - \mathbf{M}\ddot{\hat{\mathbf{U}}} - \mathbf{C}_c'\dot{\hat{\mathbf{U}}} - \mathbf{M}_c'\hat{\mathbf{U}}$), because inclusion of static correction modes generates more accurate gross body motion and constraint reaction loads as well as elastic displacements (Yoo and Haug, 1986). In this case, it can be shown that the mode-acceleration solutions can be expressed as

$$\tilde{\mathbf{U}} = \hat{\mathbf{U}}_n + (\mathbf{K}^{-1} - \mathbf{K}_m^{-1})\mathbf{F}' - (\mathbf{K}^{-1} - \mathbf{K}_m^{-1})(\mathbf{C}_c'\dot{\hat{\mathbf{U}}} + \mathbf{M}_c'\hat{\mathbf{U}}) - \mathbf{K}^{-1}\mathbf{M}\ddot{\hat{\mathbf{U}}}_a \quad (39)$$

where $\hat{\mathbf{U}}_n$ is the nodal displacement vector that is obtained by the kept vibration normal modes; and $\ddot{\hat{\mathbf{U}}}_a$ is the nodal acceleration vector that is obtained by the residual attachment modes.

In order to clearly explain that the MAM can still improve the mode-displacement solutions even though some residual attachment modes are used when obtaining the mode-displacment solutions, consider only one residual attachment mode associated with an applied concentrated force, say $\mathbf{F}_a(t)$, in one of the nodal degrees of freedom. The residual attachment mode corresponding to this force is then defined as the displacement vector that is a solution of the following static equation:

$$\Psi = (\mathbf{K}^{-1} - \mathbf{K}_m^{-1})\mathbf{F}'' \quad (40)$$

where $\mathbf{F}''$ is the unit nodal force vector corresponding to the applied concentrated force vector $\mathbf{F}_a(t)$ (Yoo and Haug, 1986). The second term in the right side of Eq. (39) can be decomposed as

$$(\mathbf{K}^{-1} - \mathbf{K}_m^{-1})\mathbf{F}' = (\mathbf{K}^{-1} - \mathbf{K}_m^{-1})\mathbf{F}_a + (\mathbf{K}^{-1} - \mathbf{K}_m^{-1})(\mathbf{F}'_c + \mathbf{F}'_i) \quad (41)$$

Since the residual attachment modal coordinate is the same as the time-varying applied concentrated force (Craig and Chang, 1977), the first term in the right side of Eq. (41) can be rewritten as

$$(\mathbf{K}^{-1} - \mathbf{K}_m^{-1})\mathbf{F}_a = \Psi\mathbf{F}_a = \hat{\mathbf{U}}_a \quad (42)$$

Since $\hat{\mathbf{U}} = \hat{\mathbf{U}}_n + \hat{\mathbf{U}}_a$, Eq. (39) can be written as

$$\tilde{\mathbf{U}} = \hat{\mathbf{U}} + \Delta\mathbf{U} \quad (43)$$

where

$$\Delta\mathbf{U} = (\mathbf{K}^{-1} - \mathbf{K}_m^{-1})(\mathbf{F}'_c + \mathbf{F}'_i - \mathbf{C}_c'\dot{\hat{\mathbf{U}}} - \mathbf{M}_c'\hat{\mathbf{U}}) - \mathbf{K}^{-1}\mathbf{M}\ddot{\hat{\mathbf{U}}}_a \quad (44)$$

Therefore, the mode-acceleration solutions improves the mode-displacement solutions by the addition of the static correction terms $\Delta\mathbf{U}$ that come from the constraint reaction loads $\mathbf{F}'_c$, from the inertial loads $\mathbf{F}'_i$ generated from the gross body motion of a flexible body, from the coupling loads ($-\mathbf{C}_c'\dot{\hat{\mathbf{U}}} - \mathbf{M}_c'\hat{\mathbf{U}}$), and from the distributed inertial loads ($-\mathbf{K}^{-1}\mathbf{M}\ddot{\hat{\mathbf{U}}}_a$) resulting from the residual attachment modal accelerations.

When including static correction modes, usually only some of the concentrated constraint reaction or applied loads are used to define residual attachment modes, because the number of static correction modes may become large if the set of nodal coordinates along which unit forces or torques are applied in order to define residual attachment modes contains all the nodal coordinates at which forces appear (Yoo and Haug, 1986). Therefore, we usually neglect the static correction effects from the neglected concentrated loads, from the inertial loads generated from the gross body motion of a flexible body, and from the distributed applied loads. The static correction vectors in Eq. (44), however, contains the neglected loads acting on the flexible body. Thus the mode-acceleration solutions still can improve the mode-displacement solutions that are obtained by using a combination of vibration normal modes and residual attachment modes that are defined only for some of the concentrated loads.

The basic procedures of the proposed MAM are applied to the dynamic simulation of a flexible multibody system as follows. First, obtain time histories of gross body motion, constraint reaction loads, and elastic deformation by approximating the elastic deformation field of a flexible component by a truncated set of modes, and by solving a coupled set of system equations of motion. Second, compute the nodal force vectors ($\mathbf{F}' - \mathbf{M}\ddot{\hat{\mathbf{U}}} - \mathbf{C}_c'\dot{\hat{\mathbf{U}}} - \mathbf{M}_c'\hat{\mathbf{U}}$) and solve the quasi-static equilibrium equation of Eq. (37) to obtain improved elastic displacements or dynamic stresses by compensating for the effect of neglected higher frequency vibration normal modes and undefined static correction modes. The hybrid method (Yim, 1990) used basically the same procedures to obtain more accurate dynamic stresses without any theoretical explanation. The explanations made in this paper show why the hybrid method generates better dynamic stresses than does the modal stress superposition method (Liu, 1987). Moreover, the theory explained so far also improves the elastic displacements.

4. EFFICIENT SOLUTION OF MODE-ACCELERATION EQS.

From Eq. (37), the mode-acceleration equations can be written as

$$\mathbf{K}\tilde{\mathbf{U}} = (\mathbf{F}_i' - \mathbf{M}\ddot{\hat{\mathbf{U}}} - \mathbf{C}_c'\dot{\hat{\mathbf{U}}} - \mathbf{M}_c'\hat{\mathbf{U}}) + \mathbf{F}_a + \mathbf{F}_c' \quad (45)$$

Even though the mode-acceleration equations are static linear equations, the solution of these equations is expensive because the load vector in the right side of Eq. (45) contains numerous time-varying loads. This section presents an efficient method of solving the mode-acceleration equations, based on the procedure proposed by Ryu et al. (1992). First, the nodal force vector is represented as a sum of space-dependent and time-dependent terms. Second, if a set of nodal loads associated with a unit value of a time-dependent term is applied statically to the finite element structural analysis model, a stress or an elastic displacement field associated with this time-invariant load is obtained at the node of interest. This stress or elastic displacement field is defined as a stress or an elastic displacement influence coefficient associated with the time-dependent term. Third, the total resultant stress or elastic displacement histories are generated by multiplying the actual magnitude of the time-dependent term with the stress or elastic displacement influence coefficient and then by summing over all load cases. Note in this procedure that even though a flexible body is moving in space, only one configuration of the structural analysis model is needed because of the local representation of kinematic variables and loads, and because modal coordinates are employed as generalized coordinates. This solution scheme is very efficient because of the superposition principle applied to the time-invariant structural analysis model. The following subsections summarize the solution procedure explained above for the inertial, constraint reaction, and applied forces.

4.1 Dynamic Stress and Elastic Displacement by Inertial Force

In the deformed configuration, D'Alembert inertial forces on a differential mass m^p (= ρdV) at point p in Fig. 1 can be expressed as

$$\mathbf{f}^p = \mathbf{f}_r^p + \mathbf{f}_d^p \tag{46}$$

Forces in Eq. (46) are defined as

$$\mathbf{f}_r^p = -m^p(\ddot{\mathbf{r}} + \tilde{\omega}\tilde{\omega}\mathbf{s}_0^p + \dot{\tilde{\omega}}\mathbf{s}_0^p) \tag{47}$$

$$\mathbf{f}_d^p = -m^p\{(\tilde{\omega}\tilde{\omega} + \dot{\tilde{\omega}})\Phi^p\mathbf{a} + 2\tilde{\omega}\Phi^p\dot{\mathbf{a}} + \Phi^p\ddot{\mathbf{a}}\} \tag{48}$$

where Φ^p is the (3 x m) translational modal matrix defined at point p. Note that $\mathbf{f}_r^p$ represents inertial forces resulting from the gross body motion, while $\mathbf{f}_d^p$ represents inertial force resulting from the elastic deformation. Therefore, $\mathbf{F}_i'$ in Eq. (45) is a nodal force vector generated from $\mathbf{f}_r^p$, while $-(\mathbf{M}\ddot{\hat{U}} + \mathbf{C}_c'\dot{\hat{U}} + \mathbf{M}_c'\hat{U})$ is a nodal force vector generated from $\mathbf{f}_d^p$.

The D'Alembert inertial forces in Eq. (47) can be rewritten as a combination of space-dependent and time-dependent terms as

$$\mathbf{f}_r^p = -m^p\begin{bmatrix} 1 & 0 & 0 & 0 & z & -y & 0 & -x & -x & y & 0 & z \\ 0 & 1 & 0 & -z & 0 & x & -y & 0 & -y & x & z & 0 \\ 0 & 0 & 1 & y & -x & 0 & -z & -z & 0 & 0 & y & x \end{bmatrix}\begin{bmatrix} \ddot{r}_x \\ \ddot{r}_y \\ \ddot{r}_z \\ \dot{\omega}_x \\ \dot{\omega}_y \\ \dot{\omega}_z \\ \omega_x^2 \\ \omega_y^2 \\ \omega_z^2 \\ \omega_x\omega_y \\ \omega_y\omega_z \\ \omega_z\omega_x \end{bmatrix} \tag{49}$$

Note that $\mathbf{s}_0^p = [x, y, z]^T$ is used in Eq. (47).

Dynamic stresses and elastic displacements induced by the inertial forces $\mathbf{f}_r^p$ can be obtained as follows. First, twelve quasi-static structural analyses generate stress and elastic displacement influence coefficients $\hat{\mathbf{S}}_i^r$ (i=1,...,12) for unit values of time-dependent terms ($\ddot{r}_x$, $\ddot{r}_y$, $\ddot{r}_z$, $\dot{\omega}_x$, $\dot{\omega}_y$, $\dot{\omega}_z$, ω_x^2, ω_y^2, ω_z^2, $\omega_x\omega_y$, $\omega_y\omega_z$, $\omega_z\omega_x$). Second, a total dynamic stress or elastic displacement history due to gross body motion is obtained by applying the superposition principle:

$$\mathbf{S}^r = \ddot{r}_x\hat{\mathbf{S}}_1^r + \ddot{r}_y\hat{\mathbf{S}}_2^r + \ddot{r}_z\hat{\mathbf{S}}_3^r + \dot{\omega}_x\hat{\mathbf{S}}_4^r + \dot{\omega}_y\hat{\mathbf{S}}_5^r + \dot{\omega}_z\hat{\mathbf{S}}_6^r + \omega_x^2\hat{\mathbf{S}}_7^r + \omega_y^2\hat{\mathbf{S}}_8^r + \omega_z^2\hat{\mathbf{S}}_9^r + \omega_x\omega_y\hat{\mathbf{S}}_{10}^r + \omega_y\omega_z\hat{\mathbf{S}}_{11}^r + \omega_z\omega_x\hat{\mathbf{S}}_{12}^r \tag{50}$$

Dynamic stresses and elastic displacements induced by inertial forces $\mathbf{f}_d^p$ due to deformation can be calculated in a similar way. First, the forces can be expressed as

$$\mathbf{f}_d^p = -m^p\sum_{i=1}^{m}\mathbf{B}_i\phi_i^p \tag{51}$$

where m is the number of modal coordinates and ϕ_i^p is the (3 x 1) translational mode shape vector defined at point p. Term $\mathbf{B}_i$ in Eq. (51) is the (3 x 3) time-dependent matrix, which can be expressed as

$$\mathbf{B}_i = (\tilde{\omega}\tilde{\omega} + \dot{\tilde{\omega}})a_i + 2\tilde{\omega}\dot{a}_i + \ddot{a}_i\mathbf{I} \tag{52}$$

where matrix $\mathbf{I}$ is a (3 x 3) identity matrix.

Second, with the flexible inertial loads distributed at all nodes, given in Eq. (51), quasi-static structural analyses generate the stress or elastic displacement influence coefficients $\hat{\mathbf{S}}_{ijk}^d$ for the unit values of each term $(\mathbf{B}_i)_{jk}$ (i = 1, 2, ... m and j & k = 1, 2, 3). For example, $\hat{\mathbf{S}}_{111}^d$ is obtained in the case that $(\mathbf{B}_1)_{11}$ = 1 and all other elements of matrix $\mathbf{B}$ are zero. Therefore, the total dynamic stress or elastic displacement history from these deformational inertial forces is obtained as

$$\mathbf{S}^d = \sum_{i=1}^{m}\sum_{j=1}^{3}\sum_{k=1}^{3}(\mathbf{B}_i)_{jk}\hat{\mathbf{S}}_{ijk}^d \tag{53}$$

According to Eq. (53), a total of 9m sets of finite element structural analyses (FEA) and superpositions are required for the computation of dynamic stresses or elastic displacements from the inertial forces due to deformation.

4.2. Dynamic Stress and Elastic Displacement by Applied and Constraint Reaction Loads

Dynamic stresses or elastic displacements can also be caused by constraint reaction loads ($\mathbf{F}_c'$), externally applied body loads ($\mathbf{f}_b$), and applied concentrated loads ($\mathbf{F}_a$). If these loads are expressed in the body reference frame, each stress or elastic displacement influence coefficient can be computed for each unit magnitude load in each coordinate direction. For the applied body force, the stress or elastic displacement influence coefficient is computed in the same way as has been done for the inertial translational load because both are distributed loads. For applied concentrated and constraint reaction loads, a stress or elastic displacement influence coefficient is computed by the unit magnitude load in each coordinate direction, and then the stress or elastic displacement is obtained by multiplication with the actual magnitude of the load.

After the stress and elastic displacement influence coefficient associated with each unit load is defined, the total dynamic stress or elastic displacement history $\mathbf{S}^f$ resulting from applied and constraint reaction loads is obtained as

$$\mathbf{S}^f = \sum_i^{naf}(F_{axi}\hat{\mathbf{S}}_{axi}^f + F_{ayi}\hat{\mathbf{S}}_{ayi}^f + F_{azi}\hat{\mathbf{S}}_{azi}^f) + \sum_j^{ncf}(F_{cxj}'\hat{\mathbf{S}}_{cxj}^f + F_{cyj}'\hat{\mathbf{S}}_{cyj}^f + F_{czj}'\hat{\mathbf{S}}_{czj}^f) + f_{bx}\hat{\mathbf{S}}_1^r + f_{by}\hat{\mathbf{S}}_2^r + f_{bz}\hat{\mathbf{S}}_3^r \tag{54}$$

for body force $\mathbf{f}_b$, applied concentrated loads $\mathbf{F}_a$, and constraint reaction loads $\mathbf{F}_c'$, where naf and ncf are number of applied and constraint reaction loads, respectively. The subscripts x, y, and z denote each body coordinate axis.

4.3. Overall Computational Procedure

Figure 2 shows the overall computational procedure and conceptual data flow for the proposed method, based on the derivations of the previous sections. First, a flexible multibody dynamic analysis generates time-dependent terms, such as gross body motion, modal coordinates, and constraint reaction loads. Note that only a truncated set of basis vectors are used to describe the elastic deformation field of a flexible body in this simulation. Second, stress or elastic displacement influence coefficients are computed in association with each unit value of time-dependent terms by quasi-static structural analyses. Last, the total dynamic stress or elastic displacement is obtained by the superposition principle.

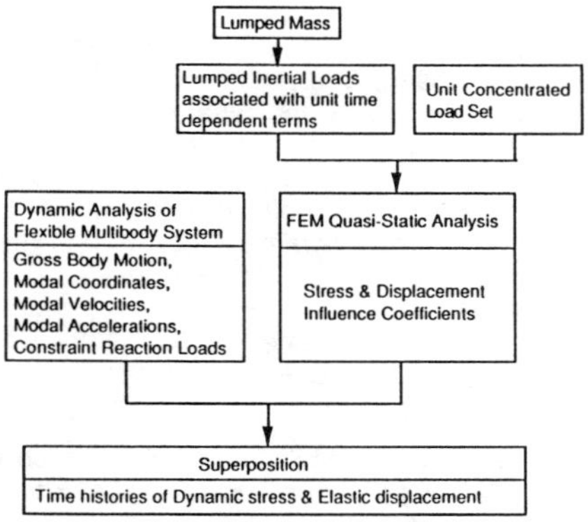

Figure 2 Overall Computational Procedure of The Proposed MAM

5. NUMERICAL EXAMPLE

In this section, the flexible planar four-bar mechanism in Fig. 3 is simulated in order to show the detailed computational procedure of the proposed MAM in flexible multibody dynamics.

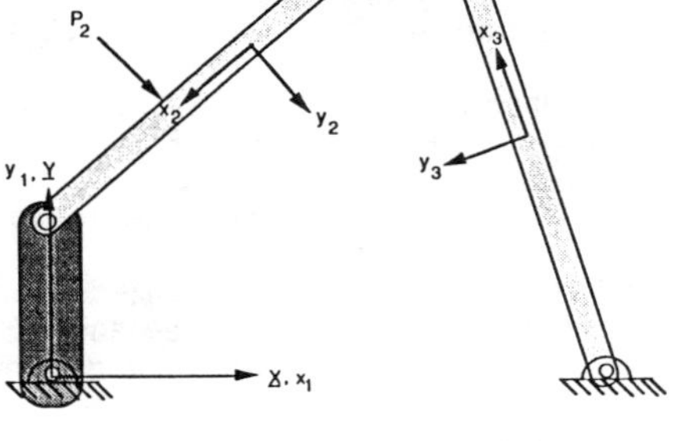

Parameters	Crank	Coupler & Follower
Body Type	Rigid Rectangular Bar	Flexible Slender Beam
Area	6.452×10^{-4} m²	4.065×10^{-5} m²
Length	1.080×10^{-1} m	2.794×10^{-1} m
Mass	1.889×10^{-1} Kg	3.080×10^{-2} Kg
Area Moment of Inertia	3.469×10^{-8} m⁴	8.673×10^{-12} m⁴
Mass Moment of Inertia about Center	1.936×10^{-4} Kg-m²	2.004×10^{-4} Kg-m²
Distance between ground pivots = 0.0254 m Material = Aluminum Gravity = 9.80665 m/sec² Mass Density = 2.712×10^{3} Kg/m³ Modulus of Elasticity = 7.10×10^{10} Pa		

Figure 3 Planar Four-Bar Crank-Rocker Mechanism

This planar four-bar crank-rocker mechanism consists of a rigid crank and two identical flexible coupler and follower links. The crank starts to rotate from the vertical position at a constant crank speed. Initial conditions on the elastic deflections and their time derivatives are assumed to be zero at the starting time. The dynamic analysis uses the first three vibration normal bending modes for each flexible beam with natural frequencies of 47.5, 190.2, and 427.9 Hz. These modes are obtained on the assumption that each beam is simply supported. In order to clearly show the improvement by the proposed MAM, a concentrated load P that is perpendicular to each flexible beam is applied.

The stress and elastic displacement influence coefficients are generated by static structural analyses with simply supported boundary conditions. Note that any boundary conditions that will give a statically determinate system can be used in static analyses because the load system is self-equilibrated. For a slender beam with planar motion in the x-y plane, the influence coefficients are defined for; (i) the gross body motions $\ddot{r}_x$, $\ddot{r}_y$, $\dot{\omega}_z$, and ω_z^2, which can be derived from Eq. (49); (ii) a longitudinal joint reaction force at the roller end; (iii) the forces resulting from deformation, which can be derived from Eq. (52); and (iv) the applied concentrated force P. Using the proposed MAM in flexible multibody dynamics, the total dynamic normal stress at point p in Fig. 4 is given by

$$\sigma^p_{xx} = \sigma^p_{xx_1} + \sigma^p_{xx_2} + \sigma^p_{xx_3} + \sigma^p_{xx_4} \qquad (55)$$

where $\sigma^p_{xx_1}$, $\sigma^p_{xx_2}$, $\sigma^p_{xx_3}$, and $\sigma^p_{xx_4}$ are normal stresses resulting from the gross body motion, joint reaction force, deformation, and concentrated applied force, respectively.

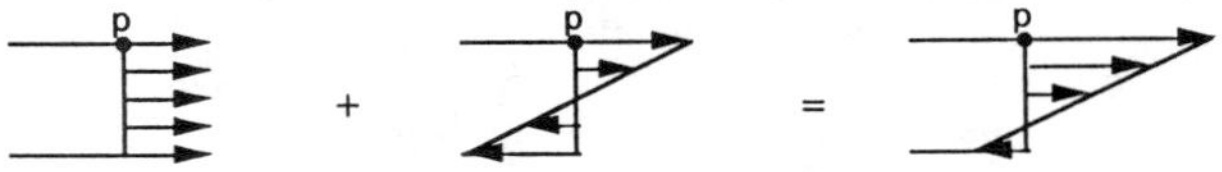

Figure 4 Normal Stress Resultant at Extreme Fiber Point p

These normal stresses are defined as

$$\sigma^p_{xx_1} = (\ddot{r}_x - g_x)\hat{S}^r_1 + (\ddot{r}_y - g_y)\hat{S}^r_2 + \dot{\omega}_z\hat{S}^r_3 + \omega_z^2\hat{S}^r_4 \qquad (56)$$

$$\sigma^p_{xx_2} = f^c_x\hat{S}^f_1 \qquad (57)$$

$$\sigma^p_{xx_3} = \sum_{i=1}^{m}(-a_i\dot{\omega}_z - 2\dot{a}_i\omega_z)\hat{S}^d_i + \sum_{i=1}^{m}(\ddot{a}_i - a_i\omega_z^2)\hat{S}^d_{i+m} \qquad (58)$$

$$\sigma^p_{xx_4} = P\hat{S}^f_2 \qquad (59)$$

where a_i is the i-th deformation modal coordinate; $\hat{S}$ are normal stress influence coefficients; and g_x, g_y, and f^c_x are x-component gravitational acceleration, y-component gravitational acceleration, and x-component constraint reaction force, respectively; and P is a concentrated applied load. A similar superposition can generate the elastic displacements in x (δ_x) and y (δ_y) directions as

$$\delta_x = (\ddot{r}_x - g_x)\hat{D}^r_1 + \omega_z^2\hat{D}^r_4 + f^c_x\hat{D}^f_1 + \sum_{i=1}^{m}(-a_i\dot{\omega}_z - 2\dot{a}_i\omega_z)\hat{D}^d_i \qquad (60)$$

$$\delta_y = (\ddot{r}_y - g_y)\hat{D}^r_2 + \dot{\omega}_z\hat{D}^r_3 + \sum_{i=1}^{m}(\ddot{a}_i - a_i\omega_z^2)\hat{D}^d_{i+m} + P\hat{D}^f_2 \qquad (61)$$

where $\hat{D}$ are elastic displacement influence coefficients. Note that even though axial modes are not used in the dynamic simulation, axial displacements can be obtained by the MAM.

On the other hand, the normal stress based on the modal stress superposition method (MDM) is defined analytically as

$$\sigma^p_{xx} = E\varepsilon^p_{xx} = -E\left(\frac{h}{2}\right)\sum_{i=1}^{m}\left(\frac{d^2\phi_i}{dx^2}\right)a_i \tag{62}$$

where ε^p_{xx} is the modal strain at point p, E is the elastic modulus, h is the height of the beam, and ϕ_i is the i-th orthonormalized bending mode shape function. Similarly, the elastic displacement in the vertical direction is obtained as

$$\delta_y = \sum_{i=1}^{m}\phi_i a_i \tag{63}$$

Notice that axial elastic displacement is zero because axial modes are not included.

Figure 5 shows the normal stresses at the upper extreme fiber of the midspan obtained by both the proposed MAM and the conventional MDM. In this simulation, the mechanism is rotating at 190.2 rpm, and P_1= 0 and P_2= 1.0 N at x_2 = 0.05588 m. In this example, the two methods give basically the same elastic displacements (not shown) but somewhat different dynamic stresses. The static correction effects of the concentrated applied and inertial loads improve the dynamic solutions because the vibration normal modes used in the dynamic simulation cannot represent the local deformation induced by these loads. Note that the improvement in the dynamic stress is clearer than that in the elastic displacement, confirming that the neglected higher frequency modes contributes more to the dynamic stresses than to the elastic displacements.

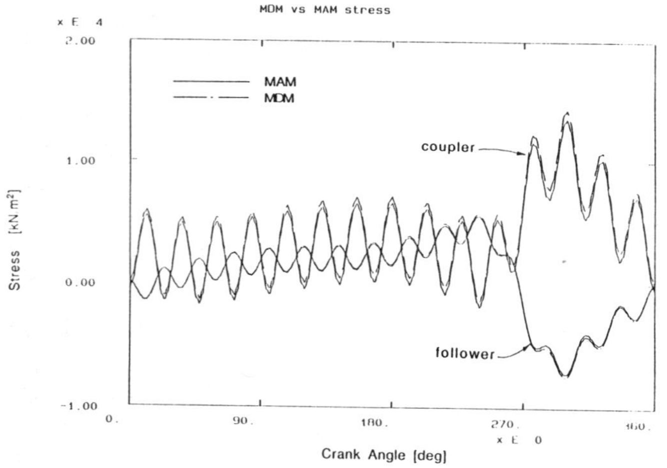

Figure 5 Dynamic Stress Histories by MAM and MDM

6. CONCLUSION

An efficient mode-acceleration method is developed for nonlinear flexible multibody dynamics. It is shown theoretically that the proposed MAM improves dynamic stresses as well as elastic displacements compared with solutions that are obtained by the MDM. Only a small amount of additional effort is needed to improve dynamic solutions by solving the mode-acceleration equations in the postprocessing stage. A numerical example of a flexible four-bar mechanism shows that the improvement in dynamic stresses is far clearer than that in elastic displacements.

REFERENCES

Bathe, K. L., 1982, *Finite Element Procedures in Engineering Analysis*, Prentice Hall, Inc. Englewood Cliffs, New Jersey.

Bisplinghoff, R. L., Ashley, H., and Halfman, R. L., 1955, *Aeroelasticity*, Addison-Wesley, Reading, MA.

Cornwell, R. E., Craig, Jr., R. R., and Johnson, C. P., 1983, "On the Application of The Mode-Acceleration Methods to Structural Engineering Problems," *Earthquake Engineering and Structural Dynamics*, Vol. 11, pp. 679-688.

Craig, Jr., R. R., 1981, *Structural Dynamics: An Introduction to Computer Methods*, John Wiley & Sons Inc., New York, NY.

Craig, Jr., R. R., and Bampton, M. C. C., 1968, "Coupling of Substructures for Dynamic Analysis," *AIAA Journal*, Vol. 6, pp. 1313-1319.

Craig, Jr. R. R., and Chang, C. J., 1977, "On the use of attachment modes in substructure coupling for dynamic analysis," *Dynamic & Structural Dynamics, AIAA/ASME 18th SDM Conference.*

Hansteen, O. E., and Bell, K., 1979, "On the Accuracy of Mode Superposition Analysis in Structural Dynamics," *Earthquake Engineering and Structural Dynamics*, Vol. 7, pp. 405-411.

Hurty, W. C., and Rubinstein, M. F., 1966, *Dynamics of Structures*, Prentice-Hall, New York.

Iman, I., Sandor, G. N., and Kramer, S. N., 1973, "Deflection and Stress Analysis in High Speed Planar Mechanisms with Elastic Links," *J. of Engineering for Industry*, Vol. 95, pp. 541-548.

Léger, P., and Wilson, E. L., 1988, "Modal Summation Methods for Structural Dynamic Computations," *Earthquake Engineering and Structural Dynamics*, Vol. 16, pp. 23-27.

Liu, T. S, 1987, "Computational Methods for Life Prediction of Mechanical Components of Dynamic Systems," *Ph. D. Dissertation*, The Univ. of Iowa.

Maddox, N. R., 1975, "On the Number of Modes Necessary for Accurate Response and Resulting Forces in Dynamic Analysis," *J. of Applied Mechanics*, pp. 516-517.

Nagarajan, S., and Turcic, D. A., 1990, "Lagrangian Formulation of the Equations of Motion for Elastic Mechanisms With Mutual Dependence Between Rigid Body and Elastic Motion. Part I & II," *J. of Dynamic Systems, Measurement, and Control*, Vol. 112, pp. 203-224.

Ryu, J., Kim, S. S., and Kim, S. S., 1992, "An Efficient Computational Method for Dynamic Stress Analysis of Flexible Multibody Systems," *Computers and Structures*, Vol. 42, No. 6, pp. 969-977.

Sadler, J. P., and Sandor, G. N., 1973, "A Lumped Parameter Approach to Vibration and Stress Analysis of Elastic Linkages," *J. of Engineering for Industry*, Vol. 95, No. 4, pp. 549-557.

Thomson, W. T., 1972, *Theory of Vibration with Applications*, Prentice-Hall, Englewood Cliffs, NJ.

Turcic, D. A., and Midha, A., 1984, "Generalized Equations of Motion for the Dynamic Analysis of Elastic Mechanism Systems," *Journal of Dynamic Systems, Measurement, and Control*, Vol. 106, pp. 243-248.

Williams, D., 1945, "Dynamic Loads in Aeroplanes Under Given Impulsive Loads with Particular Reference to Landing and Gust Loads on a Large Flying Boat," *Great Britain RAE Reports* SME 3309 and 3316.

Winfrey, R. C., 1971, "Elastic Link Mechanism Dynamics," *J. of Engineering for Industry*, Vol. 93, No. 1, pp. 268-272.

Wu, S. C., Haug, E. J., and Kim, S. S., 1989, "A Variational Approach to Dynamics of Flexible Multibody System," *J. of Mechanics of Structures and Machines*, Vol. 17, No. 1, pp. 3-32.

Yang, Z., and Sadler, J. P.. 1990, "Large-Displacement Finite Element Analysis of Flexible Linkages," *J. of Mechanical Design*, Vol. 112, pp. 175-182.

Yim, H. J., 1990, "Computational Methods for Stress Analysis of Mechanical Components in Dynamic Systems," *Ph. D. Dissertation*, The Univ. of Iowa.

Yoo, W. S., and Haug, E. J., 1986, "Dynamics of Articulated Structures, Part I: Theory," *J. of Structural Mechanics*, Vol. 14, No. 1, pp. 105-126.

AMD-Vol. 141/DSC-Vol. 37, Dynamics of Flexible Multibody Systems:
Theory and Experiment
ASME 1992

DYNAMIC SIMULATION OF FLEXIBLE MULTIBODY SYSTEMS USING VECTOR NETWORK TECHNIQUES

Marc J. Richard and Mohamed Tennich
Department of Mechanical Engineering
Laval University
Quebec, Canada

ABSTRACT:

In this paper, a new approach for the dynamic simulation of flexible multibody mechanical systems is presented. The mathematical model presented is based on the vector-network simulation method. The procedure casts, simultanously, the three dimensional Neuton-Euler equations associated with each body and the Bernoulli relationship into a symmetrical format yielding the governing differential equations of motion. The method serves as a basis for a "self-formulating" computer program which simulate system response, given only the system description as input. The method in this paper includes flexible members and is a significant extension to earlier formulation techniques. A flexible multibody aircraft during touchdown will be analyzed to demonstrate the validity of this algorithm.

1- INTRODUCTION

Engineers, who are living in a world of rapid change and extensive interaction, must continually improve their own decision-making skills or end up reacting to crises instead of controlling activity. Apprenticeships and experience are not enough. A formal and efficient technique is needed to augment the engineer's experience. The technique must be formal so that it can be learned quickly and applied directly to new situations. The technique must be efficient so that its cost does not increase in proportion to the complexity of the mechanical system. The central theme behind this work is that computer simulation is a technique that will fulfill these needs. A general purpose multibody program is a formal decision-making aid that is adaptable to the complexities and change of modern mechanical system and can be developped and communicated efficiently.

The dynamic simulation of multibody systems has been the subject of many investigations. It was used to predict the behavior and optimize the design of systems. Most of the software developed in this area was based on the rigid body motion analysis [1,9,10,11]. The advance in aeronautical mechanisms design requires higher operating velocities for lighter weight components. Representing these components by rigid bodies is not a realistic assumption. In the present investigation we present a new approach for the analysis of flexible multibody systems. This study will be limited to unidirectional flexible components. It will be based on transverse vibrations of beam represented by Bernoulli equation [14,15] and the rigid body motion will be represented by Newton-Euler equations [9,13,16]. Kinematic joints connecting the different bodies of the system will be represented by constraint equations [13]. The final mathematical model will be incorporated into the vector network algorithm which is described in the following sections. Two examples will be studied using this model.

2- VECTOR-NETWORK ALGORITHM

The vector-network is a mathematical model using graph theory as its modelling framework. It is composed of vectors which are drawn from a schematic diagram of the physical problem and a set of constitutive equations (or terminal equations), which describe the physical characteristics of the elements including flexibility in the system [1,9,10,11,13]. Essentially, the vector diagram determines the order of interconnection of the system and combined with the terminal equations, it can be shown to form a necessary and sufficient set of equations for an integral solution of the system [9]. Much of the efficiency of the vector-network method lies in the use of tree to assist in arranging the order of computation. A proper tree contains only vectors representing elements whose terminal equations can be written in the form of displacements, velocities and accelerations which are explicit functions of time. The set of remaining vectors are the complement of the tree, or the cotree. In order to maintain a general approach, all types of three-dimensional mechanical elements commonly encountered in dynamic systems will be itemized into nine basic categories and labelled according to their properties and characteristics [9,13]

N_1: mass and inertia elements
N_2: flexible-arm element
N_3: transational position/velocity driver
N_4: rotational position/velocity driver
N_5: transational force driver (external force, spring, damper)
N_6: rotational torque driver (external torque, spring, damper)
N_7: force driver of kinematic constraint
N_8: torque driver of kinematic constraint
N_9: torque coupling generated by force driver applied on flexible-arm elements

The shematic diagram of the vector network can be represented by an incidence matrix. This matrix can be transformed into the cutset and the circuit matrices noted respectively by $[A_{ij}]$ and $[B_{ij}]$. The cutset and circuit matrices will be used to formulate the mathematical model. They are related by the principle of graphic orthogonality which states [9,13] that

$$[B_{ij}] = - [A_{ij}]^T \tag{1}$$

Using these tow matrices, we can write the cutset and the circuit equations of the mechanical system represented in the vector-network formulation as

$$\{F_1 \;\; F_2 \;\; F_3 \;\; T_4\}^T = [A_{ij}]\, \{F_5 \;\; T_6 \;\; F_7 \;\; T_8 \;\; T_9\}^T \tag{2}$$

and

$$\{r_5 \;\; r_6 \;\; r_7 \;\; r_8 \;\; r_9\}^T = [B_{ij}]\, \{r_1^* \;\; r_2^* \;\; r_3 \;\; r_4\}^T \tag{3}$$

where $\{r^*\} = \{r\} + [D]^T\{u\}$ is the position of a point p of the flexible body, $\{r\}$ is the undeformed position of p, $\{u\}$ is the deformation in the body fixed axes $\{x,y,z\}$, and $[D]$ is the transforming matrix from the global to the local axes.

The transational equation of motion of a solid is given by Newton equations. It is given in a vector-network formulation by

$$[M_1]\{\dot{V}_1\} - [A_{17}]\{F_7\} = [A_{15}]\{F_5\} \tag{4}$$

where $[M_1]$ is the mass matrix, $\{\dot{V}_1\}$ is the linear acceleration of the center of mass relative to global axes, $\{F_7\}$ are the force drivers of transational constraints, $\{F_5\}$ represents the external forces, including spring and damper forces applied on the solid, $[A_{17}]$ and $[A_{15}]$ represent the cutset matrices for element number 1.

The rotational equation of motion of a solid is given by Euler equations. It can also be written in a vector-network formulation as

$$\begin{aligned}&[I_1]\{\dot{\omega}_1\} + [D_1][A_{19}][\widetilde{r_2^*}][A_{27}]\{F_7\} - [D_1][A_{18}]\{T_8\} = \\ &[D_1][A_{16}]\{T_6\} - [D_1][A_{19}][\widetilde{r_2^*}][A_{25}]\{F_5\} - [\widetilde{\omega}_1][I_1]\{\omega_1\}\end{aligned} \tag{5}$$

where $[I_1]$ is the inertia tensor of the solid, $\{\dot{\omega}_1\}$ is the angular acceleration of the solid relative to local axes, $[\widetilde{r_2^*}]$ is the anti-symmetric matrix of the flexible-arm position representing a vector product, $\{T_8\}$ is the torque driver of kinematic constraints, and $\{T_6\}$ is the external torque, including spring and damper torques applied on the solid.

In vector-network, a kinematic constraint is a vector connecting a solid to an other solid or to a position/velocity driver. The charateristic equation of a transational kinematic constraint is generated from the flexible-arm linear acceleration as

$$\begin{aligned}&-[B_{71}]\{\dot{V}_1\} + [B_{72}][\widetilde{r_2^*}]^T[B_{91}][D_1]^T\{\dot{\omega}_1\} = \\ &[B_{73}]\{\dot{V}_3\} - [B_{72}][\widetilde{V}_2]^T[B_{91}][D_1]^T\{\omega_1\}\end{aligned} \tag{6}$$

where $\{\dot{V}_3\}$ is the linear acceleration of the position velocity driver relative to the global axes, $\{\omega_1\}$ is the angular velocity of the solid, and $[\widetilde{V}_2]$ is the antisymetric matrix, representing a vector cross-product, of the flexible-arm velocity given by

$$\{V_2\} = [\widetilde{r_2^*}]\{\omega_1\} \tag{7}$$

The characteristic equation of a rotational kinematic constraint is generated from the solid angular acceleration as

$$- [B_{81}][D_1]^T\{\dot{\omega}_1\} = [B_{84}]\{\dot{\gamma}_4\} \tag{8}$$

where $\{\dot{\gamma}_4\}$ is the angular acceleration of the rotational position/velocity driver relative to the global axes.

Equations (4), (5), (6) and (8) can be assembled to generate a system of equations in the form of $[a]\{x\} = \{b\}$, given by the following equations

$$\begin{bmatrix} [M_1] & 0 & -[A_{17}] & 0 \\ 0 & [I_1] & [D_1][A_{19}][\tilde{r}_2^*][A_{27}] & -[D_1][A_{18}] \\ -[B_{71}] & [B_{72}][\tilde{r}_2^*]^T[B_{91}][D_1]^T & 0 & 0 \\ 0 & -[B_{81}][D_1]^T & 0 & 0 \end{bmatrix} \begin{Bmatrix} \dot{v}_1 \\ \dot{\omega}_1 \\ F_7 \\ T_8 \end{Bmatrix} = \begin{Bmatrix} W_1 \\ W_2 \\ W_3 \\ W_4 \end{Bmatrix} \quad (9a)$$

with

$$\{W_1\} = [A_{15}]\{F_5\} \quad (9b)$$

$$\{W_2\} = [D_1][A_{16}]\{T_6\} - [\tilde{\omega}_1][I_1]\{\omega_1\} - [D_1][A_{19}][\tilde{r}_2^*][A_{25}]\{F_5\} \quad (9c)$$

$$\{W_3\} = [B_{73}]\{\dot{V}_3\} - [B_{72}][\tilde{V}_2]^T[B_{91}][D_1]^T\{\omega_1\} \quad (9d)$$

$$\{W_4\} = [B_{84}]\{\dot{\gamma}_4\} \quad (9e)$$

This sytem of equation can be solved by a Gaussian algorithm to get the linear accelerations of the center of mass relative to the global axes, the angular accelerations of solids relative to the local axes, the force and torque drivers of kinematic constraints relative to the global axes. These accelerations can be integrated to give velocities and positions of the center of mass and the orientations of the bodies. The positions obtained can be substituted in the circuit equation to get the new positions of the cotree elements. The force and torque drivers of kinematic constraints can be susbstituted in the cutset equation to get the new force and torque applied on the solid.

The system dynamics represented by equation (9) accomodate a wide variety of nonlinear effects du to detailled geometry changes within the mechanism. The coefficient matrix, which is dependent on system displacement and time, reveals an anti-symmetric pattern attributed to the orthogonal graph proprieties existing between forces and displacements and can be perceived as a direct result of the topological structure residing within this method.

3- BEAM VIBRATIONAL RESPONSE

The vibration analysis of flexible components will be based on the model of elastic beams given by Bernoulli equation. This model neglects the effect of the transverse shear and the viscous damping of material and it is available only for small deflection.

The differential equation governing the vibration of a general elastic beam can be written in the form [5,8,14,15]

$$\frac{\partial^2}{\partial x^2}\left[E\,I(x)\frac{\partial^2 u(x,t)}{\partial x^2}\right] + \rho\,A(x)\frac{\partial^2 u(x,t)}{\partial t^2} = f(x,t) \quad (10)$$

where $u(x,t)$ is the transverse displacement, E is the modulus of elasticity, $I(x)$ is the moment of inertia, ρ is the density, $A(x)$ is the cross-section area, and $f(x,t)$ is the external length distributed force.

For a constant cross-section beam, the Bernoulli equation (10) has an exact solution composed of the product of the deflection function $X(x)$ and the time function $T(t)$. In this case, the natural frequencies of the beam can be calculated directly. The equations for natural frequencies and deflection functions are given in references [5,8,14] for different boundary conditions.

$$u(x,t) = \Sigma\, X_i(x)\, T_i(t) \quad (11)$$

$$X_i(x) = c_1 \cos \beta x + c_2 \sin \beta x + c_3 \cosh \beta x + c_4 \sinh \beta x \quad (12)$$

$$T_i(t) = a_1 \cos w_i t + a_2 \sin w_i t \quad (13)$$

$$w_i = \beta^2 \sqrt{\frac{E\,I}{\rho\,A}} \quad (14)$$

where w_i is the ith vibrational mode natural frequency, β are the solutions of natural frequency equations, a_1 and a_2 depend on initial conditions.

For a variable cross-section beam, the Bernoulli equation (10) has no exact solution and the deflection functions are given by an approximation in the form [5,8,13,14]

$$X_i(x) = a_1\, \Phi_1(x) + a_2\, \Phi_2(x) + a_3\, \Phi_3(x) + \ldots \quad (15)$$

where $\Phi_i(x)$ are continuous functions that respect the boundary conditions and the constant a_i minimize the natural frequencies of the beam. In this case we use Ritz formulation based on the balance of energy to calculate the natural frequencies of the beam. Then the generelized eigenvalues problem is given by

$$\left[[M] - w_i^2[K]\right]\{a_i\} = 0 \quad (16)$$

$$[M_{ij}] = \int_0^L I(x)\, \ddot{\Phi}_i(x)\, \ddot{\Phi}_j(x)\, dx \quad (17)$$

$$[K_{ij}] = \int_0^L \frac{\rho}{E} A(x)\, \Phi_i(x)\, \Phi_j(x)\, dx \qquad (18)$$

and the $\{a_i\}$ are the coefficient of equation (15)

We note that matrices [M] and [K] are symmetric, then we can use the Jacobi method to solve for the eigenvalues and the eigenvectors of equation (16).

A general type of support-motion involves independent displacements of individual constraint. Figure 1 shows the effects of unit translation and rotation of the support for a clamped-free beam. We define the influence function as the displacement of a generic point of the beam due to a unit displacement of a support.

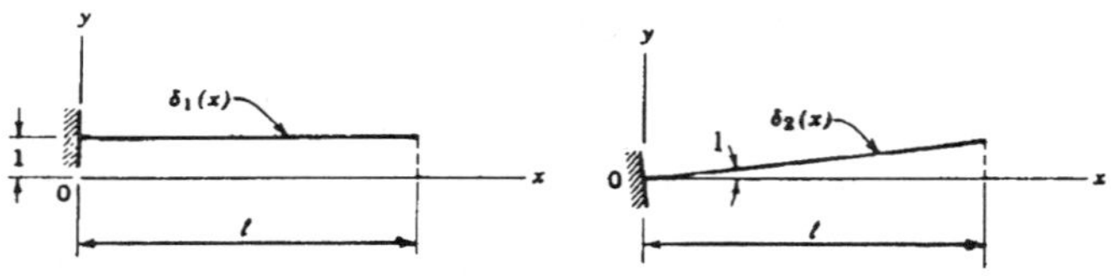

Figure 1: Clamped - free beam

When these functions are multiplied by the specified motions of the support, they produce the flexible-body motion $u_f(x,t)$ of the beam. This motion is given by [13,14]

$$u_f(x,t) = \Sigma\ \delta(x)\ u_g(t)$$

where $\delta(x)$ are the influence functions and $u_g(t)$ are the translations and rotations of the beam supports.

The differential equation of vibrational motion of a beam subjected to support motions can be derived directly from the Bernoulli equation. Figure 2 shows a clamped-free beam subjected to its support translation. We note by $u_f(x,t)$ the flexible body motion of the beam and by $u_g(t)$ the support translation and rotation. Let $u(x,t) = u_f(x,t) - u_g(t)$ represent the relative motion of the beam to its support. The Bernoulli equation, for a constant cross-section beam, can be written as [13,14]

$$E I \frac{\partial^4 u(x,t)}{\partial x^4} + \rho A \frac{\partial^2 u(x,t)}{\partial t^2} = - \rho A \frac{d^2 u_g(t)}{d t^2} \qquad (19)$$

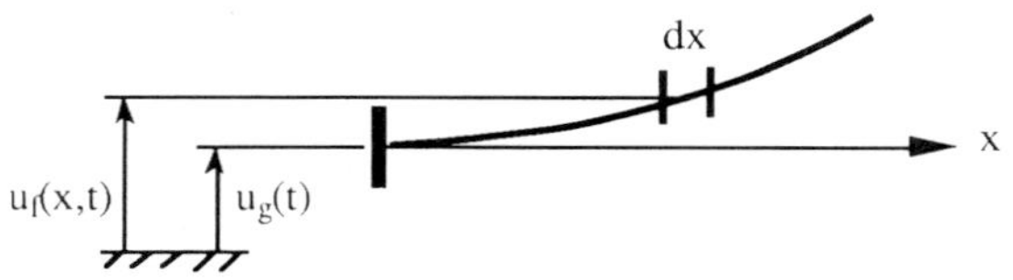

Figure 2: Support motion

This expression has the form of the differential equation of the beam forced vibration. Equation (19) will be transformed to normal coordinates by substituting expression (11) for u(x,t), multiplying by $X_j(x)$, and integrating over the length of the beam. Using the condition of orthogonality and normalization of vibrational modes we obtain

$$\ddot{T}_i(t) + w_i\, T_i(t) = \int_0^L X_i(x)\, \ddot{u}_g(t)\, dx \qquad (20)$$

where $\ddot{u}_g(t)$ is the second derivative with respect to time of the support motion, and w_i is the natural frequency of the ith vibrational mode. This expression represents a general equation of motion in normal coordinates, and the integral represents the ith normal mode load. The response of the ith vibrational mode is given by the Duhamel integral as [13,14]

$$T_i(t) = - \frac{1}{w_i} \int_0^L \delta(x)\, X_i(x)\, dx \int_0^t \ddot{u}_g(\tau) \sin(w_i (t - \tau))\, d\tau \qquad (21)$$

Substituting this expression into equation (11) gives the total vibrational response of the beam subjected to the support motion $u_g(t)$.

$$u(x,t) = - \sum_{i=1}^{\infty} \frac{X_i(x)}{w_i} \int_0^L \delta(x)\, X_i(x)\, dx \int_0^t \ddot{u}_g(\tau) \sin(w_i (t - \tau))\, d\tau \qquad (22)$$

Finally, the response of a beam subjected to external loads can be obtained in the same way as the case of support motions. This response is given below for the cases of concentrated force, moment, and distributed force.

(1) for a concentrated force q(t) applied at x_q, u(x,t) is given by,

$$u(x,t) = \sum_{i=1}^{\infty} \frac{X_i(x)\, X_{iq}}{w_i} \int_0^t \frac{q(\tau)}{\rho A} \sin(w_i (t - \tau))\, d\tau \qquad (23)$$

(2) for a concentrated moment M(t) applied at x_M, u(x,t) is given by,

$$u(x,t) = \sum_{i=1}^{\infty} \frac{X_i(x)\, X'_{iM}}{w_i} \int_0^t \frac{M(\tau)}{\rho A} \sin(w_i (t - \tau))\, d\tau \qquad (24)$$

where $X'_{iM} = \left. \frac{\partial X_i}{\partial x} \right|_{x = x_M}$

and (3) for a distributed force $Q(x,t) = f(x)\, q(t)$, $u(x,t)$ is given by,

$$u(x,t) = \sum_{i=1}^{\infty} \frac{X_i(x)}{w_i} \int_0^L f(x)\, X_i(x)\, dx \int_0^t \frac{q(\tau)}{\rho A} \sin(w_i (t-\tau))\, d\tau \qquad (25)$$

4- PRACTICAL SIMULATIONS

First, the mathematical model presented in sections 2 and 3 will be used to find the vibrational response of a variable cross-section beam. The clamped-free beam shown in figure 3, is subjected to a harmonic displacement given by

$$Y_s(t) = 0.02 \sin(w_s t) \qquad (26)$$

where t represents the time, and w_s is the frequency of the excitation.

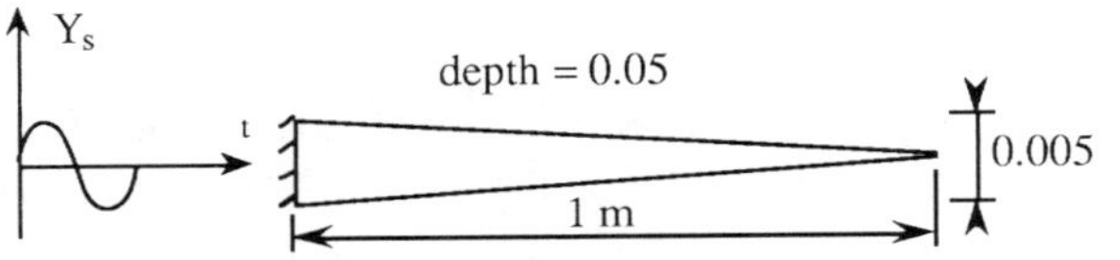

Figure 3: Support motion

The area and the inertia moment of a cross-section of the beam are given repectively by the following expressions

$$A(x) = 2.5\ 10^{-4}(1 - x) \qquad I(x) = 6.25\ 10^{-9}(1 - x)^3$$

The functions $\Phi i(x)$ used in expression (15) are given by

$$\Phi_i(x) = (-1)^{i+1} \cos\left[(i-1)\,\pi \frac{x}{L}\right]\left(\frac{x}{L}\right)^2 \qquad (27)$$

where L is the length of the beam, and x is measured from the fixed extremity. We can easily verify that these functions respect the boundary conditions of a clamped-free beam.

Figure 4 shows the vector network of the clamped free beam. Verctors 1 to 4 form the tree of the schematic diagram, and the co-tree is composed of vectors 5 to 8. The references {X,Y} and {x,y} are respectively the global and local axes and each position vector is defined as

- r_1^*: position of the center of mass 1
- r_2^*: position of the flexible-arm 2
- r_3^*: position of the flexible-arm 3
- r_4: position of driver 4
- r_5: force applied at the center of mass 1
- r_6: force applied on flexible-arm 3
- r_7: transational kinematic constraint
- r_8: rotational kinematic constraint

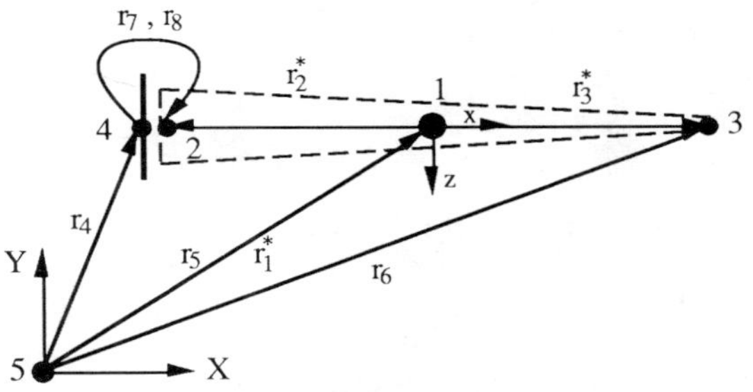

Figure 4: Shematic diagram of the beam

The incidence matrix of the shematic diagram in figure 7 is given by

$$[K] = \begin{bmatrix} 1 & -1 & -1 & 0 & 1 & 0 & 0 & 0 \\ 0 & 1 & 0 & 0 & 0 & 0 & 1 & 1 \\ 0 & 0 & 1 & 0 & 0 & 1 & 0 & 0 \\ 0 & 0 & 0 & 1 & 0 & 0 & -1 & -1 \\ -1 & 0 & 0 & -1 & -1 & -1 & 0 & 0 \end{bmatrix}$$

The cutset [A] and circuit [B] matrices are give by

$$[A] = \begin{bmatrix} 1 & 1 & 1 & 1 \\ 0 & 0 & 1 & 1 \\ 0 & 1 & 0 & 0 \\ 0 & 0 & -1 & -1 \end{bmatrix} \qquad [B] = \begin{bmatrix} -1 & 0 & 0 & 0 \\ -1 & 0 & -1 & 0 \\ -1 & -1 & 0 & 1 \\ -1 & -1 & 0 & 1 \end{bmatrix}$$

Then the cutset equations can be written as

$$\begin{aligned} \{F_1\} &= -\{F_5\} - \{F_6\} - \{F_7\} & \{T_1\} &= -\{T_8\} \\ \{F_2\} &= -\{F_7\} & \{T_2\} &= -\{T_8\} \\ \{F_3\} &= -\{F_6\} & \{T_4\} &= \{T_8\} \\ \{F_4\} &= \{F_7\} \end{aligned} \qquad (28)$$

and the circuit equations are given by

$$\begin{aligned} \{r_5\} &= \{r_1\} & \{r_7\} &= \{r_1\} + \{r_2\} - \{r_4\} \\ \{r_6\} &= \{r_1\} + \{r_3\} & \{r_8\} &= \{r_1\} + \{r_2\} - \{r_4\} \end{aligned} \qquad (29)$$

The flexible positions of elements 1, 2 and 3 are given, relative to the global axes, by

$$\{r_1^*\} = \{r_1\} + [D_1]^T\{u_1\} \qquad (30)$$

$$\{r_2^*\} = \{r_2\} + [D_1]^T\{u_2 - u_1\} \qquad (31)$$

$$\{r_3^*\} = \{r_3\} + [D_1]^T\{u_3 - u_1\} \qquad (32)$$

where the vector $\{u_j\}$ is equal to $\{0\ \ 0\ \ u_j(x,t)\}^T$, and the scalar $u_j(x,t)$ is given by

$$u_j(x,t) = -\sum_{i=1}^{6} \frac{X_i(x)}{w_i} \int_0^L X_i(x)\, dx \int_0^t -\ddot{Y}_s(\tau) \sin[w_i(t-\tau)]\, d\tau \qquad (33)$$

The equations of motion can then be written in the form

$$[M_1]\{\dot{V}_1\}-\{F_7\} = \{F_5\}+\{F_6\} \tag{34}$$

$$[I_1]\{\dot{\omega}_1\}+[D_1][\tilde{r}_2^*]\{F_7\}-[D_1]\{T_8\} = -[D_1][\tilde{r}_2^*]\{F_5 -[D_1][\tilde{r}_3^*]\{F_6\}-[\tilde{\omega}_1][I_1]\{\omega_1\} \tag{35}$$

$$\{\dot{V}_1\}-[\tilde{r}_2^*]^T[D_1]^T\{\dot{\omega}_1\} = \{\dot{V}_7\}+[\tilde{V}_2]^T[D_1]^T\{\omega_1\} \tag{36}$$

with $\{\dot{V}_7\} = \{0 \;\; \ddot{Y}_s \;\; 0\}^T$

$$[D_1]^T\{\dot{\omega}_1\} = \{0\} \tag{37}$$

The first four natural frequencies of the beam are respectively 46.6, 138.5, 298.87 and 557.64 rad/s [2,14]. The simulations were done with different values of w_s, frequency of the support motion, and the results are shown in figures 5 and 6. Figure 6 shows the variation of the maximum deflection. We note the apparence of the resonance when w_s is close to one of the natural frequencies of the beam.

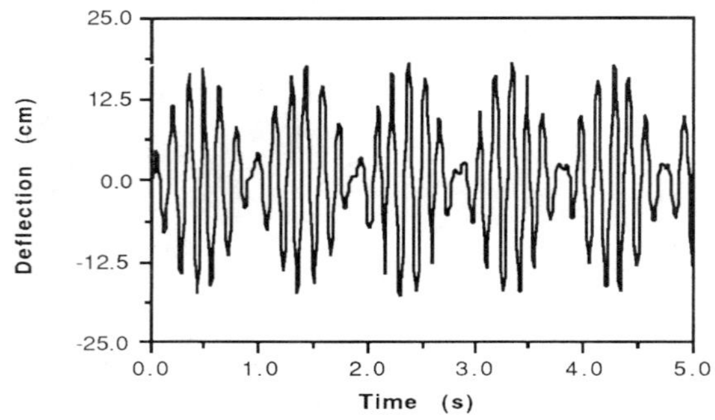

Figure 5: Response of the beam for w_s = 40 rad/s

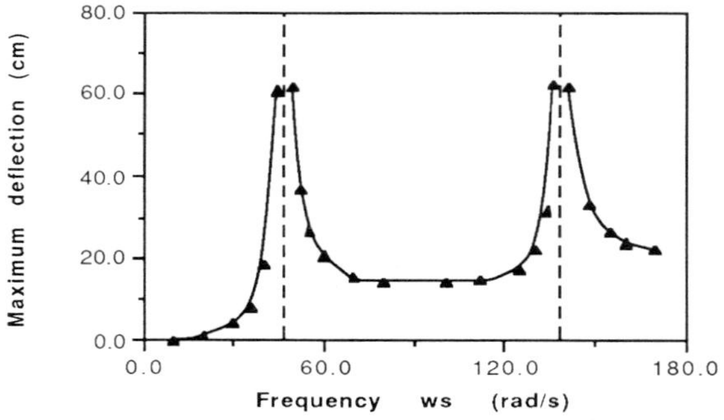

Figure 6: Variation of the maximum deflexion

As an illustration of the vector-network method developped for flexible bodies, consider the planar aircraft model shown in figure 7. The system consists of five bodies wich represents the fuselage (body 1), the tow wings (bodies 2 and 3) and the tow landing gears (bodies 4 and 5).

It is required to determine the vibrational behavior of this large scale flexible multibody aircraft during touchdown maneuvers which include impacts and rollover motion. The wings, depicted in figures 8 and 9, are modeled as elastic cantilever beams with elliptic hollow cross sections reinforced by three ribs.

Elasticity and damping of the landing gear are introduced using spring and damper elements. The fuselage is assumed to be rigid and the interconnections between the components are introduced using transational and rotational kinematic constraints [13]. Figure 10 presents the vector network of this model.

For large scale aircraft, the dynamic loads induced by the impact between the landing gear and the runway are recognized as significant factors causing vibrations, dynamic stress and fatigue damage of the aircraft structure [3]. The touchdown impact can be simulated by a vertical drop of the aircraft model. The distances between the two wheels and the runway are monitored until one of them will be zero (or negative) indicating an impact. The numerical results are plotted in figure 11 for the two wings. We note that the first impact occurs at t = 0.52 s and the maximum deflection is 18 cm. A second factor causing the fatigue damage of the aircraft structure is the irregularities of the runway surface. To simulate these irregularities we use the two independent random signals proposed by Changizi [3] and presented in figures 12 and 13. The results for the right and left wings are plotted in figures 14 and 15. We note that the maximum deflection is 2.25 cm for the right wing and 1.95 cm for the left wing.

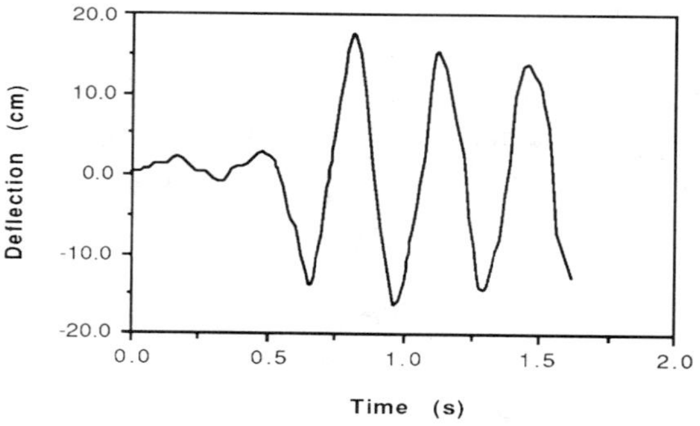

Figure 11: Wing deflection during touchdown

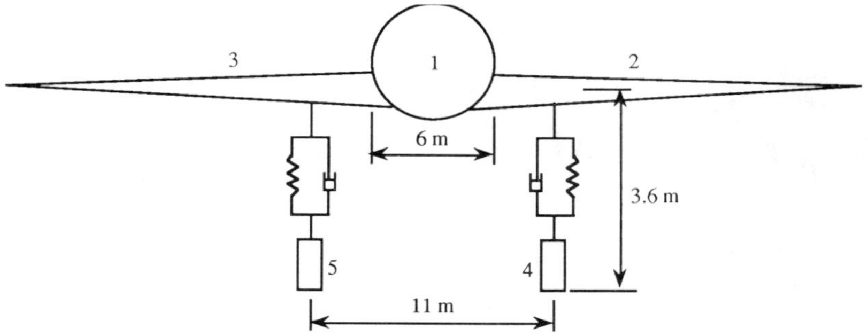

Figure 7: Aircraft model

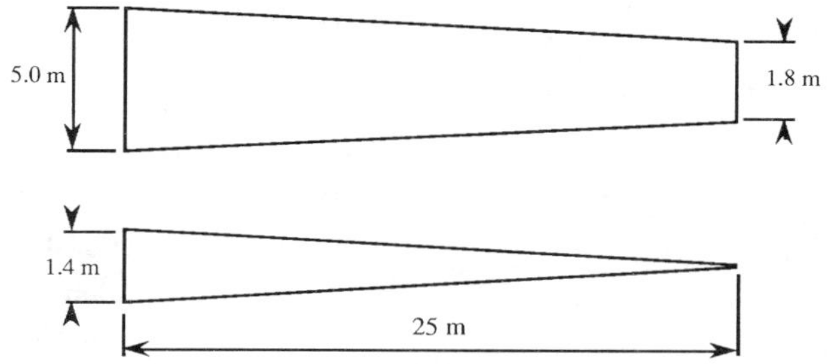

Figure 8: Wing model

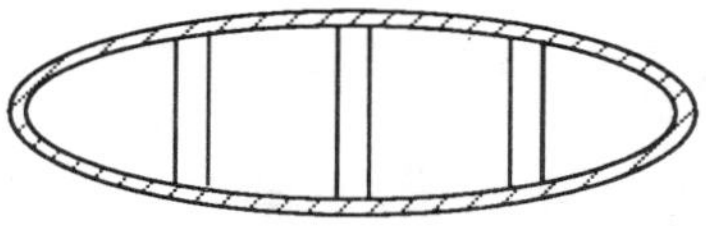

Figure 9: Wing section

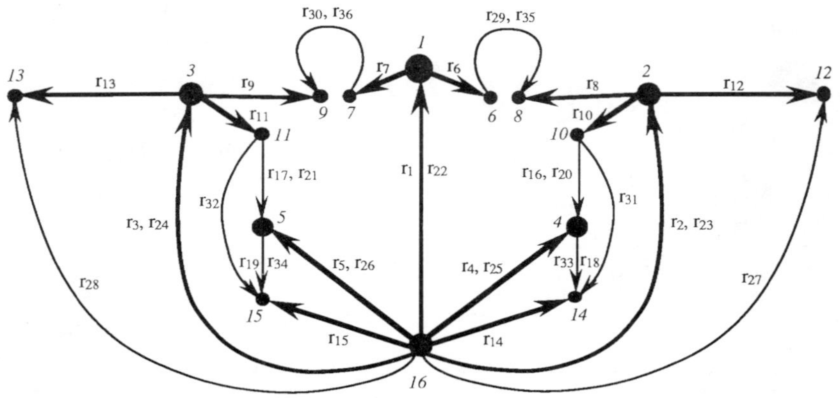

Figure 10: Aircraft schematic diagram

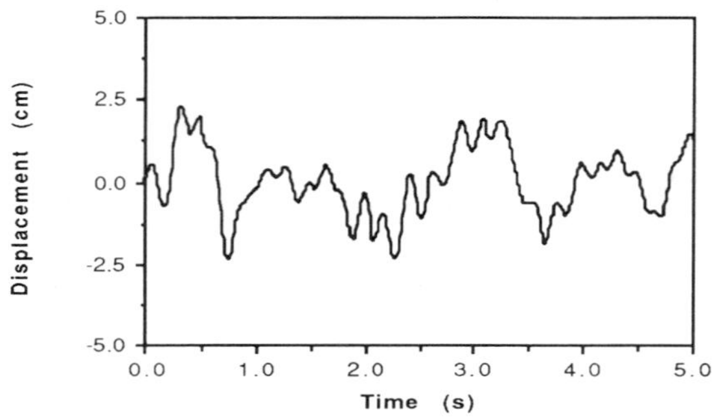

Figure 12: Runway surface for right wheel

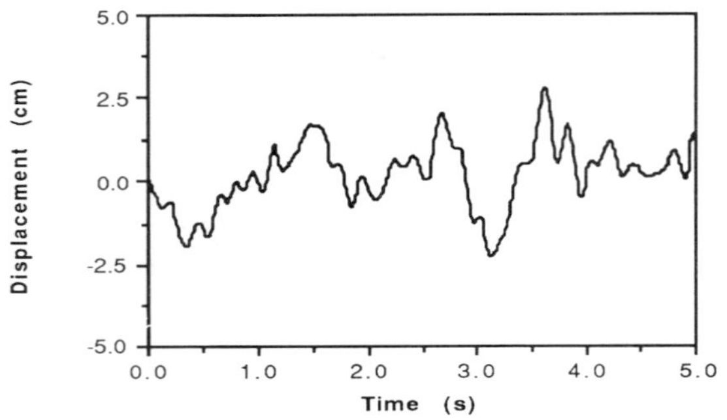

Figure 13: Runway surface for left wheel

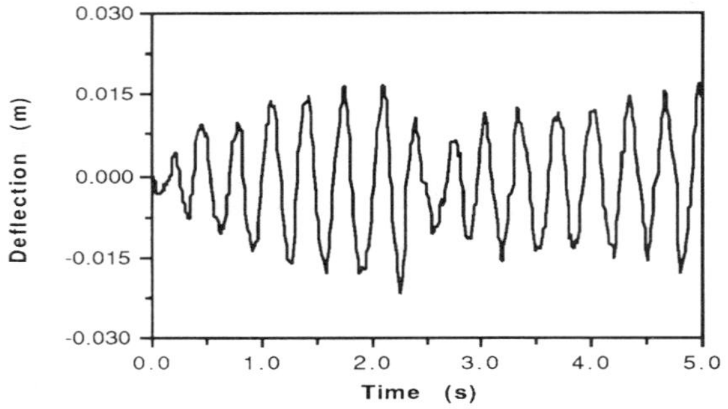

Figure 14: Right wing deflection during rolling motion

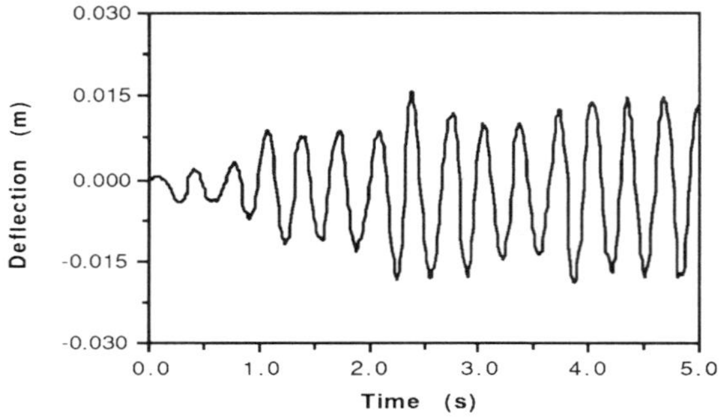

Figure 15: Left wing deflection during rolling motion

The aircraft system is a good example where the dynamical and kinematical properties of the mechanism must be respected. Note that although the simulated system is planar, all three scalar directions will be processed by the 3-D computer program but since the direction perpendicular to the plane of motion has no internal/external excitation, the scalar variables perpendicular to the motion will remain in an inactive state retaining their initialized values throughout the computation.

6- CONCLUSION

The vector-network formulation has been reviewed and extended to include flexible multibody systems. The validity of the mathematical model has been established and the general procedure for deriving the three dimensional equations of motion has been determined.

The most prominent benefit of this methodical model was its suitability for implementation on the digital computer. To demonstrate this application, a self-formulating computer program was conceived and tested. From a minimal definition of the system, the computer program is capable of automatically creating and solving the differential equations of motion following a simple substitution procedure. This computer program was used to simulate a typical aircraft during landing. However, the application of this method is not restricted to flexible systems. A variety of mechanical systems can be modelled quickly where a wide spectrum of design alternatives can be evaluated in a short period of time.

7- REFERENCES

[1] Andrews Gordon C., Richard Marc J., Anderson Ronald J., "A general vector-network formulation for dynamic systems with kinematic constraints", Mech. Mach. Theory, vol. 23, No. 3, 1988, pp. 243-256.

[2] Belvins Robert D., Formulas for natural frequency ans mode shape, Van Nostrand Reinhold Company, Toronto, 1979, 492p.

[3] Changizi Koorosh, Khulief Yehia A., Shabana Ahmed A., "Transient analysis of flexible multi-body systems. Part II: Application to aircraft landing", Computer Methods in Applied Mechanics and Engineering, vol. 54, 1986, pp. 93-110.

[4] Cranch E. T., Adler Alfred A., "Bending vibrations of variable section beams", Journal of Applied Mechanics, march 1956, pp. 103-108.

[5] Hutchison J. R., "Transverse vibrations of beams, exact versus approximate solutions", Journal of Applied Mechanics, vol. 48, december 1981, pp. 923-928.

[6] Khulief Y. A., Shabana A. A., "Dynamic analysis of constrained system of rigid and flexible bodies with intermittent motion", Journal of Mechanisms, Transmissions and Automation in Design, vol. 108, march 1986, pp. 38-45

[7] Penny J. E., Reed J. R., "An integral equation approach to the fundamental frequency of vibrating beams", Journal of Sound and Vibration, vol. 19, 1971, pp. 393-400.

[8] Rao Singiresu S., Mechanical vibrations, second edition, Addison-Wesley Publishing Company, New York, 1990, 718p.

[9] Richard Marc Joseph, Dynamic simulation of constrained three dimensional multi-body systems using vector network techniques, Ph.D. Thesis, Queen's university, Kingston, 1985, 481p.

[10] Richard M. J., Anderson R., Andrews G. C., "Generalized vector-network formulation for the dynamic simulation of multibody systems", Journal of Dynamic Systems, Mesurment, and Control, vol. 108, december 1986, pp. 322-329.

[11] Richard M. J., "Dynamic simulation of multibody mechanical systems using the vector-network model", Transaction of the CSME, vol. 12, No. 1, 1988, pp. 21-30.

[12] Shabana A. A., Patel R. D., DebChaudhury A., Ilankamban R., "Vibration control of flexible multibody aircraft during touchdown impacts", Journal of Vibration, Acoustics, Stress, and Reliability in Design, vol. 109, july 1987, pp. 270-276.

[13] Tennich Mohamed, "Simulation dynamique des systèmes de corps flexibles par la méthode réseau vectoriel", Mémoire de maîtrise, Université Laval, 1990.

[14] Timoshenko S., Young D. H., Weaver W. Jr., Vibration problems in egineering, fourth edition, John Wiley & Sons, Toronto, 1974, 521p.

[15] Thomson William T., Theory of vibration with application, second edition, Prentice-Hall, New Jersy, 1981, 493p.

[16] Wells Dare A., Lagrangian dynamics, Shaum's outline series, MaGraw-Hill, Toronto, 1967, 353p.

AMD-Vol. 141/DSC-Vol. 37, Dynamics of Flexible Multibody Systems:
Theory and Experiment
ASME 1992

SIMULTANEOUS POSITION AND FORCE CONTROL OF FLEXIBLE MANIPULATORS

Yueh-Jaw Lin and Tian-Soon Lee
Department of Mechanical Engineering
University of Akron
Akron, Ohio

ABSTRACT

This paper presents a comprehensive dynamic modeling of a flexible link manipulator by considering the shear, bending as well as rotational inertia effect of the manipulator in the dynamic formulation. In addition, the gravitational effect, which is missing in most of the flexible manipulator dynamic model, is also included in the formulation. Then, an efficient motion and force controller design method utilizing root contour analysis is proposed. With the proposed method the multi controller gains of the position/force controller for the flexible manipulator are chosen analytically, as opposed to the trial and error method. The tuning process of these controllers' gains proves the simplicity of gains selection. And the motion/force simulation results verify the effectiveness of the controllers.

INTRODUCTION

The main objective of this work is to advance the state-of-the-art in the area of motion and force control of flexible manipulators. The study is based on a flexible single-link manipulator operated in a vertical plane. The study in this work is aimed at developing an effective and systematic methodology to accurately control the motion of flexible manipulators subjected to external contact forces. Current research on flexible link manipulators has the following problems, namely, (1) most flexible manipulator models under study are either considering the manipulator operated on a horizontal plane or the shear and gravitational effects are excluded in the dynamic model, which makes it difficult to control the manipulator motion accurately, (2) the motion of the flexible manipulators being considered is usually a contact-free operation, i.e., without any contact forces, which makes it impractical as performing industrial assembly tasks, and (3) the selection of controller gains is usually by empirical methods, which makes the controller design very time consuming.

In view of the aforementioned problems the flexible link manipulator under study in this paper is constructed in such a way that it is operated vertically and in addition, subjected to a contact force at the tip. The dynamic model is comprehensively formulated by considering shear, bending, rotational inertial and gravitational effects. The inclusion of gravitational effect is considered to be necessary since most industrial robots are operated in three dimensional workspace. In the derivation of control algorithm of the flexible link manipulator, this paper also deals with external contact forces which can not be avoided when performing assembly tasks. Moreover, in this paper we propose a quick and systematic way of selecting good controller gains utilizing the root contour method (Lin and Lee, 1992). This method greatly reduces the laborious trial-and-error approach conventionally used in gains selection. The motion simulation results of the flexible single link manipulator and the most important findings with the simulation are reported to verify the effectiveness of the proposed methodology.

COMPREHENSIVE DYNAMIC MODEL

Referring to Figure 1 and applying the Hamilton's principle, we obtain dynamic equations of the arm as follows:

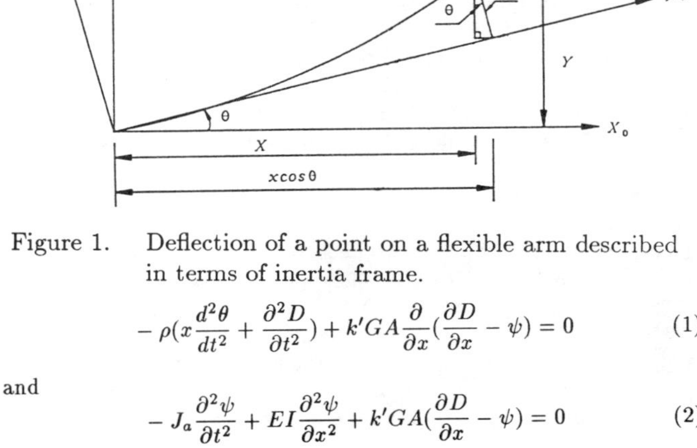

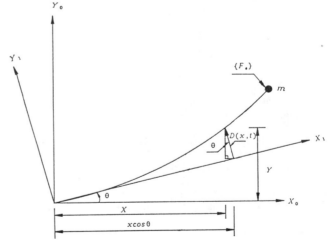

Figure 1. Deflection of a point on a flexible arm described in terms of inertia frame.

$$-\rho(x\frac{d^2\theta}{dt^2}+\frac{\partial^2 D}{\partial t^2})+k'GA\frac{\partial}{\partial x}(\frac{\partial D}{\partial x}-\psi)=0 \tag{1}$$

and

$$-J_a\frac{\partial^2\psi}{\partial t^2}+EI\frac{\partial^2\psi}{\partial x^2}+k'GA(\frac{\partial D}{\partial x}-\psi)=0 \tag{2}$$

The deflection of the arm can be written in terms of D_B as

$$EI\frac{\partial^4 D_B}{\partial x^4}+\rho\frac{\partial^2 D_B}{\partial t^2}-J_a\frac{\partial^4 D_B}{\partial x^2\partial t^2}-\frac{\rho EI}{k'GA}\frac{\partial^4 D_B}{\partial x^2\partial t^2}+\frac{\rho J_a}{k'GA}\frac{\partial^4 D_B}{\partial t^4}=-\rho x\frac{d^2\theta}{dt^2} \quad (3)$$

Because the closed form solution of Equation (3) is difficult to obtain, the approximate solution will be sought. The Galerkin's method is used to find the approximate solution. Through a laborious calculation, the forward dynamics of the generalized coordinates and joint coordinates are given by

$$[\ddot{q}] = ([M]+\rho[S][U]^T)^{-1}[([K]+\rho[S][H]^T)[q]]+ \frac{\rho}{R_1}([M]+\rho[S][U]^T)^{-1}(T-R_2cos\theta-[J_c]^T[F_e])[S] \quad (4)$$

and

$$\ddot{\theta}=\frac{(T-R_2cos\theta-[J_c]^T[F_e])}{R_1(1+\rho[U]^T[M]^{-1}[S])}+\frac{([H]^T-[U]^T[M]^{-1}[K])[q]}{(1+\rho[U]^T[M]^{-1}[S])} \quad (5)$$

CONTROLLER DESIGN

Introducing the notations W, C_g and f into the inverse dynamics equation from Equation (4) yields,

$$W\ddot{\theta}+C_g+f+[J_c]^T[F_e]=T \quad (6)$$

The control law conceived here has two parts. If the servo part is chosen as a linear combination of PD-controller operating on the position and velocity errors and P-controller operating on the force error, then we obtain the closed-loop control law given by

$$T=W[\ddot{\theta}^d+k_v(\dot{\theta}^d-\dot{\theta})+k_{fp}[J_c]^T([F_e]^d-[F_e])]+C_g+f+[J_c]^T[F_e] \quad (7)$$

Equation (7) shows that the controller indirectly controlls the Cartesian space position and force by joint space position and force. It also shows that the dynamics of gravitational effect, flexible effect and the contact force are compensated for to reduce their influences on the accuracy of the controller. It is assumed that the robots are doing assembly tasks. Thus, the contact force can be modeled as a high stiffness linear spring model.

$$[F_e]=[k_{eq}][J_c](\Delta\theta+\frac{\Delta D|_{x=L}}{L}) \quad (8)$$

where $\Delta\theta=(\theta-\theta_0)$ and $\frac{\Delta D|_{x=L}}{L}=\frac{D|_{x=L}-D_0|_{x=L}}{L}$. The characteristic equation of our system is then obtained as

$$s^2+k_vs+(k_p+k_{fp}[J_c]^T[k_{eq}][J_c])=0 \quad (9)$$

The controller gains appear in the above equation require to be tuned properly to result in suitable control actions.

SIMULATION OF MOTION SUBJECT TO FORCE

To test the effectiveness of the control law derived in the previous section, a motion simulation program has been developed to simulate the dynamic behavior of a single link flexible manipulator. The cross-sectional shape of the manipulator arm is assumed to be thin hollow round.

Figure 2 and 3 show that the overshoots of the system are reduced to zero and the position and force responses of the system are slowing down. In fact these two phenomena are corresponding to the system that behaves critically. These phenomena are due to the fact that by increasing the value of k_v will increase the damping effect of the system. Since the system takes longer time to decay to its desired value when the poles move closer to the imaginary axis, it can be expected that the responses of the system will be quite sluggish. This prediction is verified by viewing the response curves in Figures 2 and 3.

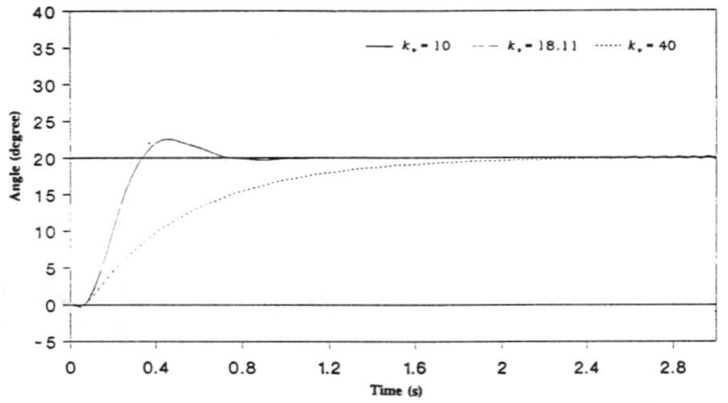

Figure 2. Closed-loop response of joint angle with $k_p = 75$, $k_{fp} = 0.045$ and $k_v = 10$.

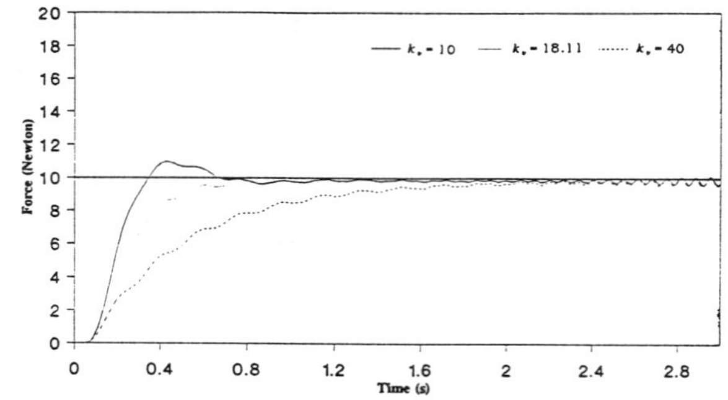

Figure 3. Closed-loop response of applied force in y-direction with $k_p = 75$, $k_{fp} = 0.045$ and $k_v = 10$.

CONCLUSION

A theoretical study for the closed-loop position and force control of a flexible link robot with an end-point payload, maneuvered in a constraint environment, has been presented. The comprehensive dynamic model along with the extended partitional control law architecture are then used to derive the control algorithm which deals with the manipulators position and contact force simultaneously. The proposed controller gains tuning method is then applied to tune these controllers' gains analytically. From the simulation results we observe that the performance of a flexible link robotic system can be predicted by investigating the roots of the system's characteristic equations. Hence, the total reliance on the trial-and-error computer simulation to evaluate its performance can be avoided or reduced. Through the force and position controller design process it is seen that it is much easier to design a multi-gains controller of a nonlinear robotic system analytically using the proposed root contour method than to design it by trial and error approach for gains selection. In addition, since the position and force controller are affecting each other during motion, the root contour method gives better insight for controller gains selection. And from the motion control simulation results, the performance of the position and force controller designed analytically using the proposed method, is quite satisfactory.

References

Lin, Y.J. and Lee T.S. (1992), "Comprehensive Dynamic Modeling and Motion/Force Control of Flexible Manipulators," *Mechatronics*, Vol. 2, pp. 129-148.

AMD-Vol. 141/DSC-Vol. 37, Dynamics of Flexible Multibody Systems:
Theory and Experiment
ASME 1992

A SINGLE-AXIS SERVOMECHANISM FOR CONTROL EXPERIMENTS INVOLVING COULOMB FRICTION, BACKLASH AND JOINT COMPLIANCE

Kwaku O. Prakah-Asante, Abu S. Islam, Daniel Walczyk, and Kevin Craig
Department of Mechanical Engineering,
Aeronautical Engineering and Mechanics
Rensselaer Polytechnic Institute
Troy, New York

ABSTRACT

An investigation of the effects of system nonlinearities, such as Coulomb friction and backlash, and undesirable structural dynamics, such as joint compliance, on the dynamic modeling and control of mechanical positioning systems is presented. The design and construction of an innovative, computer-controlled, mechanical-positioning test bed, built to facilitate investigations of the effects mentioned above, which are often neglected in dynamic system analysis and control studies, is presented. The test bed allows fully-adjustable, quantified, and well-defined Coulomb friction, backlash, joint compliance, and inertia. System identification and parameter estimation techniques are used to estimate and verify linear system parameters. Open and closed-loop simulations of the system incorporating the undesirable effects which can be exhibited by the test bed are presented and compared to experimental results. The positioning performance of the servomechanism is shown to degrade with increases in Coulomb friction, backlash, and joint compliance. Candidate control strategies to compensate for the above mentioned adverse phenomena are described and their effectiveness evaluated. Future experimental and theoretical work is described.

1. INTRODUCTION

Mechanical positioning, the need to assure that a mechanical object has a specified position and velocity at a prescribed time, is a critical element of almost all dynamic systems with significant mechanical content and there is a need for positioning systems that produce higher accuracy and faster positioning. Servomechanisms are used for linear positioning (via friction drives, rack and pinion, or ball screws) and in slewing motion (via direct drive, gear reducer, or harmonic drive). Linear positioning systems are used in CNC machines, automated machine tools, automated industrial machines, and robot prismatic joints. Slewing devices using servo-mechanisms include robot revolute joints, aircraft control-surface actuators, and gun-turret positioners. Difficulties in mechanical positioning derive from essential nonlinearities due to backlash and friction, geometric nonlinearities due to configuration, and the inherently high-order, coupled behavior of mechanical structures. Each of these problems establishes a limit to performance; when present together, which is often the case, meeting performance specifications can be a formidable challenge. A brief discussion of each of these problems follows:

(a) Dry friction (Coulomb friction and stiction), is present to some degree in almost all positioning systems. Friction introduces an element of randomness when the system is operating at low velocity and causes rapid force changes when the velocity is reversed. Because of the sticking effect and the rapid velocity reversal, dry friction cannot be linearized successfully. As the operating zone closes in on the velocity origin, the linearization progressively worsens. Research is needed in the physical phenomenon itself and its relation to positioning systems, its characterization, and its simulation. These must be linked to system mechanical design to minimize (or maximize) friction and, most importantly, to make its action as predictable as possible. Development of effective control algorithms and associated design methods must proceed in parallel with this work.

(b) Backlash, like friction, is almost always present in mechanical positioning systems and also cannot be linearized. To some extent, backlash and friction can be traded for each other, that is, if tolerances are reduced to minimize backlash, friction is increased, and vice versa. Backlash and friction are also opposing pairs from a control perspective. Whereas friction compensation often uses highcontrol gains near the velocity origin, such gains produce instabilities when backlash is present. Control of a mechanical positioning system under the influence of both friction and backlash, has to be confronted adeptly. Because impact is involved, the phenomena that take place when velocities cross an element with backlash are very complex, but understanding and characterizing them is essential to the development of control methodologies.

(c) In contemporary design of mechanical positioning systems, the mechanical system is normally considered to be rigid, or to contain isolated compliance such as explicit springs. Mechanical structures, however, are high-order systems that typically contain relatively little internal damping. As performance demands are pushed, the acceleration is increased, and the bandwidths of the controller and actuator are increased. Excitation of the high-order modes becomes inevitable and represents a limit on performance that is not predicted from low-order analysis. Methods are required for the analysis of such systems and for the design of the mechanical system in ways that will minimize the parasitic interactions between the controller and the structure, and improved control algorithms are needed for robust high-performance control, even when some residual high-order behavior is present. Existing methods for compensating for compliance should also be tested with mechanical systems under various conditions depicting real time operating conditions, and improved for optimum performance.

A considerable amount of work has been done in the modeling, identification, and control of systems in the presence of dry friction. Raboniwicz (1951) studied the transition from static to kinetic friction between stationary metal surfaces. Raboniwicz (1956, 1958) investigated stick-slip friction and also explored the intrinsic variables affecting stick-slip friction. Dahl (1976, 1977) studied and modeled pre-sliding displacement. Friction models for numerical simulation studies have been developed by Karnop (1985) and Kolston (1988). Armstrong (1988) developed ways to experimentally determine and model kinetic and viscous friction in a brush d.c. servo-motor-driven mechanism with gearing. Tustin (1947) examined the effect of backlash and non-linear friction on feedback control using a vector graphic technique similar to the modern describing function technique. Tou and Schultheiss (1953) studied the impact of static and dynamic friction on control using describing function analysis. Townsend and Salisbury (1987) studied the effects of dry friction on force control with integral feedback and also used the describing function technique for analysis. Townsend and Salisbury show that Coulomb friction could extend the system stability bounds but may lead to an input-dependent stability.

A range of control methodologies have been developed to compensate for the nonlinear effects of Coulomb friction. Gilbart and Winston (1974) developed adaptive compensation schemes to compensate for bearing friction in optical tracking telescopes. Canudas, et al. (1986), developed adaptive friction compensation for d.c. motor drives. Yang and Tomizuka (1987) developed an adaptive, pulse-width control scheme for position control of an X-Y table under the influence of stiction and Coulomb friction. Kubo, et al. (1986), Gogoussis and Donath (1987), Armstrong (1988), Craig (1988) are some examples of research efforts in robotics applications to compensate for Coulomb friction. Southward, et al. (1991) developed a nonlinear compensation force for stick-slip friction to supplement a proportional-plus-derivative control law applied to a one-degree-of-freedom mechanical system.

Researchers have also been investigating the effects of compliance on robust control. Forrest-Barlach and Babcock (1987) studied the effects of drive-train compliance and actuator dynamics in a two-degree-of-freedom manipulator and developed a position controller based on inverse dynamics. Ghorbel, et al.(1989) and Lozano, et al. (1992) presented adaptive control schemes for flexible joint manipulators.

Work has been done on the compensation techniques for backlash. As mentioned earlier, Tustin (1947) examined the effects of backlash and nonlinear friction on feedback control using a vector graphic technique similar to the describing function technique. Liversedge (1952) and Chesnut and Mayer (1955) studied the effects of backlash on closed-loop control systems and in regulating-system design. Freeman (1958) studied the effects of speed-dependent friction and backlash on system control, and Freeman (1960) investigated stabilization of control systems with backlash using a high-frequency, on-off loop.

Most of these compensation techniques were developed typically for each particular effect, friction, backlash, or compliance, without investigating the degrading effects of the other parasitic effects simultaneously (which is often the case) on system performance. Real-time tests on various mechanical positioning systems of existing compensation techniques, or new methods to determine their effectiveness are also of importance. This mechanical positioning test apparatus and associated control algorithm development hardware/software provide the tools so that various candidate control methodologies can be tried out under well-defined conditions. The objective of this research test bed is to be able to design, develop, test, and validate mechanical-positioning-system control strategies from simulations and experimental tests so that a high degree of confidence can be established in transferring the strategies to operational, mechanical-positioning systems.

2. DESIGN OF EXPERIMENTAL APPARATUS

The mechanical-system-control test apparatus (Figure 1) has been designed and built to optimize research in the design and control of multibody systems where Coulomb friction, gear backlash, and joint compliance are present. It is a multi-purpose research tool which can also be used in areas such as the design of smart sensors/actuators and control of smart structures. It consists of a spindle on which the mechanical system to be tested and controlled is mounted, e.g., a composite, flexible beam with embedded smart sensors/actuators. The mounting spindle is connected to a direct-drive servomotor. Shaft compliance, backlash (transmission hysteresis), mass moment of inertia, and Coulomb friction may be adjusted in order to examine their effects on control of the mechanical system. Sensing and control of the system is achieved by a set of incremental shaft encoders connected to an IBM PC/AT computer equipped with analog-to-digital and digital-to-analog converters. A detailed description of the apparatus follows.

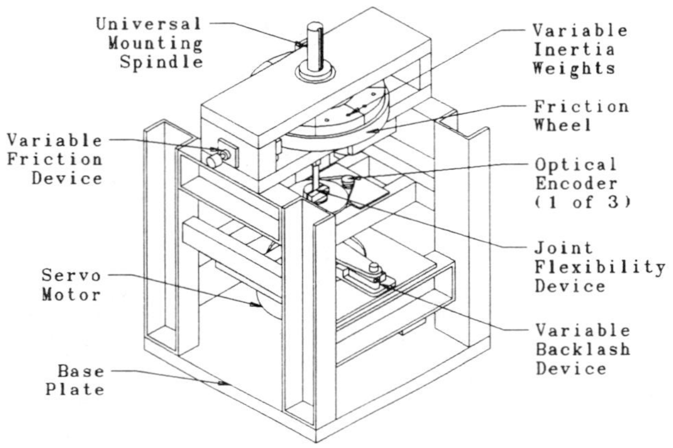

Figure 1: Single Axis Servomechanism

Mounting Spindle: The mounting spindle is a 1.0" diameter by 3.0" long vertical shaft with a 0.25" key-way. Structures mounted to the spindle are allowed a full 360 degrees of rotation. All the rotating components of the apparatus, including the mounting spindle, are supported by lightly-oiled ball bearings that introduce negligible friction into the system.

Physical Size and Mounting Requirements: The apparatus is relatively compact in size measuring only 15.0" square by 21.5" tall. The frame can be soft or hard mounted to a rigid structure or a moving base with four 3/8" diameter bolts.

Variable Friction: Adjustable Coulomb friction is achieved by a pivoted friction shoe made of nylon pressed against the outer edge of a 9.0" diameter aluminum flywheel. The constant normal force on the friction shoe is provided by a coil spring that can be adjusted in length. The normal force can be adjusted from 0 to 15 lbs. This translates to a frictional counter-torque of 0 to 20.25 in-lbs. for a frictional coefficient of 0.30 (nylon on polished aluminum).

Variable Mass Moment of Inertia: There is a set of eight steel wedges which can be bolted to the aluminum flywheel. Each wedge has a mass moment of inertia of 0.0151 lb-sec-in, which is roughly 7% of the inertia of the servomechanism itself.

Variable Shaft Compliance: A steel torsional spring with a rectangular cross section is rigidly attached to the mounting spindle. The free end of this spring can be clamped anywhere along its 2.0" length to give a compliance (reciprocal of stiffness) range of 0 to 2.32 mrad/in-lb. Higher or lower torsional compliances may be achieved by using different spring materials.

Variable Backlash: The variable backlash device consists of a compound lever arrangement which makes fine adjustments to a gap that is 5.0 inches out from the servomechanism axis of rotation. The device can be adjusted in 0.04 mrad increments up to a maximum of 4.0 mrad backlash.

Servomotor: A PMI U16M4 servomotor powered by a model AXA-180-10-20 DC servo amplifier (1800 watt maximum continuous power output) is used as a direct drive for the mechanical system. The motor is capable of producing a 15.5 ft-lb peak torque and it has a rated power output of 1.0 hp. This motor was chosen for the manipulator because of its low armature inertia, negligible electrical inductance, and low time constant.

Angular Position and Velocity Sensing: Three geared-up optical encoders with quadrature output are used to sense angular position and velocity. The resolution achieved by each encoder is 36000 pulses per spindle revolution or 0.01 degrees. As seen in Figure 2, an encoder is connected to the drive motor, to the backlash device, and to the mounting spindle. The quadrature output from the encoders is used by a four-channel quadrature card connected to the data bus of an IBM PC/AT computer. The quadrature card supplies angular position and velocity information to the computer.

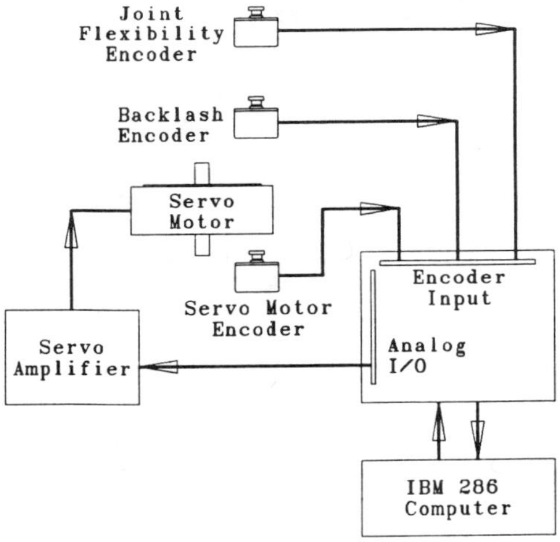

Figure 2: Servomechanism Control System

Control System: Additional analog sensors (accelerometers, strain gages, etc.) can be connected to any of the eight A/D converters linked to the computer. Control schemes for the manipulator are programmed into the computer using the *C* programming language. Digital output signals from the computer are changed to analog voltages by any of the three D/A converters. The power amp for the PMI servomotor is connected to D/A channel #1. A simplified schematic of the servomechanism control system is shown in Figure 2.

3. MATHEMATICAL MODELING, SYSTEM IDENTIFICATION, AND PARAMETER ESTIMATION OF EXPERIMENTAL APPARATUS

The equations of motion representing the dynamics of the experimental system based on a lumped-parameter model, as well as methods used for system identification and parameter estimation, are presented in this section. A free-body diagram of all of the rotating components of the servomechanism and definitions of the dynamic terms are shown in Table 1.

Table 1: Table of Parameter Values

Inertias	**J**	**0.2078 in.lb.sec^2**
	$J_0 + J_1$	**.0408 in.lb.sec^2**
	J_2	**.01670 in.lb.sec^2**
Vis.friction coeff.	**B**	**0.063 in.lb/rad/sec**
	B_1	**0.057 in.lb/rad/sec**
	B_2	**0.0066 in.lb/rad/sec**
Amplifier Const.	**K_i**	**2.0 amps/volt**
Motor Torque Const.	**K_t**	**1.818 in.lb/amp**
Spring Stiffness	**K**	**40-855.5 in.lb/rad**
Coulomb friction	**T_c**	**0-20.25 in.lb**
Backlash width	**b=H**	**0-.040 rad**

3.1 Mathematical Modeling

3.1.1 Basic System

The equations of motion for the basic system with no Coulomb friction, backlash, or compliance, is given by

$$J\ddot{\theta} + B\dot{\theta} = T_m \qquad (1)$$

where J represents the sum of the inertias (Figure 3), Θ the angle of rotation for the three lumped inertias, B the viscous friction coefficient , and T_m the motor torque. A block diagram of the basic system is shown in Figure 4. K_i and K_t are the amplifier gain and motor torque constants.

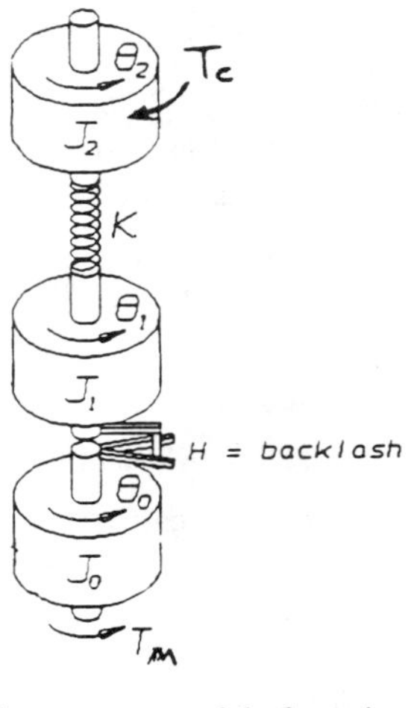

Figure 3: Lumped parameter model of rotating parts of system

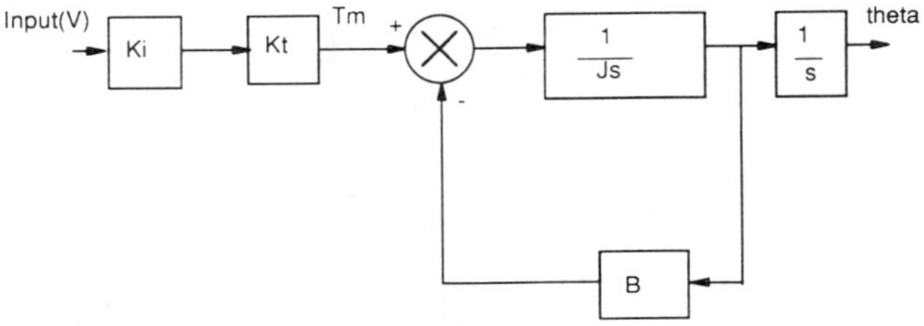

Figure 4: Block Diagram of basic system

3.1.2 System with Coulomb Friction

In the presence of Coulomb friction, Equation 1 becomes

$$J\ddot{\theta} + B\dot{\theta} = T_m - T_c \mathrm{sgn}(\dot{\theta}) \qquad (2)$$

where T_c is the motor torque required to overcome the Coulomb friction , and

$$\mathrm{sgn}(\dot{\theta}) = \begin{cases} +1 & \dot{\theta} > 0 \\ -1 & \dot{\theta} < 0 \end{cases}$$

The static friction T_s is assumed to decay to the dynamic friction T_c in the low velocity region [5]. The friction model and the block diagram of the system in the presence of nonlinear friction are also shown in Figure 5.

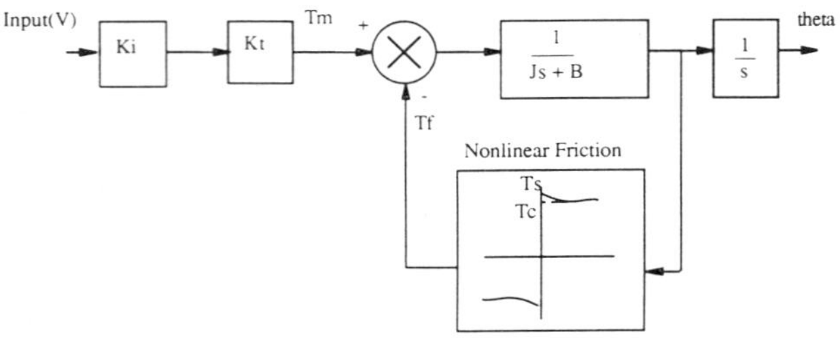

Figure 5: Friction Model and Block Diagram of System with Coulomb Friction

3.1.3 System with Backlash

The closed-loop block diagram of the system in the presence of backlash between the motor and the load shaft of Figure 2 is shown in Figure 6. Θ_{ref} is the reference position, Θ_b the actual output position and Θ the output which will exist with no backlash. The equation of motion is given by,

$$J\ddot{\theta} + B\dot{\theta} = T_m K(\theta_{ref} - \theta_b) \qquad (3)$$

where Θ_b is the output of the backlash block [2],

$$\theta_b = \theta - \frac{b}{2}\mathrm{sgn}(\dot{\theta}) \qquad (4)$$

and b is the backlash width.

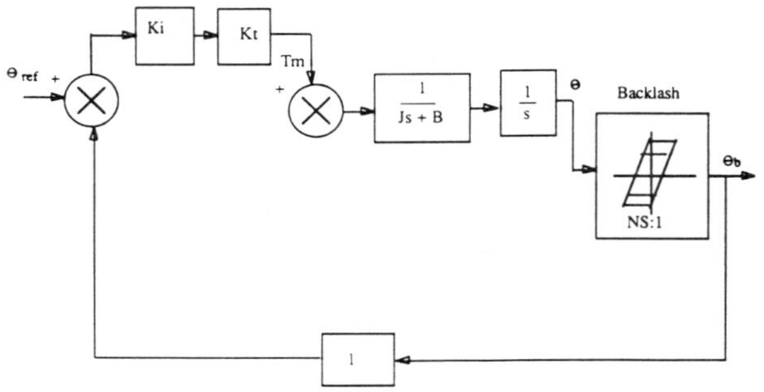

Figure 6: Closed-loop Block Diagram of System with Backlash

3.1.4 System with Compliance

In the presence of torsional compliance the governing equations of motion of the system model shown in Figure 3 is given by

$$J_1\ddot{\theta}_1 + B_1(\dot{\theta}_1 - \dot{\theta}_2) + K(\theta_1 - \theta_2) = T_m \qquad (5)$$

$$J_2\ddot{\theta}_1 + B_2(\dot{\theta}_2 - \dot{\theta}_1) + K(\theta_2 - \theta_1) = 0 \qquad (6)$$

where $J_1 = J_0 + J_1$, $\Theta_0 = \Theta_1$, B_1 and B_2 are the viscous friction coefficients of the bearings associated with J_1 and J_2 respectively.

3.2 System Identification Parameter Estimation

Non-parametric time and frequency-domain methods and parametric system identification techniques were used to estimate and verify linear system parameters based on the mathematical modeling of the system.

Time-domain step responses and bode plots of the basic system were used to verify modeling the relationship between the input torque and the output velocity and between the input torque and the output position of the physical system, to be first-order and second-order, respectively. The apriori knowledge of the model structure from the non-parametric tests was used as the underlying model structure for the parametric system identification tests.

The standard ARMAX model structure ,

$$(1 + a_1z^{-1} + a_2z^{-2} + \dots a_{na}z^{-na})\, y_{k+1} = (b_0 + b_1z^{-1} + b_2z^{-2} + \dots b_{nb-1}z^{-nb+1})u_k + (1 + c_1z^{-1} + c_2z^{-2} + \dots c_{nc}z^{-nc})v_{k+1} \qquad (7)$$

which is equivalent to

$$A(z^{-1})y_{k+1} = B(z^{-1})u_k + C(z^{-1})v_{k+1} \qquad (8)$$

where A, B, C are polynomials with degree na, nb-1, and nc, respectively, and v_k is the noise sequence, was primarily used for identification. Output data measured from the optical encoders for a white noise input after signal conditioning was used as input-output data for standard multiple regression recursive algorithms. The Approximate Maximum Likelihood (AML) recursive algorithm (Matrix_x application software), with the ARMAX model structure, identified the transfer function of the test bed in the polynomial form (equation 7).

The discrete transfer function of the system was obtained using z-transform techniques, and the corresponding s-domain transfer function was obtained using discrete to continuous time transformations. Figure 7 shows a plot of the superimposed plots of the velocity output signals from the test bed and the AML model for a 1 volt-biased white noise input. Shown in Figure 8 is the corresponding position response. The velocity and position responses of the system model compared very well with the physical system output.

The total inertia value J obtained was 0.2078 in-lb/sec^2. This agreed with the value from theoretical calculations within eight percent. The estimated viscous friction coefficient was 0.063 in-lb-sec/rad. Frequency domain analysis of the characteristics of the amplifier verified modeling its dynamics as a constant gain as recommended by the manufacturer. K_i, the amplifier gain, was estimated to be 2 amps/volt and K_t, the motor torque constant, 1.818 in-lb/amp. An approximate value of the static friction value for the motor was obtained from calculations of the minimum torque required to begin system motion with minimum input voltage.

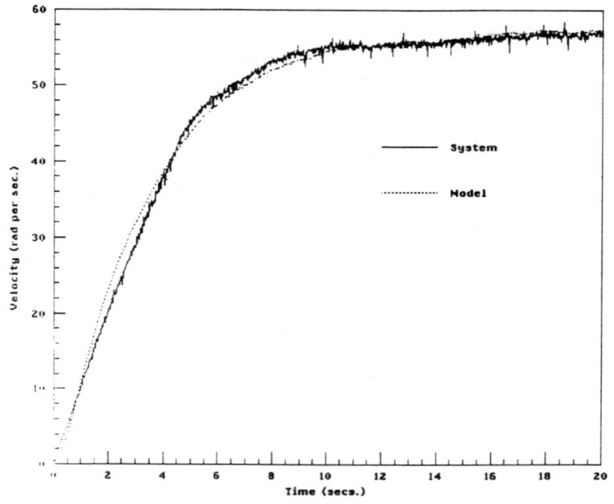

Figure 7: Plots of system/model velocity output.

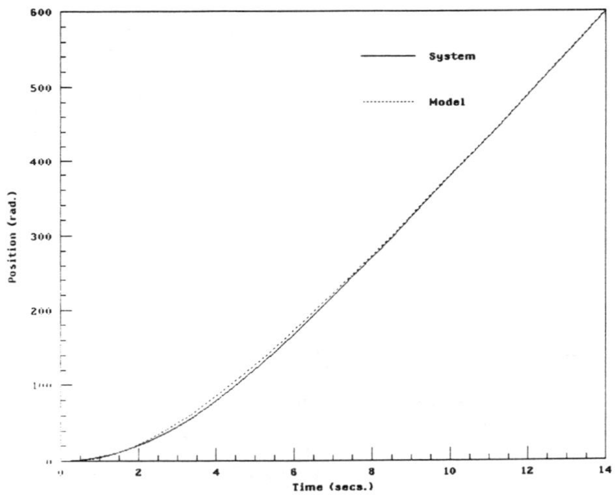

Figure 8: Superimposed Plots of system/model position output.

Estimates of the quantified values of the effects of Coulomb friction, backlash, joint compliance, and inertia, which can be applied to the system (models 3.1.2-3.1.3) are obtained as described in Section Two. Table lists the parameter values.

4. EFFECTS OF COULOMB FRICTION, BACKLASH AND JOINT COMPLIANCE ON A CONVENTIONAL CONTROLLER

This section presents simulations of the system equations developed in Section 3 with the estimated parameter values shown in Table 1. A conventional controller was designed to obtain optimum response of the basic system to a step input . The effects on the performance of the basic system with this controller when friction, backlash, and compliance are introduced, were determined.

A simulation diagram of the basic system using a linear-state-feedback PD controller is shown in Figure 9.

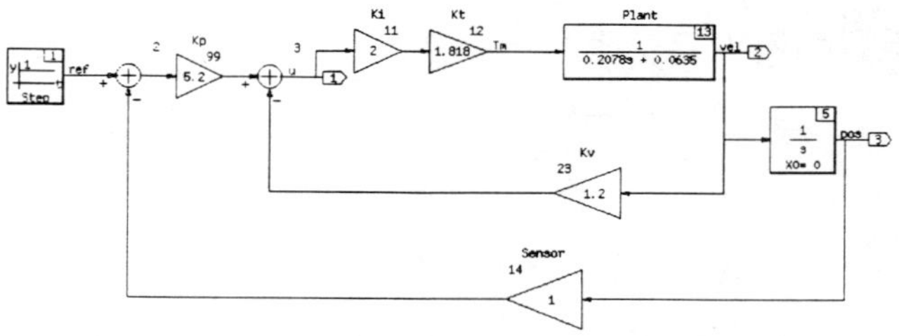

Figure 9: Simulation diagram of basic system

The proportional and derivative control gains K_p (5.2 volt/rad) and K_v (1.2 volt/rad/s), respectively, were chosen to ensure stability and good system performance. The control gains were also chosen so that the manipulating signals were below the saturation limits of the D/A converter of the test-bed. As can be seen in Figure 10, the step response of the basic system in the time domain has desirable dynamic performance essential for a typical position servo-controlled mechanical system. A fast system response with no overshoot and no steady-state error was obtained.

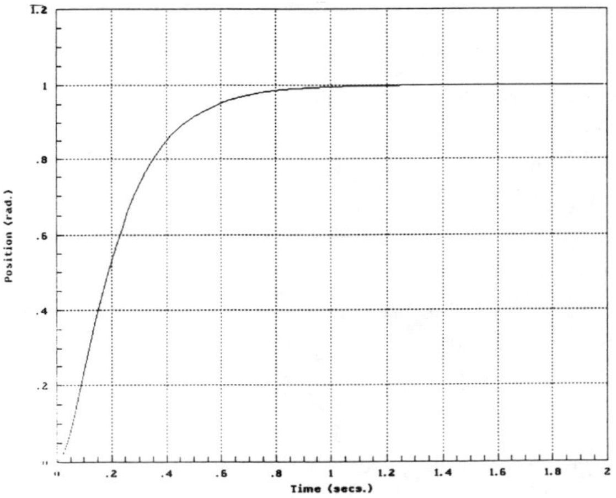

Figure 10: Step response of basic system

The simulation diagram when friction is introduced to the system (based on the mathematical model in section 3.1.2) is shown in Figure 11. The value of T_c is 1.4 in-lb.

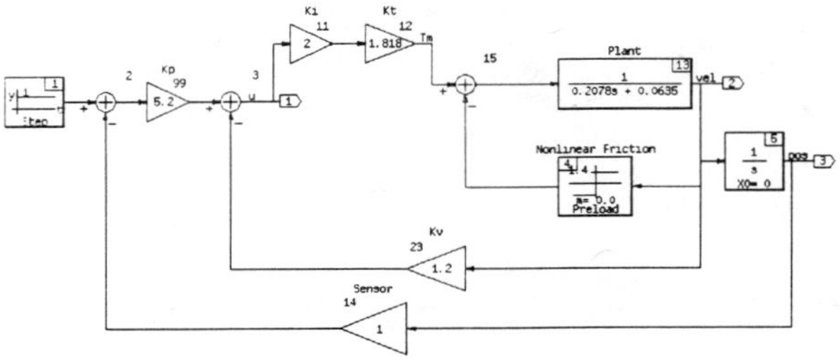

Figure 11: Simulation diagram of System with Coulomb Friction

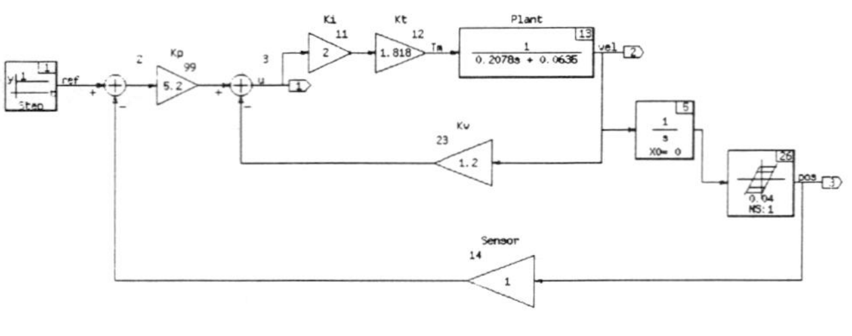

Figure 12: Simulation diagram of System with backlash (b=.04 rad)

Figure 12 shows the simulation diagram when backlash is introduced to the system.

The superimposed plots of the effects of Coulomb friction, backlash, and compliance on the basic system is presented in Figure 13. The step response plots of the velocity output from the different system models are also shown in Figure 14.

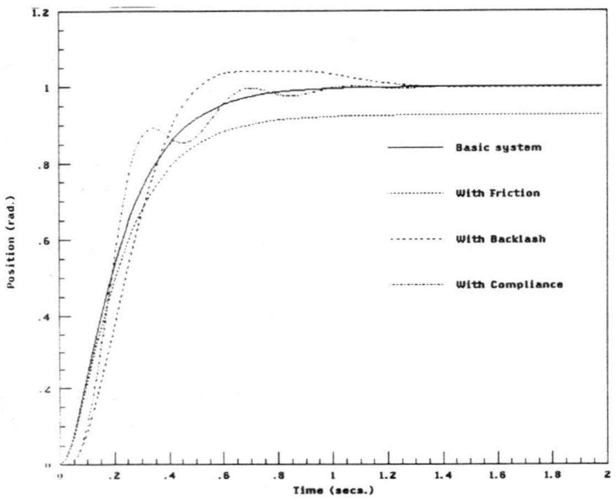

Figure 13: Superimposed position step response plots of models developed.

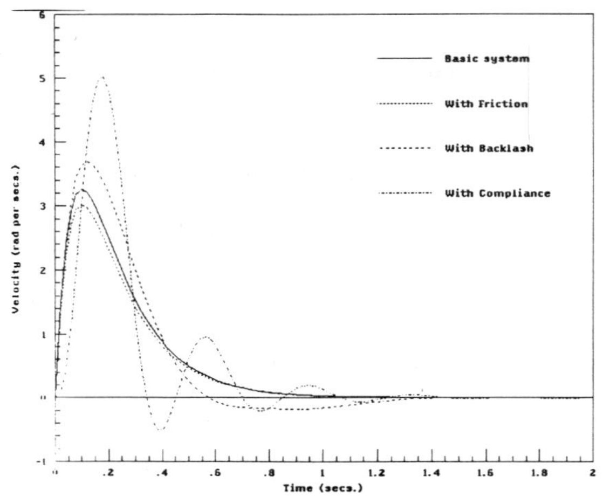

Figure 14: Velocity step response plots

The degradation of the system step response due to the nonlinear effects of Coulomb friction and backlash, and the effect of joint compliance can be observed from Figures 13 and 14. Coulomb friction introduced steady state error at the final position. Consequently, the maximum velocity is also reduced compared to the basic system. Incorporation of integral action to compensate for the steady-state error due to Coulomb friction could lead to limit cycling [16]. In the presence of backlash there is about 5% overshoot at the steady-state position compared to the system with no backlash. The velocity output is also much increased. Changes in instantaneous peak velocities, if not well accounted for in a system, could cause saturation of the controller, or with time, damage the plant structure. Joint compliance introduced oscillatory response behavior during the transient state of the position and velocity step response. This is undesirable for proper control/structure interaction. Shown in Figure 15 is the position response due to the combined effects of Coulomb friction and joint compliance. Combined effects of oscillation in the transient state due to compliance and steady state error due to Coulomb friction is observed in the simulation.

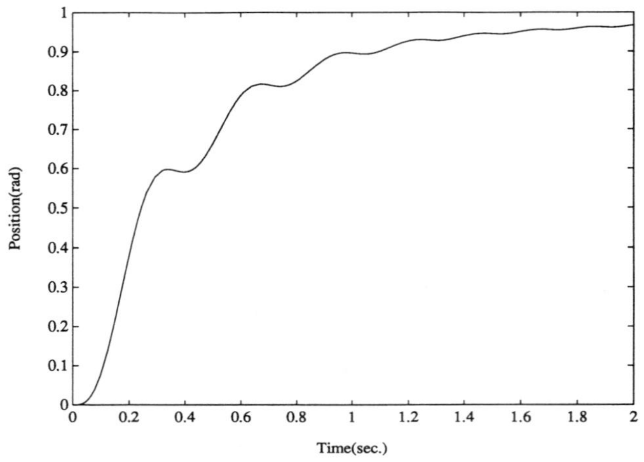

Figure 15: Position Response under the combined effects of Coulomb friction and joint compliance

The effects of Coulomb friction, backlash, and joint compliance on the position control of the experimental test bed using the conventional controller was also determined. This was compared to the observed trends in the simulations. The proportional and derivative gains, K_p and K_v, respectively, were chosen based on the simulations. A sampling frequency of 50 Hz was used. Shown in Figure 16 is the superimposed position response plots of the system for a step input of 1 rad with Coulomb friction, backlash, and joint compliance introduced. The Coulomb friction levels were about 1.4 in-lb. The amount of backlash was about .04 radians, and the stiffness K for joint compliance about 50 in-lb/rad. Trends similar to the simulation results were observed in the experimental results. Due to the limited amount of sampled data, the experimental response plots are not as smooth in comparison to the simulation plots.

Steady state error due to the effect of Coulomb friction on the step response of the test bed with the PD controller was observed (Figure 16). In the presence of backlash, the system not only demonstrated an increase in amplitude but also exhibited slight sustained oscillations (limit cycling) which was not easily observed from the simulation. Generally, if the magnitude of backlash is small compared with the accuracy limit of the system, any limit cycling occuring due to backlash in a dynamic system may be insignificant if it is stable. However, under more stringent performance specifications for precise positioning, the effect of backlash on system control becomes more important. In the presence of compliance, there is the characteristic oscillatory response of the system.

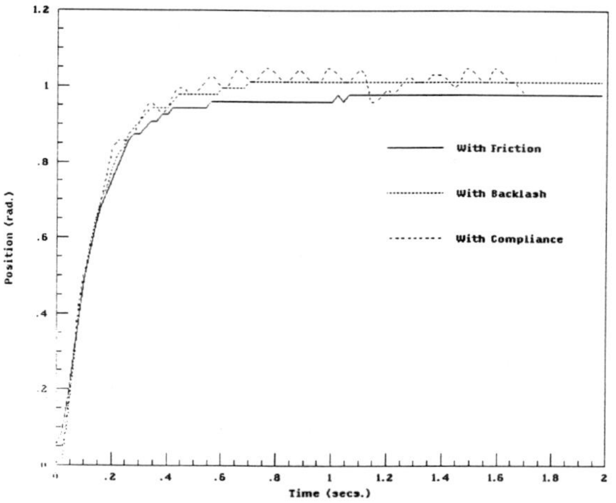

Figure 16: Real-time output responses in the presence of Coulomb Friction, backlash and joint compliance under PD control.

5. *CONTROL STRATEGIES*

Candidate control strategies to compensate for nonlinear friction, backlash, and joint compliance were investigated and a number implemented to test their effectiveness. The improvement of the performance of mechanical positioning systems from various developed compensation methods [4],[12],[13],[16],[18],[21],[28], are well acknowledged. It is however, important to know the advantages and short comings of control strategies to ensure that reliable compensation techniques are chosen or developed, and that they are well adapted to meet system performance specifications under various operating conditions. The main goal is to combine some of these strategies and others which may be developed, to improve position control of the servomechanism under the influence of the various adverse effects of Coulomb friction, backlash and joint compliance and their combined influence.

Some of the control schemes were tested by simulations and experiments. Most of the adaptive compensation schemes tend to be more reliable especially if the system parameters are not quite known . Under the influence of other adverse conditions it is prudent to ensure that parameter estimation techniques are effective. The present goal was to adapt effective control methods of minimal sophistication to control the servo-mechanism under the various 'adverse' conditions.

Nonlinear friction compensation, similar to methods in [18] and [29] to complement the linear state feedback PD controller in Section 4, was adapted and tested on the servo mechanism. The motor torque output from the control input was effectively

$$T_m = K_i K_t(-K_p(\Delta\theta) - K_v\dot{\theta}) + \hat{T}_c x$$

where

$$x = \begin{cases} \operatorname{sgn}(\dot{\theta}) & \text{for } \dot{\theta} \neq 0 \\ \operatorname{sgn}(u) & \text{for } \dot{\theta} = 0 \end{cases}$$

and u the direction of control input. The nonlinear gain

$$\hat{T}_c = K_i K_t K_c$$

where K_i and K_t are the amplifier and motor torque constants. Approximate values of friction were obtained from the calibration setup of the test bed and verified from break away torque measurements to aid in the choice of K_c. The degradation effect of Coulomb friction shown in Section 4 is compensated for using this technique. Shown in Figure (17) are results from the real-time implementation. A stable response with no steady state error was obtained.

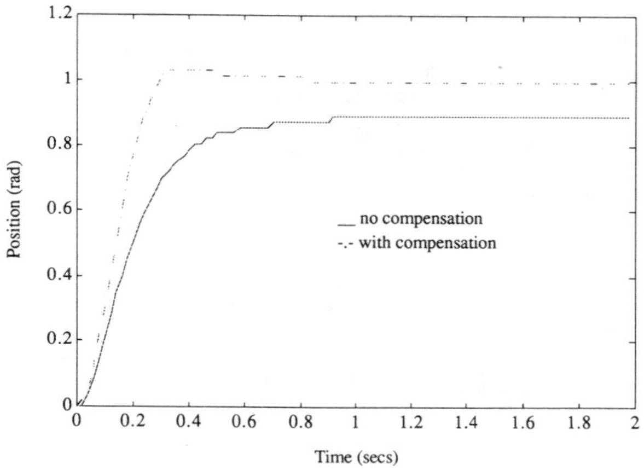

Figure 17: Real time response for Non-linear feedback compensation of friction

Dither control complementing the PD controller was also implemented to compensate for the degrading effects of Coulomb friction and backlash for the systems shown in Section 4. Dither compensation smoothing technique has been shown [6] to be effective for Coulomb friction compensation. The potential drawbacks of dither control are that the motion does not stop completely, more power is needed, and mechanical parts tend to wear more rapidly. In addition, if the dynamics of the system does not suppress the frequency of the dither signal and if it excites potential limit cycle frequencies, dithering could drive the system to instability. The control gains and frequencies were chosen to alleviate the problems associated with dither control as discussed above. In real-time implementation, once the reference position was attained the dither input was turned off. If the controller needed tuning, it was turned on, to minimize power consumption.

Typically for dither compensation [6], consider a nonlinear block connected to a linear system: w(t) is the dither signal and e is the error signal in Figure [18].

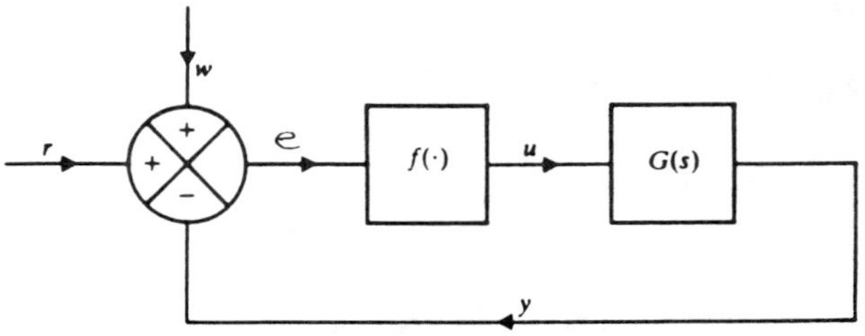

Figure 18: Introduction of dither signal

Then from the block diagram,

$$u = f(e + w)$$

and with a sinusoidal dither,

$$w(t) = W\sin(2\pi t / T_o)$$

where T_o is chosen so that the angular frequency $(2\pi t / T_o)$ is much greater than any frequency component of G(s); the dither signal introduces 'averaging' of the nonlinearity. The effective value of u being given by,

$$\hat{u} = \frac{1}{2\pi}\int_0^{2\pi} f\;(e + W\sin\theta)\,d\theta$$

The general 'averaging' effect of the dither signal results in the approximation of the nonlinearity to a linear form. Shown in Figures (19) and (20) are the simulation and real time response of friction compensation using dithering. An improved stable position response is obtained.

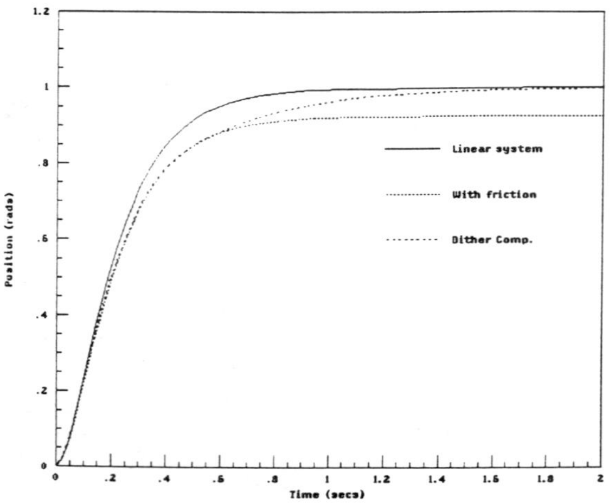

Figure 19: Simulation of friction compensation with dither.

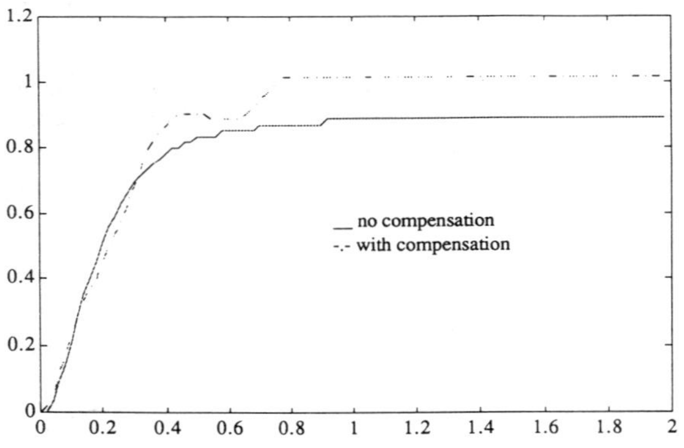

Figure 20: Real time friction compensation with dither

Shown in Figures (21) and (22) are the response plots using dither for backlash compensation. Impacts as a result of backlash were reduced substantially and there is much improved transient and steady state performance.

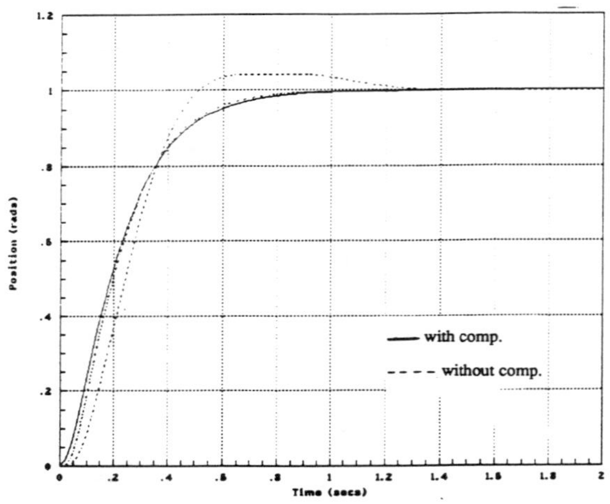

Figure 21: Simulation of backlash compensation with dither

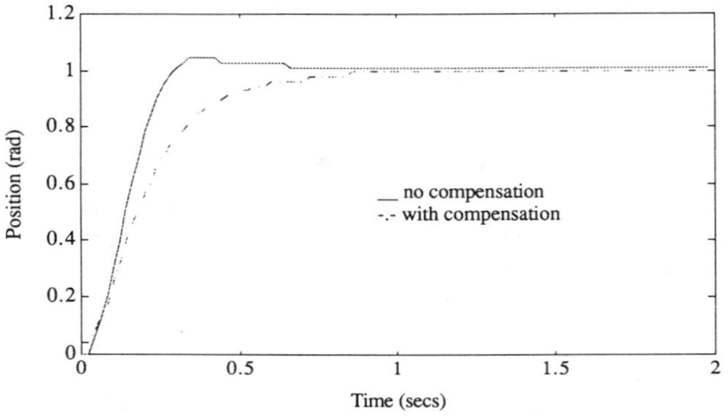

Figure 22: Real time back-lash compensation with dither

The standard linear-quadratic regulator design method, where the optimal gain matrix K is calculated so that the feedback law,

$$u = -Kx$$

minimizes the cost function,

$$J = \int (x'Qx + u'Ru)\,dt$$

for the state-equation representing the dynamics of the test-bed,

$$\dot{x} = Ax + Bu$$

was designed to compensate for the joint compliance effect. Shown in Figure (23) is a simulation result using this method. Much more improved stable response was obtained.

The nonlinear feed-back compensation method used for friction has been successfully combined with LQ compensation method for compliance to eliminate combined effect of friction and compliance. The simulation result is shown in Figure 24.

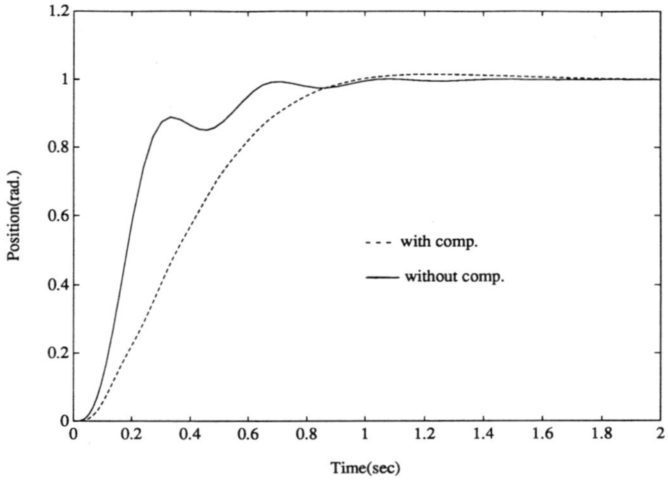

Figure 23: LQ optimal compensation for compliance

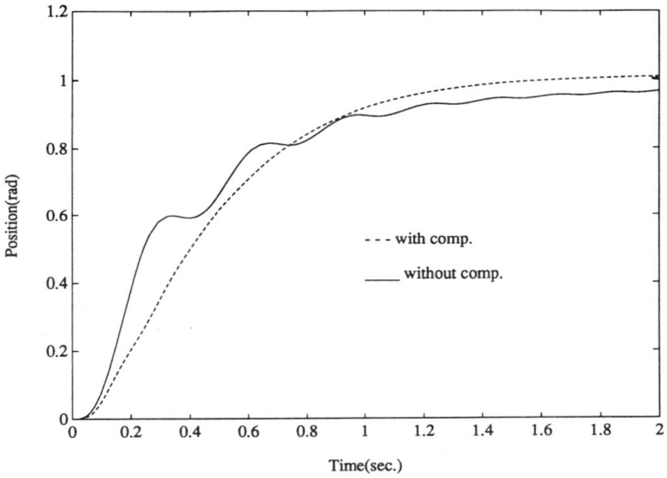

Figure 24: Combined compliance and friction compensation

6. CONCLUSIONS AND FUTURE WORK

An experimental test bed for investigating the effects of mechanical system nonlinearities, such as Coulomb friction and backlash, joint flexibility, and structural flexibility has been presented. The capabilities which can be exhibited by the test-bed make it a valuable research tool for dynamic system modeling and control studies. This has been shown through simulations and experiments. Theoretical calculations from dynamic analysis, system identification, and parameter estimation techniques were used to obtain very good system parameter values essential for simulations and for developing control algorithms. Simulations and experimental results both showed degrading effects on control system performance due to these adverse effeects. In the analysis and development of control/structure interaction techniques for optimum system performance, it is important to critically consider such adverse effects.

Control methods were adapted for the servo-mechanism, and improvement in system response was observed under the influence of Coulomb friction, backlash and joint compliance. Simulations of combined strategies for friction compensation and joint compliance showed improved performance. The real time implemented results agreed well with the results predicted by simulation.

Control algorithms are being developed for on-line identification and control of the degrading effects of friction, backlash, and joint compliance shown in Sections 4 and 5 and the effects due to their interrelations. Future work also involves mounting of a flexible composite beam on the test bed, with an embedded smart material, and developing robust control algorithms for position control of such end-effectors in the presence of adverse conditions such as the effects mentioned in this paper.

References

[1] Armstrong, B,"Friction:Experimental Determination, Modeling and Compensation," IEEE International Conference on Robotics and Automation,vol. 3, pp. 1422-1427, 1988.

[2] Atherton , Non-linear Control Engineering, Van Nostrand Reinhold, 1982.

[3] Brogan, W.L., Modern Control Theory , Prentice-Hall Inc. Englewood Cliffs N.J., 1985.

[4] Canudas,C., Astrom,K., and Braun K., "Adaptive friction compensation in dc motor drives", Proceedings of the 1986 international Conference on Robotics and Automation, pp. 1556-1561, IEEE, San Francisco, CA., April 1986.

[5] Chesnut,H., aand Mayer, R.W., "Servomechanisms and Regulating System Design", Chapman and Hall, Vol. 2,p.301, 1955.

[6] Cook, P.A., "Nonlinear Dynamical Systems", Prentice-Hall, 1986.

[7] Cosgriff, R. L., Nonlinear Control Systems, McGraw-Hill, 1958.

[8] Craig, J., "Adaptive Control of Mechanical Manipulators", Addison- Wesley, Reading, MA, 1988.

[9]Dahl, P., "Solid friction damping of mechanical vibrations", AIAA Journal,14, 1675-1682, December 1976.

[10]Dahl ,P., "Measurement of solid friction parameters of ball bearings", Proceedings of the 6th Annual Symposium on Incremental Motion, Control Systems and Devices, pp. 49-60, University of Illinois,1977.

[11] Feliu, V., Rattan, K.S., and Brown,Jr. H.B.,"Adaptive Control Of a Single-Link Flexible Manipulator In the Presence of Joint Friction and Load Changes,IEEE International Conference of Robotics and Automation, Scottsdale, pp 1036 - 1041, 1989.

[12] Forrest-Barlach, M.G., Babcock, S.M., "Inverse Dynamics Position Control of a Compliant Manipulator," Journal of Robotics and Automation, vol RA-3 No-1 pp. 75-83, 1987.

[13] Freeman, E. A., "The effect of speed dependent friction and backlash on the stability of automatic control systems," Trans. A.I.E.E., 77(II), pp. 680-691 , 1958.

[14] Freeman, E. A., " The stabilization of control systems with backlash using a high frequency on -off loop", Proc. IEEE, 107 C, pp. 150-157 , 1960.

[15] Ghorbel,Hung and Spong, "Adaptive Control of Flexible Joint Manipulators," IEEE Control Systems Magazine, vol.9,no.7,pp.9-13,Dec.1989.

[16] Gilbart, J.W. and Winston, G. C., "Adaptive Compensation for an Optical Tracking Telescope," Automatica, Vol.10., pp. 125-131, 1974.

[17] Gogouissis, A., and Donath, M., "Coulomb Friction Joint and Drive Effects in Robot in Robot Mechanisms," IEEE International Conference on Robotics and Automation, Vol. 2,pp. 828-836, 1987.

[18] Kubo, T., Anwar, G., and. Tomizuka M., "Application of Non-linear Friction Compensation to Robot Arm Control," Proceedings of IEEE International conference on Robotics and Automation, San Fransisco, pp. 722-727, 1986.

[19] Liversedge, J. H., " Backlash and Resilience within the Closed Loop of Automatic Control Systems", fro, Tustin, A. (Ed), "Automatic and Manual Control", Butterworths Scientific Publications, p.343, 1952.
[20] Ljung L., "System Identification: Theory for the User," Prentice-Hall Inc. Englewood Cliffs N.J., 1987.
[21] Lozano, R. and Brogliato, B., "Adaptive Control of Robot Manipulators with Flexible Joints," IEEE Transactions on Automatic Control, vol.37, no.2, pp.174-181, Feb 1992.
[22] Meirovitch, L.,"Methods of Analytical Dynamics", McGraw Hill, 1988.
[23] Ogata K., Modern Control Engineering, Prentice-Hall, Inc. Englewood Cliffs. N.J., 1990.
[24] Rabinowicz, E., "The Nature of the Static and Kinetic Coefficients of Friction," Journal of Applied Physics, Vol.22, No. 11, November,pp. 1373-1379, 1951.
[25] Rabinowicz, E., "Stick and slip", Scientific America, 194(5),109-118, 1956.
[26]Rabinowicz,E., "The intrinsic variables affecting the stick-slip process", Proceedings of the Physical Society of London, pp. 668-675, 1958 Volume 71.
[27]Rabinowicz, E., Friction and Wear of Materials, John Wiley and Sons, 1965.
[28]Shen,C. and Wang, H., "Nonlinear compensation of a second and third-order system with dry friction, IEEE Transactions on Applications and INdustry, 83, 128-136, 1964.
[29] Southward S.C.,Radcliff C.J. and MacCluer C.R., "Robust Nonlinear Stick-Slip Friction Compensation,"Journal of Dynamic Systems, Measurement, and Control,Vol. 113, pp.639 -645, 1991.
[30] Townsend, W. T., and Salisbury, J. K., 1987, "The Effect of Coulomb Friction and Stiction on Force Control," IEEE International Conference on Robotics and Automation, Vol. 2, pp. 883-889, 1987.
[31] Tou J., and Schultheiss, P.M.,"Static and Sliding Friction in Feedback Systems," Journal of Applied Physics, Vol. 24 No. 9. pp. 1210-1217, 1953.
[32] Tustin, A., "The effects of Backlash and of Speed-Dependent Friction on the Stability of Closed-Cycle Control Systems," Journal of the Institute of Electrical Engineers, Vol. 94(2A), pp. 143-151,1947.
[33] Yang S., Tomizuka M., "Adaptive Pulse Width Control for Precise Positioning Under Influence of Stiction and Coulomb Friction," IEEE American Control Conference, Vol 1, pp. 188-193, 1987.

AMD-Vol. 141/DSC-Vol. 37, Dynamics of Flexible Multibody Systems:
Theory and Experiment
ASME 1992

A STUDY OF THE APPLICATION OF ADAPTIVE DECENTRALIZED CONTROL TO A MULTIBODY SPACE STRUCTURE

George T. Flowers and Madheswaran Manikkam
Department of Mechanical Engineering
Auburn University
Auburn, Alabama

Vipperla B. Venkayya
Wright Laboratory (WL/FIBRA)
Wright-Patterson AFB, Ohio

Abstract

Adaptive strategies are a topic of current interest for the decentralized control of flexible, multibody structural systems. A decentralized adaptive strategy has been proposed by Gavel and Siljak that considers a system to consist of several known subsystems with unknown interconnections. The current work investigates the application of this technique to the vibration control of a two body flexible truss structure. Simulation studies were performed to illustrate the implementation and performance of this strategy for reference model tracking for a two body truss structure. The results of these studies are presented and discussed.

Nomenclature

A_i = disconnected linear state matrix for i^{th} subsystem
A_{mi} = decentrally stabilized linear state matrix for i^{th} subsystem
B_i = input matrix for i^{th} subsystem
$\bar{d}_i$ = interconnection function for i^{th} subsystem
$\bar{e}_i$ = error vector for i^{th} subsystem $(\bar{x}_i - \bar{x}_{mi})$
$\bar{f}_1$, $\bar{f}_2$ = interconnection forces
G_i = positive definite n_i x n_i matrix
H_i = solution matrix for Lyapunov equation
i = subsystem index
J_i = linear local feedback control law for i^{th} subsystem
l_i = number of outputs for subsystem i
m_i = number of interconnection forces i
n_i = number of states for subsystem i
N = number of subsystems
P_i = interconnection matrix for subsystem i
p_i = number of control inputs for subsystem i
Q_i = output matrix
q_{ij} = j^{th} generalized coordinate for flexible vibration of subsystem i
R_i = positive definite p_i x p_i matrix
$\bar{x}_i$ = state vector for i^{th} subsystem
$\bar{x}_{mi}$ = desired state vector for i^{th} subsystem
$\bar{u}_i$ = local control input acting on subsystem i
$\bar{v}_i$ = interconnection vector for subsystem i
Z_i = adaptive feedback control law for i^{th} subsystem
α_i = adaptive gain for i^{th} subsystem
θ_1, θ_2 = angular position of trusses 1 and 2 respectively
γ_i, σ_i = scalar parameters for adaptive controller for i^{th} subsystem
$\dot{()}$ = differentiation with respect to time
$[\,]$ = matrix quantity
$\bar{(\,)}$ = indicates a vector quantity
$(\,)_{nom}$ = indicates a nominal quantity use for linearized analyses

Introduction

Typically, control strategies are developed assuming that the controller has access to all the states of the system (or at least their estimates). For many large scale systems, such a formulation is not practical. Adequate control may require too many states for it to be practical to deal with all simultaneously or the system may be geographically extended such that communication of states back to a central controller is prohibitively expensive and involves excessive time delays. The solution to this problem has been to develop decentralized control techniques that are based only upon local states which are convenient subsets of the total system states. This field has advanced rapidly in recent years. Such techniques have been shown to be effective for many large scale systems, including power distribution networks, telecommunication networks, and large, flexible structures.

There has been a great deal of recent interest in the decentralized control of flexible structures. For decentralized control design, a flexible structure can be modelled as a collection of flexible appendages attached to a flexible support structure. Knowledge of the components (support structure and appendages) is typically fairly complete and can usually be modelled quite adequately using linear mode shapes and frequencies. However, knowledge of the interconnection dynamics is typically not so complete. The interconnections may be nonlinear (a pinned interconnection with friction, for example) and a great deal of uncertainty may exist with regard to adequate models (both with regard to functional form and parametric configuration). Therefore, it is natural to address the issue of decentralized control for a flexible structure in terms of substructures subjected to external disturbances due to uncertain interconnection forces.

The current work presents a simulation study of a two body space structure using adaptive decentralized control. The structure shown in Figure 1 is used as the simulation test bed. It consists of a two body system, each body consisting of a flexible truss structure. The model is meant to simulate a space platform with a tracking antenna. Kinematical nonlinearities are retained in the model. The goal of the current work is to assess the applicability of an adaptive decentralized controller design developed by Gavel and Siljak (1989) for the motion control of this structure. The controller algorithm and the results of the computer simulation are presented below.

Previous Work

An area of current rapid development is the field of adaptive control theory. Researchers in the area of decentralized control have recently begun to investigate and apply adaptive techniques. Since such methods typically result in lower closed loop feedback gains, they appear to offer significantly improved noise rejection characteristics as compared to fixed–gain strategies. A number of investigators have made important contributions to this field. Among them are Hmamed and Radouane (1983), Ioannou and Kokotović (1983), Ioannou and Reed (1988), and Siljak (1991). Much of this work has been directed toward using model reference adaptive controllers (MRAC) for the decentralized control of unknown subsystems as if they were disconnected from each other. Gavel and Siljak (1989) describe an adaptive method for designing a decentralized control strategy for a system with known subsystems that will be stable regardless of the magnitude of the interconnection forces. In order to understand this approach, let us first start with the following mathematical description for an interconnected dynamical system.

The dynamics of the interconnected subsystems may be described as:

$$\dot{\bar{x}}_i = A_i\bar{x}_i + B_i\bar{u}_i + P_i\bar{v}_i \tag{1}$$

$$\bar{y}_i = C_i\bar{x}_i \qquad i \in N$$

$$\bar{w}_i = Q_i\bar{x}_i$$

where $\bar{x}_i(t)$, $\bar{u}_i(t)$, $\bar{y}_i(t)$ are the state input and output of subsystem i, and $\bar{v}_i(t)$ and $\bar{w}_i(t)$ are the interconnection inputs and outputs of subsystem i. Each pair (A_i, B_i) and (A_i, C_i) are respectively controllable and observable.

For the purposes of this development it is necessary to formulate some boundedness relation for the unknown interconnection forces. Toward this end, it is assumed that the interconnection inputs and outputs, $\bar{v}_i$ and $\bar{w}_i$, are related as

$$\bar{v}_i = \bar{d}_i(t, w) \tag{2}$$

In addition, $\bar{d}_i(t,w)$ must satisfy the conic sector bounds

$$\| \bar{d}_i(t,w) \| \leq \sum_{j=1}^{N} \xi_{ij} \| \bar{w}_j \|, \ i \in N \tag{3}$$

Note that it is not necessary that the numbers ξ_{ij} be known, but it is required that they be fixed values.

It is also necessary to impose the requirement that the matrices P_i can be factored as

$$P_i = B_i P_i' \qquad i \in N \tag{4}$$

for some constant matrices P_i'. This is essentially a requirement that the interconnection forces be directly reachable by the actuator forces. Again, it is not necessary that the matrices P_i' be known.

The control objective is for each of the closed loop subsystems to track stable isolated reference models. A typical approach for selecting appropriate reference models consists of first selecting the local feedback gain matrices, J_i, to stabilize the local isolated subsystems. This results in state matrices for the closed loops subsystems of the form:

$$A_{m,i} = A_i - B_i J_i, \quad i \in N \tag{5}$$

The resulting reference models are governed by the relations:

$$\dot{\bar{x}}_{mi} = A_{mi}\bar{x}_{mi} + B_i\bar{r}_i$$

$$\bar{y}_{mi} = C_i\bar{x}_{mi} \qquad i \in N \tag{6}$$

where $\bar{r}_i$ are bounded, piecewise continuous vector functions of time.

An adaptive local control law can be formulated in the following manner. First, define

$$Z_i = R_i^{-1} B_i^T H_i, \quad i \in N \tag{7}$$

where H_i is the symmetric positive definite solution to the matrix Lyapunov equation

$$A_{mi}^T H_i + H_i A_{mi} = -G_i, \quad i \in N \tag{8}$$

G_i is any symmetric, positive definite n_i x n_i matrix. A solution to (8) is guaranteed to exist since A_{mi} is stable.

The resulting feedback control law is then

$$\bar{u}_i = -J_i\bar{x}_i - \alpha_i(t)Z_i\bar{e}_i + \bar{r}_i, \quad i \in N \tag{9}$$

where $\alpha_i(t)$ is a time variable scalar adaptation gain at time $t > t_o$ and $\bar{e}_i(t) = \bar{x}_i(t) - \bar{x}_{mi}(t)$. The adaptation law for $\alpha_i(t)$ is chosen as:

$$\dot{\alpha}_i = \gamma_i \bar{e}_i^T Z_i^T R_i Z_i \bar{e}_i - \sigma_i \alpha_i, \; \alpha_i(t_0) > 0, \; i \in N \qquad (10)$$

where γ_i and σ_i are given positive scalars.

Gavel and Siljak (1989) demonstrated this adaptive feedback scheme and showed that the resulting behavior is globally bounded. A properly designed adaptive controller of the form represented above can be made to be stable for any value of interconnection coupling strength.

Current Work

Of concern in the present study is the application of the controller strategy described above to the vibration suppression of a large flexible space structure. In order to examine the issues relevant to the implementation of this strategy and the performance of the resulting controller, a simulation study was performed. The details of this study are presented below.

Simulation Model

The simulation model consists of a two body flexible truss structure as shown in Figure 1. The structure is comprised of a boom truss and an antenna. The base support is fixed. The node locations are shown in Figure 2. The material for each structure is steel. The area of each spar is 1.0 in^2.

The equations of motion for this model were developed based upon the assumption that each body can be modelled as a linearly elastic structure. The resulting equations of motion are of the following form.

$$M_1\{\ddot{\bar{z}}_1\} + [(N_1)(N_2)]\{(\ddot{\bar{z}}_1)^T(\ddot{\bar{z}}_2)^T\}^T + K_1\{\bar{z}_1\} + \Delta_1\{\bar{f}_1\} = E_{11}\{\bar{u}_1\} + E_{12}\{\bar{u}_2\} \qquad (11)$$

$$\{\bar{z}_1\} \equiv (q_{11} \;\; . \;\; . \;\; q_{1k} \;\; \theta_1)^T$$

$$M_2\{\ddot{\bar{z}}_2\} + N_2^T\{\ddot{\bar{z}}_1\} + K_2\{\bar{z}_2\} + \Delta_2\{\bar{f}_2\} = E_{22}\{\bar{u}_2\} \qquad (12)$$

$$\{\bar{z}_2\} \equiv (q_{21} \;\; . \;\; . \;\; q_{2k} \;\; \theta_2)^T$$

where M_1, M_2, N_1, N_2, E_1, E_2, $\bar{f}_1$, and $\bar{f}_2$ are nonlinear functions of the states.

We select

$$\bar{r}_1 = \begin{pmatrix} 0 \\ 0 \\ 1.0\cos(10t) \end{pmatrix} \qquad (13)$$

$$\bar{r}_2 = \begin{pmatrix} 0 \\ 0 \\ 1.0\cos(10t) \end{pmatrix} \qquad (14)$$

as the reference signals.

The elastic effects are modelled using the first two bending modes for each substructure. The effects of the higher elastic modes on the dynamics of the reduced order system can be considered to be included in the (uncertain) interconnection forces, $\bar{f}_1$ and $\bar{f}_2$. The model is fully documented in Flowers and Venkayya (1992).

Controller Implementation

The decentralized adaptive controller strategy discussed above was implemented for the two body flexible structure model. The objective is to investigate the effectiveness of this approach for reference model tracking for a flexible multibody system.

For the purpose of controlling the vibration of each substructure, it is first necessary to decide on appropriate locations for actuator placement. In the current study, actuators will be applied as shown in Figure 1. The actuator forces are applied (as couples) at the nodes with bold arrows. The numbers associated with each pair of arrows indicates the corresponding row of the controller input vector. The directions of the actuator forces are indicated by the directions of the arrows in relation to the associated substructure.

For the dynamic simulation, the full nonlinear models are used. However, the linearized decoupled state equations are used for design purposes. It is important to note that the decoupled subsystems are nonlinear, but are not time dependent. The nonlinear effects occur in nonlinear functions of the flexural vibrations or angular velocities. The nonlinearities in the respective M_i matrices occur only in functions of the flexural vibration amplitudes. Since it is desirable to maintain these quantities at low levels, if possible, it is clear that these expressions can be linearized about a nominal state of zero. As a result, the linearized decoupled systems have constant coefficients regardless of the angular position of the substructure.

The linearized, decoupled state equations are of the form:

$$\begin{pmatrix} \ddot{\bar{z}}_1 \\ \dot{\bar{z}}_1 \end{pmatrix} = \begin{pmatrix} 0 & -(M_1^{-1}K_1)_{\rm nom} \\ I & 0 \end{pmatrix} \begin{pmatrix} \dot{\bar{z}}_1 \\ \bar{z}_1 \end{pmatrix} + \begin{pmatrix} (M_1^{-1}E_{11})_{\rm nom} \\ 0 \end{pmatrix} \begin{pmatrix} \bar{u}_1 \\ 0 \end{pmatrix} \qquad (15)$$

$$\begin{pmatrix} \ddot{\bar{z}}_2 \\ \dot{\bar{z}}_2 \end{pmatrix} = \begin{pmatrix} 0 & -(M_2^{-1}K_2)_{\rm nom} \\ I & 0 \end{pmatrix} \begin{pmatrix} \dot{\bar{z}}_2 \\ \bar{z}_2 \end{pmatrix} + \begin{pmatrix} (M_2^{-1}E_{22})_{\rm nom} \\ 0 \end{pmatrix} \begin{pmatrix} \bar{u}_2 \\ 0 \end{pmatrix} \qquad (16)$$

For the systems of (15) and (16), it is should be noted that the B_1 and B_2 matrices are of the form

$$(B_1)_{\rm nom} = ((\hat{B}_1)_{\rm nom}^T \;\; [0])^T$$

$$(B_2)_{\rm nom} = ((\hat{B}_2)_{\rm nom}^T \;\; [0])^T.$$

where

$$(\hat{B}_1)_{\rm nom} = (M_1^{-1}E_{11})_{\rm nom}$$

$$(\hat{B}_2)_{\rm nom} = (M_2^{-1}E_{22})_{\rm nom}$$

For this study, we choose:

$$(\hat{B}_1)_{\rm nom} = \begin{pmatrix} -0.596 & 0.0 & 1.21 \\ 0.0 & -0.125 & 0.0 \\ 0.0009 & 0.0 & 1.00 \end{pmatrix}$$

$$(\hat{B}_2)_{nom} = \begin{pmatrix} 0.131 & 0.0 & -5.11 \\ 0.0 & -0.0584 & 0.0 \\ 0.0097 & 0.0 & 1.00 \end{pmatrix}$$

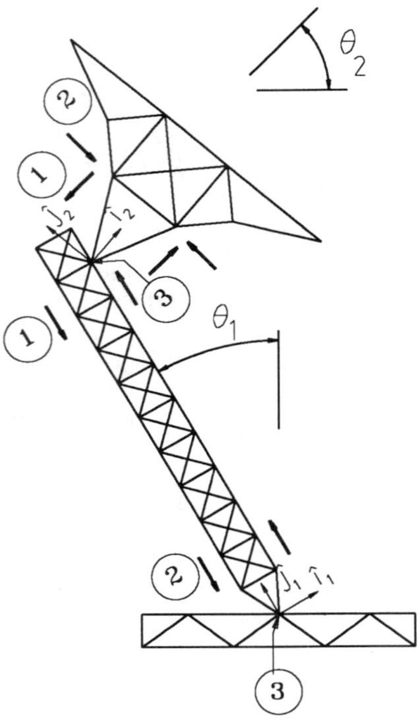

Figure 1 Two body truss structure

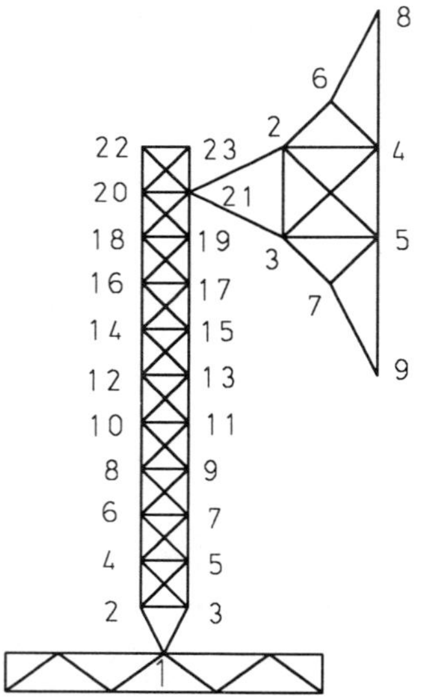

Figure 2 Node locations for two body truss structure

Linear Controller Design

The controller design consist of a linear part that stabilizes the local, disconnected subsystems and the nonlinear, adaptive controller. For the purposes of this study, the linear controllers are optimal controllers selected using the standard LQR procedure. Each truss structure is considered as a separate substructure. The resulting control laws are:

$$J_1 = \begin{pmatrix} -1.55 & 0.0 & 1.53 & -0.521 & 0.0 & 0.562 \\ 0.0 & -1.33 & 0.0 & 0.0 & -0.049 & 0.0 \\ 0.594 & 0.0 & 1.61 & 0.596 & 0.0 & 0.827 \end{pmatrix}$$

$$J_2 = \begin{pmatrix} 0.108 & 0.0 & 0.503 & -0.0678 & 0.0 & 0.166 \\ 0.0 & -1.25 & 0.0 & 0.0 & -0.162 & 0.0 \\ -0.621 & 0.0 & 3.39 & 1.40 & 0.0 & 0.986 \end{pmatrix}$$

Adaptive Controller Design

Two adaptive controller designs will be considered. For both cases,

$$R_1 = R_2 = \begin{pmatrix} 1 \times 10^{-3} & 0 & 0 \\ 0 & 1 \times 10^{-4} & 0 \\ 0 & 0 & 1 \end{pmatrix}$$

For the first case,

$$G_1 = G_2 = \begin{pmatrix} 1 & 0 & 0 & 0 & 0 & 0 \\ 0 & 1 & 0 & 0 & 0 & 0 \\ 0 & 0 & 1 & 0 & 0 & 0 \\ 0 & 0 & 0 & 1 & 0 & 0 \\ 0 & 0 & 0 & 0 & 1 & 0 \\ 0 & 0 & 0 & 0 & 0 & 1 \end{pmatrix}$$

The matrices that result are:

$$H_1 = \begin{pmatrix} 1.00 & 0.0 & -0.901 & 0.621 & 0.0 & -0.513 \\ 0.0 & 5.32 & 0.0 & 0.0 & 0.390 & 0.0 \\ -0.901 & 0.0 & 1.53 & -0.463 & 0.0 & 1.02 \\ 0.621 & 0.0 & -0.463 & 1.47 & 0.0 & -0.311 \\ 0.0 & 0.390 & 0.0 & 0.0 & 6.900 & 0.0 \\ -0.513 & 0.0 & 1.02 & -0.311 & 0.0 & 1.77 \end{pmatrix}$$

$$H_2 = \begin{pmatrix} 0.573 & 0.0 & 2.69 & -0.349 & 0.0 & 0.895 \\ 0.0 & 10.7 & 0.0 & 0.0 & 0.278 & 0.0 \\ 2.69 & 0.0 & 15.2 & -2.52 & 0.0 & 5.02 \\ -0.349 & 0.0 & -2.52 & 1.69 & 0.0 & -0.486 \\ 0.0 & 0.280 & 0.0 & 0.0 & 19.2 & 0.0 \\ 0.895 & 0.0 & 5.02 & -0.486 & 0.0 & 3.32 \end{pmatrix}$$

For the second case,

$$G_1 = G_2 = \begin{pmatrix} 5 & 0 & 0 & 0 & 0 & 0 \\ 0 & 5 & 0 & 0 & 0 & 0 \\ 0 & 0 & 1 & 0 & 0 & 0 \\ 0 & 0 & 0 & 1 & 0 & 0 \\ 0 & 0 & 0 & 0 & 1 & 0 \\ 0 & 0 & 0 & 0 & 0 & 1 \end{pmatrix}$$

The matrices that result are:

$$H_1 = \begin{pmatrix} 2.26 & 0.0 & -1.34 & 0.425 & 0.0 & -0.785 \\ 0.0 & 17.3 & 0.0 & 0.0 & 0.390 & 0.0 \\ -1.34 & 0.0 & 1.95 & -0.0521 & 0.0 & 1.24 \\ 0.425 & 0.0 & -0.0521 & 2.70 & 0.0 & -0.15 \\ 0.0 & 0.390 & 0.0 & 0.0 & 22.3 & 0.0 \\ -0.785 & 0.0 & 1.24 & -0.15 & 0.0 & 1.89 \end{pmatrix}$$

$$H_2 = \begin{pmatrix} 1.37 & 0.0 & 5.24 & -1.39 & 0.0 & 1.63 \\ 0.0 & 38.1 & 0.0 & 0.0 & 0.278 & 0.0 \\ 5.24 & 0.0 & 29.2 & -8.95 & 0.0 & 8.71 \\ -1.39 & 0.0 & -8.95 & 5.82 & 0.0 & -1.94 \\ 0.0 & 0.278 & 0.0 & 0.0 & 68.6 & 0.0 \\ 1.63 & 0.0 & 8.71 & -1.94 & 0.0 & 4.40 \end{pmatrix}$$

Simulation Results

First, let us examine the dynamical behavior of this system without the adaptive controller. Figures 3 and 4 present the time responses for the flexible and rigid body motion of the boom truss and the antenna with the adaptive controller parameters, γ_i and σ_i, set to zero. The oscillations are clearly unstable. The system responses diverge very rapidly from those of the reference model. Note that each isolated substructure is stable. However, an eigenanalysis of a linearized model of the complete system with just the linear controller indicates that it is unstable. This effect is due to the impact of the interconnections on the system behavior. The dynamics of the antenna acting through the interconnection (at boom truss node 21) on the dynamics of the boom truss and vice–versa serve to cause unstable oscillations.

Figures 5 and 6 present the time responses for the flexible and rigid body motion of the boom truss and the antenna for the indicated parameters. These responses correspond to the first case of the adaptive controller described in the previous section. For the boom truss the ability of the actual system to track the desired behavior is acceptable for the both the rigid body motion and the flexible vibration. There are some discrepancies for the flexible vibrations, but those are relatively small. In addition, the rigid body motion and the first vibrational mode of the antenna are well controlled by the adaptive controller. However, the second vibrational mode of the antenna does not converge to the associated reference model behavior. Instead, there is an offset from the desired trajectory. This is a potential problem that is inherent in the controller design. The adaptive controller does not guarantee that the tracking error will tend to zero but only that it will be bounded.

Figures 7 and 8 present the time responses for the flexible and rigid body angular motion of the boom truss and the antenna for the indicated parameters. These responses are for the second case of the adaptive controller. For this case, the ability of the actual system to track the desired behavior is acceptable for both the rigid body motion and for both modes of the flexible vibration. In fact, the model tracking for the boom truss is almost exact from the starting time. The responses of the antenna are also quite good. So it can be seen that the design of an acceptable controller is very dependent on the selection of G_i.

The time responses shown in this study are for large initial conditions in order to better assess the effect of the nonlinearities. The tracking behavior that results is good for relatively small values of γ_i and σ_i. A consideration of this performance leads to the following concern. One potentially significant problem with the control strategy is the fact that the magnitude of the control gains depend on the square of the errors. If the states are small (as they typically are for structural vibration) then the adaptive gains increase relatively slowly for modest values of γ_i. For example if the state errors are of the order 10^{-3}, when squared these values are of the order 10^{-6}. A consideration of equation (9) indicates that it would take a considerable amount of time for the adaptive controller gains to grow appreciably for such behavior if the values of γ_i are moderate quantities. This results in the requirement of high values for the γ_i's for good performance for low amplitude vibration suppression in a structural system. Such high gains may serve to amplify noise and lead to excessive controller forces when the motion is of relatively high amplitude.

Conclusion

A study of the implementation and performance characteristics of an adaptive, decentralized control scheme for a two body flexible structure has been presented. This approach has considered each body to be a separate substructure. The primary source of interconnection forces is the inertial coupling between the two substructures acting through a pinned interconnection. The control strategy is shown to provide good reference model tracking for all states of the test structure.

Some of the more significant conclusions and suggestions that have been drawn from this study are as follows.

1. The proof that the control law represented by (7), (9), and (10) is such that the state tracking error, $\bar{e}_i$, is globally ultimately bounded [Gavel and Siljak (1989)] is performed using a Lyapunov approach. Such methods typically produces stability boundaries that are very conservative. Alternative, less conservative strategies that may be more optimal in terms of the distribution of actuator energy, should be investigated.
2. In order to obtain better small amplitude vibration suppression it would be of interest to investigate the use of alternate adaptive gain dynamics (a function of the absolute values of the e_i's for example) that would have more favorable response characteristics for small amplitude motion.
3. It is important to note that in the Gavel–Siljak algorithm it is possible to weight certain states using the R_i and G_i matrices (as was done in the example studies). However, it is required that these matrices be positive definite and symmetric. This requirement coupled with the requirements on H_i make the task of designing a specific overall weighting for certain states a cumbersome, trial and error procedure.

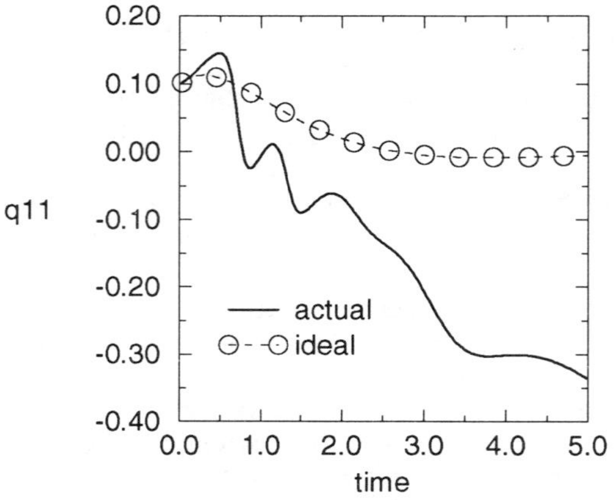

Figure 3.a Time response for boom truss.
($\gamma 1 = 0, \gamma 2 = 0, \sigma 1 = 0, \sigma 2 = 0$)

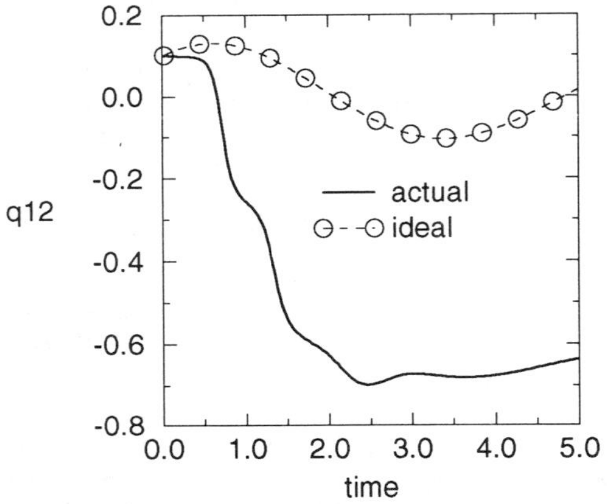

Figure 3.b Time response for boom truss.
($\gamma 1 = 0, \gamma 2 = 0, \sigma 1 = 0, \sigma 2 = 0$)

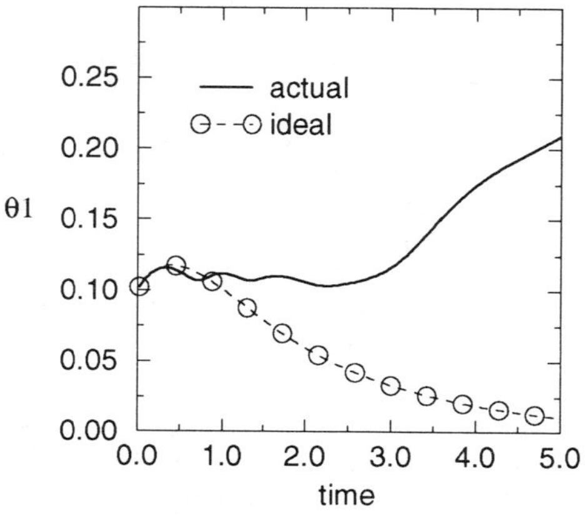

Figure 3.c Time response for boom truss.
($\gamma 1 = 0, \gamma 2 = 0, \sigma 1 = 0, \sigma 2 = 0$)

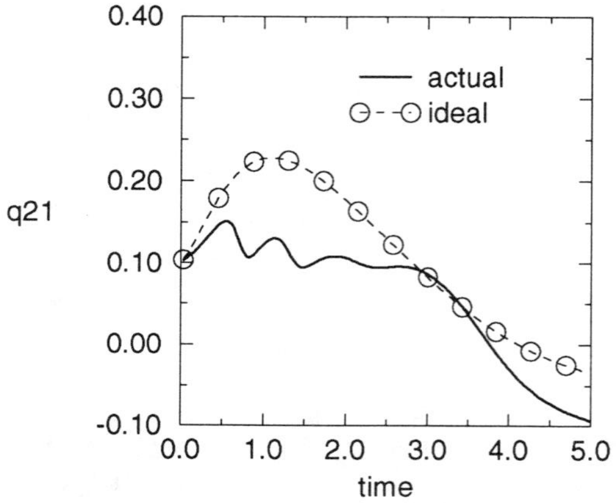

Figure 4.a Time respone for antenna.
($\gamma 1 = 0, \gamma 2 = 0, \sigma 1 = 0, \sigma 2 = 0$)

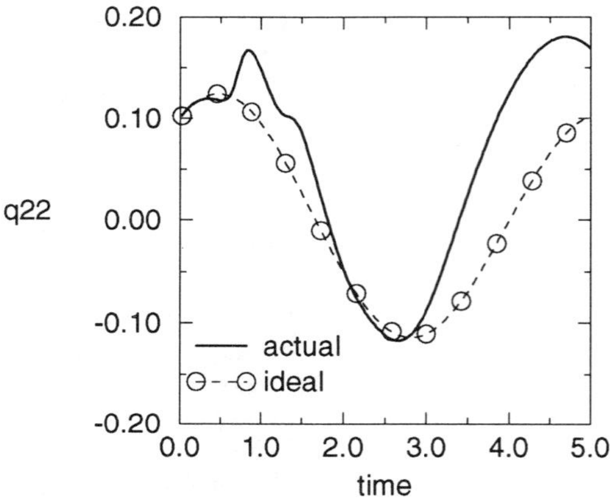

Figure 4.b Time response for antenna.
($\gamma 1 = 0, \gamma 2 = 0, \sigma 1 = 0, \sigma 2 = 0$)

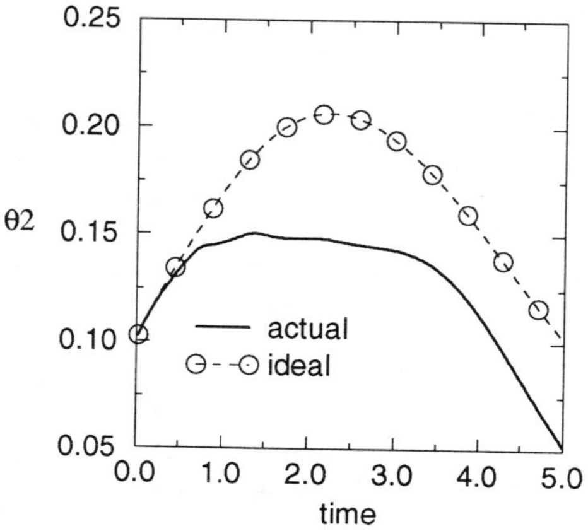

Figure 4.c Time response for antenna.
($\gamma 1 = 0, \gamma 2 = 0, \sigma 1 = 0, \sigma 2 = 0$)

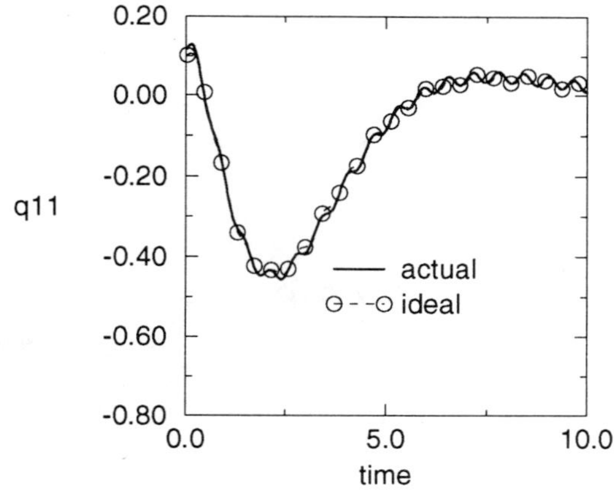

Figure 5.a Time response for boom truss for case 1 adaptive controller.
($\gamma 1 = 10, \gamma 2 = 10, \sigma 1 = 0.1, \sigma 2 = 0.1$)

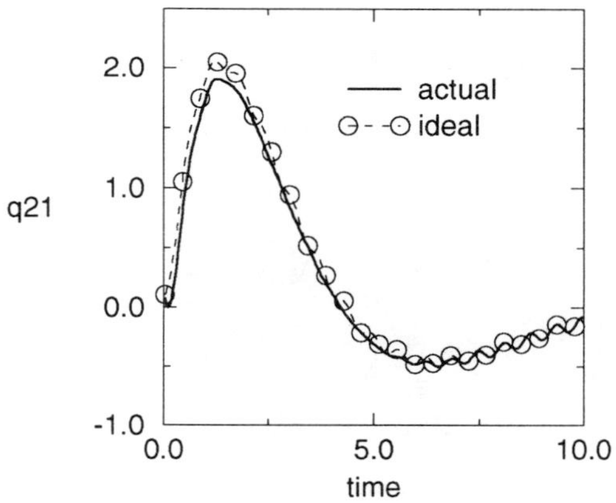

Figure 6.a Time response for antenna for case 1 adaptive controller.
($\gamma 1 = 10, \gamma 2 = 10, \sigma 1 = 0.1, \sigma 2 = 0.1$)

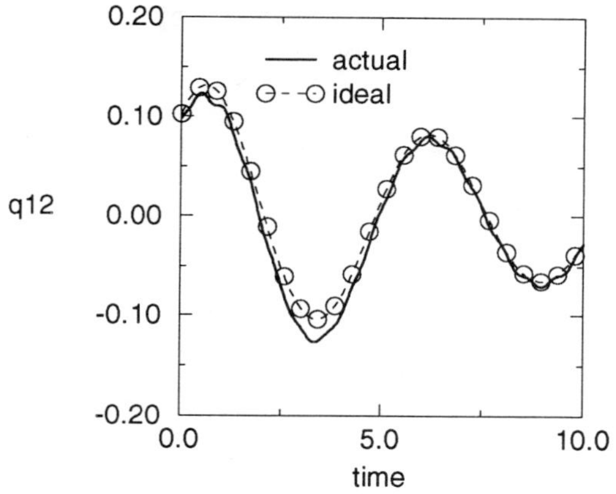

Figure 5.b Time response for boom truss for case 1 adaptive controller.
($\gamma 1 = 10, \gamma 2 = 10, \sigma 1 = 0.1, \sigma 2 = 0.1$)

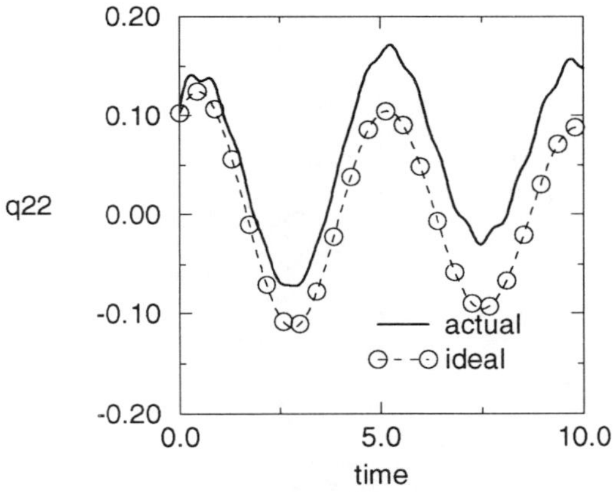

Figure 6.b Time respose for antenna for case 1 adaptive controller.
($\gamma 1 = 10, \gamma 2 = 10, \sigma 1 = 0.1, \sigma 2 = 0.1$)

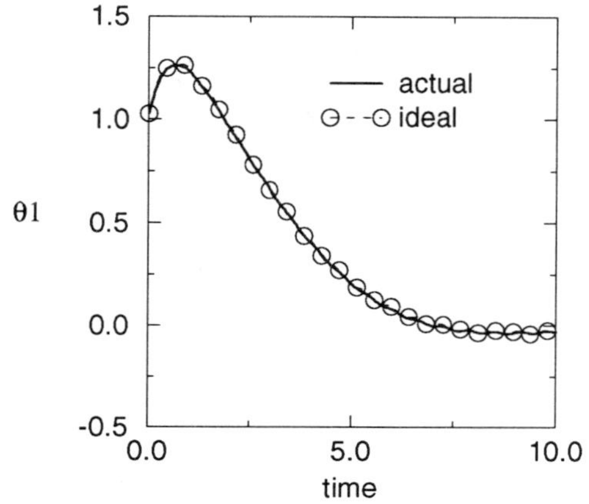

Figure 5.c Time response for boom truss for case 1 adaptive controller.
($\gamma 1 = 10, \gamma 2 = 10, \sigma 1 = 0.1, \sigma 2 = 0.1$)

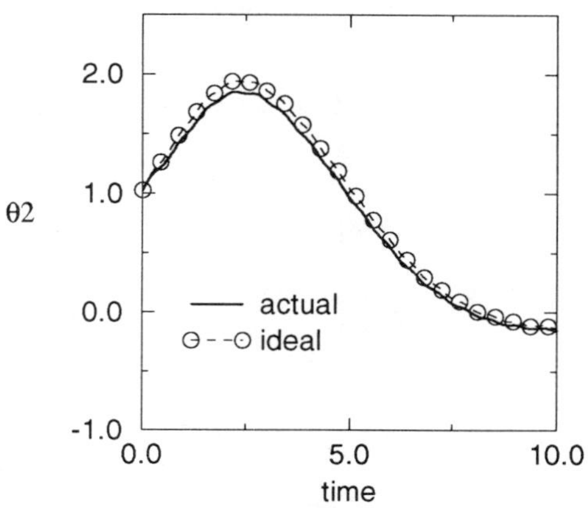

Figure 6.c Time response for antenna for case 1 adaptive controller.
($\gamma 1 = 10, \gamma 2 = 10, \sigma 1 = 0.1, \sigma 2 = 0.1$)

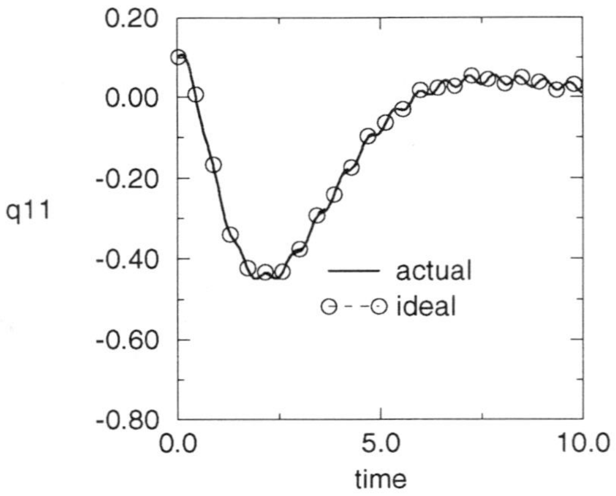

Figure 7.a Time response for boom truss for case 2 adaptive controller.

($\gamma 1 = 10, \gamma 2 = 10, \sigma 1 = 0.1, \sigma 2 = 0.1$)

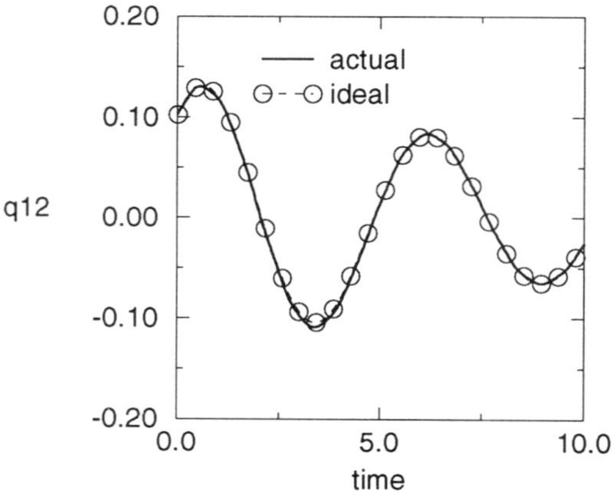

Figure 7.b Time response for boom truss for case 2 adaptive controller.

($\gamma 1 = 10, \gamma 2 = 10, \sigma 1 = 0.1, \sigma 2 = 0.1$)

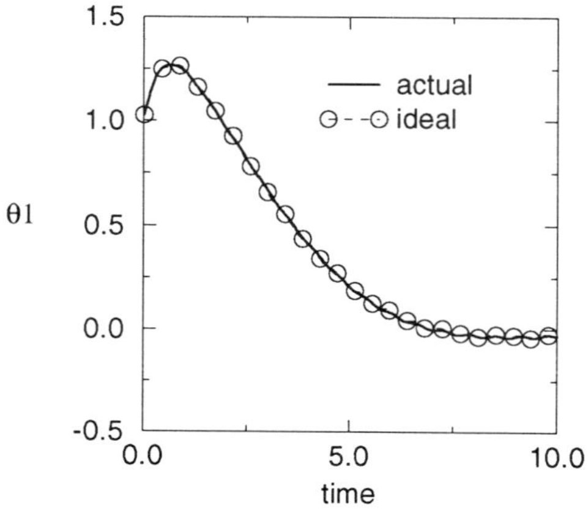

Figure 7.c Time response for boom truss for case 2 adaptive controller.

($\gamma 1 = 10, \gamma 2 = 10, \sigma 1 = 0.1, \sigma 2 = 0.1$)

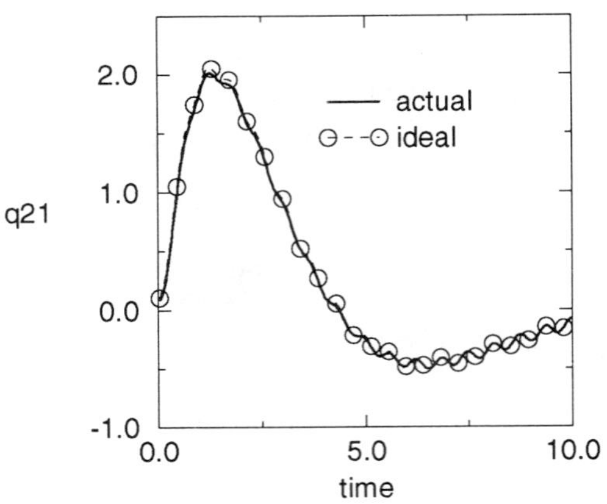

Figure 8.a Time response for antenna for case 2 adaptive controller.

($\gamma 1 = 10, \gamma 2 = 10, \sigma 1 = 0.1, \sigma 2 = 0.1$)

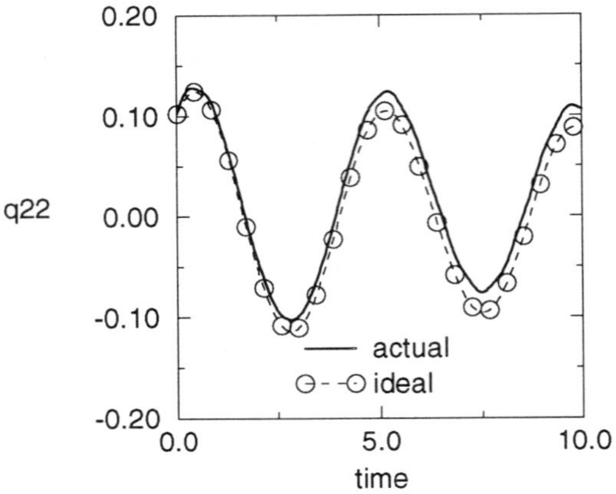

Figure 8.b Time response for antenna for case 2 adaptive controller.

($\gamma 1 = 10, \gamma 2 = 10, \sigma 1 = 0.1, \sigma 2 = 0.1$)

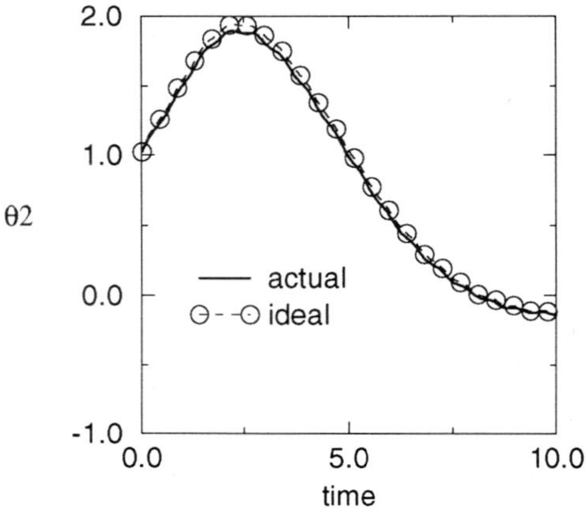

Figure 8.c Time response for antenna for case 2 adaptive controller.

($\gamma 1 = 10, \gamma 2 = 10, \sigma 1 = 0.1, \sigma 2 = 0.1$)

4. Structural systems are inherently infinite DOF systems. For the application of this algorithm to structural systems, the behavior of residual modes is a matter of critical importance. As stated in (4), a basic requirement for the guaranteed boundedness of the tracking error is that the interconnection forces be directly reachable by the controller forces. An alternative approach may be to guarantee, if possible, the stability or at least the boundedness of the vibration associated with the residual modes. For implementation of this controller on critical structures, it is very important that this issue be adequately addressed.

Acknowledgements

This work was sponsored by the Air Force Office of Scientific Research under AFOSR RIP Subcontract 92–73.

The authors are grateful to S. C. Sinha of Auburn University for his comments and suggestions.

References

Flowers, G.T., and Venkayya, V.B., *Adaptive Decentralized Control of Flexible Multibody Structures*, presented at the "AIAA/ USAF/ NASA /OAI Symposium on Multidisciplinary Analysis and Optimization," Sept. 21 –23, 1992.

Gavel, D.T., and Siljak, D.D., *Decentralized Adaptive Control: Structural Conditions for Stability*, IEEE Transactions on Automatic Control, Vol. 34, No. 4, April 1989, pp 413–426.

Gantmacher, F.R., The Theory of Matrices, Chelsea Publishing Company, New York, 1977.

Hmamed, A., and L. Radouane, *Decentralized Nonlinear adaptive feedback stabilization of Large-scale interconnected systems*, IEE Proceedings, Vol. 130-D, 1983 PP. 57–62.

Ioannou, P.A., and P.V. Kokotović, Adaptive Systems with Reduced Models, Springer, New York, 1983.

Ioannou, P.A., and P.V. Kokotović, *Robust Design of Adaptive Control*, IEEE Transactions on Automatic Control, Vol. AC–29, 1984, pp. 202–211.

Ioannou, P.S., *Decentralized Adaptive Control of Interconnected Systems*, IEEE Transactions on Automatic Control, Vol. AC–31, No. 4, April, 1986, pp. 291–298.

Ioannou, P.A., and J.S. Reed, *Discrete-time Decentralized Control*, Automatica, Vol. 24, 1988, pp. 419–421.

La Salle, J.P., and Lefschetz, S.Stability of Lyapunov's Direct Method with Applications, Academic Press, Inc., New York, 1961.

Mao, Cheng–Jyl, and Lin, Wei–Song, *On-Line Decentralized Control of Interconnected Systems With Unmodelled Nonlinearity and Interaction*, Proceedings of the 1989 Automatic Controls Conference, Vol. 1, pp. 248–252.

Siljak, D.D., *Parameter Space Methods for Robust Control Design: A Guided Tour*, IEEE Transactions on Automatic Control, Vol. 34, No. 7, July 1988, pp. 674–688.

Siljak, D.D., Decentralized Control of Complex Systems, Academic Press, Inc., New York, 1991.

Singh, M.D., Decentralised Control, North–Holland Publishing Company, Amsterdam, 1981.

Wiens, G.J., and Sinha, S.C.,*On the Application of Liapunov's Direct Method to Discrete Dynamic Systems with Stochastic Parameters*, Journal of Sound and Vibration, Vol. 94, No. 1, 1984 pp. 19–31.

Walker, J.A.,*On the Stability of Linear Discrete Dynamical Systems*, Journal of Applied Mechanics, Vol. 37, June 1970, pp. 271–275.

Walker, J.A.,*On the Application of Liapunov's Direct Method to Linear Lumped–Parameter Elastic Systems*, Journal of Applied Mechanics, Vol. 41, March 1974, pp. 278–284.

AMD-Vol. 141/DSC-Vol. 37, Dynamics of Flexible Multibody Systems:
Theory and Experiment
ASME 1992

A METHODOLOGY FOR MODELING HYBRID PARAMETER MULTIPLE BODY SYSTEMS

Alan A. Barhorst
Department of Mechanical Engineering
Texas Tech University
Lubbock, Texas

Louis J. Everett
Department of Mechanical Engineering
Texas A&M University
College Station, Texas

Abstract

In this paper, the Hybrid Parameter Multiple Body Modeling problem is addressed. Presented is a systematic methodology for deriving equations of motion for these highly complex systems. The methodology is founded in variational principles, but uses vector algebra to eliminate tedium. The method is applicable to even the most complex problems which include non-holonomic constraints, closed loop chains, and contact/impact. Such problems are not collectively nor effectively handled with other methods.

The variational nature of the methodology allows rigorous equation formulation providing not only the complete nonlinear, hybrid, differential equations, but also the boundary conditions. This can be significant especially since boundary conditions for such problems are often beyond intuition. One novelty of the method lies in the fact that its variational nature is transparent to the user. This enables it to be effectively used by researchers having a working knowledge of Lagrange's method. Users of the method need not understand variational calculus, however understanding the method's derivation does require it.

The methodology is formulated in the constraint-free subspace of the system's configuration space, thus Lagrange multipliers are not needed for constrained systems, regardless of the constraint type (holonomic or nonholonomic). The dimension of distributed parameter bodies and their interconnections is general.

In this paper the methodology is derived starting from d'Alembert's principle. The method is then demonstrated by modeling a distributed cantilever beam with tip mass/inertia attached to a rigid rotating base. Results from other researchers will be compared to the method of this paper. Although such a simple problem does not demonstrate the method's inherent utility, it is an excellent introduction.

Introduction

Presented in this paper is a systematic approach to modeling HPMB systems [1, 2]. The methodology is rigorously founded in the variational realm of mechanics. It has most of the attributes of the analytical approach (i.e. Hamilton's principle), yet eliminates most of the pitfalls, such as the need to use Lagrange multipliers for constraints, and excessive algebra. The method also incorporates one of the most useful qualities of Newton's method, namely, vector algebra. It does not retain any of the hindering features, such as the need to include constraint forces. The implementation of the method is straightforward, yet great rigor is retained. Analyst familiar with Kane's form of the Gibb's-Appell equations [3, 4, 5, 6], will find the method straightforward to implement.

The motivation for this "new" method was the need to easily derive complete models of hybrid parameter systems [1, 7, 8, 9, 10, 11]. Although the method is still relatively mathematically intense (compared to an equal number of rigid bodies), it has a predisposition for symbolic manipulation.

A hypotheses for this work is that a simple method may make it possible to bring HPMB modeling out of the academic domain and into use by product designers. Another hypothesis is that a simple (ultimately an automated) method will make it possible for researchers to rapidly regenerate models based on new continuum assumptions. For example, a key problem with HPMB modeling is trying to match a model to an existing system. Often one assumes certain connectivity, constructs a model, then compares its output to experimental measurements. If the two differ, one attempts to modify the model and reconstruct it. Having an easy to apply method, facilitates this procedure. Furthermore since the method generates the minimal set of equations along with the boundary conditions, it should be invaluable in this model matching process. Another feature of having the minimal set of equations for a given system, is the avoidance of differential-algebraic equations (usually used to describe HPMB systems) and their inherent solution difficulties [12].

The basis of the method is d'Alembert's principle [13] applied to a hybrid system of arbitrarily connected bodies. D'Alembert's principle for a generic particle is (see figure 1):

$$^{\mathcal{N}}\delta{}^{o}\vec{r}^{dm} \cdot \left[d\vec{F} - \frac{^{\mathcal{N}}d}{dt}\left({}^{o}\vec{v}_{\mathcal{N}}^{dm}\, dm \right) \right] = 0 \tag{1}$$

Where $^{\mathcal{N}}\delta(\cdot)$ is the variation of $(\cdot)$ with respect to the Newtonian frame $\mathcal{N}$ [14]. Ordinarily one thinks of a variation as a change in a scalar quantity, however, the extension of the concept to vectors is used as well. The Newtonian frame has origin at point o, and dm is the element of mass. The term $d\vec{F}$ is the differential resultant force (non-constraint) on dm. The term $\frac{^{\mathcal{N}}d}{dt}$ represents differentiation with respect to time in the Newtonian frame and ${}^{o}\vec{v}_{\mathcal{N}}^{dm}$ is the velocity of dm with respect to o, as seen in $\mathcal{N}$. A complete description of the notation is included in the appendix.

An important, and somewhat novel, relationship utilized in this work is the relative variation [14, 15]. This relationship can be written as:

$$^{A}\delta\vec{v} = {}^{B}\delta\vec{v} + {}^{A}\delta\vec{\theta}^{B} \times \vec{v} \tag{2}$$

Equation 2 describes a means by which one can relate the variation of vector $\vec{v}$ as seen in frame A (i.e., $^{A}\delta\vec{v}$), to the variation of $\vec{v}$ as seen in frame B (i.e., $^{B}\delta\vec{v}$). The term $^{A}\delta\vec{\theta}^{B}$ is an infinitesimal virtual rotation of frame B as seen in frame A. Equation 2 allows multiple reference frames to be employed in the analysis.

Everett realized the utility of equation 2 and used it with Hamilton's principle [16], other researchers used it with d'Alembert's principle (applied to rigid bodies) [15, 17, 18]. Equation 2 will be shown to be a key element in the derivation that follows in the next section. Another useful fact is that infinitesimal (virtual) rotations follow an addition theorem [1] similar to the angular velocity addition theorem [5].

D'Alembert's principle, as presented in equation 1, accompanied by equation 2, along with Kane's partial velocities [5] and the definition (to be expounded on below) of pseudo-generalized coordinates and pseudo-generalized speeds, allow the present method to be developed. The method is variational in nature but the variations are transparent to the modeler, similar to the relationship Lagrange's equations has to Hamilton's principle. The result is a systematic, easily applied methodology. The usually laborious task of performing variations is reduced to vector differentiation and scalar partial differentiation. The method produces complete, minimal, hybrid parameter sets of differential equations along with boundary conditions.

Derivation

Consider the schematic representation of an arbitrarily connected, arbitrarily constrained, system of hybrid parameter mechanical bodies depicted in figure 1. Attached to each body B, is a coordinate frame with origin at b_o.

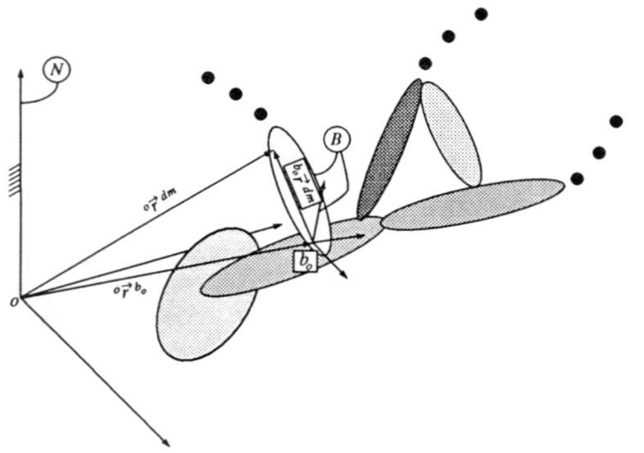

Figure 1: A General Hybrid Parameter System of Bodies

The letter B will be interchangeably used for the body and the frame. The Newtonian frame of reference $\mathcal{N}$ has an origin at o. The position (relative to o) of a differential mass element dm of the body is located by the vector ${}^{o}\vec{r}^{dm}$. Utilizing the vector triangle shown in figure 1, ${}^{o}\vec{r}^{dm}$ can be written as:

$$ {}^{o}\vec{r}^{dm} = {}^{o}\vec{r}^{b_o} + {}^{b_o}\vec{r}^{dm} \tag{3} $$

where ${}^{o}\vec{r}^{b_o}$ and ${}^{b_o}\vec{r}^{dm}$ are defined as shown.

The total mass of the system of bodies is m and can be written as the sum of each body's mass, $m = \sum_r m_r + \sum_f m_f$, where the indices r and f range over all the rigid and flexible bodies. Integrating equation 1 over the total system mass then splitting into rigid and flexible bodies results in:

$$ 0 = \sum_r \int_{m_r} {}^{\mathcal{N}}\delta^{o}\vec{r}^{dm_r} \cdot \left[d\vec{F}_r - \frac{{}^{\mathcal{N}}d}{dt}\left({}^{o}\vec{v}_{\mathcal{N}}^{dm_r} dm_r\right)\right] + \sum_f \int_{m_f} {}^{\mathcal{N}}\delta^{o}\vec{r}^{dm_f} \cdot \left[d\vec{F}_f - \frac{{}^{\mathcal{N}}d}{dt}\left({}^{o}\vec{v}_{\mathcal{N}}^{dm_f} dm_f\right)\right] \tag{4} $$

The absolute velocity ${}^{o}\vec{v}_{\mathcal{N}}^{dm}$ of the generic differential mass can be expressed, utilizing equation 3, as:

$$ \begin{aligned} {}^{o}\vec{v}_{\mathcal{N}}^{dm} = \frac{{}^{\mathcal{N}}d}{dt}{}^{o}\vec{r}^{dm} &= \frac{{}^{\mathcal{N}}d}{dt}{}^{o}\vec{r}^{b_o} + \frac{{}^{B}d}{dt}{}^{b_o}\vec{r}^{dm} + {}^{\mathcal{N}}\vec{\omega}^{B} \times {}^{b_o}\vec{r}^{dm} \\ &= {}^{o}\vec{v}_{\mathcal{N}}^{b_o} + {}^{b_o}\vec{v}_{B}^{dm} + {}^{\mathcal{N}}\vec{\omega}^{B} \times {}^{b_o}\vec{r}^{dm} \end{aligned} \tag{5} $$

where ${}^{\mathcal{N}}\vec{\omega}^{B}$ is the angular velocity of frame B as seen in frame $\mathcal{N}$. The absolute variation ${}^{\mathcal{N}}\delta^{o}\vec{r}^{dm}$ of the generic differential mass can be expressed, utilizing equations 2 and 3, as:

$$ {}^{\mathcal{N}}\delta^{o}\vec{r}^{dm} = {}^{\mathcal{N}}\delta^{o}\vec{r}^{b_o} + {}^{B}\delta^{b_o}\vec{r}^{dm} + {}^{\mathcal{N}}\vec{\delta\theta}^{B} \times {}^{b_o}\vec{r}^{dm} \tag{6} $$

Rigid Bodies

Consider the rth typical rigid body. Its contribution to equation 4 is:

$$ \int_{\Omega_r} {}^{\mathcal{N}}\delta^{o}\vec{r}^{dm_r} \cdot \left[\vec{F}_r - \frac{{}^{\mathcal{N}}d}{dt}\left(\rho_r {}^{o}\vec{v}_{\mathcal{N}}^{dm_r}\right)\right] d\Omega_r \tag{7} $$

where densities are assumed to exist such that $d\vec{F}_r = \vec{F}_r\, d\Omega_r$ and $dm_r = \rho_r\, d\Omega_r$. Ω_r is an appropriate spatial domain. Utilizing equations 3, 5, 6, vector triple product identities, and the definition of the vector to the center of mass, ${}^{b_{or}}\vec{r}^{*r}$, the following holds [5, 15, 1]:

$$ \begin{aligned} \int_{\Omega_r} {}^{\mathcal{N}}\delta^{o}\vec{r}^{dm_r} \cdot \left[\vec{F}_r - \frac{{}^{\mathcal{N}}d}{dt}\left(\rho_r {}^{o}\vec{v}_{\mathcal{N}}^{dm_r}\right)\right] d\Omega_r &= {}^{\mathcal{N}}\delta^{o}\vec{r}^{b_{or}} \cdot \left[\vec{F}_r - m_r {}^{o}\vec{a}_{\mathcal{N}}^{*r}\right] + \\ {}^{\mathcal{N}}\vec{\delta\theta}^{B_r} \cdot \Big[\vec{T}_{b_{or}} - {}^{b_{or}}\vec{r}^{*r} \times m_r {}^{o}\vec{a}_{\mathcal{N}}^{b_{or}} - \vec{\vec{I}}_{b_{or}} \cdot {}^{\mathcal{N}}\vec{\alpha}^{B_r} &- {}^{\mathcal{N}}\vec{\omega}^{B_r} \times \vec{\vec{I}}_{b_{or}} \cdot {}^{\mathcal{N}}\vec{\omega}^{B_r}\Big] \end{aligned} \tag{8} $$

where $\vec{\vec{I}}_{b_{or}}$ is the inertia dyadic [5, 15, 1]. The other individual terms are defined in the appendix. The terms containing ${}^{B}\delta^{b_o}\vec{r}^{dm}$ in equation 6 are zero because the body is rigid.

The constraint forces and torques are those which maintain the kinematic constraints. The forces and torques appearing in equation 8 are those which are not constraint forces. Equation 8 is the rigid body contribution to equation 4.

Flexible Bodies

Now consider the fth typical flexible body. Its contribution to equation 4 is:

$$ \int_{\Omega_f} {}^{\mathcal{N}}\delta^{o}\vec{r}^{dm_f} \cdot \left[\vec{F}_f - \frac{{}^{\mathcal{N}}d}{dt}\left(\rho_f {}^{o}\vec{v}_{\mathcal{N}}^{dm_f}\right)\right] d\Omega_f \tag{9} $$

As before, it is assumed that force and mass densities exist such that $d\vec{F}_f = \vec{F}_f\, d\Omega_f$ and $dm_f = \rho_f\, d\Omega_f$. Utilizing equation 6, and the vector triple product identities, integral 9 can be rewritten as:

$$ \begin{aligned} \int_{\Omega_f} {}^{\mathcal{N}}\delta^{o}\vec{r}^{dm_f} \cdot \left[\vec{F}_f - \frac{{}^{\mathcal{N}}d}{dt}\left(\rho_f {}^{o}\vec{v}_{\mathcal{N}}^{dm_f}\right)\right] d\Omega_f = \\ {}^{\mathcal{N}}\delta^{o}\vec{r}^{b_{of}} \cdot \int_{\Omega_f} \left[\vec{F}_f - \frac{{}^{\mathcal{N}}d}{dt}\left(\rho_f {}^{o}\vec{v}_{\mathcal{N}}^{dm_f}\right)\right] d\Omega_f \\ + \int_{\Omega_f} {}^{B_f}\delta^{b_{of}}\vec{r}^{dm_f} \cdot \left[\vec{F}_f - \frac{{}^{\mathcal{N}}d}{dt}\left(\rho_f {}^{o}\vec{v}_{\mathcal{N}}^{dm_f}\right)\right] d\Omega_f \\ + {}^{\mathcal{N}}\vec{\delta\theta}^{B_f} \cdot \int_{\Omega_f} {}^{b_{of}}\vec{r}^{dm_f} \times \left[\vec{F}_f - \frac{{}^{\mathcal{N}}d}{dt}\left(\rho_f {}^{o}\vec{v}_{\mathcal{N}}^{dm_f}\right)\right] d\Omega_f \end{aligned} \tag{10} $$

If the applied body forces $\vec{F}_f$ occur in regions and at specific points, then the total force per unit volume (domain) for the differential element, in deformed coordinates, is (see [19, 20]):

$$ \vec{F}_f = \mathcal{H}_f \vec{F}_{hf} + \mathcal{D}_f \vec{F}_{df} + \tau_{fij,i} \hat{b}_{fj} \tag{11} $$

where $\mathcal{H}_f$ is the Heaviside step function and $\mathcal{D}_f$ is the Dirac delta function. It is assumed that body forces encompassing the entire domain will be included by setting $\mathcal{H}_f = 1$. The Heaviside step and Dirac delta functions are defined for the regions of application of their associated forces. The usual index notation is used for the stress tensor τ_{fij}. The derivatives defined with comma notation are with respect to deformed coordinates.

In this derivation it is assumed that the continuum is of a multipolar nature. Thus the following differential equations is needed to reduce expressions that arise in the derivation of the method. Namely [19]:

$$ \mu_{fij,i} \hat{b}_{fj} + \mathcal{H}_f \vec{T}_{hf} + \mathcal{D}_f \vec{T}_{df} + \epsilon_{ijk} \tau_{fjk} \hat{b}_{fi} = 0 \tag{12} $$

where μ_{fij} is the couple stress tensor and $\mathcal{H}_f \vec{T}_{hf}$ and $\mathcal{D}_f \vec{T}_{df}$ are torques, similar to the forces described above, applied to the material.

The operations used to reduce equation 10 consist of a mix of vector and tensor calculus. The operations are interesting but are not necessary for the present demonstration. The operations are performed in detail in [1, 2]. The operations, alluded to above, allow equation 10 to be rewritten as:

$$
\begin{aligned}
\int_{\Omega_f} {}^{\mathcal{N}}\delta^{o}\vec{r}^{dm_f}\cdot\left[\vec{F}_f-\frac{{}^{\mathcal{N}}d}{dt}\left(\rho_f{}^{o}\vec{v}_{\mathcal{N}}^{dm_f}\right)\right]d\Omega_f &= \\
&{}^{\mathcal{N}}\delta^{o}\vec{r}^{b_{of}}\cdot\int_{\partial\Omega_f}\left(\mathcal{H}_f\vec{F}_{hf}+\mathcal{D}_f\vec{F}_{df}\right)d\sigma_f \\
&+{}^{\mathcal{N}}\delta^{o}\vec{r}^{b_{of}}\cdot\int_{\Omega_f}\left[\mathcal{H}_f\vec{F}_{hf}+\mathcal{D}_f\vec{F}_{df}-\frac{{}^{\mathcal{N}}d}{dt}\left(\rho_f{}^{o}\vec{v}_{\mathcal{N}}^{dm_f}\right)\right]d\Omega_f \\
&+\int_{\Omega_f}\Big\{-\delta\bar{V}_f+2\mathcal{H}_f\vec{T}_{hf}\cdot{}^{B_f}\delta\vec{\theta}^{I_f}+2\mathcal{D}_f\vec{T}_{df}\cdot{}^{B_f}\delta\vec{\theta}^{I_f} \\
&\qquad+{}^{B_f}\delta^{b_{of}}\vec{r}^{dm_f}\cdot\left[\mathcal{H}_f\vec{F}_{hf}+\mathcal{D}_f\vec{F}_{df}-\frac{{}^{\mathcal{N}}d}{dt}\left(\rho_f{}^{o}\vec{v}_{\mathcal{N}}^{dm_f}\right)\right]\Big\}d\Omega_f \\
&+\int_{\partial\Omega_f}\Big[{}^{B_f}\delta^{b_{of}}\vec{r}^{dm_f}\cdot\left(\mathcal{H}_f\vec{F}_{hf}+\mathcal{D}_f\vec{F}_{df}\right) \\
&\qquad+2\mathcal{H}_f\vec{T}_{hf}\cdot{}^{B_f}\delta\vec{\theta}^{I_f}+2\mathcal{D}_f\vec{T}_{df}\cdot{}^{B_f}\delta\vec{\theta}^{I_f}\Big]d\sigma_f \\
&+{}^{\mathcal{N}}\delta\vec{\theta}^{B_f}\cdot\int_{\partial\Omega_f}\left[{}^{b_{of}}\vec{r}^{dm_f}\times\left(\mathcal{H}_f\vec{F}_{hf}+\mathcal{D}_f\vec{F}_{df}\right)+\mathcal{H}_f\vec{T}_{hf}+\mathcal{D}_f\vec{T}_{df}\right]d\sigma_f \\
&+{}^{\mathcal{N}}\delta\vec{\theta}^{B_f}\cdot\int_{\Omega_f}\Big[{}^{b_{of}}\vec{r}^{dm_f}\times\Big(\mathcal{H}_f\vec{F}_{hf}+\mathcal{D}_f\vec{F}_{df}- \\
&\qquad\frac{{}^{\mathcal{N}}d}{dt}\left(\rho_f{}^{o}\vec{v}_{\mathcal{N}}^{dm_f}\right)\Big)+\mathcal{H}_f\vec{T}_{hf}+\mathcal{D}_f\vec{T}_{df}\Big]d\Omega_f \qquad (13)
\end{aligned}
$$

where $\bar{V}_f$ is the strain energy density of the flexible body. The symbol ${}^{B_f}\delta\vec{\theta}^{I_f}$ represents the virtual rotation of the "internal distribution" of coordinate frames. Equation 13 can be thought of as a component of a statement of the principle of virtual work for the body [20].

Now examine the strain energy term in equation 13. The strain energy density function is assumed to be of the form:

$$\bar{V}_f=\bar{V}_f\left(\tilde{u}_{fi},\tilde{u}_{fi,j},\tilde{u}_{fi,jk}\right) \qquad (14)$$

where $\tilde{u}_{fi}=\tilde{u}_{fi}\left(r_{fj},t\right)$. The r_{fj} are orthogonal curvilinear coordinates. With equation 14 and curvilinear coordinates in mind, the following holds:

$$
\begin{aligned}
\int_{\Omega_f}\delta\bar{V}_f\,d\Omega_f &= \int_{\partial\Omega_f}(\hat{n}\cdot\hat{e}_{fk})h_k\frac{\partial\bar{V}_f}{\partial\tilde{u}_{fi,jk}}\delta\tilde{u}_{fi,j}\,d\sigma_f \\
&+\int_{\partial\Omega_f}(\hat{n}\cdot\hat{e}_{fj})\left[h_j\frac{\partial\bar{V}_f}{\partial\tilde{u}_{fi,j}}-\frac{h_j}{h_1h_2h_3}\frac{\partial}{\partial r_{fk}}\left(h_1h_2h_3\frac{\partial\bar{V}_f}{\partial\tilde{u}_{fi,jk}}\right)\right]\delta\tilde{u}_{fi}\,d\sigma_f \\
&+\int_{\Omega_f}\left[\frac{\partial\bar{V}_f}{\partial\tilde{u}_{fi}}-\frac{1}{h_1h_2h_3}\frac{\partial}{\partial r_{fj}}\left(h_1h_2h_3\frac{\partial\bar{V}_f}{\partial\tilde{u}_{fi,j}}\right)\right. \\
&\qquad\left.+\frac{1}{h_1h_2h_3}\frac{\partial^2}{\partial r_{fj}\partial r_{fk}}\left(h_1h_2h_3\frac{\partial\bar{V}_f}{\partial\tilde{u}_{fi,jk}}\right)\right]\delta\tilde{u}_{fi}\,d\Omega_f \qquad (15)
\end{aligned}
$$

where h_1, h_2 and h_3 are the curvilinear coordinate scale factors. $(\hat{n}\cdot\hat{e}_{fj})$ is the jth component of the unit outward normal vector for $\partial\Omega_f$, and $\hat{e}_{fj}$ is the curvilinear triad.

Evaluating ${}^{B_f}\delta^{b_{of}}\vec{r}^{dm_f}$ from equation 13, with consideration for the curvilinear coordinates, gives:

$$
\begin{aligned}
{}^{B_f}\delta^{b_{of}}\vec{r}^{dm_f} &= \frac{\partial^{b_{of}}\vec{r}^{dm_f}}{\partial\vec{\tilde{u}}_f}\frac{\partial\vec{\tilde{u}}_f}{\partial\tilde{u}_{fi}}\delta\tilde{u}_{fi} \\
&= \frac{\partial^{b_{of}}\vec{r}^{dm_f}}{\partial\vec{\tilde{u}}_f}h_i\,\delta\tilde{u}_{fi}\,\hat{e}_{fi} \\
&= h_i\,\delta\tilde{u}_{fi}\,\hat{e}_{fi} \qquad (16)
\end{aligned}
$$

where the triad $\hat{e}_{fi}$ are a curvilinear coordinate axis set associated with the body frame B_f. Similarly the virtual rotations are written as:

$$
\begin{aligned}
{}^{B_f}\delta\vec{\theta}^{I_f} &= \frac{1}{2}\vec{\nabla}\times\delta\vec{\tilde{u}}_f \\
&= \frac{1}{2}\vec{\nabla}\times\frac{\partial\vec{\tilde{u}}_f}{\partial\tilde{u}_{fi}}\delta\tilde{u}_{fi} \\
&= \frac{1}{2}\vec{\nabla}\times h_i\,\delta\tilde{u}_{fi}\,\hat{e}_{fi} \qquad (17)
\end{aligned}
$$

D'Alembert's principle for this system of bodies can now be assembled.

Final Form of d'Alembert's Principle

The final form of d'Alembert's principle, equation 4, is composed of equations 8 and 13, with consideration of equations 16 and 17. Assembling the equations gives:

$$
\begin{aligned}
0=\int_m {}^{\mathcal{N}}\delta^{o}\vec{r}^{dm}\cdot\left[d\vec{F}-\frac{{}^{\mathcal{N}}d}{dt}\left({}^{o}\vec{v}_{\mathcal{N}}^{dm}dm\right)\right] &= \\
&\sum_r\Big\{{}^{\mathcal{N}}\delta^{o}\vec{r}^{b_{or}}\cdot\left[\vec{F}_r-m_r{}^{o}\vec{a}_{\mathcal{N}}^{*r}\right] \\
+{}^{\mathcal{N}}\delta\vec{\theta}^{B_r}\cdot\Big[\vec{T}_{b_{or}}-{}^{b_{or}}\vec{r}^{*r}\times m_r{}^{o}\vec{a}_{\mathcal{N}}^{b_{or}}-\vec{I}_{b_{or}}\cdot{}^{\mathcal{N}}\vec{\alpha}^{B_r}-{}^{\mathcal{N}}\vec{\omega}^{B_r}\times\vec{I}_{b_{or}}\cdot{}^{\mathcal{N}}\vec{\omega}^{B_r}\Big]\Big\} \\
+\sum_f\Big\{{}^{\mathcal{N}}\delta^{o}\vec{r}^{b_{of}}\cdot\Big\{\int_{\Omega_f}\left[\mathcal{H}_f\vec{F}_{hf}+\mathcal{D}_f\vec{F}_{df}-\frac{{}^{\mathcal{N}}d}{dt}\left(\rho_f{}^{o}\vec{v}_{\mathcal{N}}^{dm_f}\right)\right]d\Omega_f \\
+\int_{\partial\Omega_f}\left(\mathcal{H}_f\vec{F}_{hf}+\mathcal{D}_f\vec{F}_{df}\right)d\sigma_f\Big\} \\
+{}^{\mathcal{N}}\delta\vec{\theta}^{B_f}\cdot\Big\{\int_{\Omega_f}\Big({}^{b_{of}}\vec{r}^{dm_f}\times\Big[\mathcal{H}_f\vec{F}_{hf}+\mathcal{D}_f\vec{F}_{df} \\
-\frac{{}^{\mathcal{N}}d}{dt}\left(\rho_f{}^{o}\vec{v}_{\mathcal{N}}^{dm_f}\right)\Big]+\mathcal{H}_f\vec{T}_{hf}+\mathcal{D}_f\vec{T}_{df}\Big)d\Omega_f \\
+\int_{\partial\Omega_f}\left[{}^{b_{of}}\vec{r}^{dm_f}\times\left(\mathcal{H}_f\vec{F}_{hf}+\mathcal{D}_f\vec{F}_{df}\right)+\mathcal{H}_f\vec{T}_{hf}+\mathcal{D}_f\vec{T}_{df}\right]d\sigma_f\Big\} \\
+\int_{\Omega_f}\Big\{\Big[\mathcal{H}_f\vec{F}_{hf}+\mathcal{D}_f\vec{F}_{df}-\frac{{}^{\mathcal{N}}d}{dt}\left(\rho_f{}^{o}\vec{v}_{\mathcal{N}}^{dm_f}\right)\Big]\cdot h_i\hat{e}_{fi} \\
+\vec{\nabla}\times\left(\mathcal{H}_f\vec{T}_{hf}+\mathcal{D}_f\vec{T}_{df}\right)\cdot h_i\hat{e}_{fi} \\
-\Big[\frac{\partial\bar{V}_f}{\partial\tilde{u}_{fi}}-\frac{1}{h_1h_2h_3}\frac{\partial}{\partial r_{fj}}\left(h_1h_2h_3\frac{\partial\bar{V}_f}{\partial\tilde{u}_{fi,j}}\right) \qquad (18) \\
+\frac{1}{h_1h_2h_3}\frac{\partial^2}{\partial r_{fj}\partial r_{fk}}\left(h_1h_2h_3\frac{\partial\bar{V}_f}{\partial\tilde{u}_{fi,jk}}\right)\Big]\Big\}\delta\tilde{u}_{fi}\,d\Omega_f \\
+\int_{\partial\Omega_f}\Big\{\left(\mathcal{H}_f\vec{F}_{hf}+\mathcal{D}_f\vec{F}_{df}\right)\cdot h_i\hat{e}_{fi}+2\left(\mathcal{H}_f\vec{T}_{hf}+\mathcal{D}_f\vec{T}_{df}\right)\cdot\frac{h_k}{h_1h_2h_3}\epsilon_{kji}\,h_{i,j}\,\hat{e}_{fk} \\
-(\hat{n}\cdot\hat{e}_{fj})\Big[h_j\frac{\partial\bar{V}_f}{\partial\tilde{u}_{fi,j}}-\frac{h_j}{h_1h_2h_3}\frac{\partial}{\partial r_{fk}}\left(h_1h_2h_3\frac{\partial\bar{V}_f}{\partial\tilde{u}_{fi,jk}}\right)\Big]\Big\}\delta\tilde{u}_{fi}\,d\sigma_f \\
+\int_{\partial\Omega_f}\Big[\left(\mathcal{H}_f\vec{T}_{hf}+\mathcal{D}_f\vec{T}_{df}\right)\cdot\frac{h_kh_i^2}{h_1h_2h_3}\epsilon_{kji}\,\hat{e}_{fk}-(\hat{n}\cdot\hat{e}_{fk})h_k\frac{\partial\bar{V}_f}{\partial\tilde{u}_{fi,jk}}\Big]\delta\tilde{u}_{fi,j}\,d\sigma_f
\end{aligned}
$$

Choice of Variations

Presented in this section is a discussion of how the variational vector quantities in equation 18 can be chosen.

First consider a system with N discrete degrees of freedom. Let q_j, $j=1,2,3,\cdots,N$ be a conventional set of independent generalized coordinates used to describe the configuration of the system. Also, define N generalized speeds (quasi-coordinates) u_l to serve as an alternate set of coordinates as:

$$u_l=\sum_{m=1}^{N}Y_{lm}\dot{q}_m+Z_l \qquad l=1,2,3,\cdots,N \qquad (19)$$

where matrix Y is chosen non-singular. The original variables can be found as:

$$\dot{q}_l=\sum_{m=1}^{N}W_{lm}u_m+X_l \qquad l=1,2,3,\cdots,N \qquad (20)$$

If there are M simple nonholonomic constraint relationships then the M dependent u_m can be expressed as:

$$u_m=\sum_n A_{mn}u_n+B_m \qquad m\in\left\{I_d^M\right\},\ n\in\left\{I_i^{N-M}\right\} \qquad (21)$$

where $\{I_d^M\}$ means the set of all M indices associated with dependent u's. $\{I_i^{N-M}\}$ means the set of all $N - M$ indices associated to the independent u's. Equations 19–21 can always be written for a system of rigid bodies.

Now consider a system containing some hybrid parameter (continuously elastic) bodies. The system's kinematics will be described using the distributed coordinates of the elastic bodies where the connections are made. Figures 2 and 3 show a typical flexible body, with connections, and a typical connection in and on the body, respectively. In each region there will be a

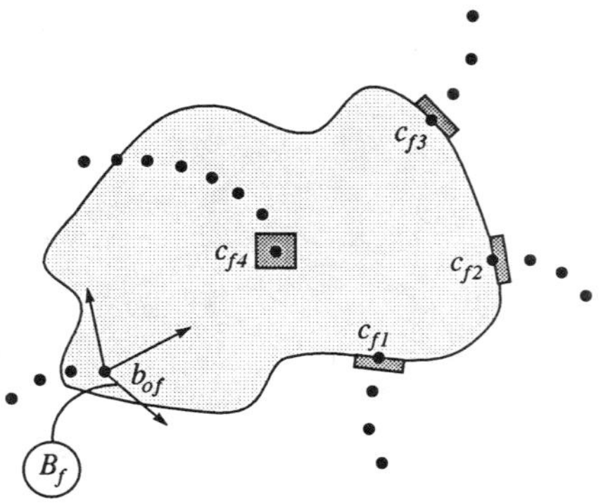

Figure 2: The fth Flexible Body and Connections

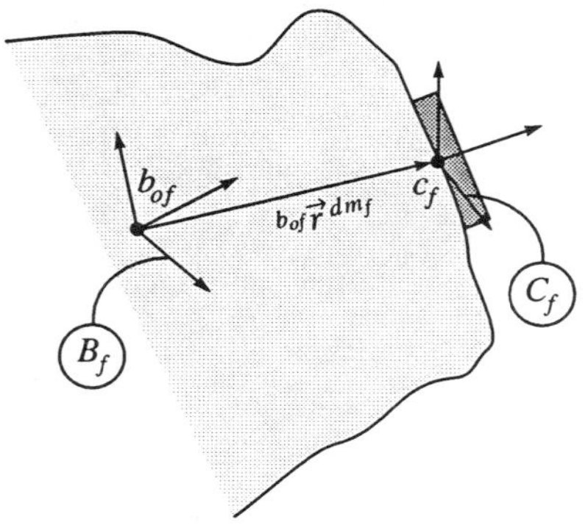

Figure 3: A Typical Connection in the fth Flexible Body

"special" point (by choice) or a point of "significance" (e.g. part of a hinge) which is labelled c_f. There is a point c_f at every connection between bodies. Also, attach a frame C_f to each point c_f. The region around each point of connection will be called the region of connection. Regions of connection for c_f which lie inside the body will be labeled Ω_{c_f}. Regions associated with c_f which lie on the surface of the body will be labeled $\partial\Omega_{c_f}$. The symbol ∂ does not imply the regions are in any way related.

It is assumed that the regions of connection are stiff enough that elastic behavior of the flexible body in the region of connection is a known function of the motion of c_f. The significance of this assumption is that the region of connection is free to translate and rotate, but not distort. The negligible distortion assumption is justified by the fact that two flexible bodies connected by finite regions will be significantly distortion free. A flexible body connected to a rigid body in a finite region will be "infinitely" stiff relative to distortion. The form of these functional relationships are:

$$\widetilde{u}_{fi} = \mathcal{L}(\widetilde{u}_{fi}(c_f,t), \widetilde{u}_{fi,j}(c_f,t)) \tag{22}$$

and

$$\widetilde{u}_{fi,j} = \mathcal{M}(\widetilde{u}_{fi,j}(c_f,t)) \tag{23}$$

in the regions of connection. $\mathcal{L}$ and $\mathcal{M}$ are known functions depending on the shape of the connections.

Assign a pseudo-generalized coordinate, q_p, to each of the flexural displacements of all of the points of connection c_f. Also assign a pseudo-generalized coordinate, to each of the relationships $\frac{\partial \widetilde{u}_{fi}}{\partial r_{fj}}$ $(i \neq j)$, for each c_f (the slopes) which are defined. The terminology "pseudo-generalized" coordinate is justified if one gains familiarity with the present method and reviews the discussion of pseudo-coordinates found in Kane [5].

If the location of the connection moves on the body, like the point of contact for a rolling thin flexible disk, then the pseudo-generalized coordinates may look like:

$$q_p = q_p(q_n, t) \qquad n \in \{I^N \setminus p\},\, p \in \{I^P\} \tag{24}$$

where $\{I^N \setminus p\}$ is the set of indices (N is the number of coordinates) without ($\setminus$) the index associated to the pth coordinate. The above pseudo-coordinate implies that:

$$\dot{q}_p = \frac{\partial q_p}{\partial q_n}\dot{q}_n + \frac{\partial q_p}{\partial t} \qquad n \in \{I^N \setminus p\},\, p \in \{I^P\} \tag{25}$$

where P is the number of pseudo-coordinates. Also, $\{I^P\} \subset \{I^N\}$. Equation 25 is a nonholonomic "condition." It is called a "condition" because (unlike a constraint) it does not lower the system's degrees of freedom due to the term $\frac{\partial q_p}{\partial t}$. If $\dot{q}_p$ is formed as in equation 25, but is associated with $\frac{\partial q_p}{\partial t}$, then equation 19 will maintain its form if the coefficients $\frac{\partial q_p}{\partial q_n}$ are absorbed in the appropriate Y matrix element.

Without loss of generality, and for ease of gathering like variations, assign pseudo-generalized speeds as follows:

$$u_p = \frac{\partial q_p}{\partial t} \tag{26}$$

with $\dot{u}_p$ defined as:

$$\dot{u}_p = \frac{\partial^2 q_p}{\partial t \partial q_n}\dot{q}_n + \frac{\partial^2 q_p}{\partial t^2} \qquad n \in \{I^N \setminus p\},\, p \in \{I^P\} \tag{27}$$

Now equation 19 can be written as:

$$\begin{Bmatrix} u_r \\ u_p \end{Bmatrix} = \begin{bmatrix} Y + \frac{\partial q_p}{\partial q_r} & 0 \\ 0 & I \end{bmatrix} \begin{Bmatrix} \dot{q}_r \\ \dot{q}_p \end{Bmatrix} + Z \tag{28}$$

The lower partition is not a kinematic differential equation, it is merely an identity. Equation 21 is still valid, with the realization that the pseudo-u's will usually (if present) be independent. There are cases in which it may be desirable to allow the pseudo-u's to be dependent (e.g. some closed chain mechanisms, see [1, 21]). The nonholonomic constraint relationship, equation 21, will be generated from a generally nonlinear constraint relationship between the generalized coordinates (pseudo and regular). Thus the pseudo-coordinates and pseudo-speeds are completely determined in terms of the independent coordinates and speeds. This fact will be utilized below.

Look at δq_p, namely:

$$\delta q_p = \frac{\partial q_p}{\partial q_n}\delta q_n + \delta' q_p \qquad n \in \{I^N \setminus p\},\, p \in \{I^P\} \tag{29}$$

where $\frac{\partial q_p}{\partial q_n}\delta q_n$ is the variation of q_p due to the variations in the other coordinates, q_n (pseudo and otherwise). $\delta' q_p$ is the variation in q_p due to either $\delta\widetilde{u}_{fi}(c_f,t)$ or $\delta\widetilde{u}_{fi,j}(c_f,t)$, $(i \neq j)$. At first glance this appears to be a misuse of the chain rule, but is justified if one considers the function being varied is itself free to be varied.

The variations δq_r and $\delta' q_p$ have been formulated such that they are independent for all r and p. The index r stands for regular generalized coordinates, while p stands for pseudo-generalized coordinates. Since δq_r and $\delta' q_p$ are an independent set of variations, a new set can be formed from a linear combination of the old. Judiciously choosing the same coefficients as in equation 28 gives:

$$\begin{Bmatrix} \delta u_r \\ \delta u_p \end{Bmatrix} = \begin{bmatrix} Y + \frac{\partial q_p}{\partial q_r} & 0 \\ 0 & I \end{bmatrix} \begin{Bmatrix} \delta q_r \\ \delta' q_p \end{Bmatrix} \tag{30}$$

The column matrix Z, in equation 28, does not contribute to the variational relationship because it depends on time and not the q's. The variations in the dependent generalized speeds (nonholonomic constraint) can be related to the variations in the independent generalized speeds as follows:

$$\delta u_m = \sum_n A_{mn}\delta u_n \qquad m \in \{I_d^M\},\, n \in \{I_i^{N-M}\} \tag{31}$$

The column matrix B, from equation 21, does not contribute to the variational relationship for the same reasons that Z does not contribute to equation 30.

With the generalized speeds, u's, formed as shown in equations 21 and 28, the correct and judicious representation of the variations ${}^{\mathcal{N}}\delta{}^{o}\vec{r}^{\,b_o}$ and ${}^{\mathcal{N}}\vec{\delta\theta}^{B}$ can be expressed as:

$$ {}^{\mathcal{N}}\delta{}^{o}\vec{r}^{\,b_o} = \sum_{n=1}^{N-M} \frac{\partial{}^{o}\vec{v}_{\mathcal{N}}^{\,b_o}}{\partial u_n}\delta u_n \tag{32}$$

and

$$ {}^{\mathcal{N}}\vec{\delta\theta}^{B} = \sum_{n=1}^{N-M} \frac{\partial{}^{\mathcal{N}}\vec{\omega}^{B}}{\partial u_n}\delta u_n \tag{33}$$

The coefficients $\frac{\partial{}^{o}\vec{v}_{\mathcal{N}}^{\,b_o}}{\partial u_n}$ and $\frac{\partial{}^{\mathcal{N}}\vec{\omega}^{B}}{\partial u_n}$ are Kane's partial velocities [5]. They are equivalent to relationships by Gibbs and Appell [6], through the "cancellation of dots" identity [22].

Further Analyses of Terms from Inter-Body Connections. If equations 32 and 33 are utilized in equation 18 there will be boundary (region of connection) terms of the form (suppressing time dependence for convenience of notation):

$$\int_{\partial\Omega_f} f_{fi}\,\delta\tilde{u}_{fi}\,d\sigma_f + g_{fi}\,\delta\tilde{u}_{fi}(c_f) + \int_{\partial\Omega_f} h_{fi}\,\delta\tilde{u}_{fi,j}\,d\sigma_f + k_{fi}\,\delta\tilde{u}_{fi,j}(c_f) \tag{34}$$

with integration over the region of connection, or equivalently:

$$\int_{\partial\Omega_f} \left(f_{fi} + \mathcal{D}_{c_f}\,g_{fi}\right)\delta\tilde{u}_{fi}\,d\sigma_f + \int_{\partial\Omega_f}\left(h_{fi} + \mathcal{D}_{c_f}\,k_{fi}\right)\delta\tilde{u}_{fi,j}\,d\sigma_f \tag{35}$$

where $\mathcal{D}_{c_f}$ is the Dirac delta for the connection point c_f. The terms f_{fi} and h_{fi} are contributions from active forces in the region. This equation describes a "spike" traction force and torque at the point of "significance." Everywhere else in the region of connection the displacements and slopes are known via equations 22 and 23. The term g_{fi} is the resultant force and k_{fi} the resultant torque that are generated automatically via the variational principle. Define an effective regional force g'_{fi} such that:

$$\int_{\partial\Omega_f} g'_{fi}\,d\sigma_f = g_{fi} \tag{36}$$

and a torque k'_{fi} such that:

$$\int_{\partial\Omega_f} k'_{fi}\,d\sigma_f = k_{fi} \tag{37}$$

where the equivalent distributions take the region's orientation into account via equations 22 and 23. Using these relationships, equation 35 can be rewritten as:

$$\int_{\partial\Omega_f}\left(f_{fi} + g'_{fi}\right)\delta\tilde{u}_{fi}\,d\sigma_f + \int_{\partial\Omega_f}\left(h_{fi} + k'_{fi}\right)\delta\tilde{u}_{fi,j}\,d\sigma_f \tag{38}$$

The variations $\delta\tilde{u}_{fi}$ and $\delta\tilde{u}_{fi,j}$ can now be treated as independent and arbitrary on $\partial\Omega_f$.

For regions of connection in the domain of a flexible continuum body there will be terms of the form:

$$\int_{\Omega_f} F_{fi}\,\delta\tilde{u}_{fi}\,d\Omega_f + G_{fi}\,\delta\tilde{u}_{fi}(c_f) + K_{fi}\,\delta\tilde{u}_{fi,j}(c_f) \tag{39}$$

or equivalently:

$$\int_{\Omega_f}\left(F_{fi} + \mathcal{D}_{c_f}\,G_{fi}\right)\delta\tilde{u}_{fi}\,d\Omega_f + \int_{\Omega_f}\mathcal{D}_{c_f}\,K_{fi}\,\delta\tilde{u}_{fi,j}\,d\Omega_f \tag{40}$$

G_i is the resultant force and K_i the resultant torque that are generated automatically. As was done above, if an equivalent regional force and torque are defined such that:

$$\int_{\Omega_f}\mathcal{H}_{c_f}\,G'_{fi}\,d\Omega_f = G_{fi} \tag{41}$$

and:

$$\int_{\Omega_f}\mathcal{H}_{c_f}\,K'_{fi}\,d\Omega_f = K_{fi} \tag{42}$$

considering equations 22 and 23. Now equation 40 can be rewritten as:

$$\int_{\Omega_f}\left(F_{fi} + \mathcal{H}_{c_f}\,G'_{fi}\right)\delta\tilde{u}_{fi}\,d\Omega_f + \int_{\Omega_f}\mathcal{H}_{c_f}\,K'_{fi}\,\delta\tilde{u}_{fi,j}\,d\Omega_f \tag{43}$$

where $\mathcal{H}_{c_f}$ is the Heaviside step function for the region of connection. Integrating the last term, and using Gauss's integral relationship gives:

$$\int_{\Omega_f}\left(F_{fi} + \mathcal{H}_{c_f}\,G'_{fi} - \frac{1}{h_1h_2h_3}\frac{\partial}{\partial r_{fj}}\left(h_1h_2h_3\,\mathcal{H}_{c_f}\,K'_{fi}\right)\right)\delta\tilde{u}_{fi}\,d\Omega_f \tag{44}$$

The $\delta\tilde{u}_{fi}$ can also be treated as independent and arbitrary in the entire domain, Ω_f, of the body.

Equations of Motion

The equations of motion can now be extracted from equation 18 with the variations chosen with equations 32, 33, 38, and 44 in mind. Since the variations are arbitrary or known on the boundary, and equation 18 is zero, the coefficients of the variations must be zero and/or certain boundary displacement and slope conditions must be prescribed. The result is the following set of hybrid parameter differential equations and boundary conditions.

Ordinary Differential Equations. For each $n \in \{\{I_i^{N-M}\}\setminus\{I^P\}\}$ (i.e. for each independent regular generalized speed), the following first order (in time) ordinary differential equation, in u_n, holds:

$$\begin{aligned} 0 \;=\; & \sum_r \left\{\frac{\partial{}^{o}\vec{v}_{\mathcal{N}}^{\,b_{or}}}{\partial u_n}\cdot\left[\vec{F}_r - \vec{I}_r\right] + \frac{\partial{}^{\mathcal{N}}\vec{\omega}^{B_r}}{\partial u_n}\cdot\left[\vec{T}_r - \vec{J}_r\right]\right\} \\ + & \sum_f \left\{\frac{\partial{}^{o}\vec{v}_{\mathcal{N}}^{\,b_{of}}}{\partial u_n}\cdot\left[\vec{F}_f - \vec{I}_f\right] + \frac{\partial{}^{\mathcal{N}}\vec{\omega}^{B_f}}{\partial u_n}\cdot\left[\vec{T}_f - \vec{J}_f\right]\right\} \end{aligned} \tag{45}$$

The individual terms are defined as follows:

$$\begin{aligned} \vec{F}_r &= \text{Resultant of all non-constraint forces.} \\ \vec{I}_r &= m_r{}^{o}\vec{a}_{\mathcal{N}}^{\,*r} \\ \vec{T}_r &= \text{Moment of all non-constraint forces about the point } b_{or}. \\ \vec{J}_r &= {}^{b_{or}}\vec{r}^{\,*r}\times m_r{}^{o}\vec{a}_{\mathcal{N}}^{\,b_{or}} + \vec{\vec{I}}_{b_{or}}\cdot{}^{\mathcal{N}}\vec{\alpha}^{B_r} + {}^{\mathcal{N}}\vec{\omega}^{B_r}\times\vec{\vec{I}}_{b_{or}}\cdot{}^{\mathcal{N}}\vec{\omega}^{B_r} \\ \vec{F}_f &= \int_{\Omega_f}\left(\mathcal{H}_f\vec{F}_{hf} + \mathcal{D}_f\vec{F}_{df}\right)d\Omega_f + \int_{\partial\Omega_f}\left(\mathcal{H}_f\vec{F}_{hf} + \mathcal{D}_f\vec{F}_{df}\right)d\sigma_f \\ \vec{I}_f &= \int_{\Omega_f}\frac{{}^{\mathcal{N}}d}{dt}\left(\rho_f{}^{o}\vec{v}_{\mathcal{N}}^{\,dm_f}\right)d\Omega_f \\ \vec{T}_f &= \int_{\Omega_f}\left[{}^{b_{of}}\vec{r}^{\,dm_f}\times\left(\mathcal{H}_f\vec{F}_{hf} + \mathcal{D}_f\vec{F}_{df}\right) + \mathcal{H}_f\vec{T}_{hf} + \mathcal{D}_f\vec{T}_{df}\right]d\Omega_f \\ &\quad + \int_{\partial\Omega_f}\left[{}^{b_{of}}\vec{r}^{\,dm_f}\times\left(\mathcal{H}_f\vec{F}_{hf} + \mathcal{D}_f\vec{F}_{df}\right) + \mathcal{H}_f\vec{T}_{hf} + \mathcal{D}_f\vec{T}_{df}\right]d\sigma_f \\ \vec{J}_f &= \int_{\Omega_f}{}^{b_{of}}\vec{r}^{\,dm_f}\times\frac{{}^{\mathcal{N}}d}{dt}\left(\rho_f{}^{o}\vec{v}_{\mathcal{N}}^{\,dm_f}\right)d\Omega_f \end{aligned} \tag{46}$$

The kinematic differential equations associated with equation 45 are the upper partition of the matrix equation 28.

Elastic Partial Differential Equations. This section gives the field and boundary conditions for each elastic body, f. The field equations, distinct for each $\tilde{u}_{fi} \in \Omega_f \setminus \Omega_{c_f}^{d}$ $(i = 1, 2, 3)$, are:

$$\begin{aligned} &\left[\mathcal{H}_f\vec{F}_{hf} + \mathcal{D}_f\vec{F}_{df} - \frac{{}^{\mathcal{N}}d}{dt}\left(\rho_f{}^{o}\vec{v}_{\mathcal{N}}^{\,dm_f}\right)\right]\cdot h_i\,\hat{e}_{fi} + \vec{\nabla}\times\left(\mathcal{H}_f\vec{T}_{hf} + \mathcal{D}_f\vec{T}_{df}\right)\cdot h_i\,\hat{e}_{fi} \\ &\quad - \left[\frac{\partial\bar{V}_f}{\partial\tilde{u}_{fi}} - \frac{1}{h_1h_2h_3}\frac{\partial}{\partial r_{fj}}\left(h_1h_2h_3\frac{\partial\bar{V}_f}{\partial\tilde{u}_{fi,j}}\right)\right. \\ &\quad\quad \left. + \frac{1}{h_1h_2h_3}\frac{\partial^2}{\partial r_{fj}\partial r_{fk}}\left(h_1h_2h_3\frac{\partial\bar{V}_f}{\partial\tilde{u}_{fi,jk}}\right)\right] \\ &\quad + \mathcal{H}_{c_f}\,G'_{fi} - \frac{1}{h_1h_2h_3}\frac{\partial}{\partial r_{fj}}\left(h_1h_2h_3\,\mathcal{H}_{c_f}\,K'_{fi}\right) = 0 \end{aligned} \tag{47}$$

where index notation is used with i, j, and k. The expression $\Omega_f \setminus \Omega^d_{c_f}$ means the domain minus the regions of connection lying in the domain which have specified displacements (kinematically dependent pseudo-coordinates). G'_{fi} is defined in equation 41, with each G_{fi} defined as:

$$G_{fi} = \text{RHS}(45) \tag{48}$$

for each pseudo-generalized speed u_n associated with each $\widetilde{u}_{fi,t}(c_f)$, for every $\Omega^i_{c_f}$ in the domain. $\Omega^i_{c_f}$ implies the regions of connection that are kinematically independent (as discussed above). RHS(45) means the right-hand side of equation 45 is the formula ($G_{fi} \neq 0$ in general). Similarly, K'_{fi} is defined in 42 with each K_{fi} defined as:

$$K_{fi} = \text{RHS}(45) \tag{49}$$

for each pseudo-generalized speed u_n associated with each $\widetilde{u}_{fi,jt}(c_f)$, for every $\Omega^i_{c_f}$ in the domain.

On portions of $\partial\Omega_f$ that have specified tractions, the following boundary conditions hold for each $\widetilde{u}_{fi}$ $(i = 1, 2, 3)$:

$$\left(\mathcal{H}_f \vec{F}_{hf} + \mathcal{D}_f \vec{F}_{df}\right) \cdot h_i \hat{e}_{fi} + 2\left(\mathcal{H}_f \vec{T}_{hf} + \mathcal{D}_f \vec{T}_{df}\right) \cdot \frac{h_k}{h_1 h_2 h_3} \epsilon_{kji} h_{i,j} \hat{e}_{fk}$$
$$-(\hat{n} \cdot \hat{e}_{fj}) \left[h_j \frac{\partial \bar{V}_f}{\partial \widetilde{u}_{fi,j}} - \frac{h_j}{h_1 h_2 h_3} \frac{\partial}{\partial r_{fk}} \left(h_1 h_2 h_3 \frac{\partial \bar{V}_f}{\partial \widetilde{u}_{fi,jk}} \right) \right] = 0 \tag{50}$$

where index notation conventions are used with i, j, and k. Also, for each $\widetilde{u}_{fi,j}$ $(i = 1, 2, 3)$ we have:

$$\left(\mathcal{H}_f \vec{T}_{hf} + \mathcal{D}_f \vec{T}_{df}\right) \cdot \frac{h_k h_i^2}{h_1 h_2 h_3} \epsilon_{kji} \hat{e}_{fk} - (\hat{n} \cdot \hat{e}_{fk}) h_k \frac{\partial \bar{V}_f}{\partial \widetilde{u}_{fi,jk}} = 0 \tag{51}$$

again with index notation.

On portions of the boundary where connections are made, the following boundary conditions hold for each $\widetilde{u}_{fi} \in \partial\Omega^i_{c_f}$ $(i = 1, 2, 3)$:

$$\left(\mathcal{H}_f \vec{F}_{hf} + \mathcal{D}_f \vec{F}_{df}\right) \cdot h_i \hat{e}_{fi} + 2\left(\mathcal{H}_f \vec{T}_{hf} + \mathcal{D}_f \vec{T}_{df}\right) \cdot \frac{h_k}{h_1 h_2 h_3} \epsilon_{kji} h_{i,j} \hat{e}_{fk}$$
$$+g'_{fi} - (\hat{n} \cdot \hat{e}_{fj}) \left[h_j \frac{\partial \bar{V}_f}{\partial \widetilde{u}_{fi,j}} - \frac{h_j}{h_1 h_2 h_3} \frac{\partial}{\partial r_{fk}} \left(h_1 h_2 h_3 \frac{\partial \bar{V}_f}{\partial \widetilde{u}_{fi,jk}} \right) \right] = 0 \tag{52}$$

where index notation is used with i, j, and k. The term $\partial\Omega^i_{c_f}$ represents all boundary connections with independent pseudo-coordinates and speeds. This is where forcing terms (other than g'_{fi}) arise from active elements at the connection. Also, for each $\widetilde{u}_{fi,j} \in \partial\Omega^i_{c_f}$ $(i = 1, 2, 3)$ we have:

$$\left(\mathcal{H}_f \vec{T}_{hf} + \mathcal{D}_f \vec{T}_{df}\right) \cdot \frac{h_k h_i^2}{h_1 h_2 h_3} \epsilon_{kji} \hat{e}_{fk} + k'_{fi} - (\hat{n} \cdot \hat{e}_{fk}) h_k \frac{\partial \bar{V}_f}{\partial \widetilde{u}_{fi,jk}} = 0 \tag{53}$$

again, where the forcing terms (other than k'_{fi}) are from active elements at the connection, and index notation is used on i, j, and k. The term g'_{fi}, in equation 52, is defined in equation 36, with each g_{fi} defined as:

$$g_{fi} = \text{RHS}(45) \tag{54}$$

for each pseudo-generalized speed u_n associated with each $\widetilde{u}_{fi,t}(c_f)$, for every $\partial\Omega^i_{c_f}$ (independent pseudo-coordinates) on the boundary. Similarly, k'_{fi}, in equation 53, is defined in 37 with each k_{fi} defined as:

$$k_{fi} = \text{RHS}(45) \tag{55}$$

for each pseudo-generalized speed u_n associated to each $\widetilde{u}_{fi,jt}(c_f)$, for every $\partial\Omega^i_{c_f}$ on the boundary.

On portions of the boundary where displacements are known, we have:

$$\widetilde{u}_{fi} = \hat{\widetilde{u}}_{fi} \text{ and } \widetilde{u}_{fi,j} = \hat{\widetilde{u}}_{fi,j} \qquad \text{(known)} \tag{56}$$

The functions $\hat{\widetilde{u}}_{fi}$ and $\hat{\widetilde{u}}_{fi,j}$ are determined as follows.

For connections that are made in the domain Ω_f of a flexible body in which the associated pseudo-coordinates and speeds are dependent on other

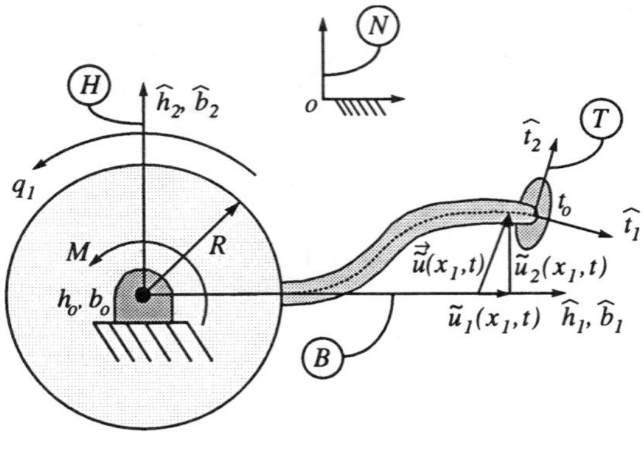

Figure 4: Cantilever Beam Attached to a Rotating Base

system coordinates and speeds, the region $\Omega^d_{c_f}$ must be extracted from the domain of Ω_f. For this boundary, namely $\partial(\Omega_f \setminus \Omega^d_{c_f})$, the displacements and slopes are given by the constraint relationships for the pseudo-coordinates and speeds and the functional forms $\mathcal{L}$ and $\mathcal{M}$ of equations 22 and 23. Because $\Omega^d_{c_f}$ has mass and is moving (in general), the integrals in equation 45 are over the entire domain Ω_f.

On normal boundary ($\partial\Omega_f$) regions of connection in which the pseudo-coordinates and speeds are dependent ($\partial\Omega^d_{c_f}$), the functions $\hat{\widetilde{u}}_{fi}$ and $\hat{\widetilde{u}}_{fi,j}$ are evaluated as follows. They can be found from the constraint relationship for the pseudo-coordinates and speeds and the functional forms 22 and 23 explicit for the region in question.

For portions of the normal boundary ($\partial\Omega_f$) where displacements are specified (prescribed), $\hat{\widetilde{u}}_{fi}$ and $\hat{\widetilde{u}}_{fi,j}$ are easily determined.

To ensure frame B_f rides with the body, (either $\partial\Omega_f$ or Ω_f) let:

$$\widetilde{u}_{fi} = \widetilde{u}_{fi,j} = 0 \qquad (i \neq j) \tag{57}$$

in regions where the frame is attached. This eliminates the rigid body solutions of the partial differential equations, accounted for in the analysis. Special considerations are discussed in Appendix B of [1].

Comments

In this section the derivation of a general vector based methodology for developing equations of motion for hybrid parameter multiple body systems was presented. By developing the methodology in the general form, and absorbing the algebra usually associated with hybrid systems in the development, allows an analyst to use the methodology without great effort. The steps a user of this technique must take are: a) perform kinematic analysis, b) realize active forces and torques, c) form the strain energy density, d) perform partial differentiation, and e) dot multiply vector quantities. The differentiation usually involves simple expressions.

The utility of the method, along with a discussion of other methods, will be demonstrated in the next section.

Cantilever Beam on a Rotating Base

Preamble

In this section the equations of motion for a cantilever beam, with a tip mass/inertia, attached to a rotating base will be derived. This system was chosen because: a) it is non-trivial due to its distributed mass and elasticity, b) its planar nature allows for heuristic equation verification, and c) Hamilton's principle can be readily applied to it. This example will demonstrate some of the qualities of the new methodology, such as: a) its systematic nature, b) its resulting closed form equations, and c) automatic boundary condition generation. No claim to the efficacy of the continuum model will be made, however, the reader is referred to a recent workshop proceedings for a discussion of this problem [23].

System Model

The system considered is shown in figure 4. The coordinate frame marked $\mathcal{N}$ is the Newtonian frame, o is the origin of frame $\mathcal{N}$. H is the hub frame located at the center of the circular hub, origin h_o. The domain of the beam is one dimensional, the independent coordinate is x_1 measured (in the beam coordinate frame) from the root of the beam along the undeformed neutral axis of the beam. The "special" point of the beam coincides with h_o and is labeled b_o. Attached to b_o is the frame B aligned with frame H. There is a torque of magnitude M applied to the hub. The coordinate measuring angular position of the hub is q_1. The hub has mass m_h, and inertia dyadic $\vec{\vec{I}}_{h_o}$ about h_o, its radius is R. The deflection of the beam is measured with $\vec{\widetilde{u}}(x_1,t)$ which has components $\widetilde{u}_1(x_1,t)\hat{b}_1$, measuring elongation, and $\widetilde{u}_2(x_1,t)\hat{b}_2$ measuring deflection. The beam has mass per unit length ρ_B, its length is L, cross sectional area A, area moment of inertia I, and Young's modulus E. The body T attached to the tip of the beam has mass m_t, and inertia dyadic $\vec{\vec{I}}_{t_o}$ about the center of gravity of the body t_o.

In order to utilize the formulas presented as equations 45–57 several items need to be calculated. The first items to be calculated are the angular velocities of the hub, frame B, and the tip mass T. They are:

$$ {}^{\mathcal{N}}\vec{\omega}^{H} = \dot{q}_1\hat{h}_3 = u_1\hat{h}_3 \tag{58} $$

and

$$ {}^{\mathcal{N}}\vec{\omega}^{B} = \dot{q}_1\hat{b}_3 = u_1\hat{b}_3 \tag{59} $$

and

$$ {}^{\mathcal{N}}\vec{\omega}^{T} = (u_1+\dot{q}_4')\,\hat{t}_3 = (u_1+u_4')\,\hat{t}_3 \tag{60} $$

where ${}^{\mathcal{N}}\vec{\omega}^{H}$ is for the hub, ${}^{\mathcal{N}}\vec{\omega}^{B}$ is for the beam frame, and ${}^{\mathcal{N}}\vec{\omega}^{T}$ is for the tip mass. The prime ($'$) denotes pseudo-coordinates such as $q_4' = \frac{\partial \widetilde{u}_2}{\partial x_1}(L,t)$.

Appropriate angular accelerations are also required. The relationship for the hub and beam is:

$$ {}^{\mathcal{N}}\vec{\alpha}^{B} = \dot{u}_1\hat{h}_3 \tag{61} $$

and

$$ {}^{\mathcal{N}}\vec{\alpha}^{T} = (\dot{u}_1+\dot{u}_4')\,\hat{t}_3 \tag{62} $$

for the tip mass T.

The absolute velocities of the points h_o, b_o and t_o ("special" points) are also required. This system is rotating about h_o so:

$$ {}^{o}\vec{v}_{\mathcal{N}}^{h_o} = {}^{o}\vec{v}_{\mathcal{N}}^{b_o} = 0 \tag{63} $$

and

$$ {}^{o}\vec{v}_{\mathcal{N}}^{t_o} = (u_2' - u_1q_3')\,\hat{b}_1 + (u_3' + u_1\,(R+L+q_2'))\,\hat{b}_2 \tag{64} $$

The pseudo-coordinates are defined as: $q_2' = \widetilde{u}_1(L,t)$ and $q_3' = \widetilde{u}_2(L,t)$. The pseudo-speeds are defined as: $u_2' = \dot{q}_2'$ and $u_3' = \dot{q}_3'$.

The absolute acceleration of the differential beam element dm can be written as:

$$ \begin{aligned} {}^{o}\vec{a}_{\mathcal{N}}^{dm} = &\left(\frac{\partial^2\widetilde{u}_1}{\partial t^2} - 2u_1\frac{\partial\widetilde{u}_2}{\partial t} - \dot{u}_1\widetilde{u}_2 - (R+x_1+\widetilde{u}_1)\,u_1^2\right)\hat{b}_1 \\ &+\left(\frac{\partial^2\widetilde{u}_2}{\partial t^2} + 2u_1\frac{\partial\widetilde{u}_1}{\partial t} + (R+x_1+\widetilde{u}_1)\,\dot{u}_1 - \widetilde{u}_2u_1^2\right)\hat{b}_2 \end{aligned} \tag{65} $$

The acceleration of t_o is:

$$ \begin{aligned} {}^{o}\vec{a}_{\mathcal{N}}^{t_o} = &\left(\dot{u}_2' - 2u_1u_3' - \dot{u}_1q_3' - (R+L+q_2')\,u_1^2\right)\hat{b}_1 \\ &+\left(\dot{u}_3' + 2u_1u_2' + (R+L+q_2')\,\dot{u}_1 - q_3'u_1^2\right)\hat{b}_2 \end{aligned} \tag{66} $$

The strain energy density function for the beam is also required. Assuming coupling between deflection and elongation is pertinent, the following will be used. Namely:

$$ \bar{V} = \frac{1}{2}\left\{EA\left(\frac{\partial\widetilde{u}_1}{\partial x_1} + \frac{1}{2}\left(\frac{\partial\widetilde{u}_2}{\partial x_1}\right)^2\right)^2 + EI\left(\frac{\partial^2\widetilde{u}_2}{\partial x_1^2}\right)^2\right\} \tag{67} $$

The torque applied to the hub is:

$$ \vec{T}_{h_o} = M\hat{h}_3 \tag{68} $$

The methodology also requires the calculations of the "preferred directions" for the variations, namely the partial speeds. They are, by inspection (of equations 58, 59, 60, 63, and 64), represented as:

	${}^{o}\vec{v}_{\mathcal{N}}^{h_o}$	${}^{o}\vec{v}_{\mathcal{N}}^{b_o}$	${}^{o}\vec{v}_{\mathcal{N}}^{t_o}$	${}^{\mathcal{N}}\vec{\omega}^{H}$	${}^{\mathcal{N}}\vec{\omega}^{B}$	${}^{\mathcal{N}}\vec{\omega}^{T}$
$\frac{\partial}{\partial u_1}$	0	0	$-q_3'\hat{b}_1 + (R+L+q_2')\,\hat{b}_2$	$\hat{h}_3$	$\hat{b}_3$	$\hat{t}_3$
$\frac{\partial}{\partial u_2'}$	0	0	$\hat{b}_1$	0	0	0
$\frac{\partial}{\partial u_3'}$	0	0	$\hat{b}_2$	0	0	0
$\frac{\partial}{\partial u_4'}$	0	0	0	0	0	$\hat{t}_3$

(69)

The equations of motion can now be written down.

The ordinary differential equation, governing angular displacement of the hub, comes from equation 45. The formula is:

$$ \begin{aligned} 0 = &\frac{\partial\,{}^{\mathcal{N}}\vec{\omega}^{H}}{\partial u_1}\cdot\left[\vec{T}_H - \vec{J}_H\right] + \frac{\partial\,{}^{\mathcal{N}}\vec{\omega}^{B}}{\partial u_1}\cdot\left[\vec{T}_B - \vec{J}_B\right] \\ &+\frac{\partial\,{}^{o}\vec{v}_{\mathcal{N}}^{t_o}}{\partial u_1}\cdot\left[\vec{F}_T - \vec{I}_T\right] + \frac{\partial\,{}^{\mathcal{N}}\vec{\omega}^{T}}{\partial u_1}\cdot\left[\vec{T}_T - \vec{J}_T\right] \end{aligned} \tag{70} $$

Upon substitution of the terms defined above, the final form is:

$$ \begin{aligned} 0 = &\left\{I_3 + J_3 + \int_0^L \rho_B\left[(R+x_1+\widetilde{u}_1)^2 + \widetilde{u}_2^2\right]dx_1 + m_t\left({q_3'}^2 + (R+L+q_2')^2\right)\right\}\dot{u}_1 \\ &+\int_0^L \rho_B\left[\frac{\partial^2\widetilde{u}_2}{\partial t^2}(R+x_1+\widetilde{u}_1) + 2u_1\frac{\partial\widetilde{u}_1}{\partial t}(R+x_1+\widetilde{u}_1) - \frac{\partial^2\widetilde{u}_1}{\partial t^2}\widetilde{u}_2 + 2u_1\frac{\partial\widetilde{u}_2}{\partial t}\widetilde{u}_2\right]dx_1 \\ &-M + m_t\left[\dot{u}_3'(R+L+q_2') + 2u_1u_2'(R+L+q_2') - \dot{u}_2'q_3' + 2u_1u_3'q_3'\right] + J_3\dot{u}_4' \end{aligned} \tag{71} $$

The inertia for the hub is I_3 and for the tip mass J_3.

The field equation governing elongation, from equation 47 with $i = 1$, is:

$$ 0 = \frac{\partial}{\partial x_1}\left(\frac{\partial\bar{V}}{\partial\widetilde{u}_{1,1}}\right) - \rho_B\,{}^{o}\vec{a}_{\mathcal{N}}^{dm}\cdot\hat{b}_1 \tag{72} $$

with boundary condition, from equation 52 with $i = 1$:

$$ \frac{\partial\bar{V}}{\partial\widetilde{u}_{1,1}} = \frac{\partial\,{}^{o}\vec{v}_{\mathcal{N}}^{t_o}}{\partial u_2'}\cdot\left[\vec{F}_T - \vec{I}_T\right] \tag{73} $$

at $x_1 = L$ and $\widetilde{u}_1 = 0$ at $x_1 = 0$. Performing the required differentiation results in:

$$ \begin{aligned} 0 = &\frac{\partial}{\partial x_1}\left[EA\left(\frac{\partial\widetilde{u}_1}{\partial x_1} + \frac{1}{2}\left(\frac{\partial\widetilde{u}_2}{\partial x_1}\right)^2\right)\right] \\ &-\rho_B\left(\frac{\partial^2\widetilde{u}_1}{\partial t^2} - 2u_1\frac{\partial\widetilde{u}_2}{\partial t} - \dot{u}_1\widetilde{u}_2 - (R+x_1+\widetilde{u}_1)\,u_1^2\right) \end{aligned} \tag{74} $$

replacing equation 72. While equation 73 is replaced with:

$$ EA\left(\frac{\partial\widetilde{u}_1}{\partial x_1} + \frac{1}{2}\left(\frac{\partial\widetilde{u}_2}{\partial x_1}\right)^2\right) = -m_t\left(\dot{u}_2' - 2u_1u_3' - \dot{u}_1q_3' - (R+L+q_2')\,u_1^2\right) \tag{75} $$

at $x_1 = L$.

The field equation governing deflection, from equation 47 with $i = 2$, is:

$$ 0 = \frac{\partial}{\partial x_1}\left(\frac{\partial\bar{V}}{\partial\widetilde{u}_{2,1}}\right) - \frac{\partial^2}{\partial x_1^2}\left(\frac{\partial\bar{V}}{\partial\widetilde{u}_{2,11}}\right) - \rho_B\,{}^{o}\vec{a}_{\mathcal{N}}^{dm}\cdot\hat{b}_2 \tag{76} $$

with boundary condition, from equations 52 and 53 with $i = 2$:

$$ \begin{aligned} \frac{\partial\bar{V}}{\partial\widetilde{u}_{2,1}} - \frac{\partial}{\partial x_1}\left(\frac{\partial\bar{V}}{\partial\widetilde{u}_{2,11}}\right) &= \frac{\partial\,{}^{o}\vec{v}_{\mathcal{N}}^{t_o}}{\partial u_3'}\cdot\left[\vec{F}_T - \vec{I}_T\right] \\ \frac{\partial\bar{V}}{\partial\widetilde{u}_{2,11}} &= \frac{\partial\,{}^{\mathcal{N}}\vec{\omega}^{T}}{\partial u_4'}\cdot\left[\vec{T}_T - \vec{J}_T\right] \end{aligned} \tag{77} $$

at $x_1 = L$ and

$$ \widetilde{u}_2 = \widetilde{u}_{2,1} = 0 \tag{78} $$

at $x_1 = 0$. Performing the required differentiation results in:

$$ \begin{aligned} 0 = &\frac{\partial}{\partial x_1}\left[EA\left(\frac{\partial\widetilde{u}_1}{\partial x_1} + \frac{1}{2}\left(\frac{\partial\widetilde{u}_2}{\partial x_1}\right)^2\right)\frac{\partial\widetilde{u}_2}{\partial x_1}\right] - \frac{\partial^2}{\partial x_1^2}\left(EI\frac{\partial^2\widetilde{u}_2}{\partial x_1^2}\right) \\ &-\rho_B\left(\frac{\partial^2\widetilde{u}_2}{\partial t^2} + 2u_1\frac{\partial\widetilde{u}_1}{\partial t} + (R+x_1+\widetilde{u}_1)\,\dot{u}_1 - \widetilde{u}_2u_1^2\right) \end{aligned} \tag{79} $$

replacing equation 76. While equation 77 is replaced with:

$$EA\left(\frac{\partial\widetilde{u}_1}{\partial x_1}+\frac{1}{2}\left(\frac{\partial\widetilde{u}_2}{\partial x_1}\right)^2\right)\frac{\partial\widetilde{u}_2}{\partial x_1}-\frac{\partial}{\partial x_1}\left(EI\frac{\partial^2\widetilde{u}_2}{\partial x_1^2}\right)=$$

$$-m_t\left(\dot{u}_3'+2u_1u_2'+(R+L+q_2')\,\dot{u}_1-q_3'u_1^2\right)$$

$$EI\frac{\partial^2\widetilde{u}_2}{\partial x_1^2}=-J_3(\dot{u}_1+\dot{u}_4') \tag{80}$$

at $x_1 = L$. Initial conditions for $q_1, u_1 = \dot{q}_1, \widetilde{u}_1, \frac{\partial\widetilde{u}_1}{\partial t}, \widetilde{u}_2$ and $\frac{\partial\widetilde{u}_2}{\partial t}$ must also be specified.

Discussion

Examples found in papers by Lee and Junkins [11] and Choura, et al. [24] along with many papers in [23] can be used as comparison to the example presented above. In [11] an explicit Lagrangian approach is used to model the system with a point mass at the tip. In [24] Hamilton's principle and Rayleigh beam assumptions are used to model the system with no tip mass. Although the algebra required to present the equations of motion for the example herein is relatively extensive, it is believed that the total labor costs should compare favorably to the methods in [9, 10, 11, 24]. If one applies Hamilton's principle directly to the example problem, the advantages of the method presented in this work will become evident. The utilization of vector algebra, as the kinematic medium, should also compare favorably to the scalar based methods presented in [9, 10, 11, 24], when many bodies are considered. It is not demonstrated in this paper, but nonholonomic constraints can be easily absorbed with the new technique [1, 21], whereas in [9, 10, 11, 24] Lagrange multipliers are necessary.

Summary

Presented in this paper is an explicit formulation for modeling hybrid parameter multiple body systems. The technique is variational in nature and is formulated in the system's constraint-free subspace of the configuration space. Thus holonomic and nonholonomic systems are managed without the use of Lagrange multipliers [1, 21]. The spatial continuum dimension is general as well as the connections between system constituent bodies. The methodology is applied systematically to any system at hand. Vector algebra is utilized, which makes the use of multiple coordinate frames natural.

The methodology was applied to a planar cantilever beam (with tip mass/inertia) attached to a rotating base. In a systematic presentation of the technique, the complete hybrid model for the beam system was presented.

One of the advantages of using the methodology is in the realm of rapid regeneration of the equations of motion for a particular system. If an investigator decided to make some changes to the continuum model, the technique's systematic nature make this a much simpler task. The technique is easily applied to problems that are constrained and/or have higher dimensional continuum bodies. The technique is also extendible to continuum assumptions such as in a Timoshenko beam. Discussion of these related topics is presented in [1].

REFERENCES

[1] Barhorst, A. A., *On Modeling the Dynamics of Hybrid Parameter Multiple Body Mechanical Systems*, Ph. D. Dissertation, Texas A & M University, College Station, Texas, 1991.

[2] Barhorst, A. A., and Everett, L. J., "Modeling Hybrid Parameter Multiple Body Systems: A Different Approach," Submitted to *The International Journal of Nonlinear Mechanics*, April 1, 1992.

[3] Gibbs, J. W., *The Scientific Papers of J. Willard Gibbs, Volume II: Dynamics, Etc.*, Dover, New York, 1961.

[4] Desloge, Edward A., *Classical Mechanics*, John Wiley & Sons, New York, 1982.

[5] Kane, T. R., and Levinson, D. A., *Dynamics Theory and Applications*, McGraw-Hill, New York, 1985.

[6] Desloge, E. A., "Relationship Between Kane's Equations and the Gibbs-Appell Equations," *J. Guidance, Control, and Dynamics*, Vol. 10, No. 1, 1987, pp. 120–122.

[7] Lips, K. W., and Singh, R. P., "Obstacles to High Fidelity Multibody Dynamics Simulation," *Proc. of the American Control Conference*, Vol. 1, June 1988, pp. 587–594.

[8] Likins, P. W., "Multibody Dynamics: A Historical Perspective," *Proc. of the Workshop on Multibody Simulation*, JPL D-5190, Vol. 1, April 1988, pp. 10–24.

[9] Meirovitch, Leonard, "State Equations of Motion for Flexible Bodies in Terms of Quasi-Coordinates," *Proc. IUTAM/IFAC Symposium on Dynamics of Controlled Mechanical Systems*, Zurich, 1988, Springer-Verlag, Berlin, 1989, pp. 37–48.

[10] Low, K. H., and Vidyasagar, M., "A Lagrangian Formulation of The Dynamic Model for Flexible Manipulator Systems," *ASME Journal of Dynamic Systems, Measurement and Control*, Vol. 110, June 1988, pp. 175–181.

[11] Lee, S., and Junkins, J. L., "Explicit Generalizations of Lagrange's Equations for Hybrid Coordinate Dynamical Systems," Submitted to *AIAA J. Guidance, Control and Dynamics*, March 1991.

[12] Petzold, L., "Differential/Algebraic Equations Are Not ODE's," *SIAM J. Sci. Stat. Comput.*, Vol. 3, No. 3, 1982, pp. 367–384.

[13] Lanczos, Cornelius, *The Variational Principles of Mechanics*, The University of Toronto Press, Toronto, 1970.

[14] Everett, L. J., and McDermott, M. Jr., "The Use of Vector Techniques in Variational Problems," *ASME Journal of Dynamic Systems, Measurement and Control*, Vol. 108, No. 2, June 1986, pp. 141–145.

[15] Wittenberg, Jens, *Dynamics of Systems of Rigid Bodies*, B. G. Teubner, Stuttgart, 1977.

[16] Everett, L. J., "An Alternative Algebra for Deriving Equations of Motion of Compliant Manipulators," *J. Robotic Systems*, Vol. 5, No. 6, Dec 1988, pp. 553–566.

[17] Sol, E. J., *Kinematics and Dynamics of Multibody Systems, A Systematic Approach to Systems with Arbitrary Connection*, Dr. Ir. Thesis, Technische Hogeschool Eindhoven, Holland, 1983.

[18] Haug, E. J., and McCullough, M. K., "A Variational-Vector Calculus Approach to Machine Dynamics," *J. Mechanisms, Transmissions, and Automation in Design*, Vol. 108, March 1986, pp. 25–30.

[19] Malvern, L. E., *Introduction to The Mechanics of Continuous Medium*, Prentice-Hall, Englewood Cliffs, New Jersey, 1969.

[20] Reddy, J. N., *Energy and Variational Methods in Applied Mechanics*, John Wiley & Sons, New York, 1984.

[21] Barhorst, A. A., and Everett, L. J., "Obtaining the Minimal Set of Hybrid Parameter Differential Equations for Mechanisms," ASME Design Engineering Technical Conference, September 13-16, 1992, Phoenix AZ.

[22] Junkins, John L., *Optimal Spacecraft Rotational Maneuvers*, Elsevier Scientific Publishing Co., New York, 1985.

[23] Man, G. and Laskin, R., *Proc. of the Workshop on Multibody Simulation*, JPL D-5190, Vol. 1–4, April 1988.

[24] Choura, S., Jayasuriya, S., and Medick, M. A., "On the Modeling, and Open Loop Control of a Thin Flexible Beam," *ASME J. Dynamic Systems, Measurement and Control*, Vol. 113, No. 1, March 1991, pp. 26–33.

Appendix

This appendix discusses the notation used in the paper. The symbols not included here are obvious given this list and the context.

$\hat{b}_i$: unit vector $(i = 1, 2, 3)$.

$\hat{n}_i$: unit vector $(i = 1, 2, 3)$ in Newtonian frame, sometimes designates outward pointing surface unit vector.

${}^{A}\vec{\omega}^{B}$: angular velocity of frame B relative to frame A. Any uppercase superscript letter denotes a frame of reference.

${}^{A}\vec{\alpha}^{\,B}$: angular acceleration of frame B relative to frame A.

${}^{a}\vec{r}^{\,b}$: a vector from point a (tail) to point b (head). Any lower case superscript letter denotes points.

${}^{A}\delta^{a}\vec{r}^{\,b}$: variation of ${}^{a}\vec{r}^{\,b}$ with respect to frame A.

$\frac{{}^{A}d}{dt}$: differentiation with respect to frame A.

$\frac{{}^{A}d}{dt}\left({}^{a}\vec{r}^{\,b}\right) = {}^{a}\vec{v}_{A}^{\,b}$: velocity of point b relative to point a as seen in frame A. Note: ${}^{a}\vec{v}_{A}^{\,b}$ could also represent a nonholonomic velocity vector that is not a differential of a position vector.

$\frac{{}^{A}d}{dt}\left({}^{a}\vec{v}_{A}^{\,b}\right) = {}^{a}\vec{a}_{A}^{\,b}$: acceleration of point b relative to point a as seen in frame A.

$\tilde{u}_i$: components ($i = 1,2,3$) of displacement field variable. The symbol $\tilde{}$ is used to denote field variables.

$\tilde{\vec{u}}$: displacement vector field.

$\tilde{V}$: strain energy density function (scalar).

i, j, k, l: as subscript indices, the usual tensor subscript notation is implied. That is: summation on repeated indices, comma notation for derivatives, etc.

r, f: subscripts used to denote body type, i.e. rigid, and flexible.

$\mathcal{H}, \mathcal{D}$: Heaviside step function and Dirac delta function, respectively. They are defined for the spatial domain in question.

$\vec{F}, \vec{T}$: forces and torques (may be spatially distributed).

$\vec{F}_r$: resultant of all non-constraint forces acting on body r.

m_r: mass of body r.

${}^{o}\vec{a}_{\mathcal{N}}^{\,*r}$: absolute acceleration of the center of mass of body r.

$\vec{T}_{b_{or}}$: resultant torque of all non-constraint forces about b_{or} including applied couples.

${}^{b_{or}}\vec{r}^{\,*r}$: position of the center of mass with respect to point b_{or}.

${}^{o}\vec{a}_{\mathcal{N}}^{\,b_{or}}$: absolute acceleration of the point b_{or}.

$\vec{\vec{I}}_{b_{or}}$: Inertia dyadic of body r about the point b_{or}.

${}^{\mathcal{N}}\vec{\omega}^{\,B_r}$: angular velocity of body or frame B_r.

${}^{\mathcal{N}}\vec{\alpha}^{\,B_r}$: angular acceleration of body or frame B_r.

AMD-Vol. 141/DSC-Vol. 37, Dynamics of Flexible Multibody Systems:
Theory and Experiment
ASME 1992

EFFICIENT MODELLING OF DOCKING AND BERTHING DYNAMICS FOR FLEXIBLE SPACE STRUCTURES

Juan Jose Gonzalez Vallejo
Sener Tecnica Industrial
Y Naval, S.A.
Las Arenas, Vizcaya, Spain

Javier Edgar Benavente
Dynacs Engineering Co., Inc.
Palm Harbor, Florida

Gonzalo Taubmann
Sener Tecnica Industrial Y Naval, S.A.
Las Arenas, Vizcaya, Spain

ABSTRACT

Currently envisioned space based infrastructures will rely heavily on berthing and docking maneuvers during both the build-up and mission phases of operations. Integrated systems level analysis tools that combine robotics, controls, guidance and multi-flexible-body simulation techniques in an easy to use format are needed to understand the subsystem coupling that can be expected on-orbit.

Recent advances in several technologies, including computer software and hardware technologies, have now made it possible to simulate the detailed contact dynamics associated with docking and berthing of large, flexible space structures in conjunction with other major sub-systems on a relatively low cost workstation.

This paper presents the SENDAP simulation environment as a tool capable of analysing flexible space structures with many varied sub-systems, including, but not limited to, attitude control systems, manipulator joint controllers, docking mechanisms, etc., all in the presence of contact forces generated in docking or berthing operations. The overall objective of this paper is to show, using an applicable, complex example, the varied, systems level analysis capabilities SENDAP has for modelling flexible space structures.

The paper is organized as follows: First, a description of SENDAP and its general capabilities is presented. Next, a discussion of contact modelling within SENDAP is presented along with results from a series of simulations made with SENDAP to validate the contact routines. Finally, SENDAP's system level analysis capabilities are exercised via simulation of a complex berthing and mating maneuver between a multi-flexible-body Space Station Freedom (SSF) model and the European Attached Pressurized Module (APM) via the Canadian Space Station Remote Manipulator System (SSRMS).

THE SENDAP ENVIRONMENT

SENDAP is an integrated set of software tools designed to simulate the multi-body dynamics and control issues related to berthing and docking operations. Though designed with the docking and berthing operation in mind, SENDAP is in no way restricted to analysis of this type. The SENDAP environment is, in reality, a generic dynamics and control simulation tool with a capability to analyse and simulate contact dynamics.

SENDAP is unique in the fact that the environment allows the user to take on a "systems level" perspective and analysis methodology in an efficient, user friendly manner. SENDAP's primary features include:

* A symbolic, case specific FORTRAN code generator
* Order (N), frontal-like dynamics solution algorithms for generic, open-loop topology multi-body flexible or rigid structures
* Contact dynamics algorithms
* Multiple system capability with each system consisting of multiple bodies and controllers
* Multiple digital and continuous controller capability
* MSC and COSMIC NASTRAN interface
* Inverse kinematics module for robot manipulators
* An interactive 3-D graphic animation tool
* An interactive 2-D plot package
* An interactive problem preparation tool
* Orbital environment, including gravity gradient, orbit dynamics and aerodynamic drag
* A full set of theoretical sensors and actuators typical of spacecraft operations

* User defined software interfaces

The SENDAP time history simulation consists of two independent modules, which together solve a spacecraft's system equations of motion. The "shell" module comprises the generic portion. It contains all of the sensors, actuators, orbital environment and control interfaces, that is, all routines that are independent of the spacecraft's topology.

The second module is termed the "core" and it performs the actual dynamic computations for a specific system topology. The core module is generated symbolically by the FORTRAN code generator, DAPGEN. DAPGEN generates the FORTRAN code to solve case specific system equations of motion cast in the form of Kane's Dynamical Equations [1]. Furthermore, Kane's equations are written in a recursive, frontal-like solution algorithm. The result is a set of case specific equations of motion who's computational burden is proportional to "N", where N is the number of rigid and flexible degrees of freedom (DOF) in the system. Traditional numerically based simulations have a computational burden proportional to N^3. Additional details concerning SENDAP's computational algorithms can be found in [2].

SENDAP's Order (N) algorithm allows it to efficiently handle complex spacecraft system analysis with structural flexibility modelled. In addition, SENDAP's efficiency makes the simulation of "stiff" systems feasible. Stiff systems, such as the contact dynamics class of problems, require small integration steps, on the order of 10-4 seconds.

Coupling the small time step requirement with high DOF systems makes traditional numerical simulation impractical from a "wall clock" point of view. SENDAP was designed to overcome these limitations and make the simulation of complex contact dynamics problems practical.

MODELLING CONTACT DYNAMICS IN SENDAP

The original concept for SENDAP called for the contact dynamics routines to be generic in nature, that is, major software changes should not be required for evolving contact dynamic modelling needs.

This requirement dictated the development of "geometric contact primitives" within SENDAP's contact routines. The geometric contact primitives are basic shapes that can be used to build more complex contacting structures. Examples of the primitives are cylinders, planar rings, torus and hemispheres.

The advantage of this approach is that the basic software that detects contact between any two primitives never changes. The engineer need only specify the position and orientation of each contact primitive that describes his contact model. Everything else is handled by the SENDAP environment.

A second advantage is that basic geometric primitives are easy to display in a graphic, 3-D animation tool. As a collection of primitives composing a simulated system, the animation becomes an excellent conceptualization tool that is invaluable to the engineer.

The basic approach to modelling contact dynamics in SENDAP is as follows: two bodies in the simulated system are identified as the target and chase bodies. Each of these bodies is further identified with a contact dynamics reference node. All primitives defining the contact geometry for the chase body (CH) are defined relative to the chase body contact reference node and all primitives defining the contact geometry for the target body (TA) are defined relative to the target body contact reference node.

Besides the target and chase bodies, the SENDAP contact dynamics software allows the definition of three other types of bodies. The first is the "AR" body (AR for Attenuation Ring), the second is the "ND" body (ND for Non-Docking body). The third is the "IB" body (IB for Interference Body). The AR body allows

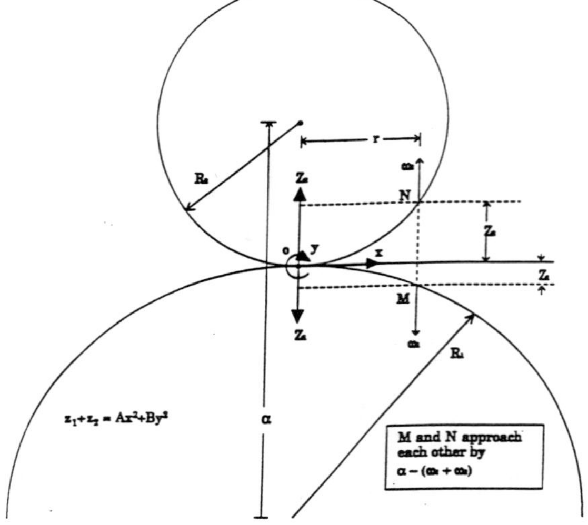

Figure 1: Two Arbitrary Bodies in Contact

for a third independent contacting body in the simulation and the ND bodies are for animation purposes only. IB bodies allow detection of contact between bodies without generating contact forces.

Associated with each primitive is a material identification number. Properties defined for each material are its Young's Modulus, Poisson's Ratio, coefficient of friction, coefficient of material damping and apparent material frequency. In addition to defining a material for each primitive, the engineer must specify which other primitives each primitive can make contact with. This specification is based on physical constraints and has been implemented for contact search efficiency.

The SENDAP contact dynamics software computes contact forces based on a Hertzian contact model [3]. Figure 1 represents two bodies of arbitrary geometry in contact. By neglecting higher order terms, the surfaces near the point of contact can be expressed by the equations:

$$Z_1 = A_1x^2 + A_2xy + A_3y^2$$
$$Z_2 = B_1x^2 + B_2xy + B_3y^2$$

By appropriate choice of axis, the distance between two points M and N can be expressed as:

$$Z_1 + Z_2 = Ax^2 + By^2$$

where A and B are constants depending on the magnitudes of the principal curvature of the contacting surfaces and the angles between the normals at the point of contact.

If we let E_i and ν_i be the Young's Modulus and Poisson's Ratio for the contacting materials, then

$$k_i = \frac{1-\nu^2}{\pi E_i}$$

and

$$(k_1+k_2)\int\int \frac{qdA}{r} = \alpha - Ax^2 - By^2$$

where qdA is the pressure acting on an infinitely small element of the surface of contact and r is the distance from the point under consideration. Hertz found that an elliptical pressure distribution satisfies this integral; hence,

$$q_0 = \frac{3}{2}\frac{P}{2\pi ab}$$

$$a = m\left[\frac{3\pi}{4}\frac{P(k_1+k_2)}{(A+B)}\right]^{\frac{3}{2}}$$

$$b = n\left[\frac{3\pi}{4}\frac{P(k_1+k_2)}{(A+B)}\right]^{\frac{3}{2}}$$

where q_o is the maximum pressure between the two bodies and a and b are the magnitudes of the semi-axis of the elliptical contact surface. m and n are parameters based on A and B. The equations for a and b can be solved simultaneously for P, and; hence, the contact force.

CONTACT VALIDATION: THE DBS-FE MODEL

During the first quarter of 1991, Matra of France performed a series of docking tests on the Docking/Berthing System Front End Model. The tests consisted of a hardware in the loop docking device driven by a numerical "environment" simulation. The empirical data from these contact tests form the basis for the validation of SENDAP's contact dynamics routines. This section briefly describes the SENDAP models developed to simulate the DBS-FE Matra simulations and validate the contact routines.

Figure 2 shows the general topology utilised throughout the simulation cases ran to validate the contact software. In general, a 7 body system with 22 DOF was simulated, except for cases with 3 latches, where a 6 body system with 21 DOF was simulated. Figure 3 shows a SENDAP animated representation of the Hermes/Columbus Docking System (HCDS), which is a derivitive of and very similar to the DBS-FE docking device developed by Sener.

Figures 4, 5 and 6 compare simulation results from SENDAP to the Matra experimental contact results for a typical test run. Compared are the relative axial displacement and velocity between the chase and target vehicles in addition to the axial contact force applied to the target vehicle.

In general, SENDAP's contact computations agreed favorably with the Matra test results in all comparison cases. This increased the overall confidence in the SENDAP environment in general and the contact computation routines in particular.

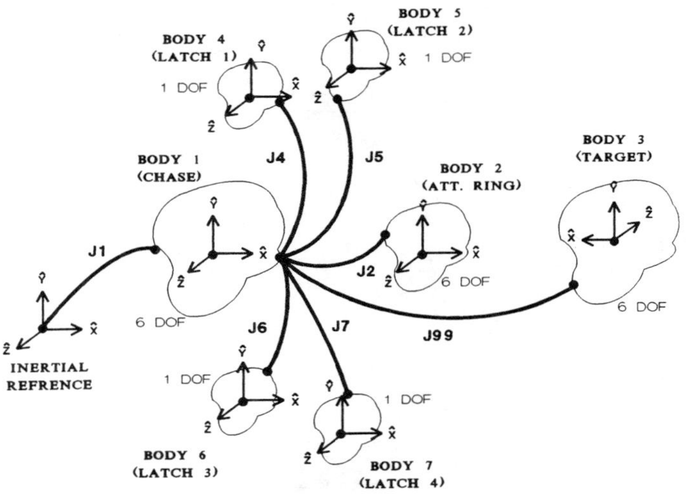

Figure 2: General Topology for DBS-FE Model

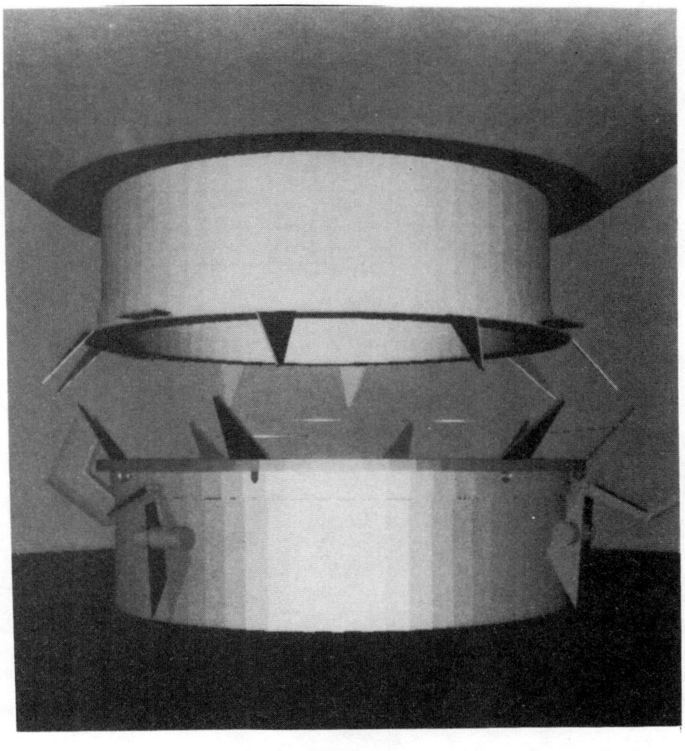

Figure 3: Photograph of a SENDAP Animation of the HCDS

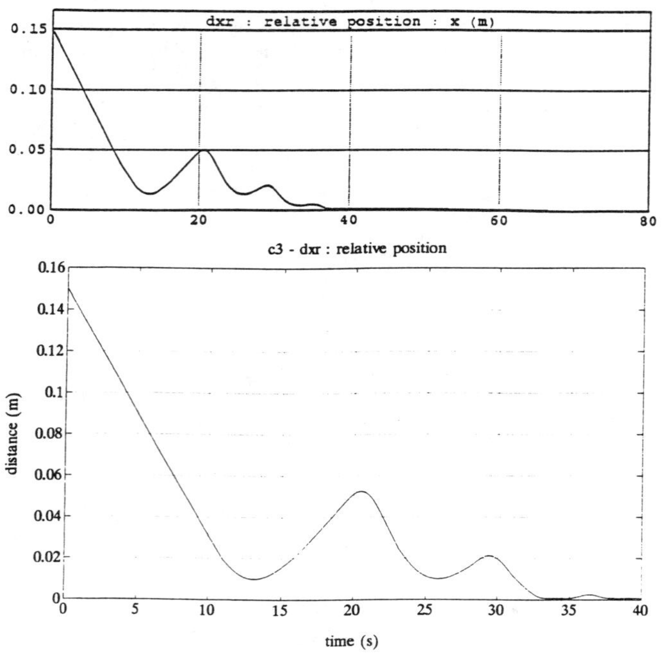

Figure 4: Relative Axial Displacement (m), Experimental and SENDAP

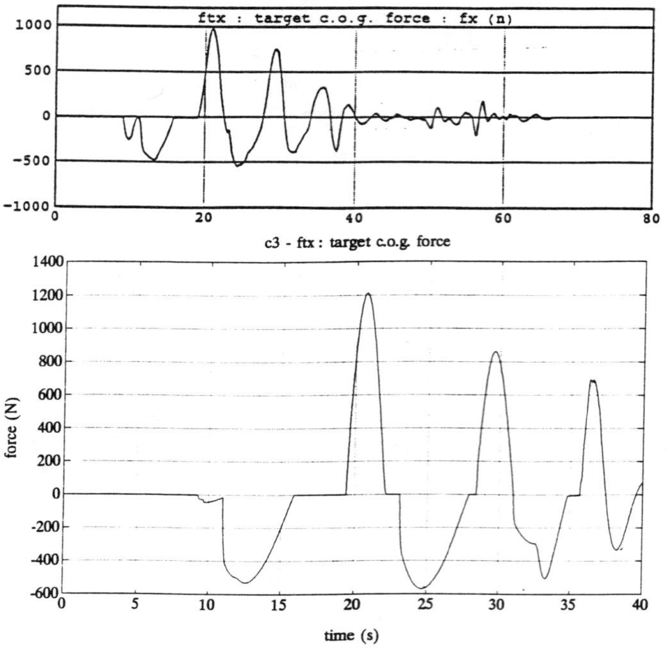

Figure 6: Relative Axial Force on Target (N), Experimental and SENDAP

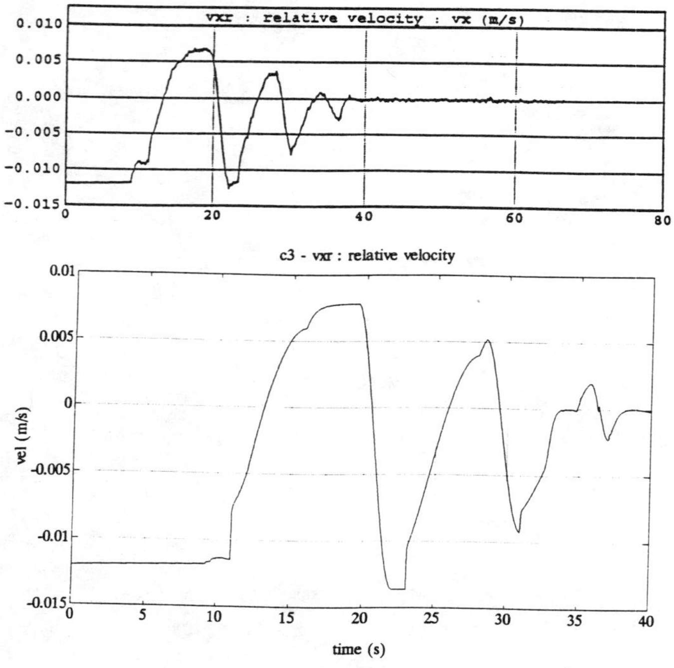

Figure 5: Relative Axial Velocity (m/s), Experimental and SENDAP

FLEXIBLE BODY MODEL WITH CONTACT: A COMPLEX EXAMPLE

Details regarding how SENDAP utilizes flexible body data can be found in [2]; however, in general, SENDAP accepts component modal data in the form of mode shapes and slopes and nodal body data from the MSC and COSMIC NASTRAN finite element programs. Data from other finite element programs may be used, but the SENDAP environment includes interfaces to these two popular finite element programs.

To demonstrate the overall capabilities of the SENDAP environment, we have chosen a non-trivial example: the berthing and mating of the APM to the SSF by the SSRMS.

This is a non-trivial example for several reasons:

* The simulation includes a large number (36) of rigid and flexible bodies
* The simulation contains a total of 149 rigid and flexible degrees of freedom
* Digital user defined controllers implement detailed representations of the SSRMS joint servo controllers for all seven SSRMS articulated joints
* Detailed models of the four independent latch mechanisms for the SSF/APM docking port, including the closed loop topologies and the step motor drivers, are represented through continuous controllers
* A full contact dynamics model between the SSF and APM representing the Common Berthing Mechanism (CBM)
* Inverse kinematics generation for the SSRMS joint servo controller commands

Figure 7 shows in some detail the scope of the topology for the SSF/SSRMS/APM simulation. The Space Station Freedom model is based on a COSMIC NASTRAN component model of the PDR MB-15, circa 1st quarter, 1991. The component model of the SSF consists of 11 bodies: a rigid core (to facilitate placement of docking port node locations, SSRMS attachment points, etc.), 2 flexible alpha booms attached outboard of the core body and 8 solar panels attached outboard of the alpha booms. The first 12 modes of each of the 10 flexible component bodies for the SSF model were retained for the SENDAP simulation.

Although this SSF model is no longer valid for on-going analysis work on the SSF project, it fully demonstrates the capabilities of SENDAP in modelling the station as newer, more up-to-date, flexible body models become available.

The SSRMS model consists of 8 rigid bodies, each with a single rotational degree of freedom except the first body, which is locked to the SSF core body. The last SSRMS body is locked to a rigid body model of the APM. The contact model for the APM and SSF is based on the CBM.

The berthing and mating operation consisted of the following simulated sequences:

* The system is initialized with the SSRMS base body attached to the Japanese Experimental Module (JEM). The SSRMS end effector is grappling the APM. The APM's docking port is poised approximately 1 meter from the SSF docking port along the axial docking port axis.

* Inverse kinematics commands for the SSRMS joint rates are generated to move the APM from its initial position to a "capture" position near the CBM latches. This maneuver is performed with two saw-tooth rate commands on the SSRMS joints and will place the APM within the capture zone in about 50 seconds.

* Once the APM's docking port is within the capture zone, the latches are activated and the APM is "pulled" into a mating position. During latch activation, the SSRMS is placed in its brakes-on mode along with 0 commanded rates. It is at this time that the contact routines are utilized to compute the dynamics at the docking ports.

* Following full latch closure, the simulation ends. Total simulation time is 70 seconds.

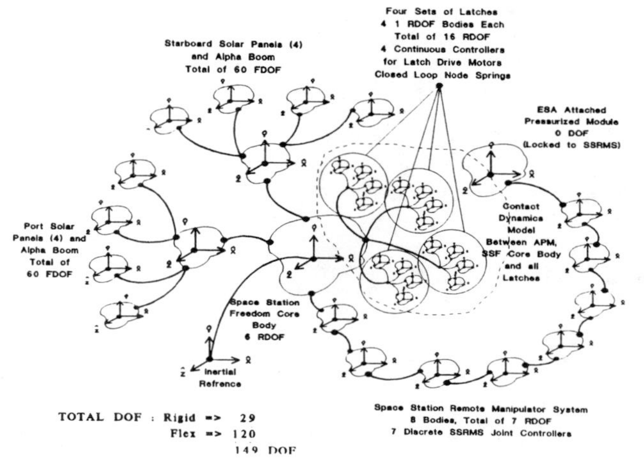

Figure 7: SSF/APM/SSRMS Berthing Operation Topology

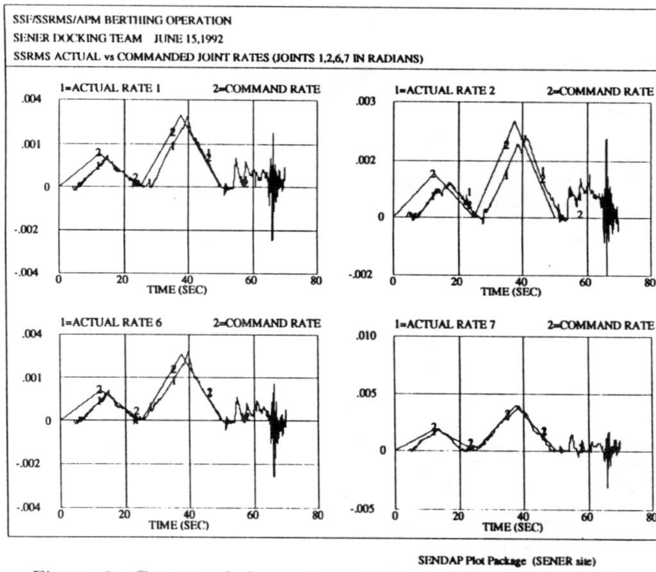

Figure 8: Commanded vs. Actual Selected SSRMS Joint Rates (rads/s)

Figure 8 shows the commanded verse actual joint rates on four of the seven SSRMS joint servo controllers. Notice that the commands were in the form of two saw-toothed rate commands over a 50 second period. The commanded rate after 50 seconds is zero on all of the SSRMS joints; however, the effects of the latches "pulling" in the APM and the subsequent contact cause what appear to be impulsive rates at the joints. Responses for the other three joint servos are similar.

Figures 9 and 10 show the relative translational kinematics between the docking ports of the SSF and APM. The values are expressed in SSF core body coordinates. Notice the persistence of relative motion between the SSF and APM. The latch models implemented include the relatively soft springs of the latch mechanisms with a considerable capability to back drive. In addition, we have not simulated the eventual lockdown between the two mating spacecraft. It was also noted that the SSRMS joint servos continued a force fight during latch capture of the APM.

The effects of the berthing and mating operation on highly flexible components of the SSF are shown in Figures 11 and 12. These figures show the modal response of the first eight modes of one of the flexible solar panels. Shown are the modal coordinates, which form the flexible portion of the system's state vector within SENDAP. The modal coordinates combined with the mode shapes and slopes result in physical modal deformations (see [2]); hence, the modal coordinates give an indication of a finite distributed body's deformation.

As can be seen in the plots, the slower modes are excited by the disturbances caused by the motion of the SSRMS as well as the docking port contact forces. This is especially true of the lowest frequency modes simulated. The solar panel's higher frequency modes are predominantly excited by the higher frequency contact disturbance. Table 1 lists the radian frequency of the modes shown in Figures 11 and 12. The data in Table 1 was extracted from the COSMIC NASTRAN data for the MB-15 component models of the SSF solar panels.

Figures 13 and 14 show the contact forces and torques experienced by the APM at the docking port. Figures 15 and 16 show the constraint forces and torques on the first SSRMS body attached to the JEM. Information similar to that contained in these four plots can be invaluable to structural designers working on systems similar to the Space Station or smaller in scope.

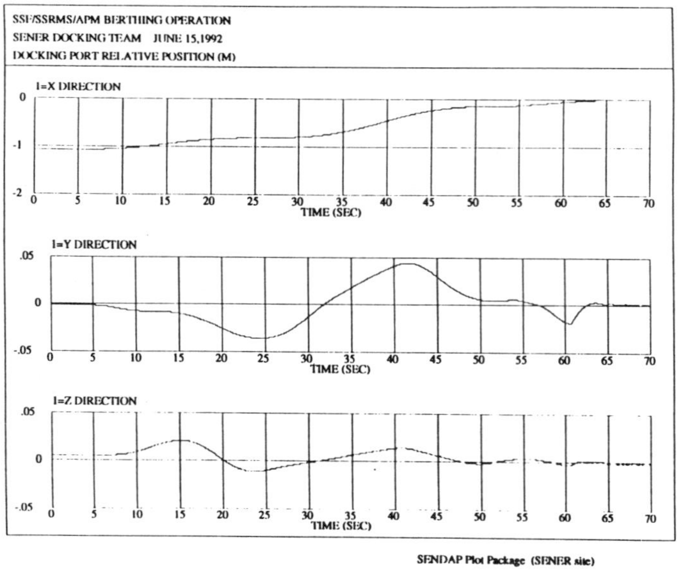

Figure 9: Relative Docking Port Position (m)

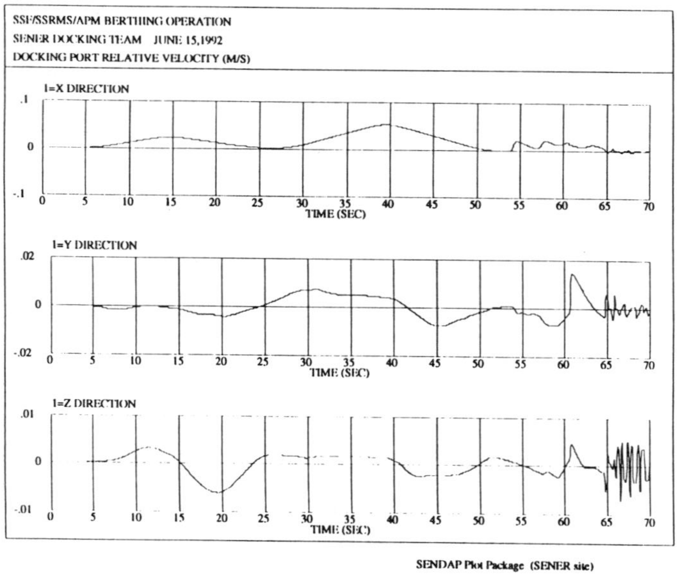

Figure 10: Relative Docking Port Velocity (m/s)

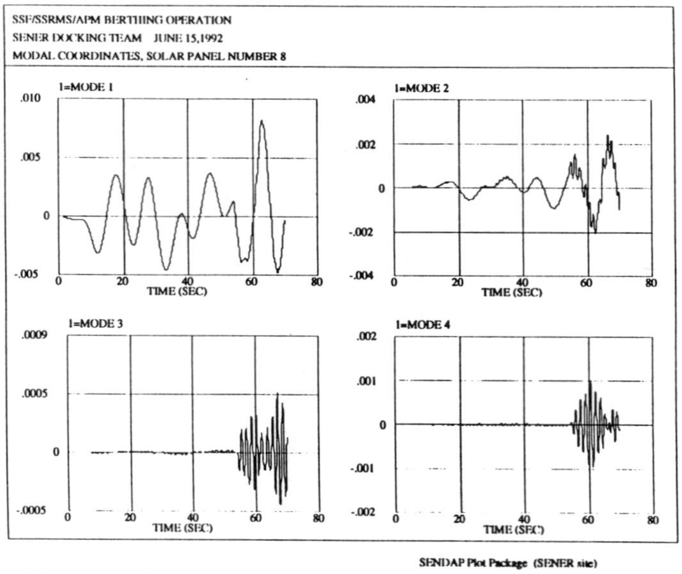

Figure 11: Modal Coordinate, Solar Panel 8, Modes 1-4 (non-dimensional)

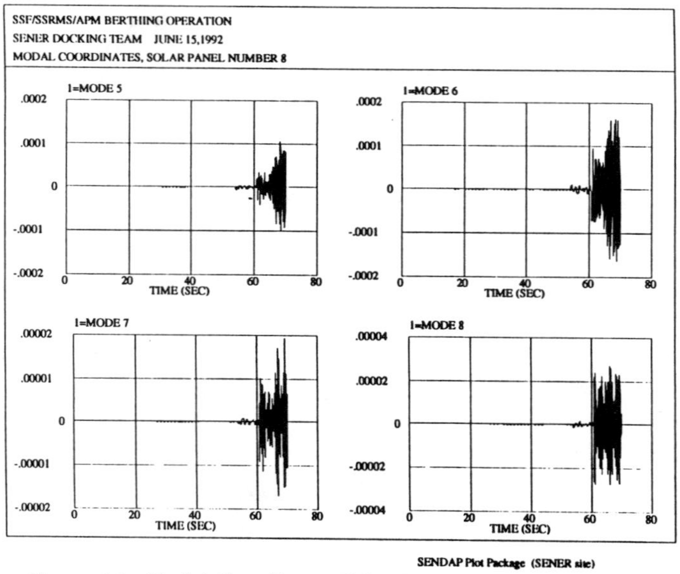

Figure 12: Modal Coordinate, Solar Panel 8, Modes 5-8 (non-dimensional)

Real Eigenvalues		
Mode No.	Eigen-Value	Radian Freq.
1	.397	.630
2	.397	.630
3	15.60	3.95
4	15.60	3.95
5	122.44	11.07
6	122.44	11.07
7	471.72	21.72
8	471.72	21.72

Table 1: COSMIC NASTRAN Eigenvalue Output for SSF Solar Panel No. 8

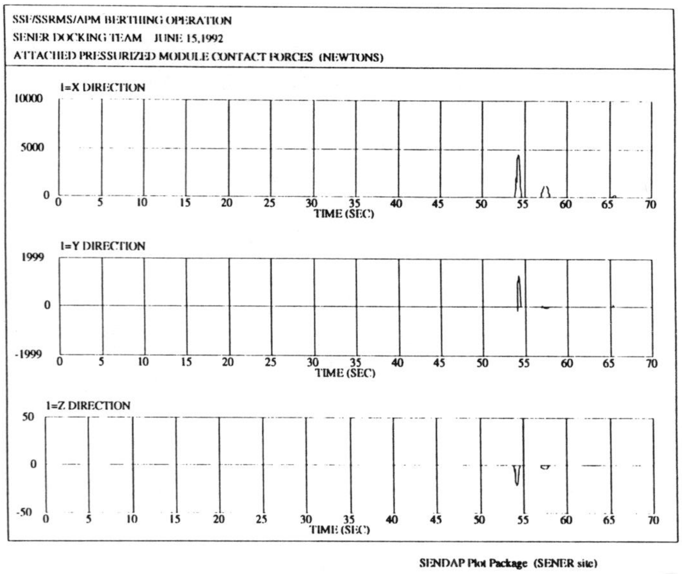

Figure 13: APM Docking Port Contact Force (N)

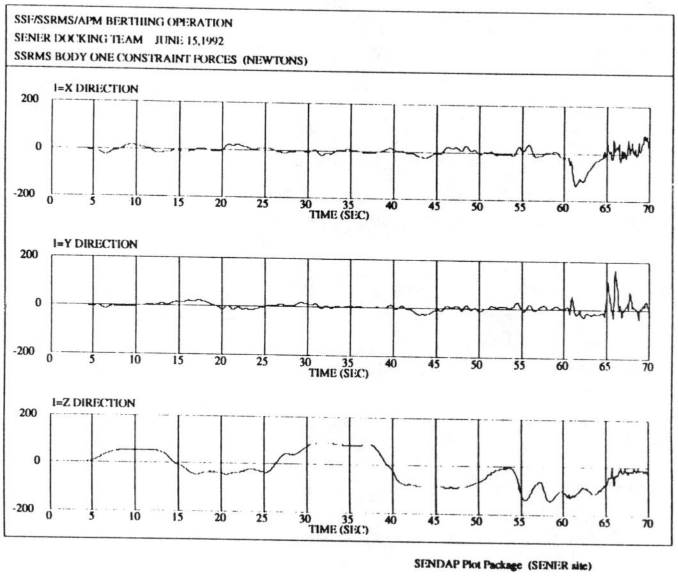

Figure 15: SSRMS Body 1 Constraint Force (N)

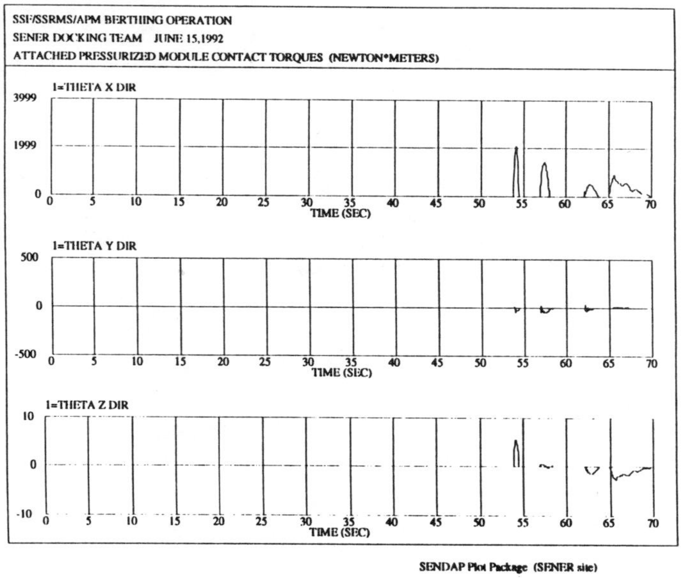

Figure 14: APM Docking Port Contact Torques (N*m)

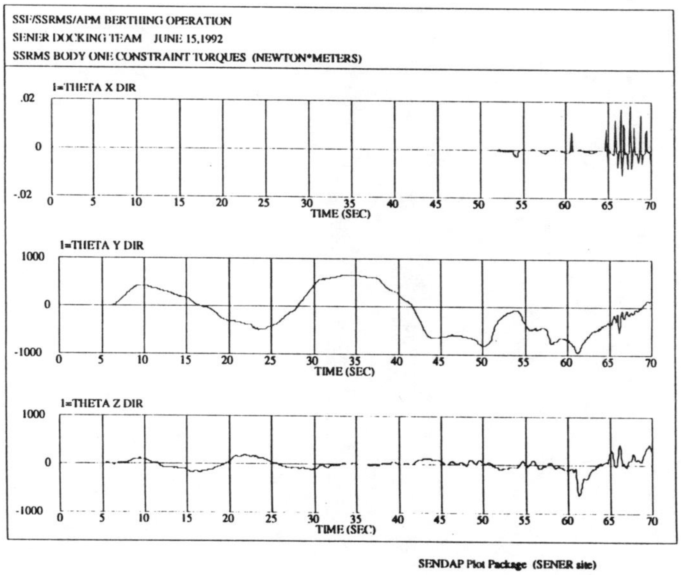

Figure 16: SSRMS Body 1 Constraint Torques (N*m)

SUMMARY

This paper has presented an integrated digital simulation environment for modelling contact dynamics with rigid and flexible multi-body systems. In addition, through a large, complex example, this paper has clearly shown the capabilities of the simulation environment to model mission operations consisting of several major systems and sub-systems that are currently planned as part of future, international space based infrastructures.

ACKNOWLEDGEMENTS

The authors wish to thank the staff members of Sener, S.A. and Dynacs, Inc. for the valuable contributions of expertise and effort that made the SENDAP development effort a successful reality. Special thanks go to Juan Jose Uriarte (Sener) for his efforts in validation and exercise of the SENDAP environment and Anil Singh (Dynacs) for his hard work and persistence.

References

[1] Kane, Likens and Levinson, *Spacecraft Dynamics*, McGraw-Hill, 1983

[2] *SENDAP Theory Manual*, Dynacs Engineering Company, Inc, 1991

[3] Timoshenko and Goodier, *Theory of Elasticity*, McGraw-Hill, 1934

[4] *SENDAP Docking Device User's Manual*, Dynacs Engineering Company, Inc., 1991

[5] *SENDAP User's Manual*, Dynacs Engineering Company, Inc., 1991

AMD-Vol. 141/DSC-Vol. 37, Dynamics of Flexible Multibody Systems:
Theory and Experiment
ASME 1992

CONSTRAINT VIOLATION IN CONCURRENT RANGE SPACE METHODS FOR TRANSIENT DYNAMIC ANALYSIS

Ramesh G. Menon, Andrew J. Kurdila, and Thomas W. Strganac
Department of Aerospace Engineering
Texas A&M University
College Station, Texas

1 INTRODUCTION

Over the past three years, the authors have demonstrated that alternative, nonrecursive order N formulations for the transient dynamic analysis of multibody systems can be achieved using range space methods . These formulations of multibody dynamics have been shown to be rapidly convergent for linear (Kurdila,1991),(Kurdila and Menon, 1992) and nonlinear (Menon and Kurdila, 1991) structural systems with a large number of degrees of freedom. Inasmuch as these methods induce concurrency via the popular method of subdomain decomposition, it can be easily shown that they are amenable to implementation on the forthcoming class of highly parallel architectures (Menon and Kurdila, 1991), (Menon and Kurdila, 1992). This class of formulations can be distinguished from other variants of subdomain decomposition in that redundant degrees of freedom are retained, and the coupling of substructures is achieved by explicitly calculating the coupling forces in an order N computational cost. Following nomenclature introduced in the constrained quadratic optimization literature (Gill, 1981), the former set of subdomain decomposition methods that eliminate redundant coordinates *a priori* and utilize a minimal coordinate formulation are referred to as nullspace methods, as opposed to the range space methods discussed in this paper. One potential drawback to the range space formulation is the well known constraint violation drift associated with redundant formulations of transient multibody dynamics. This paper introduces a family of constraint violation correction algorithms that retain the sequential order N computational cost of the range space formulation, while ensuring simulation fidelity. The hybrid methods introduced are based upon penalty multibody formulations and augmented Lagrangian formulations. Explicit bounds on measures of the constraint violation are presented as well as a qualitative Lyapunov analysis of the attractors of the constraint violation in phase space. Several completed numerical examples of the approach are provided on the nCUBE2, including simulations of the Space Station Freedom and the NASA Langley CSI Evolutionary model. Current research is ongoing to extend the methodology to applications in aeroservoelastic control and the control of distributed parameter systems.

2 GOVERNING EQUATIONS

We now present the dynamic equations for a multi-body system comprised of numerous sub-structures. This decomposition of the system into small substructures can be accomplished by a sub-domain decomposition scheme. This presents an avenue to perform computational load balancing on multiprocessors, which is the task of mapping a given problem among the available processors. An automated procedure to carry out subdomain decomposition to quasi-optimally map an arbitrary structure onto a given number of processors is presented in a forthcoming paper by the authors (Kurdila,1992). We adopt a redundant coordinate formulation so as to be able to use a non-recursive procedure to compute the accelerations of bodies in the system. The rationale behind using a redundant formulation, is that the additional cost of using a larger number of coordinates can be more than compensated by the potential for computing the dynamics of each body in parallel. When absolute coordinates are used, the inertial acceleration of a body in a chain in not dependent on computations of accelerations of adjacent bodies. The coupling arises only through constraints between the bodies which can be computed in parallel at each constraint interface. This method should be carefully contrasted with concurrent recursive methods in (Bae and Haug, 1988).

Let $\boldsymbol{q}$ represent the vector of generalized coordinates for the system. If the constraints between the bodies are holonomic, they can in general be expressed as a vector function of the generalized coordinates and time.

$$\Phi(q, t) = 0 \tag{1}$$

Differentiating equation (1) with respect to time we obtain

$$C(q, t)\dot{q} + \Phi_t = 0 \tag{2}$$

where C is the constraint Jacobian matrix defined as follows:

$$C(q, t) \equiv \left[\frac{\partial \Phi}{\partial q}\right] \tag{3}$$

and

$$\Phi_t = \frac{\partial \Phi}{\partial t} \tag{4}$$

If there are non-holonomic constraints (of the equality type), then they are generally functions of the generalized velocities. These velocities may either be the time derivatives of $\boldsymbol{q}$ or they may be based upon a quasi-coordinate formulation. Nevertheless, if the velocities appear linearly in the non-holonomic constraints, they can be cast in the form of equation (2) and our formulation is still applicable.

The governing equations describing the motion of the multi-body system can now be written as:

$$M(q)\ddot{q} = f(q, \dot{q}, t) + C^T\lambda \tag{5}$$

In equation (5), M is the generalized mass matrix and λ is the vector of Lagrange multipliers representing the constraint forces and torques. f consists of all the other terms including centrifugal, stiffness, damping, external forces and torques. The final form of the constraint equations to be used in the formulation is used by taking the second derivative with respect to time of equation (1):

$$C\ddot{q} + \dot{C}\dot{q} + \dot{\Phi}_t = 0. \tag{6}$$

There are many different approaches to solving the system of equations given by (5) and (6) (Kurdila and Kamat, 1990). This paper deals with parallel preconditioned conjugate gradient methods. One means of classifying these methods is based on whether they use a minimal or redundant coordinate system. Examples of minimal, or nullspace, methods are given in (Axelsson, 1984) and

(Hughes, 1983). We explicitly solve for the Lagrange multipliers using the range space solution method. As described in Kurdila (1991), the range space method has many advantages over the null space method for systems composed of block diagonal coefficient matrices. In the null space method, it is necessary to compute the null space basis for the constraint Jacobian and the system coefficient matrices need to be transformed to the null space. As a result, the coefficient matrices may lose their block diagonal properties and become dense or banded. In the range space method, the system coefficient matrix retains its strictly block diagonal structure and hence the factorization of the individual blocks can occur asynchronously in parallel. Also, because of the physical motivation for the preconditioner, it is rapidly convergent (Kurdila, 1991) and (Kurdila and Menon, 1992).

Rewriting equation (5) by inverting the mass matrix and pre-multiplying by the constraint Jacobian, we obtain:

$$C\ddot{q} = CM^{-1}f + CM^{-1}C^T\lambda \qquad (7)$$

Substituting for $C\ddot{q}$ from equation (6) we obtain

$$(CM^{-1}C^T)\lambda = -CM^{-1}f + e(q, \dot{q}, t) \qquad (8)$$

where

$$e(q, \dot{q}, t) = -\dot{C}\dot{q} - \dot{\Phi}_t \qquad (9)$$

We assume that there are no redundant constraints in the system. As a result the rows of the constraint Jacobian matrix are linearly independent. We further assume that in the case of deformable bodies, the sub-structures are obtained through a consistent finite element approximation so that the mass matrices are not singular. In the case of rigid bodies, we require that there be no massless links in the system. These assumptions assure that the constraint metric, $CM^{-1}C^{T}$, is not singular. These assumptions are valid for a large class of structures, including examples presented in this paper.

In the following sections (3), (4) and (5) we present order N concurrent constraint stabilization methods that are amenable to implementation with the range space formulation. The details of the concurrent formulations will be presented in the final manuscript. Some preliminary numerical results are given at the end of this abstract.

3 BAUMGARTE'S METHOD

This is perhaps the single most widely used method for constraint stabilization (Baumgarte, 1983). It is attractive in its simplicity and the ease with which it can be implemented. Motivated by feedback control theory we add terms proportional to the constraint violations and their derivatives to equation (6):

$$\ddot{\Phi} + 2\xi\omega\dot{\Phi} + \omega^2\Phi = 0 \qquad \xi, \omega > 0 \quad (10)$$

where ζ and ω are positive constants to control the constraint violations. The only change that this introduces in our governing equations is the modification of the e term in equation (9):

$$e(q, \dot{q}, t) = -\dot{C}\dot{q} - \dot{\Phi}_t - 2\zeta\omega\dot{\Phi} - \omega^2\Phi \qquad (11)$$

Since the stabilization involves only the modification of equation (9), it is implementable in the concurrent preconditioned conjugate gradient scheme. As a result this preserves the order N formulation. Despite its simplicity, Baumgarte's method has the following disadvantages:

- Picking ξ and ω are problem dependent

and as a result difficult to automate.

- There are no *a priori* bounds on constraint violation.
- There is no vehicle to tackle accumulated constraint violations.

4 PENALTY METHOD

By using a modified form of the classical penalty method to enforce the constraints between the sub-structures , we obtain the penalized system of equations given by:

$$(M + \beta C^T C)\ddot{q} \qquad (12)$$
$$= f - C^T(\beta \dot{C}\dot{q} + \beta\dot{\Phi}_t + \mu\dot{\Phi} + \alpha\Phi)$$

where β, μ and α are the penalties on the inertial, damping and stiffness terms respectively. We now have a system of equations in as many unknowns as the number of unconstrained coordinates.

We can prove the following theorem on the time domain constraint violation behavior of the penalized system, when the system is natural and if the initial conditions are consistent with the constraints (Kurdila and Junkins, 1991).

Theorem:

The constraint violation in the penalized system satisfies

$$\|\dot{\Phi}_\varepsilon\|^2 + \|\Phi_\varepsilon\|^2 \qquad (13)$$
$$\leq \frac{2E_0}{min(\sigma_{min}(\alpha), \sigma_{min}(\beta))}$$

where $\sigma min(.)$ denotes the minimum singular value of the argument.

The penalty method requires the solution of equation (12), which can be implemented using a concurrent preconditioned conjugate gradient procedure for the solution of the accelerations.

5 AUGMENTED LAGRANGIAN METHOD

Recently there has been considerable interest in the application of augmented Lagrangian methods to multibody formulations (Bayo, 1988), (Serna, 1988). The fundamental scheme can be stated as follows:

$$\lambda_0 = 0$$
$$M(q)\ddot{q}_n = f(q, \dot{q}, t) \qquad (14)$$
$$-C^T\frac{1}{\varepsilon}(\ddot{\Phi}_n + 2\xi\omega\dot{\Phi} + \omega^2\Phi) - C^T\lambda_n$$
$$\lambda_{n+1} = \lambda_n + \frac{1}{\varepsilon}(\ddot{\Phi}_n + 2\xi\omega\dot{\Phi} + \omega^2\Phi)$$

where $\xi, \omega \geq 0$ are the penalty damping ratio and penalty frequency respectively. Although the augmented Lagrangian method has asymptotic convergence properties presented in (Glowinski, 1989), it is not sufficient for practical simulation and control generation. In other words, for achieving practical computational efficiency, the method should be able to converge to a specified tolerance within a finite number of iterations. While this philosophy has been used (Bayo, 1991) to realize what are claimed to be real time simulation methods, the effect of finite number of iterations on the constraint violation behavior of the multibody simulation has not appeared in the literature to date. Motivated by these factors, we investigate the behavior of the constraint violations when a constant number of iterations are used per time step.

Theorem:

The constraint violation measure decreases by order of ε with each additional iteration, when a constant number of iterations are used at each time step

$$\left|\ddot{\Phi} + 2\xi\omega\dot{\Phi} + \omega^2\Phi\right|_{i+1} \qquad (15)$$

$$\leq \frac{\varepsilon}{\upsilon}\left|\ddot{\Phi}_i + 2\xi\omega\dot{\Phi} + \omega^2\Phi\right|_i$$

where ν is the smallest non-zero eigen value of M-1CTC.

The above equation describes the behavior of the constraint violation function with each additional iteration. Therefore we can state that the constraint violation measure decreases by order of ε with each additional iteration, since ν is typically much larger than ε. This statement applies to the quasi static process that we have considered, *viz*., the process at each time step.

The evolution of the system, with a fixed number of iterations at each time step is given by the following time domain constraint violation bound.

Theorem:

The energy in the constraint violation shrinks by O(ε) with every additional iteration.

$$(\dot{\Phi}(t)^2 + \omega^2\Phi(t)^2)_{i+1} \qquad (16)$$

$$\leq \frac{\varepsilon}{\upsilon}(\dot{\Phi}(t)^2 + \omega^2\Phi(t)^2)_i$$

The advantages of using the augmented lagrangian method is that:

- *a priori* time domain constraint violation bounds as a function of number of fixed iterations are available.
- Accumulation of constraint violation can be handled by allowing variable number of iterations.
- Since only small penalty values need be used to achieve desired accuracy, the method results in well conditioned formulation.
- It is readily implementable in a concurrent preconditioned conjugate gradient method similar to the solution process in the penalty method.

6 RESULTS

Figure (1) shows the order *N* computational cost for the range space formulation with preconditioned conjugate gradient (PCG) solution when various hybrid constraint stabilization schemes are used. The cost for the augmented Lagrangian method is also shown for reference. Figure (2) shows their corresponding concurrent cost when implemented on an nCUBE 2. The concurrent cost approaches the constant time paradigm of the order *N* scheme. Additional results in aeroservoelasticity and distributed control synthesis are currently under investigation.

7 REFERENCES

1. Kurdila, A.J., Menon, R.G., and Sunkel, J.W., "A Nonrecursive Order *N* Method for Simulating Multibody Dynamics", Presented at the 32 SDM Structures, Structural Dynamics and Materials Conference, Baltimore, MD, April, 1991.
2. Kurdila, A.J., Menon, R.G., and Sunkel,

J.W., "Nonrecursive Order *N* Preconditioned Conjugate Gradient / Range Space Method for Multibody Dynamics", accepted for publication in AIAA Journal of Guidance, Control and Dynamics, 1992.

3. Menon, R.G., and Kurdila, A.J., "Concurrent Simulation Method for Linear Multibody Dynamics", Presented at the First U.S. National Congress on Computational Mechanics, Chicago, IL, July 20-24, 1991.

4. Menon, R.G., and Kurdila, A.J., "A Nonrecursive Order *N* Method for Simulation of Nonlinear Multibody Systems", presented at the First U.S. National Congress on Computational Mechanics, Chicago, IL, June 20-24, 1991.

5. Menon, R.G., and Kurdila, A.J., "Subdomain Decomposition Methods and Computational Control for Multibody Dynamical Systems", to be presented at the Symposium on High-Performance Computing for Flight Vehicles, Washington, D.C., December 7-9, 1992.

6. Gill, P.E., Murray, W., and Wright, M.H., Practical Optimization, Academic Press, London, 1981.

7. Kurdila, A.J., "Subdomain Decomposition using an Entropy Maximizing Principle", preprint.

8. Bae, D., and Haug, E.J., "A Recursive Formulation for Constrained Mechanical System Dynamics: Part III. Parallel Processor Implementation", Mechanics of Structures and Machines, Vol. 16, No.2, pp. 249-269, 1988.

9. Kurdila, A.J., and Kamat, M.P., "Concurrent Multiprocessing Methods for Calculating Null Space and Range Space Bases for Multi-body Simulation", AIAA Journal, Vol. 28 No. 7, July 1990.

10. Axelsson, O., and Barker, V.A., "Finite Element Solution of Boundary Value Problems", Academic Press, Inc., Orlando, 1984.

11. Hughes, T.J.R., Levit, I., and Winget, J., "An Element-by-Element Solution Algorithm for Problems of Structural and Solid Mechanics", Computer Methods in Applied Mechanics and Engineering, Vol. 36, No. 2, pp. 241-254, 1983.

12. Baumgarte, J.W., "A New Method of Stabilization for Holonomic Constraints", Journal of Applied Mechanics, Vol. 50, pp. 869-870, 1983.

13. Kurdila, A.J., Junkins, J.L. and Menon, R.G., "Linear Substructure Synthesis via Lyapunov Stable Penalty methods", Finite Elements in Analysis and Design, no. 10, pp. 101-123, December 1991.

14. Bayo, E. *et al*, "A Modified Lagrangian Formulation for the Dynamic Analysis of Constrained Mechanical Systems", Computer Methods in Applied Mechanics and Engineering, vol.71, pp. 183-195, 1988.

15. Serna, M.A. and Bayo, E., "Numerical Implementation of Penalty Methods for the Analysis of Elastic Mechanisms", Trends and Developments in Mechanisms, Machines and Robotics - 1988, pp. 449-456, ASME, New York.

16. Bayo, E., Garcia de Jalon, J., Avello, A., and Cuadrado, J.,"An Efficient Computational Method for Real Time Multibody Dynamic Simulation in Fully Cartesian Coordinates", Computer Methods in Applied Mechanics and Engineering, vol.92, pp. 377-395, 1991.

17. Glowinski, R. and Le Tallec, P., Augmented Lagrangian and Operator-Splitting Methods in Nonlinear Mechanics, Society for Industrial and Applied Mathematics, Philadelphia, 1989.

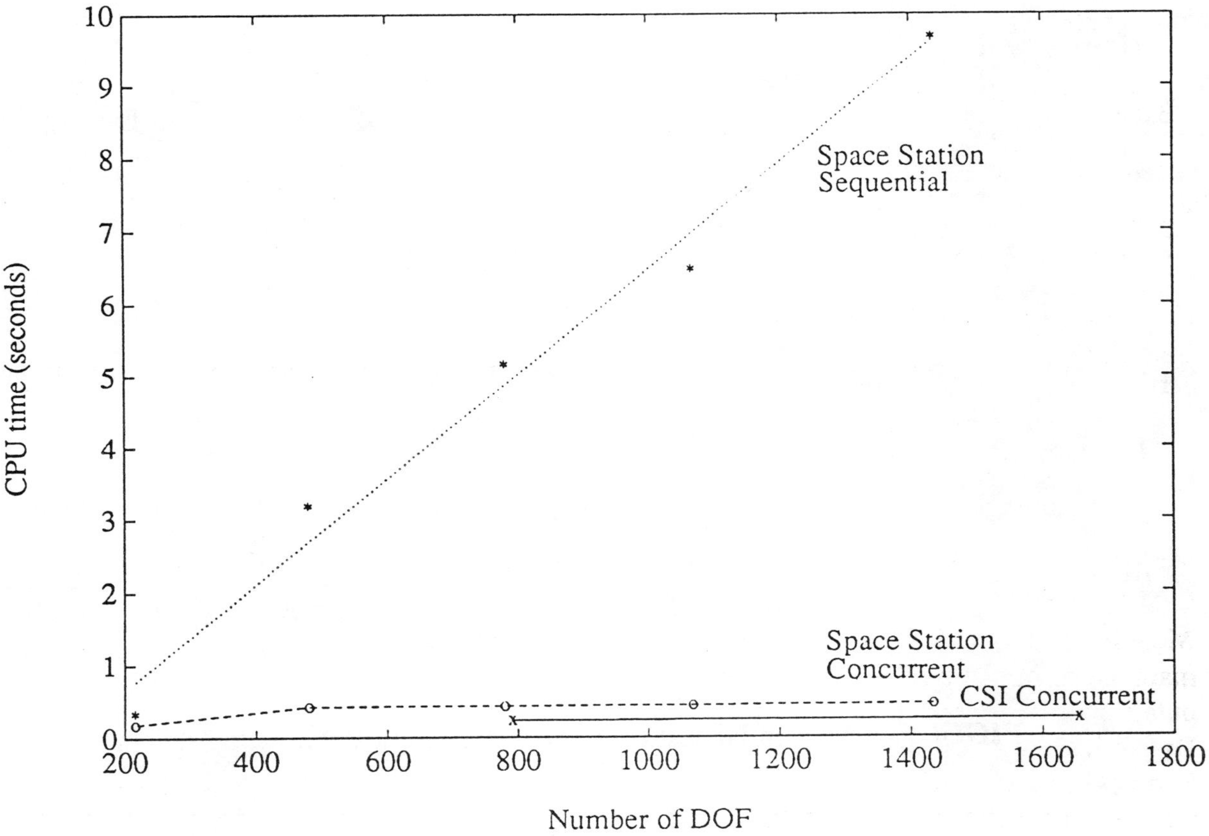

FIGURE 1: Order *N* sequential cost on a VAX 8650 and concurrent cost on an nCUBE2 for a 13 substructure Space Station Model and the CSI Model.

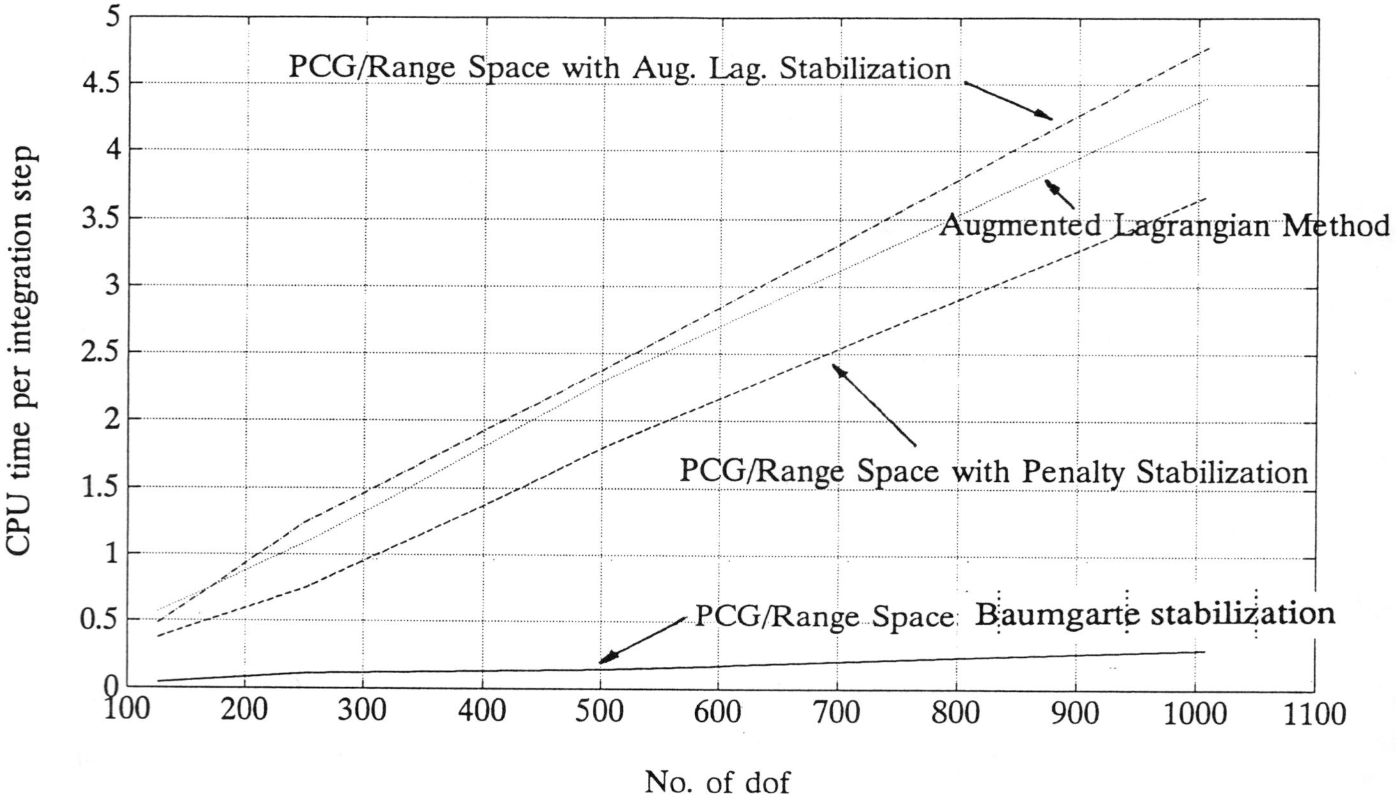

FIGURE 2: Sequential cost for various constraint stabilization schemes.

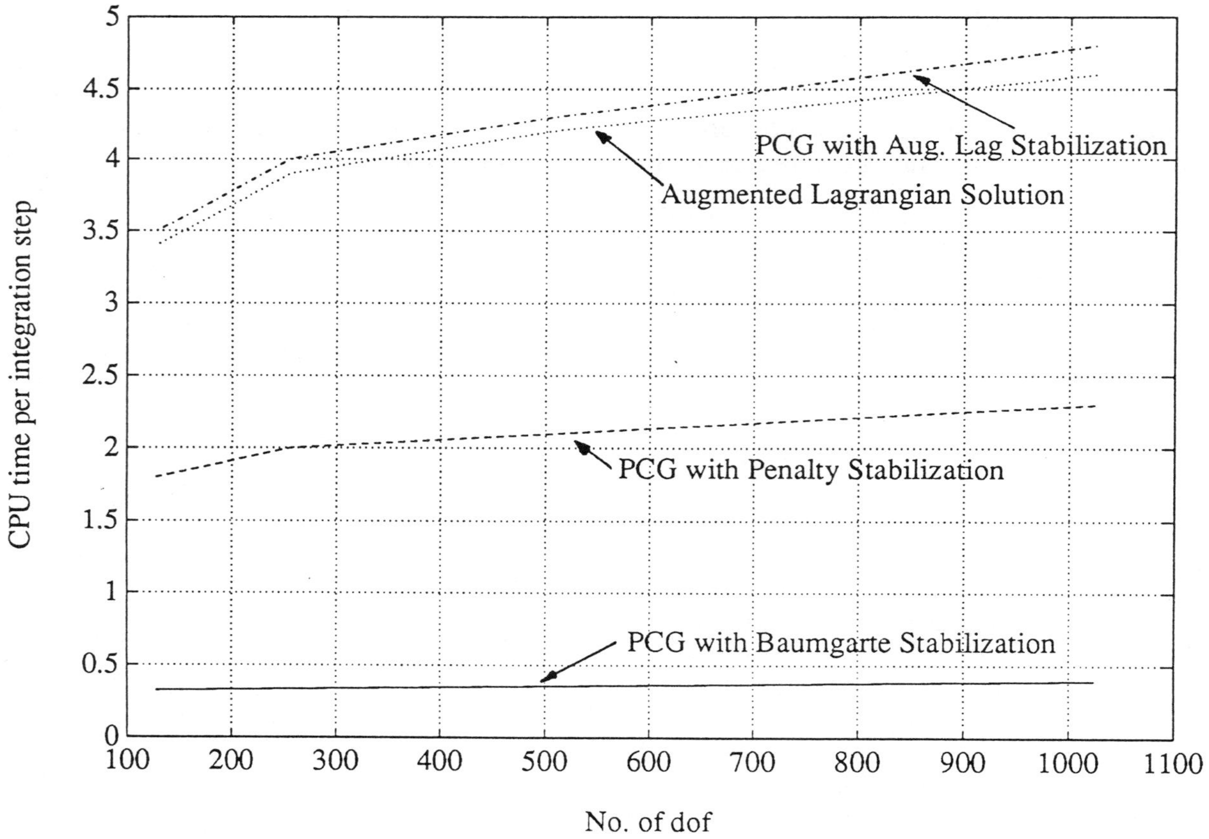

FIGURE 3: Concurrent implementation on nCUBE2 of constraint stabilization methods.

AMD-Vol. 141/DSC-Vol. 37, Dynamics of Flexible Multibody Systems:
Theory and Experiment
ASME 1992

ON THE GEOMETRIC STIFFNESS MATRICES IN FLEXIBLE MULTIBODY DYNAMICS

Sivakumar S. K. Tadikonda and H. T. Chang
Dynacs Engineering Company Incorporated
Palm Harbor, Florida

ABSTRACT

The geometric stiffness matrix for a flexible body in a multibody configuration is due to two sets of forces (and torques). One set is due to the self motions, and this has been addressed in the literature. Another and equally important set of forces (and torques) on a body is due to the inertia forces and torques of all of its outboard bodies. The second set of forces are transmitted only through the hinge connections. Thus, this set requires the consideration of only up to six additional inertia forces in setting up the differential stiffness matrix. Simulation results presented indicate that, for a flexible body consisting of outboard bodies, one may only consider the second set of forces without loss of accuracy, and thus significantly reduce the computational burden. This approach is especially suitable for multibody systems such as the Shuttle Remote Manipulator System (SRMS), where the payloads are several orders of magnitude larger than the SRMS link masses.

1 INTRODUCTION

The flexibility of a body is described mathematically using the method of assumed modes in several multibody dynamics formulations - see Singh *et al* (1985) for example. It has been realized that this approach also leads to spurious loss of stiffening in the modal equations (Baruh and Tadikonda, 1989). For flexible bodies such as beams and plates, this spurious loss of stiffness can be annulled by considering the "foreshortening effect" (Baruh and Tadikonda, 1989), also known as the "shortening of projection" (Meirovitch, 1980). Another method, but yielding identical results, consists of considering second order terms in the strain-displacement relationships. When the finite element method is used to model the structural flexibility of an arbitrarily shaped body, the use of the differential stiffness matrix is shown to account for these geometric nonlinear effects (Cook, 1985). This result was exploited by Banerjee and Dickens (1990) for a single flexible body to account for the geometric nonlinearity effects arising due to the self-motion of a body by considering the body inertia forces as the set of inertia loads acting on the system, and was later extended to multi-body systems by Banerjee and Lemak (1991). This results in the computation and use of 21 generalized stiffness matrices in the body modal equations.

Using the nomenclature of Singh *et al* (1985), a body in a tree-topology can be categorized either as a leaf body or a branch body. A leaf body is defined as the last body in a branch. Thus, a leaf body contains bodies only inboard of it and leading to the base body ("root"), but does not contain any bodies outboard of it. A branch body contains both inboard and outboard bodies (see Fig.1).

Now consider an articulated two link manipulator from Baruh and Tadikonda (1989) as shown in Figure 2. Link 1 is connected to the ground by Joint 1 with two rotational degrees of freedom (DOF's) and Links 1 and 2 are connected by a revolute joint, Joint 2. One axis of rotation of Joint 1 is parallel to Joint 2 revolute axis, while the other is in an orthogonal direction. Each link is a flexible beam of uniform cross-section, having uniform stiffness and mass properties. Let the mass densities, masses and lengths of links 1 and 2 be denoted by $\rho_1, \rho_2, m_1, m_2, L_1$ and L_2, respectively. The geometric nonlinearity effects for Link 2 are due to self motion, as explained by Banerjee and Dickens (1990). However, the geometric nonlinearity effects for Link 1 are those due to its self motion, as well as those due to the rigid-body motion of Link 2. The modal equations derived for a two link manipulator by Baruh and Tadikonda (1989) are presented in the following section to demonstrate these two contributions.

2 A TWO LINK MANIPULATOR EXAMPLE

Let a body frame $\mathbb{B}_j = \left[\underline{b}_{j1}, \underline{b}_{j2}, \underline{b}_{j3}\right]^T$ be associated with link $j\,(j = 1,2)$ such that $\underline{b}_{j1}$ points along the beam axis of the undeformed beam. The rotational degree of freedom between Links 1 and 2 is about the $\underline{b}_{23}$ axis. The planar rigid-body motion of the manipulator is defined by the variables θ_1 and θ_2 and the nonplanar rotation is defined by ψ, as shown in Fig.2. The two rotational DOF's between Link1 and the ground are described by the variables ψ about the $\underline{N}_3$ axis, followed by $(\theta_1 - 90^o)$ about the $\underline{b}_{13}$ axis. The body frame is located at the joint connecting to the ground for Link 1 while the body frame for Link 2 is located at the middle of the undeformed link, with $\underline{b}_{23}$ parallel to $\underline{b}_{13}$. The flexibility in the links is assumed to be small, and the small elastic displacements are measured with respect to the respective body frames. Let the elastic transverse displacements $v_j\,(j = 1,2)$ occur in the $\underline{b}_{j1}\underline{b}_{j2}$ plane, with the subscript indicating the body association.

The position vector of an elemental mass dm at a configuration point P_1 on Link 1 is given by

$$\underline{R}_1(P_1,t) = (x + u_1(P_1,t))\,\underline{b}_{11} + v_1(P_1,t)\,\underline{b}_{12} \tag{1}$$

where, $u_1(P_1,t)$ is the shortening of projection due to the transverse elastic displacement v_1 and can be shown to have the form (Baruh and Tadikonda, 1989):

$$u_1(P_1,t) \approx -\frac{1}{2}\int_o^x \left(\frac{\partial v_1}{\partial \sigma}\right)^2 d\sigma \tag{2}$$

Differentiation of Eq.(1) with respect to time to obtain the velocity and its use in the computation of the kinetic energy for Link 1 can be shown to result in a term:

$$\frac{1}{2}\int_0^{L_1} \rho_1 x\left(\dot\theta_1^2 + \dot\psi^2\cos^2\theta_1\right)\left(\int_0^x \left(\frac{\partial v_1}{\partial \sigma}\right)^2 d\sigma\right) dx \tag{3}$$

If the foreshortening term u_1 in Eq.(1) is neglected, this term can be accounted for through the potential energy contribution due to the centrifugal force. After interchanging the order of integration and simplification, Eq.(3) can be written as (Meirovitch,1980):

$$V_1(t) = \frac{1}{2}\left(\dot\theta_1^2 + \dot\psi^2\cos^2\theta_1\right)\int_o^{L_1} \frac{\rho_1}{2}\left(L_1^2 - x^2\right)\left(\frac{\partial v_1}{\partial x}\right)^2 dx \tag{4}$$

Using the method of assumed modes, the transverse elastic displacements, $v_i\,(i = 1,2)$, are expressed as:

$$v_i(P_i,t) = \sum_{j=1}^{NM_i} \phi_{ij}(P_i)\,\eta_j^{(i)}, \qquad i = 1,2 \tag{5}$$

where, ϕ_{ij} represents the jth assumed mode shape for body i, $\eta_j^{(i)}$ is its amplitude and $NM_i\,(i = 1,2)$ are the number of modes used in the expansion for i^{th} link. Substituting the above expansion into Eq. (4), the potential energy function can be obtained as:

$$V_1(t) = \frac{1}{2}\left(\dot\theta_1^2 + \dot\psi^2\cos^2\theta_1\right)\sum_{i=1}^{NM_1}\sum_{j=1}^{NM_1} h_{ij}^{(1)}\eta_i^{(1)}\eta_j^{(1)} \tag{6}$$

where,

$$h_{ij}^{(1)} \triangleq \frac{1}{2}\int_o^{L_1} \rho_1\left(L_1^2 - x^2\right)\phi'_{1i}\phi'_{1j}\,dx \tag{7}$$

The virtual work expression then is

$$\delta W = -\delta V_1 = -\left(\dot\theta_1^2 + \dot\psi^2\cos^2\theta_1\right)\sum_{i=1}^{NM_1}\sum_{j=1}^{NM_1} h_{ij}^{(1)}\eta_i^{(1)}\delta\eta_j^{(1)} \tag{8}$$

Now, consider the effect of Link 2 motions on Link 1. Following the approach of Banerjee and Dickens (1990), we first compute the inertia force of Link 2, assuming that both Links are rigid. It can be shown that (subscript r indicates that elastic motion is neglected)

$$\begin{aligned} -\int \ddot{\underline{R}}_{2r}dm \cdot \underline{b}_{11} \;=\;& m_2 L_1\dot\theta_1^2 + m_2\frac{L_2}{2}\ddot\theta_2\sin(\theta_2 - \theta_1) \\ &+ m_2\frac{L_2}{2}\dot\theta_2^2\cos(\theta_2 - \theta_1) \\ &+ m_2\dot\psi^2\left(L_1\sin\theta_1 + \frac{L_2}{2}\sin\theta_2\right)\sin\theta_1 \end{aligned} \tag{9}$$

This inertia force is transmitted to Link 1 through the hinge between the two links. The potential energy function due to this impressed force and the shortening of projection at the tip, extending the approach of Meirovitch (1980), is:

$$V_2(t) = \frac{1}{2}\left(\underline{f}_{Tr}^{2*}\cdot\underline{b}_{11}\right)\int_o^L \left(\frac{\partial v_1}{\partial x}\right)^2 dx \tag{10}$$

where,

$$\underline{f}_{Tr}^{2*} = -\int \ddot{\underline{R}}_{2r}\,dm \tag{11}$$

is the rigid body translational inertia force of Link 2. Substitution of the assumed modes expression in Eq. (10) yields

$$V_2(t) = \frac{1}{2}\left(\underline{f}_{Tr}^{2*}\cdot\underline{b}_{11}\right)\sum_{i=1}^{NM_1}\sum_{j=1}^{NM_1} G_{ij}^{(1)}\eta_i^{(1)}\eta_j^{(1)} \tag{12}$$

where,

$$G_{ij}^{(1)}(L_1) \triangleq \int_o^{L_1} \phi'_{1i}\phi'_{1j}\,dx \tag{13}$$

The virtual work expression due to this potential energy function (for a given impressed force) then is:

$$\delta W = -\delta V_2 = -\left(\underline{f}_{Tr}^{2*}\cdot\underline{b}_{11}\right)\sum_{i=1}^{NM_1}\sum_{j=1}^{NM_1} G_{ij}^{(1)}\eta_j^{(1)}\delta\eta_i^{(1)} \tag{14}$$

Thus, the geometrically nonlinear term results in two distinct terms in the modal equations - one term due to self motions and another due to the motions of outboard bodies. The modal force contributions due to the outboard body motions then are:

$$f_i = -\left(\underline{f}_{Tr}^{2*}\cdot\underline{b}_{11}\right)\sum_{j=1}^{NM_1} G_{ij}^{(1)}\eta_j^{(1)} \tag{15}$$

The equations of motion presented by Baruh and Tadikonda (1989) for this two link configuration were derived using the kinematic relationships in Eqs. (1-2), and through a Lagrangian approach. That is, Eqs. (1) and (2) are differentiated with respect to time, retaining the second order term u_1, the kinetic and potential energy functions were formed and the required partial derivatives were taken. The resulting modal equations for Link 1 (Eq. (53) in Baruh and Tadikonda, 1989) are presented below:

$$
\begin{aligned}
&\textstyle\sum_{\ell=1}^{NM_1}[m_{i\ell}^{(1)} + m_2\phi_{1i}(L_1)\phi_{1\ell}(L_1)]\ddot{u}_{1j} \\
&- \textstyle\sum_{p=1}^{NM_2}[m_2\phi_{1i}(L_1)\phi_{2p}(-\frac{L_2}{2})cos(\theta_2-\theta_1)]\ddot{u}_{2p} \\
&+ [f_{1i} + m_2 L_1\phi_{1i}(L_1)]\ddot{\theta}_1 \\
&+ m_2\left[\tfrac{L_2}{2}\phi_{1i}(L_1)cos(\theta_2-\theta_1) + \underline{\left\{\tfrac{L_2}{2}\textstyle\sum_{\ell=1}^{NM_1} G_{ib}^{(1)}(L_1)u_{1\ell}\right.}\right. \\
&+ \left.\left.\phi_{1i}(L_1)\textstyle\sum_{p=1}^{NM_2}\phi_{2p}(-\frac{L_2}{2})u_{2p}\right\} sin(\theta_2-\theta_1)\right]\ddot{\theta}_2 \\
&+ \left[\textstyle\sum_{\ell=1}^{NM_1}\left(m_{i\ell}^{(1)} + m_2\phi_{1i}(L_1)\phi_{1\ell}(L_1)\right)u_{1\ell}\right. \\
&- \left. m_2\phi_{1i}(L_1)\textstyle\sum_{p=1}^{NM_2}\phi_{2p}(-\frac{L_2}{2})u_{2p}\right] cos\theta_1\ddot{\psi} \\
&+ \textstyle\sum_{\ell=1}^{NM_1}\left[k_{i\ell}^{(1)} + h_{i\ell}^{(1)}(\dot{\theta}_1^2 + \dot{\psi}^2\sin^2\theta_1) - m_{i\ell}^{(1)}(\dot{\theta}_1^2 + \dot{\psi}^2\cos^2\theta_1)\right. \\
&+ \underline{m_2 G_{i\ell}^{(1)}(L_1)\{L_1\dot{\theta}_1^2 + \tfrac{L_2}{2}\dot{\theta}_2^2\cos(\theta_2-\theta_1) + g\cos\theta_1\}} \\
&+ \underline{m_2 G_{i\ell}^{(1)}(L_1)(L_1 sin\theta_1 + \tfrac{L_2}{2}sin\theta_2)sin\theta_1\dot{\psi}^2} \\
&- \left. m_2\phi_{1i}(L_1)\phi_{1\ell}(L_1)(\dot{\theta}_1^2 + \dot{\psi}^2 cos^2\theta_1)\right] u_{1\ell} \\
&+ D_i \;=\; F_{1i}, \qquad i = 1, 2, \cdots NM_1
\end{aligned}
\tag{16}
$$

where,

$$m_{i\ell}^{(1)} \;=\; \int_o^{L_1} \rho_1\phi_{1i}\phi_{1\ell}dx \tag{17}$$

$$f_{1i}^{(1)} \;=\; \int_o^{L} \rho_1 x\phi_{1i}dx \tag{18}$$

$$k_{i\ell}^{(1)} \;=\; \int_o^{L} EI_1\phi_{1i}''\phi_{1\ell}''dx \tag{19}$$

D_i contains the Coriolis, centrifugal and gravity terms, and F_{1i} represent the modal forces.

Note that the underlined terms in Eq.(16) are precisely those terms obtained by substituting Eqs.(9) and (11) into (15). As explained earlier, these terms arise due to the rigid-body inertia forces of bodies outboard of Link 1. The terms containing $h_{ij}^{(1)}$ in Eq.(16) are exactly those in Eq.(8). Thus, we now see that the geometrically nonlinear "shortening of projection" of link 1, due to the transverse flexible motion of the link, gives rise to two sets of terms in the modal equations - one set containing the inertia force contributions due to self motions, and a second set containing the inertia force contributions due to its outboard bodies. Since the second set of forces is transmitted only through the hinge connections, it is much simpler to deal with than those due to self motions. The formulation of Banerjee and Lemak (1991) accounts for the h_{ij} terms in the above equations for any arbitrarily shaped body, but G_{ij} terms cannot be obtained using the same approach. To complement the contribution of Banerjee and Lemak (1991), a formulation that accounts for the geometric nonlinearity effects on an arbitrarily shaped flexible body in a multibody chain, with the body flexibility modeled using the finite element method, is presented next. Simulation results are then presented using trusses that are representative of space structures as the links, in the manipulator shown in Fig.2.

3 GENERAL FORMULATION

Consider a generic body, Body j, in a tree topology. A body fixed frame $\mathbb{b}_j = \left[\, \underline{b}_{j1}\underline{b}_{j2}\underline{b}_{j3} \,\right]^T$ is associated with Body j. Let the vector locating the origin of this body frame with respect to an inertial frame be denoted by $\underline{R}_f^j$. The body is assumed to be linearaly elastic. The position vector locating an elemental mass dm on this body, in the inertial frame, is given by

$$\underline{R}^j = \underline{R}_f^j + \underline{r}^j + \underline{u}^j \tag{20}$$

where, $\underline{r}^j$ is the vector locating dm in $\mathbb{b}_j$ in the undeformed configuration and $\underline{u}^j$ defines the elastic deformation vector of dm in $\mathbb{b}_j$. Using the method of assumed modes, $\underline{u}^j$ can be expressed as

$$\underline{u}^j\left(\underline{r}^j, t\right) = \sum_{i=1}^{NM_j} \underline{\phi}_i^j\left(\underline{r}^j\right)\eta_i^j(t) \tag{21}$$

where, $\underline{\phi}_i^j$ are a set of assumed mode shapes and η_i^j denote their time varying amplitudes.

Differentiation of Eq. (20) yields the velocity expression

$$\dot{\underline{R}}^j = \dot{\underline{R}}_f^j + \underline{\omega}^j \times \left(\underline{r}^j + \underline{u}^j\right) + \overset{o}{\underline{u}}^j \tag{22}$$

where, $\underline{\omega}^j$ is the angular velocity of $\mathbb{b}_j, \overset{o}{\underline{u}}^j$ represents the deformational velocity of dm relative to $\mathbb{b}_j$ and has the form:

$$\overset{o}{\underline{u}}^j = \sum_{i=1}^{NM_j} \underline{\phi}_i^j\left(\underline{r}^j\right)\dot{\eta}_i^j(t) \tag{23}$$

The open dot in Eqs. (22) and (23) denotes differentiation in the local reference frame.

In a tree-topology, the number of bodies equals the number of joints connecting these bodies, with each joint permitting rigid body relative motion between the bodies connected by it. Let the number of degrees of freedom (DOF) permitted across joint i be denoted by NJ_i. The total number of DOF for tree topology then is given by

$$N = \sum_{i=1}^{NB}(NJ_i + NM_i) \tag{24}$$

where, NB denotes the number of bodies. Let $\nu_k\,(k = 1, \ldots N)$ represent a set of generalized speeds. Then, Eq. (22) can always be written as (Singh *et al*, 1985):

$$\dot{\underline{R}}^j = \sum_{i=1}^{N}\underline{V}_i^j\nu_i + \underline{V}_t^j \tag{25}$$

In the above, $\underline{V}_i^j$ is the coefficient of the generalized speed ν_i for body j. Similarly, $\underline{\omega}^j$ can be written as:

$$\underline{\omega}^j = \sum \underline{\omega}_i^j\nu_i + \underline{\omega}_t^j \tag{26}$$

The generalized speeds $\nu_i\,(i = 1, \ldots N)$ can be selected as the linear and angular rates corresponding to the joint DOF, and the modal rates, $\dot{\eta}_i^j$ (Singh *et al*, 1985).

Differentiation of Eq. (22) yields the acceleration expression:

$$
\begin{aligned}
\ddot{\underline{R}}^j \;=\; & \ddot{\underline{R}}_f^j + \dot{\underline{\omega}}^j \times \left(\underline{r}^j + \underline{u}^j\right) + \underline{\omega}^j \times \left(\underline{\omega}^j \times \left(\underline{r}^i + \underline{u}^j\right)\right) \\
& + 2\underline{\omega}^j \times \overset{o}{\underline{u}}^j + \overset{oo}{\underline{u}}^j
\end{aligned}
\tag{27}
$$

Equivalently, one also has, from Eq. (25),

$$\ddot{\underline{R}}^j = \sum_{i=1}^{N} \underline{V}_i^j \dot{\nu}_i + \sum_{i=1}^{N} \dot{\underline{V}}_i^j \nu_i + \dot{\underline{V}}_t^j \tag{28}$$

The equations of motion are obtained, using Kane's method, from

$$f_i^* + f_i = 0, \qquad i = 1, \ldots N \tag{29}$$

where,

$$f_i^* = -\sum_{j=1}^{NB} \int \underline{V}_i^j \cdot \ddot{\underline{R}}^j dm \tag{30}$$

and

$$f_i = \sum_{j=1}^{NB} \int \underline{V}_i^j \cdot d\underline{f} \tag{31}$$

The generalized inertia forces and the generalized active forces associated with DOF i are represented by f_i^* and f_i, respectively. The external forces as well as body elastic forces are denoted by $d\underline{f}$ in Eq.(31). The modal equations of motion of body j can be obtained using Eq.(29) in the form:

$$\sum_{\ell=1}^{N} M_{i\ell}\dot{\nu}_\ell + \sum_{\ell=1}^{NM_j} K_{i\ell}^j \eta_\ell^j + C_i = f_i', \tag{32}$$

$$f_i' = f_i + \sum_{\ell=1}^{NM_j} K_{i\ell}^j \eta_\ell^j \tag{33}$$

for i corresponding to the modal DOF of body j. The elements of the stiffness matrix due to body flexibility are denoted by $K_{i\ell}^j$ above, and C_i contains the remaining terms in the inertia force.

Now consider the translational inertia force of body j that is purely due to the rigid body motion, with link flexibility neglected. To this end, Eqs. (22) and (27) are modified as:

$$\dot{\underline{R}}_r^j = \dot{\underline{R}}_{fr}^j + \underline{\Omega}^j \times \underline{R}_r^j \tag{34}$$

and

$$\ddot{\underline{R}}_r^j = \ddot{\underline{R}}_{fr}^j + \underline{\Omega}^j \times \underline{r}^j + \left(\underline{\Omega}^j \times \underline{r}^j\right) \tag{35}$$

The subscript r is used to indicate that only the rigid body motion is considered and $\underline{\Omega}^j$ is the rigid body angular velocity of body j. In terms of the generalized speeds ν_i, these equations become:

$$\dot{\underline{R}}_r^j = \sum_{i=1}^{N} \underline{W}_i^j \nu_i + \underline{W}_t^j \tag{36}$$

$$\underline{\Omega}^j = \sum_{i=1}^{N} \underline{\Omega}_i^j \nu_i + \underline{\Omega}_t^j \tag{37}$$

and

$$\ddot{\underline{R}}_r^j = \sum_{i=1}^{N} \underline{W}_i^j \dot{\nu}_i + \sum_{i=1}^{N} \dot{\underline{W}}_i^j \nu_i + \dot{\underline{W}}_t^j \tag{38}$$

The coefficients $\underline{W}_i^j$ and $\underline{\Omega}_i^j$ appearing in Eqs. (36) and (37) are defined as:

$$\begin{aligned} \underline{W}_i^j &= \underline{V}_i^j \quad \text{for } i \text{ corresponding to joint DOF} \\ &= \underline{O} \quad \text{for } i \text{ corresponding to elastic DOF} \end{aligned} \tag{39}$$

and

$$\begin{aligned} \underline{\Omega}_i^j &= \underline{\omega}_i^j \text{ for } i \text{ corresponding to joint DOF} \\ &= \underline{O} \text{ otherwise} \end{aligned} \tag{40}$$

The rigid body translational inertia force of an elemental mass dm on body j is:

$$d\underline{f}_{Tr}^{j*} = -\ddot{\underline{R}}_r^j dm \tag{41}$$

and that of body j is:

$$\begin{aligned} \underline{f}_{Tr}^{j*} &= -\int \ddot{\underline{R}}_r^j dm \\ &= -\left[\, m_j \ddot{\underline{R}}_{fr}^j + \dot{\underline{\Omega}}^j \times m_j \underline{r}_c^j + \underline{\Omega}^j \times \left(\underline{\Omega}^j \times m_j \underline{r}_c^j\right) \,\right] \end{aligned} \tag{42}$$

where,

$$\underline{r}_c^j = \frac{1}{m_j} \int \underline{r}^j dm \tag{43}$$

Alternately, from Eq.(38) we have

$$\int \ddot{\underline{R}}_r^j dm = \sum_{i=1}^{N} \underline{P}_i^j \dot{\nu}_i + \underline{Q}^j \tag{44}$$

where,

$$\underline{P}_i^j = \int \underline{W}_i^j dm \tag{45}$$

and

$$\underline{Q}^j = \int \left(\sum_{i=1}^{N} \dot{\underline{W}}_i^j \nu_i + \dot{\underline{W}}_t^j \right) dm \tag{46}$$

Thus, the inertia forces due to all bodies in the branch outboard of hinge k are given by

$$\left(\underline{f}_{Tr}^*\right)_k = -\sum_{j=k}^{NBO_k} \int \ddot{\underline{R}}_r^j dm \tag{47}$$

$$= -\sum_{i=1}^{N} \left(\sum_{j=k}^{NBO_k} \underline{P}_i^j \right) \dot{\nu}_i - \sum_{j=k}^{NBO_k} \underline{Q}^j \tag{48}$$

The number of bodies in the branch outboard of hinge k is denoted by NBO_k in Eq.(48). However, since a hinge permitting relative translational motion across it does not transmit force along that direction to its inboard body, one needs

$$\left(\underline{f}_{Tr}^*\right)_k' = \left(\underline{f}_{Tr}^*\right)_k - \sum_{p=1}^{NT_k} \left[\left(\underline{f}_{Tr}^*\right)_k \cdot \underline{\ell}_p^{L(k)} \right] \underline{\ell}_p^{L(k)} \tag{49}$$

where, NT_k denotes the number of translational DOF across joint k and $\underline{\ell}_p$ is the pth Euler axis. Note that the term on the left hand side of Eq.(49) represents the contact force at the joint due to the outboard bodies, if all the bodies were rigid.

Now consider the rotational inertia forces of an element of mass dm on body j, neglecting flexibility. To this end, we first compute the elemental angular momentum:

$$dH_o = \underline{R}_r^j \times \dot{\underline{R}}_r^j dm \tag{50}$$

The rotational inertia force of this elemental mass then is:

$$d\underline{f}_{Rr}^{j*} = -\frac{d}{dt}\left(\underline{R}_r^j \times \dot{\underline{R}}_r^j dm \right) \tag{51}$$

The body angular momentum vector for body j, associated solely with rigid body motion then is:

$$\underline{H}_{or}^{j} = \int \underline{R}_r^j \times \dot{\underline{R}}_r^j dm \qquad (52)$$

Substituting Eq.(36) into the above, we obtain

$$\underline{H}_{or}^{j} = \sum_{i=1}^{N} \underline{C}_i^j \nu_i + \underline{D}^j \qquad (53)$$

where,

$$\underline{C}_i^j = \int \underline{R}_r^j \times \underline{W}_i^j dm \qquad (54)$$

and

$$\underline{D}^j = \int \underline{R}_r^j \times \underline{W}_t^j dm \qquad (55)$$

The rate of change of angular momentum, obtained by differentiating Eq.(53) as

$$\dot{\underline{H}}_{or}^{j} = \sum_{i=1}^{N} \underline{C}_i^j \dot{\nu}_i + \sum_{i=1}^{N} \dot{\underline{C}}_i^j \nu_i + \dot{\underline{D}}^j, \qquad (56)$$

is used to obtain the rotational inertia force of all bodies outboard of hinge k as:

$$\left(\underline{f}_{Rr}^*\right)_k = -\sum_{j=k}^{NBO_k} \dot{\underline{H}}_{or}^j \qquad (57)$$

$$= -\sum_{i=1}^{N}\left(\sum_{j=k}^{NBO_k} \underline{C}_i^j\right)\dot{\nu}_i - \sum_{j=k}^{NBO_k}\left(\sum_{i=1}^{N} \dot{\underline{C}}_i^j \nu_i + \dot{\underline{D}}^j\right) \qquad (58)$$

Once again, we need to eliminate the rotational inertia torque contributions about the rotational DOF. Thus, one needs

$$\left(\underline{f}_{Rr}^*\right)_k' = \left(\underline{f}_{Rr}^*\right)_k - \sum_{p=1}^{NR_k}\left[\left(\underline{f}_{Rr}^*\right)_k \cdot \underline{\ell}_p^{L(k)}\right]\underline{\ell}_p^{L(k)} \qquad (59)$$

where, NR_k denotes the number of rotational DOF across joint k.

The approach of Banerjee and Lemak (1991) consists of using Eqs.(41) and (51) to set up loads on each node of a finite element representation of the elastic body to obtain the initial stress state. This results in the computation of up to 21 S matrices for each elastic body. The reader is referred to Banerjee and Lemak (1991) for details. Here, we only consider the effect of outboard body motions on an elastic body.

Let the body inboard of hinge (or body) k be denoted by $L(k)$. Eqs. (48) and (58) then provide the inertia forces of all bodies outboard of hinge k, acting on body $L(k)$ through hinge k. The discussion in section 2 showed that these loads can be considered as "pre-loads" acting on body $L(k)$. The geometric nonlinearity effects on body $L(k)$ due to these pre-loads can be obtained first through a quasi-static stress analysis, followed by the setting up of the geometric stiffness matrix (Cook, 1985). The quasi-static problem for body $L(k)$ is solved as follows:

1. Lock all the degrees of freedom of joints inboard of body $L(k)$. Set $i = 1$.

2. At the location of each hinge on body $L(k)$ leading to outboard bodies, set up a unit force along $\underline{b}_{L(k)i}$ and let the loads in all the other directions and other hinge locations on body $L(k)$ be zero.

3. Solve the static problem and obtain the initial stress state

$$\begin{bmatrix} \sigma_o \end{bmatrix} = \begin{bmatrix} \sigma_{xo} & \tau_{xyo} & \tau_{xzo} \\ \tau_{xyo} & \sigma_{yo} & \tau_{yzo} \\ \tau_{zxo} & \tau_{zyo} & \sigma_{zo} \end{bmatrix} \qquad (60)$$

for each of the elements of body $L(k)$. The physical displacement $\{u\} = [u\ v\ w]^T$ of an elastic body can be represented using a finite element discretization, as

$$\{u\} = \lfloor N(x,y,z) \rfloor \{d\} \qquad (61)$$

where, $\lfloor N(x,y,z)\rfloor$ is the matrix of interpolation functions and $\{d\}$ is a column matrix representing the modal degrees of freedom. Define

$$\{\delta\} = \{u_{,x}\ u_{,y}\ u_{,z}\ v_{,x}\ v_{,y}\ v_{,z}\ w_{,x}\ w_{,y}\ w_{,z}\} \qquad (62)$$

such that

$$\{\delta\} = \lfloor G \rfloor \{d\} \qquad (63)$$

where, $\lfloor G \rfloor$ is the derivative of the shape function matrix, arranged so that Eq. (63) is valid.

4. Set up the element geometric stiffness matrix

$$[k_g]_i = \int_v \lfloor G \rfloor^T [\Gamma] \lfloor G \rfloor dv \qquad (64)$$

where,

$$\begin{bmatrix} \Gamma \end{bmatrix}_{9\times 9} = \begin{bmatrix} \sigma_o & 0 & 0 \\ 0 & \sigma_o & 0 \\ 0 & 0 & \sigma_o \end{bmatrix} \qquad (65)$$

5. Form the global geometric stiffness matrix $[K_G]_i$.

6. Set $i = i + 1$. For $i \leq 3$ this corresponds to setting up forces along the ith axis of $\mathbb{b}_{L(k)}$. The case $3 < i \leq 6$ corresponds to setting up a unit torque about the $(i-3)^{th}$ axis of $\mathbb{b}_{L(k)}$. Repeat steps 2-6.

7. The loads $A_i^{L(k)}$ $(i = 22, \ldots 27)$ at the hinge location, where,

$$A_i^{L(k)} = \underline{b}_{L(k)\ell} \cdot \left(\underline{f}_{Tr}^*\right)_k' \qquad \begin{matrix} i = 22, 23, 24 \\ \ell = (i-21) \end{matrix} \qquad (66)$$

and

$$A_i^{L(k)} = \underline{b}_{L(k)m} \cdot \left(\underline{f}_{Rr}^*\right)_m' \qquad \begin{matrix} i = 25, 26, 27 \\ m = i - 24 \end{matrix} \qquad (67)$$

can now be used along with the geometric stiffness matrices developed in step 5. If the rotatory inertia forces are negligible, the set of loads in Eq.(67) need not be considered.

8. Obtain the contributions of the geometric stiffness matrices to the ℓth modal equation of body $L(k)$ (the body directly inboard of joint k) from

$$\left(f_g^j\right)_\ell = -\sum_{i=22}^{27}\sum_{m=1}^{NM_j} S^{ji}_{\ell m}\eta^j_m A^j_i, \qquad j = L(k) \tag{68}$$

where,

$$\left[S^{ji}\right] = \left[\phi^j\right]^T [K_g]_i \left[\phi^j\right] \tag{69}$$

and $[\phi^j]$ is the modal matrix of Body j.

9. Repeat steps 1 through 8 for all the elastic branch bodies in the topology.

The modal equations for the elastic body $L(k)$ will now be modified as:

$$f_i^* + f_i + \left(f_g^j\right)_i = 0, \cdots i = 1\cdots NM_{L(k)} \tag{70}$$

where, the subscript 'i' corresponds to the modes of body $j = L(k)$. With the inclusion of the 21 S matrices defined by Banerjee and Lemak (1991) representing the effects of self motions on body stiffness, the modal equations for the body $j = L(k)$ which is inboard of joint k will now become

$$\sum_{\ell=1}^{N} M_{i\ell}\dot{\nu}_\ell + \sum_{m=1}^{27}\sum_{\ell=1}^{NM_j} S^{jm}_{i\ell}\eta^j_\ell A^j_m + \sum_{\ell=1}^{NM_j} K^j_{i\ell}\eta^j_\ell + C_i = f'_i. \tag{71}$$

for i corresponding to body j modal coordinate. Note that f'_i is as defined in Eq.(33) and does not contain the modal stiffness terms. Also, if the rotatory inertia forces are neglilgible, the matrices $S_{13}, \cdots S_{21}$ and $S_{25}\cdots S_{27}$ need not be computed.

4 ILLUSTRATIVE EXAMPLES

Two examples are presented here to illustrate the geometric nonlinear effects due to self motions and outboard body motions on a branch body.

The first example concerns the relative magnitude, and thus the relative importance of these two contributions. Let the eigenfuntions of a fixed-free beam be selected as the assumed mode shapes for Link 1 in the two link manipulator example presented in Section 2. That is,

$$\phi_i(x) = c_i\left[(\cos\beta_i x - \cosh\beta_i x) + A_i(\sin\beta_i x - \sinh\beta_i x)\right] \tag{72}$$

where, $A_i = \frac{\sin\beta_i L - \sinh\beta_i L}{\cos\beta_i L + \cosh\beta_i L}$, λ's are the solutions to the characteristic equation for a uniform fixed-free beam,

$$\cos\lambda\cos h\lambda = -1, \qquad \lambda = \beta L \tag{73}$$

and $c_i = \frac{1}{\sqrt{\rho L_1}}$ are the normalization constants satisfying

$$\int_o^L \phi_i\phi_j\rho dx = \delta_{ij}. \tag{74}$$

The mass per unit length of the beam is denoted by ρ and δ_{ij} is the kronecker delta in Eq.(74).

From Eq.(16), the coefficient of $\left(\dot\theta_1^2 + \dot\psi^2\sin^2\theta_1\right)$ in the stiffness term is

$$h^{(1)}_{ij} + m_2 L_1 G^{(1)}_{ij}(L_1) \tag{75}$$

h_{ij}	$j=1$	$j=2$	$j=3$
$i=1$	1.1933	-0.6859	-0.7921
$i=2$	-0.6859	6.4781	0.1702
$i=3$	-0.7921	0.1702	17.8624

Table 1: Stiffening Term Due to Self Motions, Using Fixed-Free Eigenfunctions

G'_{ij}	$j=1$	$j=2$	$j=3$
$i=1$	4.6478	-7.3799	3.9420
$i=2$	-7.3799	32.4172	-22.3507
$i=3$	3.9420	-22.3507	77.3046

Table 2: Stiffening Term Due to Outboard Body Motions, Using Fixed-Free Eigenfunctions

The values of the dimensionless parameters $h^{(1)}_{ij}$ and $G'_{ij} = \rho_1 L_1^2$ $G^{(1)}_{ij}(L_1)$ are shown in Tables 1 and 2, respectively. From these tables, it is clear that the latter terms are much larger. However, the second term in Eq.(75) implies that the contribution due to the "end mass" (the second link) depends on the ratio of the end mass (m_2) to the link mass $\rho_1 L_1$. Thus, it is appropriate to compare $h^{(1)}_{ij}$ and $\left(\frac{m_2 L_1}{\rho_1 L_1^2}\right) G'_{ij}$. Since the results in Table 2 can be viewed as those corresponding to a ratio of $\frac{m_2}{m_1} = 1.0$, it implies that, for other ratios, these values must be multiplied by the ratio $\left(\frac{m_2}{\rho_1 L_1}\right)$ to obtain the second term in Eq. (75). Note that both the terms in Eq. (75) will be of the same order for a mass ratio of 0.2, and as this ratio increases, it is clear that the latter term will dominate in the stiffness computations. The above analysis can be extended to modes with other boundary conditions as well.

Simulation results with two truss beams are presented next. These beams are similar to the beams designed as part of the NASA Marshall Space Flight Center Multibody Modeling, Verification and Control (MMVC) program shown in Fig.3. Each beam consists of eight bays. The beam with unequal length bays is made up of 0.4877 cm dia steel longerons and is considered for the upper link in a vertically suspended two link configuration. The bay lengths, starting from the short section, are: 15cm for bays 1 and 2, 22.5cm for bays 3 and 4, and 30.3 cm for the remaining bays. The softer truss for the lower link is made up of 0.3429 cm dia longerons. All the bays in this truss are 30.3 cm long. Both these beams have intermediate stiffeners that form an equilateral triangle with a side of 7.62 cm. The only difference between the links considered here and those shown in Fig.3 is that the lateral stiffeners in the truss links considered here form isosceles triangles as shown in Fig.4. Strength criteria for supporting self-weight as well as the payloads were considered in the design. The examples consist of spin-up and point-to-point maneuvers, without gravity loads. Structural damping is neglected.

We consider three models for comparison purposes:

1. Linear model, in which geometric nonlinearities are neglected,

2. With inertia loads (using the 21 S matrices defined by Banerjee and Lemak (1991)), and

3. With contact loads at the tip (using the upto 6 S matrices proposed here) in addition to the inertia loads.

The softer beam is used for the spin-up maneuver (prescribed rigid-body motion). Fixed-free modes are used in the simulations. The first mode of the component model of this link (without any endmass) has a natural frequency of 8.333 Hz. The 26 S matrices are generated for this body using the finite element method. A payload of 10 Kg is added as a second body in the simulation. This scenario is similar to that of the Shuttle Remote Manipulator System (SRMS) manipulating a satellite that is several times heavier than the SRMS. With the coordinate system defined as in Fig.4, spin-up is performed about the body x-axis. The spin-up profile and the y-deflection of the tip of the line of centers of the beam are shown in Fig.5. The linear model, whether the softening term is considered or omitted, gave the same results. It can be seen that the addition of the inertia load matrices does not change the results significantly either. However, the results shown with contact loads at the tip correctly capture the dominating stiffening effect due to the tip mass.

The point-to-point motion example consists of another open-loop maneuver. Referring to Fig.2, ψ is varied from 0° to 270° in 1.2 seconds while θ_1 and θ_2 are held constant at 120° and 60°, respectively. The profile for $\dot{\psi}$ is shown in Fig.6. It provides smooth acceleration and deceleration with cubic transients for $0< t \leq 0.3$ sec and $0.9< t \leq 1.2$ secs, as per the recommendation of Baruh and Tadikonda (1991), to minimize elastic motion excitation. The motor at the elbow joint has a mass of 10 Kg while the upper arm and the forearm have masses of 3.7 Kg and 1.43 Kg, respectively. The y-deflection of the tip of the centerline of the upperarm truss is shown in Fig.6. As can be seen, the linear model and the approach of Banerjee and Dickens (1990) yielded similar results. However, when the contact forces at the tip of Link 1 due to both the motor and the forearm mass are considered, the tip response is significantly different.

These simulations are carried out using a modified version of TREETOPS that incorporated all the geometric nonlinearity effects. Flexible body models are generated using COSMIC/NASTRAN.

5 CONCLUSIONS

A generic flexible branch body in a multibody chain is shown to experience two sets of "preloads" from inertia forces: one due to its own motions and the other due to the motions of bodies outboard of such a body. The contributions from the self motions of a body are modeled using the 21 geometric stiffness matrices in the literature. In addition to the above, it is shown here that the motions of outboard bodies also cause "preloads" on a flexible branch body, and that their effect can be modelled using at most 6 such generalized stiffness matrices, because the outboard body inertia forces are transmitted only through the hinge connections. In general, for a flexible body, both the terms in Eq. (75) may be needed for an accurate modeling of the contributions from the geometrically nonlinear terms. However, it is also clear that one may choose to retain only the second term in Eq. (75) for computational simplicity, without encountering the spurious loss of stiffness (or the softening effect). Retaining only the upto 6 geometric stiffness matrices presented here results in significant computational savings. This approach is especially attractive and applicable for multibody systems such as the Shuttle Remote Manipulator System (SRMS), where, the payload is several orders of magnitude larger than the SRMS link masses.

6 ACKNOWLEDGEMENT

The authors gratefully acknowledge the support of Dr. Henry Waites at the NASA Marshall Space Flight Center. Part of this research was conducted under the NASA contract No. NAS 8-38092, Multibody Modeling, Verification and Control program.

References

[1] H. Baruh and S.S.K. Tadikonda, " Issues in the Dynamics and Control of Flexible Robot Manipulators", *Journal of Guidance*, Vol. 12, No. 5, 1989.

[2] H. Baruh and S.S.K. Tadikonda, "Gibbs Phenomenon in Structural Control," *Journal of Guidance*, Vol.14, No.1, Jan-Feb 1991, pp. 51-58.

[3] A.K. Banerjee and J.M. Dickens, "Dynamics of an Arbitrary Flexible Body Undergoing Large Rotation and Translation with Small Vibration," *Journal of Guidance*, Vol.13, No.2, 1990, pp.221-227.

[4] A.K. Banerjee and M.E. Lemak, "Multi-Flexible Body Dynamics Capturing Motion-Induced Stiffness," *Journal of Applied Mechanics*, Vol.58, No.3, 1991, pp.766-775.

[5] R.D. Cook, *Concepts and Applications of Finite Element Analysis*, McGraw-Hill, 1985.

[6] L. Meirovitch, *Computational Methods in Structural Dynamics*, Sijthoff - Noordhof, The Netherlands, 1980.

[7] R.P. Singh, R.J. VanderVoort and P.W. Likins, "Dynamics of Flexible Bodies in Tree Topology - A Computer - Oriented Approach", *Journal of Guidance*, Vol. 8, No. 5, 1985.

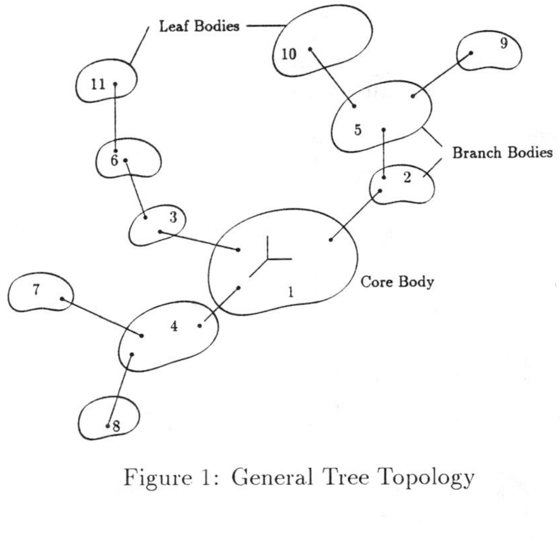

Figure 1: General Tree Topology

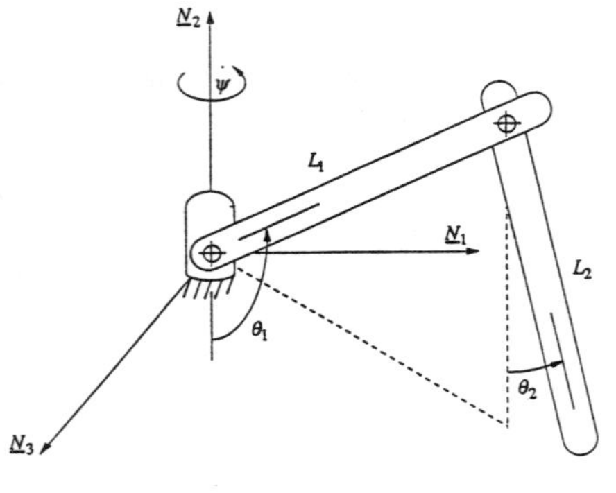

Figure 2: A 3 degree of freedom, 2 link manipulator

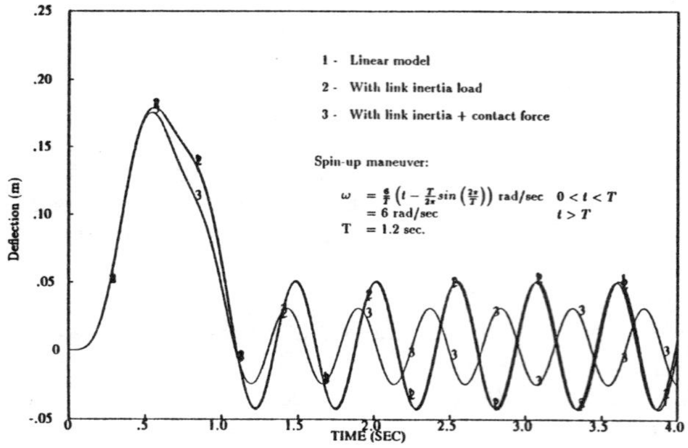

Figure 5: Tip deflection during the spin-up maneuver

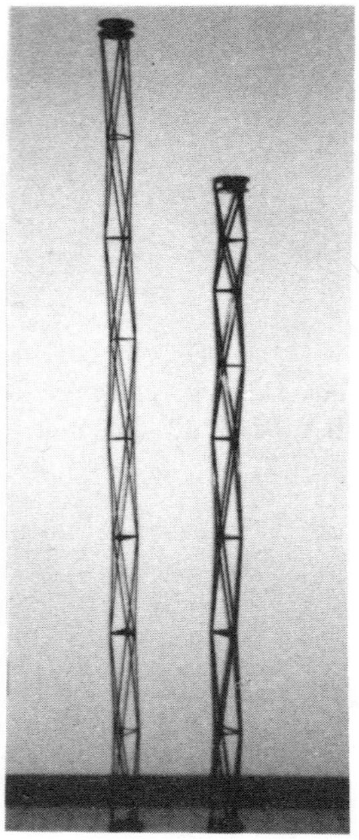

Figure 3: Truss beams designed for MMVC Experiments

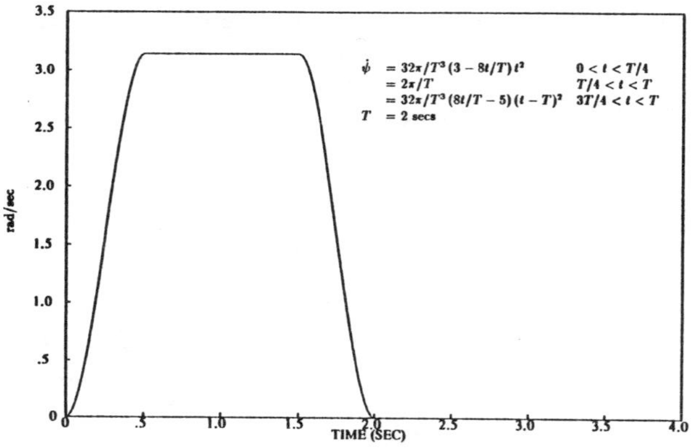

Figure 6: Prescribed angular velocity profile for $\dot{\psi}$

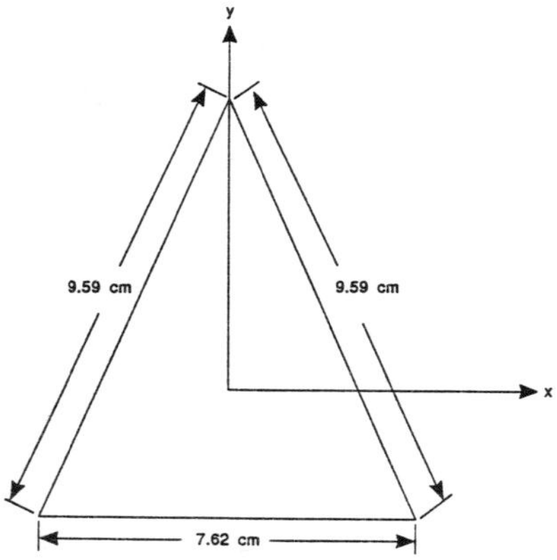

Figure 4: Coordinate frame definition

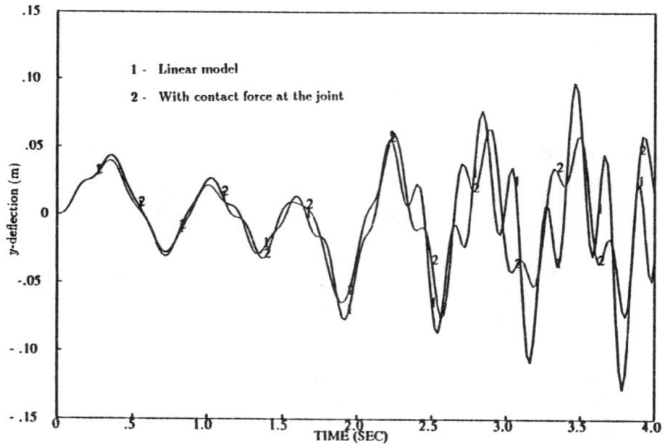

Figure 7: Tip y-deflection of the upper arm, due to the prescribed motion in Fig.6

AUTHOR INDEX

Dynamics of Flexible Multibody Systems: Theory and Experiment

Book Number: G00720